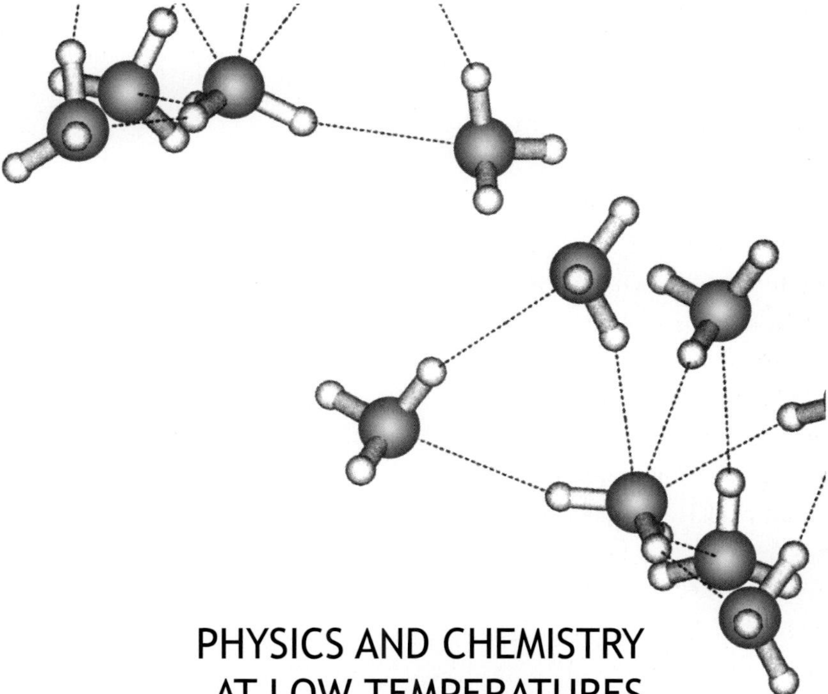

PHYSICS AND CHEMISTRY AT LOW TEMPERATURES

edited by Leonid Khriachtchev

PHYSICS AND CHEMISTRY AT LOW TEMPERATURES

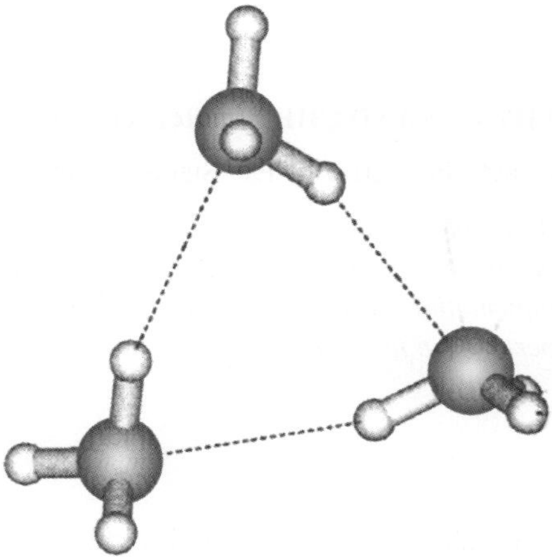

PAN STANFORD PUBLISHING

Published by

Pan Stanford Publishing Pte. Ltd.
Penthouse Level, Suntec Tower 3
8 Temasek Boulevard
Singapore 038988

Email: editorial@panstanford.com
Web: www.panstanford.com

British Library Cataloguing-in-Publication Data
A catalogue record for this book is available from the British Library.

PHYSICS AND CHEMISTRY AT LOW TEMPERATURES

ISBN 978-981-4267-51-9 (Hardcover)
ISBN 978-981-4267-82-3 (eBook)

Printed in Singapore by Mainland Press Pte Ltd.

To my daughters Vera and Ksenia, with love

Leonid

Contents

Preface

Low-temperature conditions have long attracted scientists. The reasons for this interest are both fundamental and practical. At such low (cryogenic) temperatures the physical and chemical properties of many substances undergo great change, and the behavior of matter is very different when compared with that at room temperature. Many fundamental discoveries (superconductivity, superfluidity, Bose–Einstein condensation, etc.) have been done at low temperatures. The development of thermodynamics has been essentially based on potentials of cryogenic technology. Tunneling reactions in chemistry are characterized by the low-temperature limit when the classical contribution is negligible. Remarkable advances have been made in chemical synthesis at low temperatures. Many practical applications benefit from the lack of heat and have a deep physical basis.

A number of key topics in low-temperature research are described in this book. The chapters are written by leading scientists who have strong practical experience in these research areas. The authors work in United States, Russia, Finland, Belgium, The Netherlands, Portugal, Switzerland, Poland, and Israel. They have published many papers in *Nature* and *Science*, which guarantees a broad spectrum of the described topics. Among the authors is David M. Lee, who shared the Nobel Prize in Physics in 1996 for his discovery of the superfluid properties of helium-3, one of the greatest achievements in low-temperature research. The authors provide an extensive bibliography for further reading and give many practical recommendations. This information will be hopefully useful for a broad readership, including young researchers coming to the fascinating field of low-temperature research.

Many of the authors have been active in the field of matrix-isolation spectroscopy. Discovered in the Pimentel laboratory in 1954, this technique has been widely used in various research areas. After more than five decades of matrix isolation, it is still possible to find many novel chemical species, processes, and reactions for spectroscopic investigation. The first four chapters give examples of traditional matrix-isolation research. Marilyn Jacox describes the photochemical production of free radicals, molecular ions, and other highly reactive species, their trapping in noble-gas solids for studies of their vibrational and electronic spectra, and the use of secondary photolysis for product identification. Lester Andrews uses laser ablation to provide metal atoms for reactions with small molecules upon condensation in excess argon, neon, or hydrogen for matrix infrared spectroscopic investigations. Chapter 3 (Fausto, Khriachtchev, and Hamm) describes thermal and light-induced conformational changes of small molecules in solid matrices, including attempts to understand the dynamics of the conformational transitions. Ara Apkarian and Mika Pettersson examine various aspects of the photodynamics in matrices in more detail, from the simplest systems to molecular complexes and polyatomics.

Several other chapters describe "less classical" matrix-isolation materials. Chapter 5 (Khmelenko, Lee, and Vasiliev) discusses the recent results on matrix-isolated hydrogen and deuterium atoms in impurity-helium condensates, which consist of nanoclusters of the matrix atoms or molecules (H_2, D_2, HD, and Kr), and this research utilizes temperatures as low as 0.1 K. Mario Fajardo concentrates on solid parahydrogen that offers several advantages as an isolating host due to the special "quantum solid" nature of this material, attracting our attention to a number of surprises. Helium nanodroplets described by Andrey Vilesov's team constitute an excellent medium for isolation and spectroscopy of molecules and clusters, which benefits from low temperature (0.37 K) and superfluid state of the droplets as well as the weakness of the interaction of helium atoms with embedded species. Wouter Herrebout and Benjamin van der Veken review experimental data obtained using cryogenic solutions with emphasis on red- and blue-shifting hydrogen bonds. Alexey Tsyganenko examines the interaction of molecules with solid surfaces (silica, zeolites, ice) using optical spectroscopy and electrostatic models.

The next two chapters take a look at the places where cryogenic conditions are practically realized without artificial equipment. Robert Johnson describes the chemical changes produced in ice by electrons, ions, and photons, which are relevant to the problems in astronomy and also of general interest for solid-state chemistry. Alexander Tielens and Louis Allamandola consider a number of cool interstellar processes involving simple molecules (H_2O, CH_3OH, NH_3, CO_2, and CO) in low-temperature molecular ices, which play an important role in the organic inventory of regions of star and planet formation.

Many fundamental phenomena and applications are prominently highlighted at low temperatures. For example, optical spectroscopy of single impurity molecules in low-temperature solids described by Michel Orrit and William Moerner provides a very sensitive probe of the structure and dynamics of the specific local environment around the single molecule. This chapter represents the limit of spectroscopic detection and many interesting phenomena masked by the assembly. Chapter 13 (Grochala, Khriachtchev, and Räsänen) introduces noble-gas chemistry, the area of research which sounds somewhat strange for many people. It discusses the various noble-gas compounds, starting with strongly bound systems and progressing to weak interactions. Noble-gas hydrides with the common formula HNgY are highlighted, including an argon compound HArF.

Computational methods can demonstrate the full power while describing cryogenic phenomena. The last two chapters present theoretical approaches to molecular science with an emphasis on comparison with experiments. Alexander Nemukhin and Bella Grigorenko discuss development and applications of the hybrid quantum mechanical — molecular mechanical methods for modeling properties of trapped species such as metal clusters, diatomic molecules, and noble-gas compounds. Benny Gerber and Jirka Sebek explain what can be learned about biological molecules at low temperatures. The anharmonic calculations used by them provide a deep insight into various molecular properties including intermolecular interactions.

The editor thanks all contributors for accepting his invitation to participate in the book and writing exciting stories.

Leonid Khriachtchev
Editor

CHAPTER 1

PHOTOEXCITATION OF FREE RADICALS AND MOLECULAR IONS TRAPPED IN RARE-GAS SOLIDS

Marilyn E. Jacox

Optical Technology Division, National Institute of Standards and Technology
Gaithersburg, MD 20899-8441, USA
E-mail: marilyn.jacox@nist.gov

The photochemical production of free radicals, molecular ions, and other highly reactive species, their trapping in rare-gas solids for studies of their vibrational and electronic spectra, and the use of secondary photolysis for product identification are surveyed. The application of gas-phase thermodynamic and spectroscopic data in planning and interpreting the experiments is emphasized and illustrated.

1. Introduction

Photoexcitation of a suitable precursor is widely used in matrix isolation experiments to prepare free radicals and molecular ions and to study their spectra and photolysis properties. The study of changes in the spectrum of the initial deposit when the sample is later exposed to filtered visible or ultraviolet radiation aids greatly in identifying the products obtained in attempts to generate molecules of interest, chemical reaction intermediates, free radicals, or molecular ions. When the radiation source is sufficiently intense, secondary photolysis products may appear even after the initial irradiation. Often the selection of wavelength ranges for subsequent irradiations has been quite empirical, with the principal objective to simplify the spectral analysis by sorting the product absorptions into categories corresponding to changes in intensity during a given irradiation or sequence of irradiations. However, this selection is most effective when the processes which can occur in the matrix environment as a result of various sets of irradiation conditions are considered.

This chapter is intended to provide examples of the use of photon sources both for generating free radicals and molecular ions in rare-gas matrices and for planning the strategy for subsequent irradiation of these matrix deposits.

2. Photoexcitation as a Source of Free Radicals

In the early days of matrix isolation studies, photodetachment of N_2 from diazomethane and azides using a medium-pressure mercury arc as the radiation source was believed to be one of the most promising avenues to free radical production. However, the results of these first experiments were disappointing. Among the early workers was Dolphus E. Milligan, who believed that the difficulty in using these species as sources of free radicals arose from cage recombination, which could be a major process when the barrier for the reverse recombination is low. Nevertheless, he found a few azide precursors which yielded stable free radicals. The first of these was ClN_3, which when photolyzed in an argon matrix yielded NCl.[1] Although the expected product of photolyzing CH_3N_3 in solid argon was CH_3N, that product rapidly isomerized to $CH_2=NH$.[2] A few years later, the importance of cage recombination was confirmed by Moore and Pimentel,[3] who detected the infrared absorptions of $CH_2{}^{15}N_2$ when they photolyzed a deposit of unenriched matrix-isolated diazomethane when $^{15}N_2$ had been added to the sample mixture.

In some systems, isomerization was observed upon cage recombination. Mercury-arc irradiation, for which the most intense emission is near 254 nm, led to the stabilization of HOCN in a deposit of matrix-isolated HNCO.[4] Gas-phase studies had indicated that near 254 nm the primary photodecomposition products of HNCO are NH + CO. In a supplementary study of the photolysis of an Ar:CO:HN_3 deposit, for which the primary photoproducts are NH + N_2, both HNCO and HOCN appeared. This observation suggests that in both systems initially formed NH may react with CO to form an unstable complex.

When an atom is photodetached from a molecule trapped in the solid, it may have sufficient kinetic energy to diffuse from the site of its photoproduction and to react with another suitable molecule which is also present in the solid. Photodetachment of an H atom from HI, HBr, or H_2S by 254-nm radiation occurs readily. In 1963, Milligan and Jacox succeeded in stabilizing HO_2, an important atmospheric and combustion reaction intermediate, by 254-nm photolysis of an Ar:O_2:HI sample deposit.[5] Detailed isotopic substitution studies — extremely important to positive product identification — confirmed the identification and demonstrated that the two O atoms of HO_2 are inequivalent.

The species which results from atom photodetachment often is itself a free radical. Higher-energy photolysis sources are usually required for atom photodetachment, leaving the desired product. Appropriate radiation sources include the rare-gas resonance lamps and a microwave-excited hydrogen discharge lamp, for which the design was described by Warneck,[6] which provides a high-intensity 122-nm radiation source. This lamp was first applied to matrix isolation experiments to produce NH_2, starting from an Ar:NH_3 sample,[7] and CH_3, starting from an Ar:CH_4 sample.[8] Still another example of H-atom detachment and subsequent diffusion is provided by the 122-nm photolysis of HNCO,[9] in which the predominant product was NCO rather than the HOCN stabilized at lower photon energies. In these and other vacuum-ultraviolet photolysis experiments, it is necessary to photolyze the sample as it is being deposited, in order to reduce screening of the effective radiation from the precursor molecules by the very strongly absorbing Rydberg bands of molecules in the deposit.

Cage recombination sometimes occurs even at relatively high photon energies. In studies on the vacuum ultraviolet photolysis of Ar:XCN samples (X = H, F, Cl, Br), prominent ultraviolet absorption by CN appears, but the XNC isomers of the starting molecules are also stabilized.[10]

Because free radicals have partially filled electron orbitals, many radicals have excited electronic states to which transitions can occur in the visible and near-ultraviolet spectral regions. Excitation into these states may dissociate the radical, isomerize it, or alter its reactivity with a neighboring molecule.

It is sometimes assumed that once a photodecomposition threshold is reached photolysis of the species of interest at higher energies will proceed to give the same products. However, often sufficient Franck–Condon overlap to support this generalization occurs over only a limited energy range. The first band system of HCO extends from 860 to 460 nm. The bands are diffuse because of predissociation to form H + CO. The origin of a nondiffuse band system lies near 260 nm. Because the bands in this higher-energy system are ubiquitous in hydrocarbon combustion, they have become known as the hydrocarbon flame bands. Matrix isolation experiments using photolysis sources with their major output at wavelengths shorter than 460 nm permit the stabilization of HCO and the detailed study of these hydrocarbon flame bands,[11] assisting in the vibrational assignment of this band system.

A recent example of the occurrence of reaction on photoexcitation is provided by the stabilization of NO_3 in a neon matrix in the laboratories of Jacox[12] and of Beckers and Willner.[13] The well-known $\tilde{B}\ ^2E'$ state of NO_3, which lies 663 nm above the ground state, is strongly predissociated. Both NO and NO_2 are produced. When Ne:NO and Ne:O_2 mixtures are deposited through separate jets to minimize oxidation of the NO to NO_2 and the deposit is exposed to radiation of wavelength shorter than the high energy limit of the NO_3 $\tilde{B}$ state, absorptions of NO_3 appear. The NO_3 yield can be optimized by warming the deposit to approximately 7 K for a few minutes to permit limited reorientation and diffusion of the molecules in the neon matrix. When this NO_3 is exposed to visible radiation in the 450–663-nm wavelength range of its $\tilde{B} - \tilde{X}$ transition, it is destroyed, and absorptions appear within a few cm^{-1} of the positions of the NO absorption and the band center of the infrared-inactive fundamental of O_2. This suggests the stabilization of a very weakly bound (O_2)(NO) complex in the neon matrix. Exposure of the resulting sample to radiation of wavelength shorter than the high energy limit of the NO_3 $\tilde{B}$ state causes regeneration of the NO_3. Application of this procedure has greatly facilitated the study of the behavior of the NO_3 spectrum on isotopic substitution. It has also led to the first detection in absorption of the 366-cm^{-1} degenerate deformation fundamental of NO_3, including measurements of its isotopic shifts.[13]

Photoprocesses are also exceptionally important in stabilizing rare-gas-containing uncharged molecules, a process which will be considered in greater detail in other chapters of this volume. While most of these species do not possess unpaired electrons and technically are not free radicals, matrix isolation studies of these transient species have opened a new chapter in the chemistry of the rare gases. Typically, a deposit which contains suitable precursors is exposed to ultraviolet or vacuum ultraviolet radiation, followed by annealing, which results in the appearance or growth in infrared absorption of a $HX^{+}Y^{-}$ product, where X is a rare-gas atom and Y a good electron captor, stabilized in the matrix by charge-transfer interaction. The first examples of these compounds were HXeCl, HXeBr, HXeI, and HKrCl, observed in krypton and xenon matrices by Pettersson and co-workers.[14] This work initiated an active program of research in that laboratory. A highlight in the study of this series of compounds was the stabilization and infrared study of HArF[15] and the subsequent discovery of the existence of a more stable trapping site for HArF in the argon matrix.[16] Compounds have also been studied in which the rare gas is coordinated with oxygen, sulfur, nitrogen, or carbon.[17,18]

3. Photoionization as a Source of Free Radicals and Molecular Ions

All of the photoprocesses discussed for neutral species can occur for molecular ions as well. The dependence of photoproducts on the wavelength of the irradiation becomes even more important with the extension of the energy range into the vacuum ultraviolet, necessary for ion production. This dependence is exemplified by a series of studies of the photolysis of $HCCl_3$. The 122-nm photolysis of $HCCl_3$ provided an excellent source of CCl_3.[19] However, a structured absorption at 1037 cm^{-1} was not readily assigned. With a compendium of ionization energies of small molecules[20] in hand, it was noted that the ionization energy of CCl_3 was 8.78 eV — a value that has since been revised downward to approximately 8.1 eV.[21,22] An updated online version of this compendium is currently available as the ion energetics portion of the *NIST Chemistry WebBook*, at http://webbook.nist.gov/chemistry/. When a 107-nm argon resonance lamp was substituted for the 122-nm radiation source, the cation product was instead $HCCl_2^+$.[23] The energy of the argon resonance lamp, 11.6 eV, slightly exceeds the 11.5-eV appearance energy of $HCCl_2^+$ from $HCCl_3$.[24,25]

3.1. *Discharge sampling*

Beginning in the early days of matrix isolation, many workers attempted to stabilize free radicals and molecular ions in inert matrices by passing the sample mixture through a discharge and rapidly quenching the products on the cryogenic observation surface. However, only a few simple, uncharged species were identified as a result of such experiments. Several important factors are responsible for this failure. Often, the energy released on quenching such mixtures exceeds the cooling capacity of the cell. Matrices are sufficiently rigid to inhibit molecular diffusion over only a relatively small temperature range. Diffusion becomes important at temperatures well below those at which the matrix material has a sufficient vapor pressure to be pumped away. A similar problem also occurred in the earliest studies of the vaporization of high-temperature samples. For such systems, the problem was readily solved by introducing the matrix gas downstream from the effusion cell. In addition, molecules introduced into the discharge region are extensively fragmented, resulting in mixtures in which atoms and exceptionally stable diatomics such as CN and CO predominate. Immediately after quenching, these atoms continue to diffuse and react.

3.2. *Modified discharge sampling*

A modified discharge sampling configuration, in which only the rare-gas matrix material is passed through the discharge and the molecule of interest mixed with a large excess of unexcited matrix gas is introduced downstream, has been much more successful for studies of the spectra of free radicals and molecular ions. In a recent study,[26] the yield of NH_4^+ trapped in a neon matrix was compared for various discharge configurations ranging from the conventional discharge of a Ne:NH_3 mixture during deposition to the codeposition of an undischarged Ne:H_2:NH_3 mixture with a beam of neon atoms that had been excited in a microwave discharge. By far the highest yield was obtained using the modified discharge configuration just described. Backstreaming into the discharge region is relatively unimportant in such experiments. Observations on many systems have demonstrated that the primary photoionization processes are dominated by those which require energies near or below those of the first excited states of the rare gas. The infrared spectrum of the resulting deposit almost always includes prominent absorptions of the undecomposed molecule. The extent of excitation usually is insufficient for major occurrence of secondary photoprocesses.

When argon is chosen as the rare gas, the accessible ionization range is approximately 11.5–11.8 eV. This range is primarily useful for ionizing species which have relatively low ionization energies, such as halides, unsaturated molecules, and aromatic species. Nevertheless, such modified discharges through argon have been quite useful for free radical production, as in the use of acetylene as a source of HCC[27] and of methanol to form CH_2OH.[28]

When neon is chosen, the energy range which is most important for ion production is 16.61–16.84 eV. Two of the four neon-atom levels in that range readily emit radiation. The other two, the $^3P_0$ and $^3P_2$ levels, have calculated lifetimes of many seconds, and so can transfer energy to other species by forming a collision complex, which results in Penning ionization. The energy range provided by neon is sufficiently high to ionize virtually all molecules. Successful use of this configuration for electron spin resonance studies, which are highly sensitive to species with unpaired electrons, was first reported by Knight and Steadman.[29,30]

CO_2 was selected as the test molecule to check whether excited neon atoms produced with the modified discharge configuration could provide a sufficient yield of molecular ions for infrared detection.[31] The ionization energy of CO_2 is approximately 13.8 eV, and the gas-phase band center for its ν_3 fundamental had been determined to lie at 1423.01 cm^{-1}.[32] In the test experiment, the corresponding absorption appeared at 1421.7 cm^{-1}, and studies of isotopically enriched CO_2 were consistent with the CO_2^+ identification. Moreover, the ν_3 fundamental of CO_2^- was identified at 1658.3 cm^{-1}.

Because the rare-gas matrix may effectively remove excess energy from initially formed free radicals and molecular ions, processes which predominate in gas-phase photolysis systems sometimes may not occur in rare-gas matrices. An important example of this is provided by modified discharge sampling studies of allene and propyne.[33] The first ionization energy of allene, determined by molecular beam photoelectron spectroscopy,[34] is 9.688(2) eV. Threshold photoelectron–photoion coincidence studies by Parr and co-workers[35] established thresholds and relative yields of cation products formed from allene at energies up to 20 eV. The appearance energy of $C_3H_3^+$ is 11.48(2) eV. At higher energies, that product grew rapidly, and was the predominant ion at 16.6 eV, although some allene cation, $C_3H_4^+$, persisted. Although the onset of $C_3H_2^+$ occurred at 13.5(2) eV, its yield grew slowly as the energy was increased. Similarly, the first ionization energy of propyne was determined to be 10.36 eV.[36] Threshold photoelectron–photoion coincidence studies[37] gave an appearance energy of $C_3H_3^+$ from propyne of 11.58(4) eV. At 16.6 eV, $C_3H_3^+$ was, once again, the most prominent cation, but appreciable $C_3H_4^+$ was still present. The most stable isomer of $C_3H_3^+$ was predicted to be the cyclic form, a prototypical aromatic molecule,[38,39] as was later confirmed.[40]

A modest amount of photoisomerization of allene to propyne and of propyne to allene was observed in the neon-matrix experiments using the modified discharge configuration. One set of absorptions was destroyed on irradiation of the initial deposit at wavelengths shorter than 280 nm, but persisted at somewhat shorter wavelengths when a small concentration of NO_2, a good electron captor, was added to the sample. This behavior is appropriate for a cation product. The positions of these absorptions did not correlate with those reported by Craig and co-workers[41] for polycrystalline salts of $C_3H_3^+$. Detailed isotopic substitution studies required the presence of four H atoms in the carrier molecule, which was identified as the allene cation. Calculations by Frenking and Schwarz[42] and by van der Hart[43] indicated that the potential energy surface for $C_3H_4^+$ has several minima, corresponding to the various stages in the photoisomerization of allene cation to propyne cation, as well as to the intermediate cation structures and to H-atom detachment to form *cyc*-$C_3H_3^+$. It was concluded that the lifetime of $C_3H_4^+$ is sufficiently long for collisions with the high concentration of neon atoms in the sampling region to remove the excess energy from the $C_3H_4^+$ before H-atom detachment can occur.

It sometimes happens that the ground-state structure of a cation differs from that of the corresponding uncharged molecule. For example, ionization of methyl fluoride results in the formation of H_2CFH^+, which

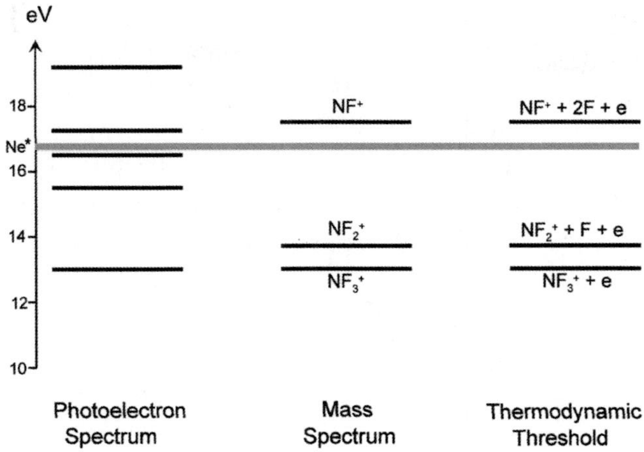

Fig. 1. Positions (eV) of the photoelectron bands of NF_3 and appearance energies (eV) of products of its photoionization.

is more stable than CH_3F^+. Such protonated species are known as ylidions. Although the methyl halide structure of the corresponding chloride and bromide is somewhat more stable than that of the ylidion, in matrix isolation experiments with the modified discharge configuration, the absorptions of both structures were present.[44]

3.3. ***Secondary photolysis***

In many systems, secondary photolysis is exceptionally important for product identification. An excellent example of this generalization is provided by studies of NF_3.[45] Mass spectrometric studies by Dibeler and Walker obtained a first ionization energy of 13.00(2) eV for NF_3 and appearance energies of 14.12(1) and 17.54(2) eV for NF_2^+ and NF^+, respectively, derived from NF_3.[46] A later photoelectron–photoion coincidence study by Rüede and co-workers required downward revision of the appearance energy for NF_2^+ from NF_3 to 13.71(2) eV.[47] These observations indicate that, using the first group of excited states of the neon atom, between 16.61 and 16.84 eV, as the energy source, NF_2^+ can be produced but NF^+ is not accessible. This result is illustrated by Fig. 1.

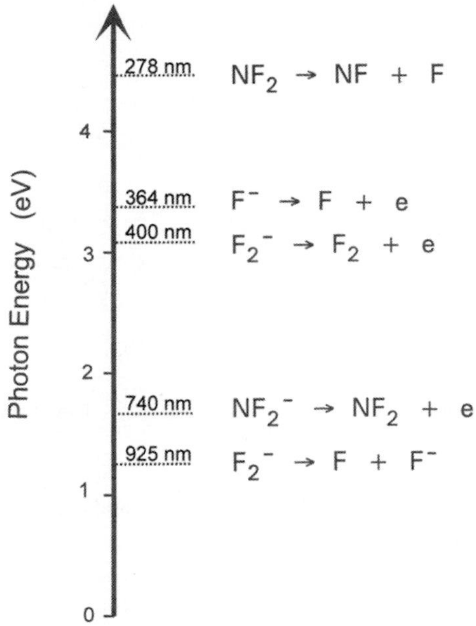

Fig. 2. Gas-phase photodissociation and photodetachment energies of the various anions expected to be formed in modified discharge experiments on Ne:NF_3 samples, together with the photodissociation energy of uncharged NF_2.

Jones and co-workers reported that the Penning ionization of NF_3 by Ne ($^3P_{0,2}$) atoms yielded 82% NF_2^+ and 18% NF_3^+.[48] Thus, the expected cation products in the modified discharge experiments on Ne:NF_3 samples are NF_2^+ and NF_3^+. The infrared spectrum of a Ne:NF_3 sample deposited without the discharge showed only the infrared absorptions of NF_3 and was unchanged on subsequent exposure of the deposit to filtered or unfiltered medium-pressure mercury-arc radiation. In contrast, infrared absorptions of uncharged NF_2 and of NF_2^+ and NF_3^+ were prominent in the spectra of Ne:NF_3 samples that had been codeposited with discharged neon.

As has been discussed in greater detail in a recent review by Jacox,[49] anions must also be present in the sample deposit in order to prevent the buildup of a high repulsive potential for the deposition of cations. Gas-phase studies of the interaction of low-energy electrons with NF_3 show that NF_3^- is no more than a minor product, but that the onset of

F^- and NF_2^- production occurs at very low electron energy.[50,51] The F^- atomic anions are relatively mobile, and can diffuse through the neon matrix and react with species trapped in the solid. The gas-phase photodissociation and photodetachment energies of the various anions expected to be present in this system are summarized, together with the photodissociation energy of uncharged NF_2, in Fig. 2. The onset of electron detachment from gas-phase NF_2^- occurs at 1.68(20) eV, which corresponds to the energy of approximately 740-nm photons.[50] Some of the F atoms can diffuse through the matrix during either the deposition or subsequent irradiations. These atoms may react to form F_2, which is immobile in the solid. F_2 is a good electron captor, with an electron affinity near 3.0 eV,[52,53] an energy which corresponds approximately to that of 400-nm radiation. The onset of electron detachment from any isolated F^- in the solid would occur at a slightly higher photon energy, corresponding to 364 nm.[54] Neon-matrix observations on many ion-production systems have shown that an excess electron energy of 1 or 2 eV is required for the detached electron to escape the site of its photoproduction.

The above generalizations receive further support from the results of anion observations for a $Ne:NF_3:NO_2$ = 1600:2:1 sample, shown in Fig. 3. Infrared absorptions of NF_2^-, as well as the familiar 1241.6-cm^{-1} absorption of NO_2^-, appeared in the initial deposit. At relatively low photon energies, the absorptions of NF_2^- decreased because of electron photodetachment, while the absorptions of uncharged NF_2 and of NO_2^-, which has a higher electron detachment threshold than does NF_2^-, grew. At higher photon energies, the NO_2^- absorption also decreased, and the NF_2 absorptions grew considerably more.

The technique has since been useful for studying molecules with very high first ionization energies. Kickel and co-workers estimated the first ionization energy of SiF_4 to be 15.29(3) eV.[55] Ionization of that molecule has a slow onset, with SiF_3^+ the predominant product both in the gas phase and in a neon matrix.[56] BF_3 has an even higher first ionization energy, 15.55(4) eV, with the appearance energy of BF_2^+ from BF_3 only slightly higher, 15.81 eV.[57] Above that energy, BF_2^+ is the predominant ion product. The spectrum of the initial sample deposit included absorptions of all three vibrational fundamentals and a combination band

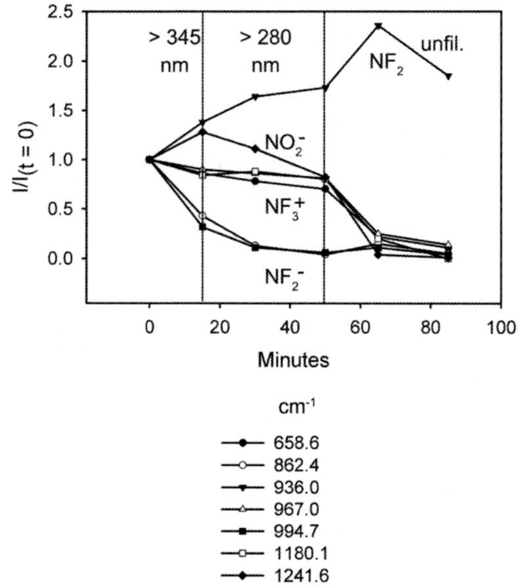

Fig. 3. Fractional changes in the intensities of various product absorptions observed in a modified discharge sampling experiment on an $Ne:NF_3:NO_2$ = 1600:2:1 experiment after subsequent irradiation with various wavelength ranges of the output of a medium-pressure mercury-arc lamp.

of BF_2.[58] Although absorptions of BF_2^+ were identified in the product spectrum, BF^+ was not, and BF was at most a minor contributor. Other product absorptions appeared, including several which had a disappearance threshold near 780 nm. The positions of this set of absorptions are appropriate for their contribution by BF_3^-. It is inferred that the electron affinity of BF_3, which has not been definitively established, has its upper bound at an energy corresponding to approximately 780 nm.

3.4. *Ion–matrix interaction*

As has been considered in greater detail in a recent review of the products of ion–molecule reactions in matrices,[49] many small molecular ions have exceptionally high reactivities. Indeed, the probability of their reaction on collision with another species often approaches unity. In contrast, the corresponding collisional reaction probability for a free radical–molecule pair is typically two or three orders of magnitude smaller. Often collisions of an ion with its uncharged counterpart lead to the stabilization of dimer ions in the matrix. Indeed, in studies on Ne:NO samples the trimer anion also contributes prominent absorptions to the spectrum.[59]

The exceptionally high reactivity of molecular ions extends to the occurrence of significant interaction with rare-gas matrices, accompanied by matrix shifts which may be as great as several hundred cm^{-1}. Several matrix properties need to be considered in order to minimize ion–matrix interactions. The most important of these are compared for inert solid rare gas and diatomic molecule matrices in Fig. 4. Because of the recent availability of helium cluster studies, helium is also included in the comparison. A high ionization energy of the matrix species reduces the extent of its charge-transfer interaction with molecular ions, which is greatest when the difference between the ionization energy of the molecule and that of the matrix is small. Almost all small molecules have first ionization energies below 16 eV. Neon has a significantly higher ionization energy than any of the other solid matrix materials. The low polarizabilities of helium and neon also suggest that they should have minimal interaction with highly polar ionic species. Proton sharing is an important cause of the large matrix shifts observed for proton stretching fundamentals of a number of protonated cations in rare-gas and molecular complexes and in matrices. Helium and neon, which have appreciably smaller proton affinities than those of most molecules,[60] excel in minimizing the importance of proton sharing, as well. Because the proton affinity of helium is only slightly smaller than that of neon, proton sharing may cause significant spectral shifts in both helium clusters and neon matrices for some protonated species derived from molecules which have relatively low proton affinities.

In 1972, Bondybey and Pimentel noted the appearance of infrared absorptions in their discharge sampling experiments using argon or krypton as the matrix material which they attributed to the trapping of interstitial hydrogen atoms.[61] The origin of the relatively high absorption intensities attributed to these interstitial atoms remained puzzling. The

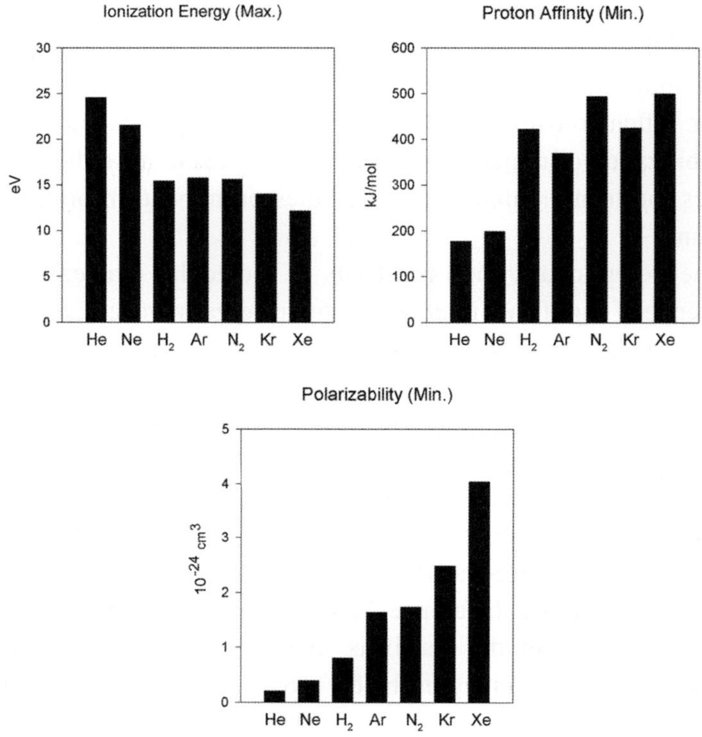

Fig. 4. Matrix properties needed to minimize ion–matrix interactions.

following year, Milligan and Jacox noted that these same absorptions had been seen in many of their 122-nm photolysis studies on matrix isolation systems which had in common the presence of an H-atom source and an atom or group of atoms that was a good electron captor and proposed their reassignment to HAr_n^+ and HKr_n^+, with $n = 2$, 4, or 6.[62] (Later studies favored $n = 2$.) In support of this reassignment, they cited flowing afterglow reaction tube observations on the argon–hydrogen system which indicated that HAr_2^+ was a dominant cation, as well as a mass spectrometric study by Chupka and Russell in which ArH^+ was demonstrated to be formed by the reaction of an Ar atom with an H atom excited to its 2 ^2S state, at 122 nm.[63] Clustering of that species with one or more additional Ar atoms in the solid would occur readily.

In an argon-matrix discharge sampling study of $HCCl_3$, Jacox observed prominent infrared absorptions of $HCCl_2^+$.[64] The photodestruction threshold of that species was near 280 nm. When the photolyzing radiation had a wavelength cutoff shorter than 260 nm, a marked increase in the concentration of HAr_2^+ resulted, which is consistent with the calculated threshold energy for proton transfer from $HCCl_2^+$ to the argon matrix. Similar experiments using a krypton matrix, in which HKr_2^+ appeared, were consistent with this mechanism. Studies which support the identification of HAr_2^+ and HKr_2^+ and extend our information on their properties have since been conducted in a number of laboratories.[65–68]

As discussed in a review by Bieske and Dopfer,[69] many protonated species form complexes with rare-gas atoms. This complexation usually results in a significant shift in the position of the fundamental contributed by the proton stretching vibration of MH^+. The magnitude of this shift depends on the difference between the proton affinity of the parent molecule, M, and that of the rare gas. The dependence on the nature of the rare-gas atom in the complex is exemplified by a series of gas-phase observations on the CH stretching fundamental of HCO^+. In its complex with a single helium atom[70] (proton affinity 177.8 kJ mol^{-1}), this fundamental appears 12.4 cm^{-1} below the gas-phase band center. For a neon atom (proton affinity 198.8 kJ mol^{-1}),[71] the shift is 42.6 cm^{-1}; for an argon atom (proton affinity 369.2 kJ mol^{-1}),[72] the shift is 273.7 cm^{-1}; and for a hydrogen molecule (proton affinity 422.3 kJ mol^{-1})[73] the shift is 249 cm^{-1}. Two proton affinities, 594 kJ mol^{-1} for complexation at the C atom and 426.3 kJ mol^{-1} for complexation at the O atom, have been reported for carbon monoxide. Correspondingly, spectra have been observed for both HCO^+ and HOC^+. All proton affinities here cited are values given by Hunter and Lias.[60]

The extent of the shift resulting from complexation of a rare gas with a protonated molecule is dependent on the number of rare-gas atoms in the complex. A detailed study for gas-phase Ar_nHCO^+ clusters, which have values of *n* from 1 to 13, has been made by Nizkorodov and co-workers.[72]

The proton affinity of the F atom is exceptionally low, 340.1 kJ mol^{-1}.[60] This value approaches that of the neon atom, suggesting that there should be a strong interaction between HF^+ and the neon atom or matrix. In an attempt to stabilize HF^+ in a neon matrix, Lugez and co-workers[74] found a shift of approximately 700 cm^{-1} for the HF^+ stretching fundamental, suggesting that the absorption was best characterized as a proton stretching vibration of $NeHF^+$.

If only cations (or anions) were to be deposited in a matrix, a high electric field would rapidly be established that would repel ions of the same sign from the deposition surface. In practice, processes such as electron flow into the deposit may occur to relieve the charge imbalance. A more detailed discussion of this phenomenon is given in a recent review.[49]

3.5. *Stabilization of anions*

The 122-nm photolysis of acetylene in rare-gas matrices proved to be an excellent source of both HCC and C_2, which was stabilized in its ground singlet state.[75] Still other absorptions in the visible spectral region that had previously been attributed to the Swan bands of triplet C_2 also appeared. These bands had vibrational spacings which differed appreciably from those measured for the Swan bands of gas-phase C_2. Subsequently, the bands were reassigned to C_2^-, formed by electron capture.[76] C_2 has an exceptionally high electron affinity, 3.27 eV.[77] The source of the photoelectrons still has not been definitively identified. One possibility is photoionization of a low-ionization-energy species such as diacetylene, the product of combination of pairs of HCC radicals formed in the intense beam of 122-nm radiation. The identification was confirmed by other experiments in which cesium atoms were added to the sample during the initial deposition, with concurrent 122-nm irradiation. Cesium has an exceptionally low ionization energy and provides a supplementary source of electrons. Fortunately, the density of the Rydberg states of C_2H_2 near 122 nm was sufficiently low to permit photoproduction of appreciable C_2. The yield of C_2^- increased greatly. Further details of the C_2^- identification are given in a recent review.[78]

That experiment demonstrated that anions can be stabilized in rare-gas matrices and paved the way for many subsequent anion identifications. Alkali metal atoms were adopted as a convenient source of electrons. The first subsequent anion identification was that of NO_2^-.[79,80] The electron affinity of NO_2, 2.273(5) eV,[81] is sufficiently high that charge-transfer interaction with the alkali metal occurs during the deposition. Often spontaneous charge-transfer interaction occurs during the codeposition of an alkali metal with another species. The electron affinity of SO_2 is 1.10 eV.[82,83] Although in the argon-matrix experiments all three vibrational fundamentals of SO_2^- were identified in the initial deposit,[84] subsequent experiments by Bencivenni and co-workers[85] provided evidence that the product was $M^+SO_2^-$, with M^+ the alkali metal cation. The problem was later avoided in experiments on a Ne:SO_2 sample using the modified discharge configuration.[86] Unlike the cations formed from alkali metal atoms, SO_2^+ cannot migrate through the matrix and form a complex with SO_2^-. In this experiment, the ν_3 fundamental of SO_2^- was shifted upward by approximately 50 cm^{-1}. Even when alkali metal atoms are codeposited with CO_2, which has an adiabatic electron affinity of –0.6 eV,[87] in an argon matrix absorptions attributed to the anion appeared.[88] In that study a new absorption attributed to CO_2^- appeared near 1600 cm^{-1}, at a position that was somewhat dependent on the alkali metal used in the experiment. The participation of the alkali metal in a complex with CO_2^- was confirmed by a modified discharge sampling experiment on a Ne: CO_2 mixture.[31] The ν_3 fundamental of CO_2^- was identified at 1658.3 cm^{-1} in the resulting neon-matrix deposit. The absorption is readily destroyed by subjecting the deposit to visible radiation. CO_2^+ is also stabilized in this system, but cannot migrate through the neon lattice.

In other experiments, an Ar:N_2O mixture was codeposited with a beam of alkali metal atoms.[80] Although new absorptions did not appear in the initial deposit, when the deposit was exposed to the full or filtered radiation of a medium-pressure mercury arc, the absorptions of N_2O gradually decreased in intensity and a prominent new absorption appeared at 1206 cm^{-1}. On prolonged irradiation this absorption was also destroyed, and the NO stretching absorptions of $(NO)_2$ grew in intensity. Isotopic substitution absorption studies supported the assignment of the new absorption to $N{=}NO_2^-$, produced from the addition of O^- to N_2O.

In 1968, Noble and Pimentel observed absorptions near 700 and 960 cm^{-1} which they attributed to an uncharged ClHCl species trapped in an

Ar:HCl sample that had been passed through a glow discharge while it was being deposited.[89] Subsequently, Milligan and Jacox observed the same two peaks with greatly enhanced intensity in the infrared spectrum of an Ar:HCl mixture that had been codeposited with a beam of potassium atoms and then exposed to 254-nm radiation.[90] This observation dictated the reassignment of the two absorptions to the $ClHCl^-$ anion. The reassignment has stood the test of time. Shortly thereafter, Bondybey and co-workers repeated the glow discharge sampling experiment using an Ar:HBr mixture and attributed two new absorptions to uncharged BrHBr.[91] However, this pair of peaks was reproduced when an Ar:HBr mixture was codeposited with a beam of alkali metal atoms and then subjected to 254-nm irradiation.[92] Coulombic interaction resulting from charge transfer facilitates anion formation when an alkali metal is present in the system. A more detailed discussion of the role of charge transfer has been given by Jacox.[93]

The stabilization of $N{=}NO_2^-$ suggested that the codeposition of an $Ar{:}CO_2{:}N_2O$ mixture with an alkali metal and subsequent mercury-arc irradiation might lead to the appearance of the important CO_3^- free radical, which is isoelectronic with $N{=}NO_2^-$. Two new CO stretching absorptions appeared in the resulting experiment. Other experiments with similar results were conducted starting instead with an $Ar{:}CO{:}O_2$ mixture.[94] However, the appearance of two appreciably separated CO stretching absorptions was at first puzzling, since CO_3^- was expected to possess D_{3h} symmetry, for which only a single, doubly degenerate CO stretching fundamental should be infrared-active. It was suggested that the apparent anion structure was distorted because of the inhomogeneous electric field associated with the alkali metal atomic cations. The argument was supported by observations on the stabilization of $M^+NO_3^-$ in matrices. Although X-ray diffraction studies had definitively established that the nitrate anion in crystalline alkali metal nitrates possesses threefold symmetry, Smith and co-workers[95] reported a splitting of similar magnitude for the degenerate NO stretching fundamental of vapors from various alkali metal nitrates trapped in inert matrices. A later neon-matrix study involving the codeposition of a $Ne{:}NO_2{:}O_2$ or a $Ne{:}NO{:}O_2$ sample with a beam of neon atoms that had been excited in a microwave discharge led to the appearance of a single prominent NO stretching absorption of NO_3^- trapped in solid neon.[96] In these experiments, molecular cations were trapped throughout the solid, providing a more homogeneous electric field than for atomic cations, which could diffuse through the lattice to sites near the molecular anions.

The possibility of electron capture in studies on ionic systems increases the number of processes which may occur. The activation barrier for electron diffusion through rare-gas matrices is low. Because electron–cation recombination is highly exothermic, the uncharged product may contribute prominent absorptions to the spectrum. In addition to the expected anion, typically impurities such as CO_2 and O_2 which are always present in low concentration provide weak electron traps. Sometimes an electron that is photodetached from a weak electron trap is captured by another species having a higher electron affinity, as exemplified by the previous discussion of the Ne:NF_3:NO_2 system.

If photodestruction of cations results from electron capture rather than from photodecomposition, the threshold and extent of cation photodestruction will depend on the properties of the anions in the system.[49] When electrons are released from weak electron traps by exposure of the sample to low-energy radiation, a fraction of the cations will be neutralized by electron capture. On prolonged irradiation, the electron source is depleted, and the concentration of cations levels off at a new, lower value. When a somewhat more energetic radiation source is used, electrons are released from slightly stronger electron traps, and the concentration of cations "tails off" at a new, lower value. Such behavior can aid in determining that a species is a cation. Addition of a good electron captor (e.g., Cl_2, NO_2) to the system may increase the energy range of stability of cations, facilitating their observation and identification. However, when such a species is added to the sample it is necessary to consider the possibility of reaction of the added species or one of its photodissociation products with other atoms or molecules in the system.

4. Summary

Photodecomposition of a free radical precursor trapped in a rare-gas matrix proceeds in the same spectral region and usually yields the same products as observed in the gas phase. However, when the energy barrier to the reverse process is low, cage recombination in the matrix suppresses product formation. Alternatively, cage recombination may result in isomerization. Occasionally, the product observed in gas-phase studies results from dissociation of an initially formed excited species. When this occurs, the matrix may collisionally deactivate the initially formed species before there is sufficient time for it to rearrange or dissociate. When photodecomposition eliminates an atom from the precursor, atomic diffusion through the solid can result in stabilization of the free radical formed by the atom detachment. The released atom sometimes reacts with another species to form a second free radical product.

Similar principles apply to the stabilization of molecular ions in rare-gas matrices, with the added generalization that electrons, like atoms, have at least limited mobility in the matrix. In ionic systems, approximate charge neutrality of the deposit must be maintained. Small molecular ions are often much more reactive than are free radicals, and may quite strongly interact even with the matrix atoms.

In studying the spectra of both free radicals and molecular ions trapped in rare-gas matrices, reference to the literature for the corresponding gas-phase processes defines energetic constraints on the system and aids greatly in selecting wavelength ranges for the study of secondary photoprocesses.

References

1. D. E. Milligan, *J. Chem. Phys.* **35**, 372 (1961).
2. D. E. Milligan, *J. Chem. Phys.* **35**, 1491 (1961).
3. C. B. Moore and G. C. Pimentel, *J. Chem. Phys.* **41**, 3504 (1964).
4. M. E. Jacox and D. E. Milligan, *J. Chem. Phys.* **40**, 2457 (1964).
5. D. E. Milligan and M. E. Jacox, *J. Chem. Phys.* **38**, 2627 (1963).
6. P. Warneck, *Appl. Opt.* **1**, 721 (1962).
7. D. E. Milligan and M. E. Jacox, *J. Chem. Phys.* **43**, 4487 (1965).
8. D. E. Milligan and M. E. Jacox, *J. Chem. Phys.* **47**, 5146 (1967).
9. D. E. Milligan and M. E. Jacox, *J. Chem. Phys.* **47**, 5157 (1967).
10. D. E. Milligan and M. E. Jacox, *J. Chem. Phys.* **47**, 278 (1967).
11. D. E. Milligan and M. E. Jacox, *J. Chem. Phys.* **51**, 277 (1969).

12. M. E. Jacox and W. E. Thompson, *J. Chem. Phys.* **129**, 204306 (2008).
13. H. Beckers, H. Willner and M. E. Jacox, *ChemPhysChem.* **10**, 706 (2009).
14. M. Pettersson, J. Lundell and M. Räsänen, *J. Chem. Phys.* **102**, 6423 (1995).
15. L. Khriachtchev, M. Pettersson, N. Runeberg, J. Lundell and M. Räsänen, *Nature* **406**, 874 (2000).
16. L. Khriachtchev, M. Pettersson, A. Lignell and M. Räsänen, *J. Am. Chem. Soc.* **123**, 8610 (2001).
17. M. Pettersson, J. Lundell and M. Räsänen, *Eur. J. Inorg. Chem.*, 729 (1999).
18. L. Khriachtchev, H. Tanskanen, J. Lundell, M. Pettersson, H. Kiljunen and M. Räsänen, *J. Am. Chem. Soc.* **125**, 4696 (2003).
19. E. E. Rogers, S. Abramowitz, M. E. Jacox and D. E. Milligan, *J. Chem. Phys.* **52**, 2198 (1970).
20. J. L. Franklin, J. G. Dillard, H. M. Rosenstock, J. T. Herron, K. Draxl and F. H. Field, *Ionization Potentials, Appearance Potentials, and Heats of Formation of Gaseous Positive Ions*, National Bureau of Standards,Washington, DC (1969).
21. J. W. Hudgens, R. D. Johnson III, R. S. Timonen, J. Seetula and D. Gutman, *J. Phys. Chem.* **95**, 4400 (1991).
22. E. S. J. Robles and P. Chen, *J. Phys. Chem.* **98**, 6919 (1994).
23. M. E. Jacox and D. E. Milligan, *J. Chem. Phys.* **54**, 3935 (1971).
24. A. S. Werner, B. P. Tsai and T. Baer, *J. Chem. Phys.* **60**, 3650 (1974).
25. J. L. Holmes, F. P. Lossing and R. A. McFarlane, *Int. J. Mass Spectrom. Ion Proc.* **86**, 209 (1988).
26. M. E. Jacox and W. E. Thompson, *Phys. Chem. Chem. Phys.* **7**, 768 (2005).
27. M. E. Jacox and W. B. Olson, *J. Chem. Phys.* **86**, 3134 (1987).
28. M. E. Jacox, *Chem. Phys.* **59**, 213 (1981).
29. L. B. Knight Jr. and J. Steadman, *J. Chem. Phys.* **77**, 1750 (1982).
30. L. B. Knight Jr., *Acc. Chem. Res.* **19**, 313 (1986).
31. M. E. Jacox and W. E. Thompson, *J. Chem. Phys.* **91**, 1410 (1989).
32. K. Kawaguchi, C. Yamada and E. Hirota, *J. Chem. Phys.* **82**, 1174 (1985).
33. D. Forney, M. E. Jacox, C. L. Lugez and W. E. Thompson, *J. Chem. Phys.* **115**, 8418 (2001).
34. Z. Z. Yang, L. S. Wang, Y. T. Lee, D. A. Shirley, S. Y. Huang and W. A. Lester Jr., *Chem. Phys. Lett.* **171**, 9 (1990).
35. A. C. Parr, A. J. Jason and R. Stockbauer, Int. *J. Mass Spectrom. Ion Proc.* **26**, 23 (1978).
36. T. Nakayama and K. Watanabe, *J. Chem. Phys.* **40**, 558 (1964).
37. A. C. Parr, A. J. Jason, R. Stockbauer and K. E. McCulloh, *Int. J. Mass Spectrom. Ion Proc.* **30**, 319 (1979).
38. J. D. Roberts, A. Streitweiser and C. M. Regan, *J. Am. Chem. Soc.* **74**, 4579 (1952).
39. F. P. Lossing, *Can. J. Chem.* **50**, 3973 (1972).
40. J. L. Holmes and F. P. Lossing, *Can. J. Chem.* **57**, 249 (1979).

41. N. C. Craig, J. Pranata, S. J. Reinganum, J. R. Sprague and P. S. Stevens, *J. Am. Chem. Soc.* **108**, 4378 (1986).
42. G. Frenking and H. Schwarz, *Int. J. Mass Spectrom. Ion Phys.* **52**, 131 (1983).
43. W. J. van der Hart, *Int. J. Mass Spectrom. Ion Proc.* **151**, 27 (1995).
44. C. L. Lugez, D. Forney, M. E. Jacox and K. K. Irikura, *J. Chem. Phys.* **106**, 489 (1997).
45. M. E. Jacox and W. E. Thompson, *J. Chem. Phys.* **102**, 6 (1995).
46. V. H. Dibeler and J. A. Walker, *Inorg. Chem.* **8**, 1728 (1969).
47. R. Rüede, H. Troxler, C. Beglinger and M. Jungen, *Chem. Phys. Lett.* **203**, 477 (1993).
48. M. T. Jones, T. D. Dreiling, D. W. Setser and R. N. McDonald, *J. Phys. Chem.* **89**, 4501 (1985).
49. M. E. Jacox, *Int. J. Mass Spectrom.* **267**, 268 (2007).
50. P. W. Harland and J. L. Franklin, *J. Chem. Phys.* **61**, 1621 (1974).
51. G. D. Sides and T. O. Tiernan, *J. Chem. Phys.* **67**, 2382 (1977).
52. J. A. Ayala, W. E. Wentworth and E. C. M. Chen, *J. Phys. Chem.* **85**, 768 (1981).
53. P. G. Wenthold and R. R. Squires, *J. Phys. Chem.* **99**, 2002 (1995).
54. C. Blondel, P. Cacciani, C. Delsart and R. Trainham, *Phys. Rev.* **A40**, 3698 (1989).
55. B. L. Kickel, E. R. Fisher and P. B. Armentrout, *J. Phys. Chem.* **97**, 10198 (1993).
56. M. E. Jacox, K. K. Irikura and W. E. Thompson, *J. Chem. Phys.* **103**, 5308 (1995).
57. V. H. Dibeler and S. K. Liston, *Inorg. Chem.* **7**, 1742 (1968).
58. M. E. Jacox and W. E. Thompson, *J. Chem. Phys.* **102**, 4747 (1995).
59. C. L. Lugez, W. E. Thompson, M. E. Jacox, A. Snis and I. Panas, *J. Chem. Phys.* **110**, 10345 (1999).
60. E. P. L. Hunter and S. G. Lias, *J. Phys. Chem. Ref. Data* **27**, 413 (1998).
61. V. E. Bondybey and G. C. Pimentel, *J. Chem. Phys.* **56**, 3832 (1972).
62. D. E. Milligan and M. E. Jacox, *J. Mol. Spectrosc.* **46**, 460 (1973).
63. W. A. Chupka and M. E. Russell, *J. Chem. Phys.* **49**, 5426 (1968).
64. M. E. Jacox, *Chem. Phys.* **12**, 51 (1976).
65. C. A. Wight, B. S. Ault and L. Andrews, *J. Chem. Phys.* **65**, 1244 (1976).
66. H. M. Kunttu and J. A. Seetula, *Chem. Phys.* **189**, 273 (1994).
67. N. Legay-Sommaire and F. Legay, *Chem. Phys. Lett.* **314**, 40 (1999).
68. T. D. Fridgen, X. K. Zhang, J. M. Parnis and R. E. March, *J. Phys. Chem.* **A104**, 3487 (2000).
69. E. J. Bieske and O. Dopfer, *Chem. Rev.* **100**, 3963 (2000).
70. S. A. Nizkorodov, J. P. Maier and E. J. Bieske, *J. Chem. Phys.* **103**, 1297 (1995).
71. S. A. Nizkorodov, O. Dopfer, M. Meuwly, J. P. Maier and E. J. Bieske, *J. Chem. Phys.* **105**, 1770 (1996).
72. S. A. Nizkorodov, O. Dopfer, T. Ruchti, M. Meuwly, J. P. Maier and E. J. Bieske, *J. Phys. Chem.* **99**, 17118 (1995).
73. E. J. Bieske, S. A. Nizkorodov, J. E. Bennett and J. P. Maier, *J. Chem. Phys.* **102**, 5152 (1995).

74. C. L. Lugez, M. E. Jacox and R. D. Johnson III, *J. Chem. Phys.* **110**, 5037 (1999).
75. D. E. Milligan, M. E. Jacox and L. Abouaf-Marguin, *J. Chem. Phys.* **46**, 4562 (1967).
76. D. E. Milligan and M. E. Jacox, *J. Chem. Phys.* **51**, 1952 (1969).
77. D. W. Arnold, S. E. Bradforth, T. N. Kitsopoulos and D. M. Neumark, *J. Chem. Phys.* **95**, 8753 (1991).
78. M. E. Jacox, *Acc. Chem. Res.* **37**, 727 (2004).
79. D. E. Milligan, M. E. Jacox and W. A. Guillory, *J. Chem. Phys.* **52**, 3864 (1970).
80. D. E. Milligan and M. E. Jacox, *J. Chem. Phys.* **55**, 3404 (1971).
81. K. M. Ervin, J. Ho and W. C. Lineberger, *J. Phys. Chem.* **92**, 5405 (1988).
82. R. J. Celotta, R. A. Bennett and J. L. Hall, *J. Chem. Phys.* **60**, 1740 (1974).
83. M. R. Nimlos and G. B. Ellison, *J. Phys. Chem.* **90**, 2574 (1986).
84. D. E. Milligan and M. E. Jacox, *J. Chem. Phys.* **55**, 1003 (1971).
85. L. Bencivenni, F. Ramondo, R. Teghil and M. Pelino, *Inorg. Chim. Acta* **121**, 207 (1986).
86. D. Forney, C. B. Kellogg, W. E. Thompson and M. E. Jacox, *J. Chem. Phys.* **113**, 86 (2000).
87. M. Knapp, O. Echt, D. Kreisle, T. D. Märk and E. Recknagel, *Chem. Phys. Lett.* **126**, 225 (1986).
88. M. E. Jacox and D. E. Milligan, *Chem. Phys. Lett.* **28**, 163 (1974).
89. P. N. Noble and G. C. Pimentel, *J. Chem. Phys.* **49**, 3165 (1968).
90. D. E. Milligan and M. E. Jacox, *J. Chem. Phys.* **53**, 2034 (1970).
91. V. Bondybey, G. C. Pimentel and P. N. Noble, *J. Chem. Phys.* **55**, 540 (1971).
92. D. E. Milligan and M. E. Jacox, *J. Chem. Phys.* **55**, 2550 (1971).
93. M. E. Jacox, *Rev. Chem. Intermed.* **2**, 1 (1978).
94. M. E. Jacox and D. E. Milligan, *J. Mol. Spectrosc.* **52**, 363 (1974).
95. D. Smith, D. W. James and J. P. Devlin, *J. Chem. Phys.* **54**, 4437 (1971).
96. D. Forney, W. E. Thompson and M. E. Jacox, *J. Chem. Phys.* **99**, 7393 (1993).

CHAPTER 2

METAL ATOM REACTIONS TO FORM NOVEL SMALL MOLECULES

Lester Andrews

Department of Chemistry, University of Virginia, P.O. Box 22904-4319, Charlottesville, Virginia, USA

E-mail: lsa@virginia.edu

Laser ablation is used to provide metal atoms for reactions with small molecules upon condensation in excess argon, neon, or hydrogen for matrix infrared spectroscopic investigations. Examples considered in this chapter include Li, Al, Ti, W, Pt, Th, and U, and the subjects are the ionic $Li^+O_2^-$ molecule; the bridged Al_2H_6 dialane; the superhydride $WH_4(H_2)_4$ complex; the $CH_2{=}TiHF$, $CCl_2{=}PtCl_2$, and $CH_2{=}ThH_2$ methylidenes; the $CH{\equiv}MoH_3$ and $HC{\equiv}UF_3$ methylidyne complexes; and the $N{\equiv}WH_3$ nitrido trihydride.

1. Introduction

Reactions of metal atoms with small molecules have provided a productive route to many new molecules of fundamental importance for their contributions to our understanding of chemical bonding, and the matrix isolation technique has contributed to this large body of information over the last half century.[1–4] Initial investigations of this type employed the more volatile alkali metal atoms for spontaneous reactions of a charge-transfer nature.[5] The less volatile transition metal atoms required more challenging higher-temperature experimental methods.[6] Laser ablation provided a very efficient and effective means of heating a very small volume element of solid material to a very high temperature sufficient for vaporization (and even ionization).[7–10] The key point here is that the laser is pulsed and focused to a spot less than 0.1 mm in diameter

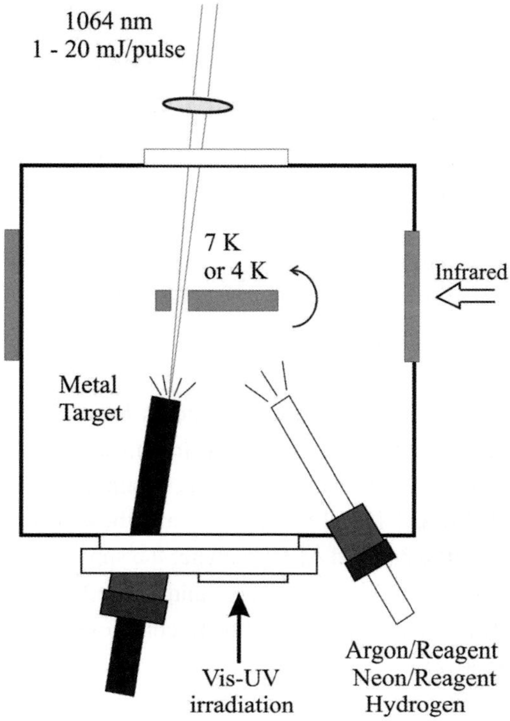

Fig. 1. Diagram of laser-ablation matrix isolation apparatus for reacting metal atoms with small molecules and trapping the products in solid argon, neon, or hydrogen for infrared spectroscopic and photochemical investigations.

so that a large amount of energy is deposited into a small volume element of sample in a very short time interval. Accordingly, the most refractory metals like tungsten and uranium can be evaporated for simple atom–molecule reactions.[2] The ablated metal atoms are translationally hot, and many are electronically excited as well. Thus, activation energy is provided to assist in the reactions of laser-ablated metal atoms with small molecules.

The present chapter describes some examples of the use of laser ablation to generate metal atoms for reactions with small molecules to make interesting new subject molecules.

2. Experimental Methods

Our matrix isolation laser-ablation apparatus shown in Fig. 1 has been described in more detail previously.[2] The Sumitomo Heavy Industries Model RDK-205D closed-cycle cryogenic cooler is a very good source of 4 K refrigeration if properly configured with substrate in thermal contact through indium gaskets and radiation shielded as completely as possible. A focused, pulsed Nd-YAG laser (1–20 mJ/pulse, 10 Hz) ablates metal atoms toward the cryogenically cooled window (4 K for hydrogen and neon or 7 K for argon matrix experiments) for codeposition and reaction with pure hydrogen, Ne/reagent, or Ar/reagent gas mixtures. Isotopically enriched reagents are employed whenever possible. The laser-ablation plume contains visible (Vis) light, which is useful to focus the laser and direct the ablated materials; however, the plume also provides ultraviolet (UV) and vacuum UV radiation to the sample and can perform photochemistry on the reagent molecule as well.[2] Infrared spectra are recorded by Nicolet Fourier-transform instruments at 0.5 cm^{-1} resolution and 0.1 cm^{-1} frequency accuracy after deposition, after subsequent irradiation by a mercury arc street lamp (Sylvania, 175 W, outer globe removed) in combination with glass filters, and after sequential annealing of the matrix sample to allow diffusion and further reaction of trapped species. We have found the common mercury arc street lamp to be a very effective, inexpensive, and readily available source of near-UV and Vis radiation.

3. Lithium and Oxygen Form the Ionic LiO_2 and LiO_2Li Molecules

The LiO_2 molecule was first prepared by Andrews as a presumed variation on the theme of the HO_2 radical, and it was discovered that the lower ionization energy of lithium compared with that of hydrogen supported a symmetrical, isosceles triangular ion-pair molecule (Li^+)(O_2^-), based on the triplet pattern in $^{16}O_2$, $^{16}O^{18}O$, $^{18}O_2$ isotopic spectra, rather than a bent, asymmetric HO_2 type of radical.[11] The lithium superoxide molecule, LiO_2, was the first polyatomic ionic molecule to be observed by inert gas matrix isolation method with distinct interionic and intraionic vibrational modes. The infrared spectrum of the LiO_2

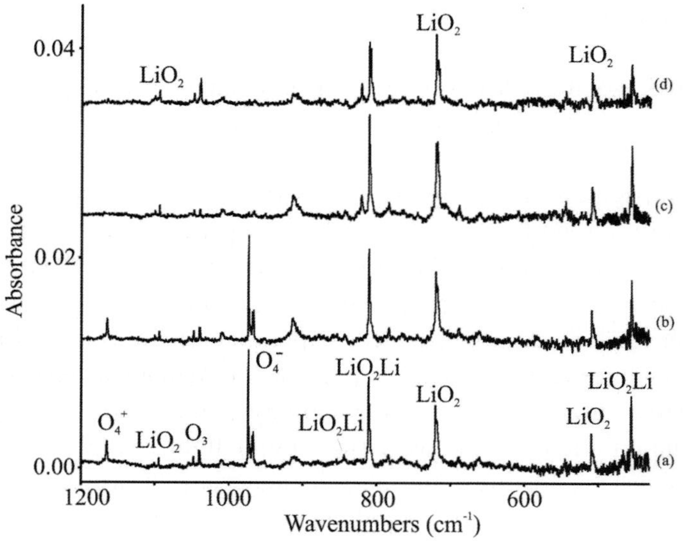

Fig. 2. Infrared spectra for laser-ablated natural lithium atom and O_2 (0.2%) reaction products in solid neon at 5 K. (a) After deposition for 60 min, (b) after >470 nm irradiation, (c) after >220 nm irradiation, and (d) after annealing to 8 K.

molecule formed spontaneously from thermal lithium atoms and oxygen molecules produced two strong interionic (Li^+–O_2^-) and one weak intraionic $(O–O)^-$ vibrational mode absorptions, but the subsequently observed Raman spectrum yielded an intense intraionic mode signal. These complementary infrared and Raman spectra substantiated the ionic model for the bonding deduced in part from the superoxide (O_2^-) frequency position (1097 cm^{-1}) of the O–O stretching mode.[12] An ionic model of polarizable ion pairs was proposed to explain the bonding and the trend in $(O–O)^-$ stretching frequencies in the analogous alkali metal family series MO_2 molecules. Subsequent theoretical calculations have confirmed the ionic model for the bonding in the LiO_2 molecule.[13]

Recently, the LiO_2 molecule has been isolated in solid neon using soft laser ablation of lithium to compare the matrix shifts as a further characterization of the nature of the bonding.[14] The infrared spectrum in Fig. 2 shows a number of product absorptions that have been assigned previously, including oxygen ionic species. The observation of such cationic species evidences vacuum UV radiation in the laser-ablation plume. Let us focus now on the three bands labeled LiO_2 at 1093.9, 719.7, and 508.9 cm^{-1}. These were shifted from the 1096.9, 698.8, and 492.4 cm^{-1} absorptions, respectively, observed 40 years ago in solid argon.[11] The neon matrix blue-shifts the two strong interionic modes by 20.9 and 16.5 cm^{-1} and red-shifts the weak intraionic mode by 3.0 cm^{-1} with respect to the argon matrix band positions. Based on earlier work with polar molecules, it is expected that the neon matrix will blue-shift the strong interionic vibrational modes of this ionic molecule.[4] The weak intraionic mode exhibits a small (0.5 cm^{-1}) lithium isotopic shift as the small infrared intensity of this mode depends on coupling with lithium motion. The electropositive lithium cation in LiO_2 is expected to bind more strongly to the more polarizable Ar matrix atom than the smaller Ne atom and to decrease the vibrational frequency as is observed.[14]

Two additional strong infrared absorptions for a lithium product species at 810.4 and 455.1 cm^{-1} using natural isotopes increase together on UV irradiation and acquire similar red satellites at 808.3 and 507.7 cm^{-1}, respectively, on annealing. These bands are blue-shifted 14.4 and 9.6 cm^{-1} by the neon matrix from the argon matrix observations, which were assigned to the motions of Li cations perpendicular and parallel, respectively, to the peroxide anion bond in the LiO_2Li molecule.[11] The above blue shifts observed for LiO_2Li are 1.8% and 2.1% of the argon matrix frequencies, which are smaller than the 3.0 % and 3.4% shifts for the analogous modes of the LiO_2 molecule. The LiO_2 molecule has a large dipole moment (5.7 D, B3LYP calculation), and accordingly it interacts with the matrix more strongly than the symmetrical ionic LiO_2Li molecule. In addition, a weaker 843.6 cm^{-1} band is appropriate for the 6(16-16)7 isotopic counterpart in natural abundance, which is blue-shifted 13.6 cm^{-1} from the argon matrix counterpart observed as the intermediate component of a triplet absorption prepared from a mixture of lithium isotopes.[11] Observation of the weaker 843.6 cm^{-1} band reaffirms that the associated neon matrix 810.4 and 455.1 cm^{-1} bands are due to the LiO_2Li molecule with two equivalent lithium atoms.

Reactions (1) and (2) are spontaneous in the argon matrix using thermal lithium atoms,[11] and these reactions proceed in the neon matrix with lithium atoms from laser ablation. Finally, the vibrational frequencies assigned to the ionic molecules LiO_2 and LiO_2Li are substantiated by computations as reported and discussed.[14]

$$Li + O_2 \rightarrow LiO_2 \quad (1)$$

$$LiO_2 + Li \rightarrow LiO_2Li \quad (2)$$

4. Aluminum and Hydrogen: The First Preparation of Dialane, Al_2H_6

Boron hydride chemistry has been investigated for a century, and a large number of distinct boron hydride compounds have been identified and characterized; however, aluminum hydride chemistry under normal conditions is limited to the nonvolatile polymeric solid trihydride $(AlH_3)_n$.[15,16] The diborane molecule is fundamentally important, and the textbook example of hydrogen (μ-hydrido) bridge bonding. Although the isostructural dialane molecule is calculated to be stable, molecular Al_2H_6 was not isolated until our work with solid hydrogen.[17,18] The failure to observe dialane earlier is even more surprising in view of the synthesis of the isostructural Ga_2H_6 molecule.[19] Alane (AlH_3) has been observed by three groups from reactions of energetic aluminum atoms with hydrogen in solid argon and characterized by infrared spectroscopy[20]; however, the concentration of AlH_3 was not sufficient to form Al_2H_6 in the more rigid argon matrix. Our successful synthesis of Al_2H_6 for the first time involved pure hydrogen as the matrix. This ensured the selective formation of the highest monohydride AlH_3, and diffusion on annealing the soft hydrogen matrix to allow dimerization to Al_2H_6. The dialane (DA) bands are produced in much greater yield in the softer hydrogen and neon than in the harder argon matrix samples.

The spectrum of the deposited sample is dominated by the strong AlH absorption at 1598.7 cm^{-1}, which is intermediate between the gas phase (1624.4 cm^{-1}) and argon matrix (1590.7 cm^{-1}) absorptions for diatomic AlH. Absorptions are also observed for AlH_3 (1883.7, 777.9, 711.3 cm^{-1}) and AlH_2 (1821.9, 1787.8 cm^{-1}), which are for the most part higher than the argon matrix values. Important new absorptions are observed at 1838.4, 1835.8, 1825.5, and 747 cm^{-1} (labeled Al_2H_4) with the hint of very weak absorptions at 1932.3, 1915.1, 1408.1, 1268.2, 835.6, 702.4,

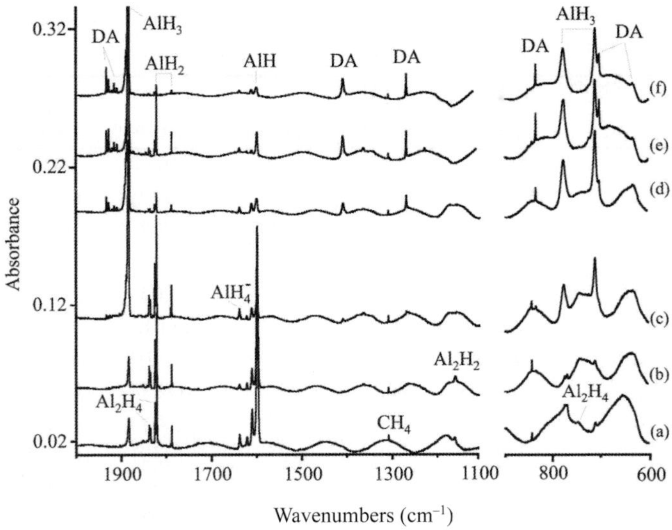

Fig. 3. Infrared spectra in the 2000–600 cm^{-1} region for laser-ablated Al codeposited with normal hydrogen at 4 K. (a) Spectrum of sample deposited for 25 min, (b) after annealing to 6.2 K, (c) after >290 nm photolysis, (d) after >220 nm photolysis, (e) after annealing to 6.5 K, and (f) after second >220 nm photolysis.

and 631.9 cm^{-1} (labeled DA) (Fig. 3a). Annealing to 6.2 K increased these bands except for AlH (Fig. 3b). Stepwise photolysis at >290 nm and >220 nm decreased the AlH band, and increased the AlH_3 absorptions and the DA band set (Figs. 3c and d). A subsequent annealing to 6.5 K increased all but the 1638 cm^{-1} band (Fig. 3e), and a final >220 nm irradiation increased the DA bands another 25% (Fig. 3f). Al atoms and D_2 were codeposited, and the spectra containing all but the weakest DA bands are described and assigned in our research papers.[17,18]

The seven-band DA set has straightforward chemistry. The strongest three bands at 1932.3, 1408.1, and 1268.2 cm^{-1} are detected on laser-ablated Al deposition with normal hydrogen (absorbance 0.001). These bands double on annealing to 6.0 K and double again on >290 nm photolysis while AlH is reduced by 90%, AlH_3 increases fourfold, and weaker DA bands join at 1915.1, 835.6, 702.4, and 631.9 cm^{-1}. Subsequent >220 nm photolysis increases the DA bands fourfold in concert and the AlH_3 bands twofold, while the next 6.5 K annealing

Table 1. Comparison of calculated and observed infrared active frequencies (cm^{-1}) for dibridged Al_2H_6 in solid hydrogen.

Sym[a]	Mode	SCF/ TZP[b]	CCSD/ DZP[c]	B3LYP[d]	BPW91[d]	Exp.[e]	H/D[f]
b_{1u}	8	2062	2047	1989	1934	1932	
		(518)	(344)	(419)	(379)	(0.069)	1.366
	9	977	954	866	808	836	
		(393)	(317)	(244)	(199)	(0.037)	1.376
	10	249	235	223	205		
		(18)	(15)	(13)	(10)	–	
b_{2u}	13	1350	1368	1292	1275	1268	
		(544)	(463)	(352)	(291)	(0.053)	1.379
	14	694	664	634	607	632	
		(353)	(328)	(263)	(230)	(0.039)	–
b_{3u}	16	2051	2024	1966	1908	1915	
		(101)	(88)	(126)	(130)	(0.023)	1.364
	17	1603	1589	1483	1431	1408	
		(1399)	(1162)	(1096)	(968)	(0.128)	1.370
	18	766	744	712	683	702	
		(890)	(684)	(648)	(575)	(0.048)	1.376

[a]Symmetry (D_{2h} point group) and mode description from Ref. 44. [b]From Ref. 44 with calculated intensities (km mol^{-1}) in parentheses. [c]From Ref. 45. [d]Ref. 18, 6-311++G** basis set. [e]In solid hydrogen, integrated intensities at maximum yield in parentheses. [f]Hydrogen/deuterium isotopic frequency ratio.

increases DA bands by 25% at the expense of AlH_3 (Fig. 3). Thus, AlH_3 is produced on photoexcitation of AlH with excess hydrogen, and DA is formed on UV photolysis along with AlH_3 and on annealing from diffusion and reaction of AlH_3. The seven DA bands are therefore assigned to dialane, Al_2H_6, which is confirmed by comparison to extensive vibrational frequency calculations at different levels of quantum theory, which are compared in Table 1.[17,18]

We discuss Table 1 for pedagogical reasons. First, it shows how the different levels of computational theory for harmonic frequencies tend to overestimate observed anharmonic frequencies. The two density functionals produce calculated frequencies that are very close to the observed values. Extrapolation through these sets of calculated frequencies to the observed values confirms the assignments to dialane and thus the first preparation of this novel molecule. Second, notice that the computed and observed infrared intensities are in very good qualitative agreement as well. Thus, the correlation between calculated and observed frequencies makes the case for the identification of this important molecule.

When the hydrogen matrix samples are annealed to 6.8 K, H_2 evaporates, molecular aluminum hydrides diffuse, aggregate and their absorptions decrease, and broad absorptions appear at 1720 (20) and 720 (20) cm^{-1}. These broad bands remain on the CsI window with decreasing absorbance until room temperature is reached. The deuterium matrix samples evaporate D_2 at about 10 K and aluminum deuterides produce a broad 1260 (20) cm^{-1} absorption, which remains on the CsI window. The frequency ratio 1720/1260 = 1.365 demonstrates that these bands are due to Al–H/Al–D vibrations. The spectrum of pure solid $(AlH_3)_n$ gives strong broad bands at 1760 and 680 cm^{-1}, solid $(AlH_3)_n$ in Nujol reveals a very strong, broad 1592 cm^{-1} band, and solid $(AlD_3)_n$ gives a corresponding 1163 cm^{-1} band.[17,18] The present broad absorptions at 1720 and 1260 cm^{-1} are therefore due to solid $(AlH_3)_n$ and $(AlD_3)_n$ on the salt window. Thus, we have made solid alane through the reaction of the elements aluminum and hydrogen to give AlH, then AlH_3, Al_2H_6, and finally the solid $(AlH_3)_n$ polymeric material. The crystal structure for this solid material shows six-coordinate aluminum with equivalent hydrogen atoms in bridge-bonding arrangements between aluminum atoms.[16] The average Al–H distance is 1.72 Å. It is interesting to note that the solid Al–H distance is intermediate between the 1.74 Å bridge and 1.58 Å terminal Al–H distances calculated[17,18] for Al_2H_6 and that the solid frequency of 1720 cm^{-1} is likewise intermediate between the terminal (1932 and 1915 cm^{-1}) and bridged (1408 and 1268 cm^{-1}) Al_2H_6 values observed here.

5. Tungsten and Hydrogen Form the $WH_4(H_2)_4$ Supercomplex

The first experimental observation of naked W atom reactions with dihydrogen found evidence for WH_4 as a major reaction product, and annealing the solid neon matrix increased a group of six sharp new absorptions, which were assigned to WH_6, formed by the spontaneous reaction of WH_4 and H_2.[21] These absorptions matched density functional

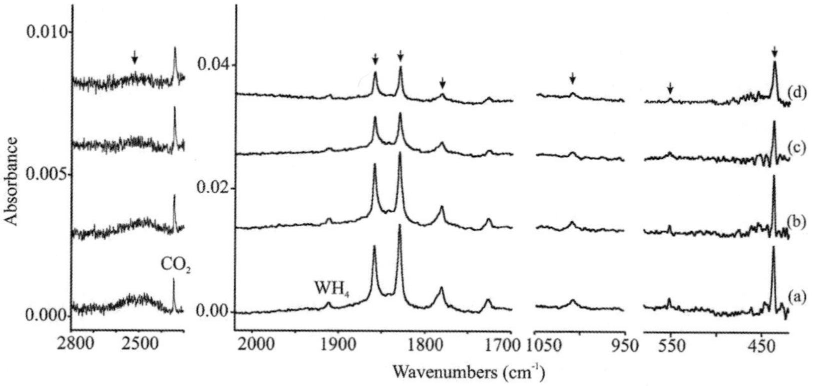

Fig. 4. Infrared spectra for the laser-ablated W atom and pure hydrogen reaction product in solid normal hydrogen at 4 K. (a) Spectrum after reagent codeposition for 30 min, (b) after annealing to 6 K, (c) after >220 nm irradiation, and (d) after annealing to 6.3 K.

calculated frequencies for the distorted C_{3v} prism WH_6 structure, which remains the highest neutral hydride and the only neutral metal hexahydride to be observed experimentally. This distorted structure is more stable than the octahedral form, which has been the subject of extensive theoretical calculations.[22] Subsequent work showed that laser-ablated W atoms react with neat normal and para-hydrogen to form the largest possible physically stable tungsten hydride species, which is the tungsten tetrahydride tetradihydrogen complex.[23]

Figure 4 illustrates infrared spectra for tungsten ablation and reaction with pure normal hydrogen during condensation at 4 K. The most prominent new absorptions are observed at 1859.1, 1830.3, and 437.2 cm^{-1} with weaker bands at 2500, 1781.6, 1007.6, and 551.5 cm^{-1} (marked with arrows). Annealing this sample to 6 K had little effect on the spectrum, while full arc irradiation (>220 nm) reduced the absorptions and a subsequent annealing to 6.3 K only sharpened them. The seven absorptions marked with arrows track together on annealing and UV irradiation, and they are thus associated with a common new product species. Note the absence of absorptions in the 1920–2020 cm^{-1} region where the strongest absorptions of WH_6 appeared in solid neon.[21] A final annealing to 7 K allowed the hydrogen matrix to evaporate and the new product to aggregate, leaving behind no infrared absorption.

Deuterium counterparts for the stronger bands were also observed and are reported in our publication.[23]

The above infrared spectra reveal diagnostic absorptions due to molecular subunits that identify this new tungsten hydride–dihydrogen complex. First, the WH_4 molecule in tetrahedral symmetry has been characterized by the triply degenerate antisymmetric W−H stretching and bending modes at 1920.5 and 525.2 cm^{-1} in solid neon,[21] and the weak new band at 1911.5 cm^{-1} in the solid hydrogen experiment can be assigned to WH_4 trapped on the surface where limited coordination can occur. The strong new absorptions at 1859.1 and 1830.3 cm^{-1} are slightly lower but still appropriate for W–H stretching modes as the 1.3951 and 1.3853 isotopic H/D frequency ratios indicate, and the new 551.5 cm^{-1} band is likewise due to an analogous H–W–H bending mode. Hence, the new product contains two or more W–H hydride bonds. Second, the new absorptions at 2500, 1781.6, 1007.6, and 437.2 cm^{-1} arise from the presence of side-bound dihydrogen molecules in this new product. The broad 2500 cm^{-1} band is characteristic of the H–H stretching mode for strongly complexed dihydrogen molecules as this mode was first observed at 2690 cm^{-1} in the important Kubas complex.[24] The 2500/1790 = 1.397 H/D isotopic ratio is in accord with that expected for an H−H stretching mode. The 1781.6 cm^{-1} band can be assigned to the antisymmetric W −(H_2) stretching mode on the basis of its 1.3892 isotopic H/D frequency ratio and its prediction from density functional calculations to fall about 80 cm^{-1} below the aforementioned highest antisymmetric W−H stretching mode at 1859.1 cm^{-1}. Such a mode was observed lower at 1570 cm^{-1} in the Kubas complex. The remaining two weak 1007.6 and strong 437.2 cm^{-1} bands are due to H_2−W−H_2 bending modes based on the prediction of such vibrational frequencies at 1065 and 414 cm^{-1} by density functional theory (DFT) calculations.[23] The higher of these has the appropriate 1.3996 H/D isotopic frequency ratio, and the lower of these may be compared with the W–(H_2) deformation mode observed near 450 cm^{-1} for the Kubas complex. This identification of $WH_4(H_2)_4$ based on the above vibrational assignments is confirmed by the close correlation of the seven strongest observed and calculated frequencies.[23]

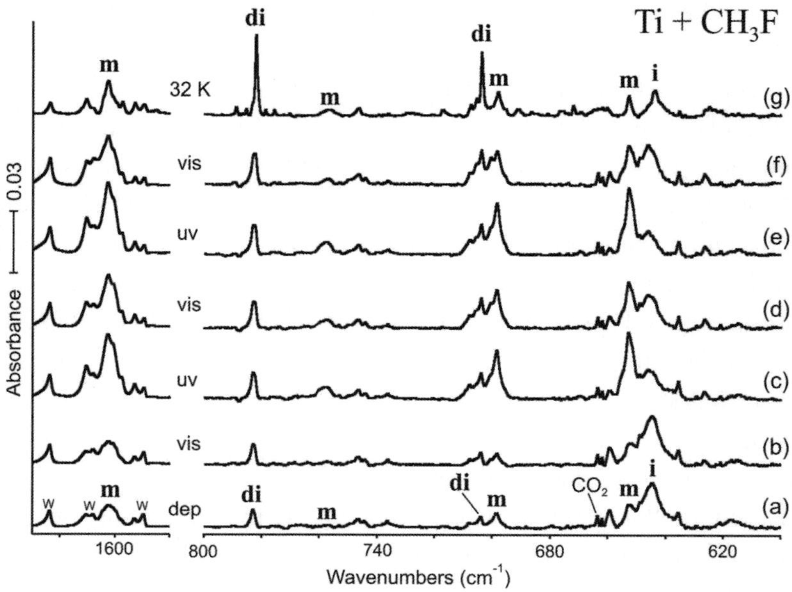

Fig. 5. Infrared spectra for laser-ablated Ti and CH_3F reaction products in solid argon at 7 K. (a) Ti and 0.2% CH_3F in argon codeposited for 1 h, (b) after Vis irradiation (>530 nm), (c) after UV irradiation (240–380 nm), (d) after a second Vis irradiation, (e) after a second UV irradiation, (f) after a third Vis irradiation, and (g) after annealing the sample to 32 K. W denotes water impurity absorptions.

Thus, we see that WH_4 is ligated by four dihydrogen molecules after preparation from the spontaneous reaction of W and neat H_2. Calculations show that the lowest energy supercomplex that can be formed is the $WH_4(H_2)_4$ species.[23] In like fashion laser-ablated Ru reacts with pure hydrogen to form an analogous $RuH_2(H_2)_4$ supercomplex.[25]

6. Ti and CH_3F Form an Agostic Methylidene Product

We were initially interested in methane activation by early transition metal atoms, but we knew from earlier experiments that the product yield would be relatively low. Hence, we reasoned that the analogous reactions with the more electron-rich methyl halides would be more favorable, and the Ti and CH_3F reaction provided our first contribution to the methylidene complex field.[2d,26]

Figure 5 shows representative spectra of Ti and CH_3F reaction products. The deposited sample reveals three sets of IR absorptions that are grouped by their behavior on sample irradiation and annealing. The first group (labeled **m** for methylidene) at 1602.8, 757.9, 698.6, and 652.8 cm^{-1} decreases on Vis light irradiation but increases markedly on UV irradiation, while the second set at 646.3 cm^{-1} (including 1105.7 and 504.3 cm^{-1}, not shown) (labeled **i** for insertion product) decreases. Note that these photochemical changes are reversible. In addition, the bands labeled **di** at 782.3 and 703.8 cm^{-1} (for dimethyl titanium difluoride) increase slightly on UV irradiation and markedly on annealing and with higher CH_3F reagent concentrations.

DFT calculations were performed, and the lowest energy products were CH_3–TiF (triplet state) and CH_2=TiHF (singlet state), with the latter 22 kcal mol^{-1} higher energy than the former. The observed frequencies for the **i** group correlated with the calculated frequencies having the higher IR intensities for CH_3–TiF, and the **m** group absorptions corresponded to the strongest calculated IR frequencies for CH_2=TiHF. The matching shifts with $^{13}CH_3F$ and CD_3F and computed isotopic frequencies described in our original paper confirm the vibrational assignments and the CH_2=TiHF molecular identification.[26]

The major 782.3 and 703.8 cm^{-1} absorptions for the **di** group reveal Ti isotopic splittings in natural abundance for the antisymmetric and symmetric stretching vibrations of a TiF_2 subunit. In addition, the associated 1385.2 and 1378.4 cm^{-1} bands are characteristic of methyl bending modes and the 566.9 cm^{-1} band (not shown) with the antisymmetric Ti–CH_3 stretching mode. The **di** bands are in excellent agreement with frequencies computed for the $(CH_3)_2TiF_2$ molecule, which is chemically analogous to the $(CH_3)_2TiCl_2$ compound synthesized earlier.[26,27]

The persistent photoreversible nature of the **i** and **m** absorptions is noteworthy. Alpha-hydrogen transfer appears to be reversible on the triplet and singlet potential energy surfaces and to involve higher energy CH_2–TiHF (triplet) and CH_3–TiF (singlet) states in spin-allowed electronic transitions as reported and reviewed.[2d,26] Starting with the global minimum energy CH_2–TiF (triplet) species, UV absorption to the higher energy CH_2–TiHF (triplet) occurs followed by relaxation in the

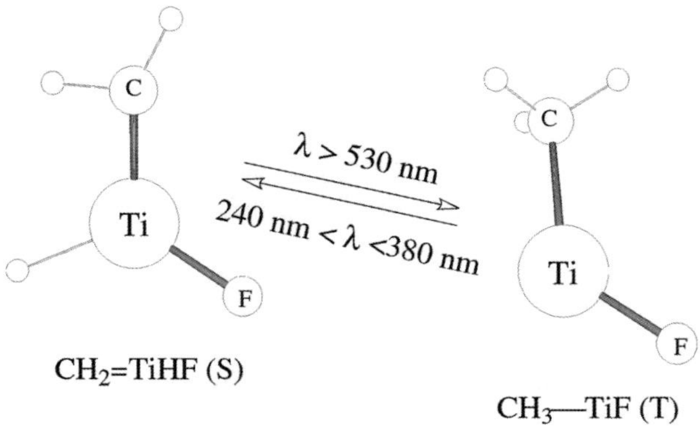

Fig. 6. Reversible photochemical α–H transfer between CH_3–TiF (triplet) and CH_2=TiHF (singlet).

argon matrix to the CH_2=TiHF (singlet) state. In the reverse process, the CH_2=TiHF (singlet) state absorbs Vis light to the higher energy CH_3–TiF (singlet) state, which is followed by matrix relaxation to the ground CH_3–TiF (triplet) state. The singlet methylidene is stable in the matrix even though it is 22 kcal mol^{-1} higher in energy than the triplet insertion product. This stability arises from the singlet transition state energy barrier (18 kcal mol^{-1}) for reverse H transfer in the ground singlet methylidene and shows the need for photo-excitation to drive the reversible α–H transfer process (Fig. 6). The computed structure for CH_2=TiHF reveals a H′–C–Ti angle (H′ is the agostic hydrogen) of 91.6° and agostic distortion of the CH_2 subgroup as illustrated by the structure. Recently, Grunenberg *et al.* have used the CH_2=TiHF complex as a model to explore the agostic bonding interaction, which is manifest in organometallic chemistry.[28]

7. Mo and CH_4 Form a Methylidyne Complex

After our successful preparation of the methylidene CH_2=TiHF, we investigated the reaction of Mo and CH_3F anticipating that the higher valence of Mo would support more α–H transfer. In addition to the expected CH_3–MoF insertion product and the methylidene CH_2=MoHF, the higher oxidation state methylidyne complex CH≡MoH_2F was

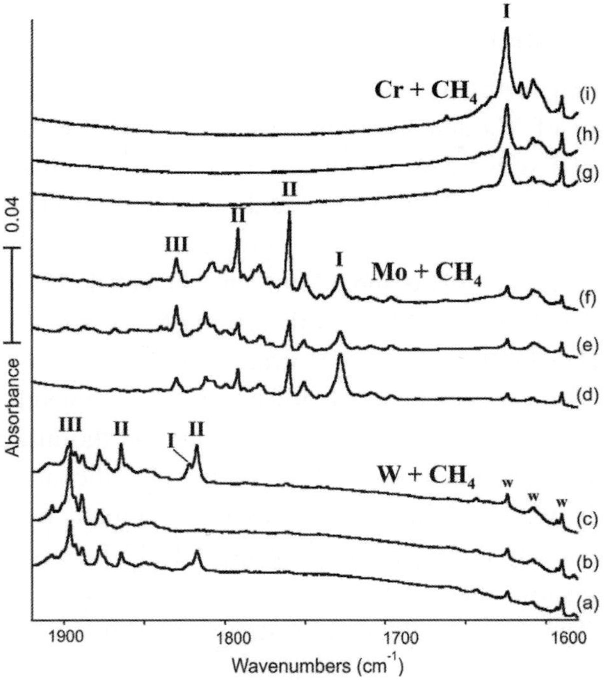

Fig. 7. Infrared spectra in the M–H stretching region for the products of laser-ablated group 6 metal atom reactions with 2% methane in argon deposited at 7 K for 1 h. (a) W, (b) after Vis irradiation (>530 nm), (c) after UV irradiation (240–380 nm), (d) Mo, (e) after Vis irradiation, (f) after UV irradiation, (g) Cr, (h) after Vis irradiation, and (i) after UV irradiation. **w** denotes water impurity absorptions.

observed.[29] Such reactions vary within the group 6 metals as Cr only forms CH_3–CrF, but W reacts to give all three products.[30] The situation is similar for methane as Cr yields only CH_3–CrH, but Mo and W react to give all three products.[31,32]

Infrared spectra for the Cr, Mo, and W reaction products with CH_4 under identical experimental conditions are compared in Fig. 7. Laser-ablated Cr, Mo, and W atoms are sufficiently energetic to overcome the activation energy for C–H insertion. Chromium gives only one product (labeled **I**), which increases on Vis and UV irradiations, but Mo and W form three products, which are interconverted by persistent reversible Vis and UV photochemical processes, as described elsewhere.[32]

The single chromium product is CH_3–CrH since CH_2=CrH_2 is too high in energy to form in these experiments. We note that very few chromium methylidene complexes are known, in contrast to the large number of Mo and W complexes.[33] The Mo case is particularly noteworthy.[31] Excited Mo activates CH_4 to give CH_3–MoH, which is characterized by a Mo–H stretching mode at 1728.0 cm^{-1} (labeled **I**). Vis irradiation promotes α–H transfer to give CH_2=MoH_2 with two Mo–H stretching modes at 1791.6 and 1759.6 cm^{-1} (labeled **II**), and a second α–H transfer to form CH≡MoH_3 with a strong MoH_3 stretching mode at 1830 cm^{-1} (labeled **III**). Next, UV irradiation transfers hydrogen back to carbon. These photochemical processes are completely reversible (three cycles performed) as shown by changes in the infrared spectra in Fig. 7 and summarized below. The product identification is based on agreement between calculated and observed isotopic frequencies, as detailed previously for each new molecule.[31] B3LYP density functional calculations show that CH≡MoH_3 has C_{3v} symmetry and is a true trihydride. The Mo–H and C≡Mo stretching, MoH_2 and Mo–C–H bending and MoH_3 rocking frequencies have also been observed for the CH MoH_3 methylidyne complex. A similar case has been made for CH≡WH_3.[32]

$$Mo(^7S)+CH_4 \xrightarrow{UV} CH_3\text{—}MoH \underset{UV}{\overset{Vis}{\rightleftarrows}} CH_2{=}MoH_2 \underset{UV}{\overset{Vis}{\rightleftarrows}} CH{\equiv}MoH_3 \tag{3}$$

The C=Mo double-bond lengths computed here for CH_2=MoHF and CH_2=MoH_2 (1.838 and 1.870 Å) are in the range of X-ray diffraction values ranging from 1.827 to 1.878 Å for several ligand-substituted Mo alkylidene complexes reviewed by Schrock.[33] The C≡Mo triple-bond lengths calculated for CH≡MoH_2F and CH≡MoH_3 (1.719 and 1.714 Å) are slightly shorter than the 1.743 and 1.754 Å values measured for larger $(AdO)_3$Mo≡CR complexes and the 1.762 Å value for $(R_1R_2N)_3$Mo≡$CPPh^-$. Finally, these methylidene and methylidyne complexes prepared by laser-ablated metal atom reactions provide simple models for the ligand-stabilized organometallic complexes important in catalytic and C–H activation reactions.[2d,33]

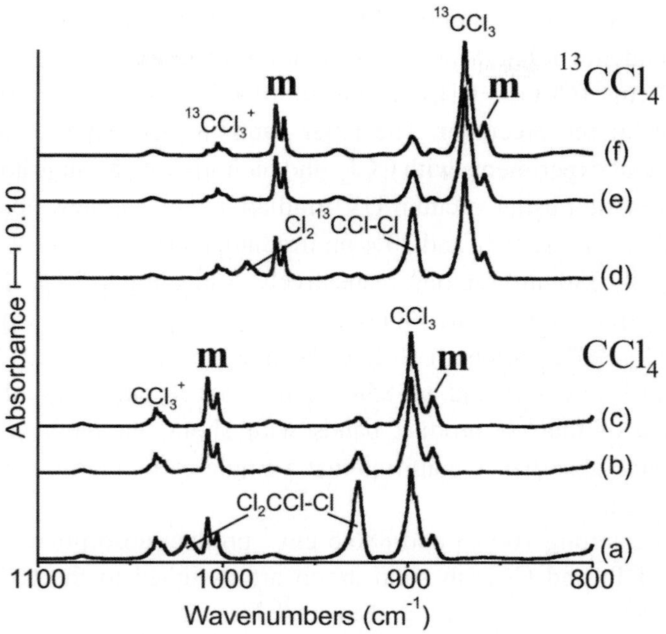

Fig. 8. Infrared spectra in the 1100–800 cm^{-1} region for the laser-ablated platinum atom and carbon tetrachloride reaction products in excess argon at 10 K. (a) Pt and CCl_4 (0.5% in argon) codeposited for 1 h, (b) after Vis ($\lambda > 420$ nm) irradiation for 20 min, and (c) after UV (240–380 nm) irradiation for 20 min. (d) Pt and $^{13}CCl_4$ (0.5% in argon, 90% enriched) codeposited for 1 h, (e) after Vis ($\lambda > 420$ nm) irradiation for 20 min, and (f) after UV (240–380 nm) irradiation for 20 min.

8. Pt and CCl_4 Form a Carbene

Platinum metal is well known as a catalyst, and a number of platinum carbene complexes and their organometallic chemistry in catalytic reactions have been explored.[34] Carbon tetrachloride reacts extensively with metal atoms, and the simple $ClC{\equiv}MoCl_3$ methylidyne product is a testament to this point.[35] In the case of platinum product energies terminate the reaction with the $CCl_2{=}PtCl_2$ carbene.[36]

Laser-ablated Pt atoms react with carbon tetrachloride in excess argon during condensation at 8 K as evidenced by two strong new absorptions at 1008.3 cm^{-1} (1003.1 cm^{-1} satellite for matrix site splitting) and

886.5 cm^{-1} (884.6 cm^{-1} shoulder for chlorine isotopic splitting) (labeled **m** for methylidene) in Fig. 8 along with bands at 1036.4 cm^{-1} (CCl_3^+), 1019.3 and 926.7 cm^{-1} ($Cl_2CCl{-}Cl$), and 898 cm^{-1} (CCl_3) produced by laser plume irradiation of the precursor. The latter bands are common to all laser-ablated metal experiments with CCl_4, and their identification follows from earlier formation using vacuum UV irradiation.[35,36] The new absorptions increased in concert 10% and 20% on irradiations in the Vis ($\lambda > 420$ nm) and UV (240–380 nm) regions, respectively. Annealing to 28 K sharpened the bands and resolved the 884.6 cm^{-1} shoulder. A similar experiment with $^{13}CCl_4$ (90% enriched) shifted the new absorptions to 971.4 cm^{-1} (966.9 cm^{-1} satellite) and to 858.5 cm^{-1} (856.1 cm^{-1} shoulder). The appearance of the ^{12}C product bands with about one-tenth of the ^{13}C product band absorbance confirms that a single carbon atom participates in these vibrational modes.

The two strong 1008.3 and 886.5 cm^{-1} product absorptions from the reaction of Pt and CCl_4 in solid argon are assigned to the $CCl_2{=}PtCl_2$ methylidene for the following reasons.[36] First, the observed bands are 1% and 2% higher than the harmonic frequencies calculated for the two strongest modes of this minimum energy product using the B3LYP density functional. Furthermore, the strongest calculated (B3LYP) absorption for the 23 kcal mol^{-1} higher energy $CCl_3{-}PtCl$ insertion product at 838 cm^{-1} is not detected. Second, the carbon-13 shifts, 36.9 and 28.0 cm^{-1}, observed for the two strong bands very nearly match the computed values, 36.6 and 28.1 cm^{-1}, as the observed 12/13 isotopic frequency ratios 1.0380 and 1.0326 are almost the same as the calculated frequency ratios, 1.0380 and 1.0334. This means that the calculation is describing the same normal modes as observed for the reaction product, and it reinforces the match in calculated frequency position and high intensities. Third, the 886.5 cm^{-1} product band and 884.6 cm^{-1} shoulder have the appropriate 9/6 relative intensity for natural abundance chlorine isotopes (35,35/35,37 statistical population) for a vibration involving two equivalent chlorine atoms. The calculation predicts a 2.0 cm^{-1} shift, and we observe a 1.9 cm^{-1} shift for this isotopic effect.

Interestingly, the 1008.3 cm^{-1} vibrational coordinate involves mostly Pt=C stretching mixed with symmetric C–Cl_2 displacement. For a pure symmetric Cl–C–Cl stretching mode with the 115.3° valence angle calculated here, the 12/13 isotopic frequency ratio would be 1.0250, which is substantially less than the observed 1.0380 value. This underscores the description of this normal mode as C vibrating back and forth between Pt and two Cl atoms. The calculated 438.9 cm^{-1} mode with only 0.4 cm^{-1} carbon-13 shift is the symmetric Cl_2–C=Pt stretching counterpart, which involves little carbon motion.

The matrix infrared spectrum with the two strongest modes calculated for CCl_2=$PtCl_2$ facilitates the identification of this simple platinum–carbene complex.

9. Uranium Methylidenes and Methylidynes

Thorium and uranium chemistry are dominated by IV and VI oxidation states, respectively,[15] so we thought it should be possible to prepare some of the above group 4 and 6 complexes using the same reactions with these laser-ablated early actinide metals. Following Ti, Zr, and Hf reactions, Th experiments were done with methane, and a group of five bands (1435.7, 1397.1 cm^{-1} ThH_2 stretch, 670.8 cm^{-1} C=Th stretch, 634.6 cm^{-1} CH_2 wag, and 458.7 cm^{-1} ThH_2 bend) were assigned to CH_2=ThH_2 based on excellent agreement with calculated frequencies.[37] The computed structure reveals comparable agostic distortion to that calculated for CH_2=HfH_2.[2d] The methyl halide series of reaction products was also investigated, and the computed CH_2=ThHX structures (X = F, Cl, Br) reveal the now established small increase in agostic distortion of the CH_2 group with increasing halogen size while the single Th–H stretching frequency increased (1380.5 to 1401.7 to 1409.5 cm^{-1}).[38]

Analogous reactions were also performed for uranium, and similar infrared spectra were observed for the CH_2=UHX complexes.[39] The U–H stretching mode for CH_2=UHF (1410.4 cm^{-1}) is higher than that for CH_2=ThHF (1380.5 cm^{-1}) just as molecular uranium hydride species are higher than their thorium counterparts. Calculations for triplet ground state CH_2=UHF are complicated by multireference character, but distorted agostic structures were found at the DFT level, which show more distortion than the thorium counterpart. The two weak U–H stretching frequencies observed at 1461 and 1425 cm^{-1} for the uranium and methane reaction product bear the same relationship to the 1410.5 cm^{-1}

band for CH_2=UHF that the two ThH_2 stretching modes for CH_2=ThH_2 do to the 1380.5 cm^{-1} band for CH_2=ThHF. The straightforward assignment of the new 1461 and 1425 cm^{-1} bands to CH_2=UH_2 is supported by DFT and CASPT2 calculations on this uranium methylidene dihydride complex in the triplet ground state.[40]

Owing to strong metal fluoride bonds, it appeared that higher oxidation state products would be favored with more heavily halogenated reagents. Thorium and fluoroform gave a strong product absorption at 521.3 cm^{-1} just above the 508 cm^{-1} precursor band, and weaker absorptions at 565.4 and 502.1 cm^{-1}. These bands increased 10% on >290 nm and 20% on >220 nm irradiations and shifted slightly with CDF_3 to 520.6 cm^{-1} and to 563.4 and 493.1 cm^{-1}. The most stable complex with this stoichiometry is triplet HC÷ThF_3, which has very weak degenerate pi bonds each populated by one electron located mostly on the carbon center.[41] This reaction proceeds through two α–F transfers after the initial insertion, which is overall 227 kcal mol^{-1} exothermic. The CHF=ThF_2 structural isomer singlet methylidene complex is 46 kcal mol^{-1} higher in energy than the final product and is an intermediate step in the reaction mechanism.

The analogous reaction with uranium produced similar product spectra. New absorptions marked with arrows in Fig. 9a are observed at 576.2, 540.2, and 527.5 cm^{-1} in the infrared spectrum recorded after the initial reaction of U and CHF_3. These bands increased by 30% on UV irradiation (λ > 290 nm) and another 20% on further UV irradiation (λ > 220 nm). The reaction with CDF_3 shown in Fig. 9d gave the same upper band, the strong band shifted to 535.9 cm^{-1}, and the lower band shifted below our region of observation. As CF_4 is less reactive, reagent codeposition produced weaker bands at 578.7 and 536.4 cm^{-1}, which increased slightly during the first irradiation (λ > 290 nm) and considerably in the second irradiation (λ > 220 nm) as shown in Figs. 9i, g, and h. These bands were stable on annealing the sample to 30 K, and they increased slightly on further UV irradiation.[42]

Based on agreement with density functional computed frequencies, the higher bands are assigned to the symmetric stretching mode of a trigonal UF_3 group and the strongest band to the degenerate antisymmetric counterpart. The lowest band is due to the degenerate

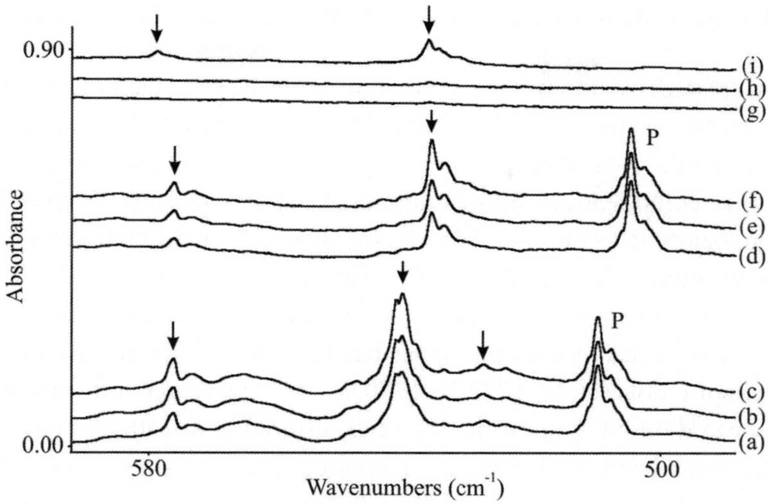

Fig. 9. Infrared spectra in the 590–490 cm^{-1} region for laser-ablated U atoms codeposited with fluoromethanes in excess argon at 8 K. (a) U and 1% CHF_3 in argon codeposited for 1 h, (b) after >290 nm irradiation, and (c) after >220 nm irradiation. (d) U and 1% CDF_3 in argon codeposited for 1 h, (e) after >290 nm irradiation, and (f) after >220 nm irradiation. (g) U and 1% CF_4 in argon codeposited for 1 h, (h) after >290 nm irradiation, and (i) after >220 nm irradiation. Precursor absorptions are labeled P.

H–U–F bending mode, which is, of course, absent in the CF_4 reaction product spectrum. Theoretical analysis shows that a triple bond is formed between carbon and uranium in these $HC{\equiv}UF_3$ and $FC{\equiv}UF_3$ methylidyne molecules.[42] Thus, the additional two valence electrons in U as compared to Th complete the formation of a relatively strong triple bond with U that is very weak with Th.

Expanding the extensive methylidene and methylidyne organometallic chemistry of the transition metals[33] to include actinide metals is a difficult challenge. This may be due in part to the fact that actinide 6*d* and 5*f* valence orbitals do not behave like transition-metal *nd* orbitals, but we find some transition-metal-like behavior in the early actinides. These investigations on Th and U methylidenes and methylidynes[37–42] provide information on the existence and nature of novel actinide–carbon multiple bonded species.

10. Tungsten Reacts with Ammonia to Produce Nitrido Trihydride

Laser-ablated W atoms were reacted with NH_3 in argon in many experiments using different substrate temperatures, laser energies, and ammonia concentrations in order to optimize product yield and to minimize complications from ammonia clusters.[43] In the W−H stretching region, bands at 1924.1 and 1917.0 cm^{-1} were observed after deposition, and they increased slightly on 20 K annealing, increased on >470 nm irradiation but decreased on 240–380 nm irradiation. A new weak band at 1092.2 cm^{-1} tracked with the upper bands on annealing and irradiation. A site-split doublet at 1772.2, 1770.9 cm^{-1} decreased on annealing, disappeared on >470 nm irradiation but restored with 240–380 nm light. A very weak band at 1188.0 cm^{-1} increased markedly on first annealing, disappeared on >470 nm irradiation, but returned on further annealing. It appears that these product species are converted into each other on photochemical rearrangements. Isotopic labeling experiments were done using $^{15}NH_3$ and ND_3 samples. The W–H stretching frequencies with $^{15}NH_3$ were essentially the same as those with NH_3, but the weak band at 1092.2 cm^{-1} shifted to 1058.4 cm^{-1} in the ammonia cluster region: these bands increased together on >470 nm irradiation. With ND_3 the upper bands shifted into the W–D stretching region at 1377.0 and 1272.3, 1271.6 cm^{-1}.

The 1188.0 cm^{-1} band, the doublet at 1772.2, 1770.9 cm^{-1}, and the 1924.1, 1917.0, and 1092.2 cm^{-1} bands have been assigned to the $W{:}NH_3$ complex, the H_2N–WH intermediate, and the $N{\equiv}WH_3$ nitrido trihydride on the basis of isotopic shifts and agreement with density functional frequency calculations in our original report.[43] The very high frequency absorption was observed at 1917.0 cm^{-1} on deposition of laser-ablated W atoms with NH_3 tracks with a weak band at 1924.1 cm^{-1} in the W–H stretching region and a weak band at 1092.2 cm^{-1} in the W−N stretching region. The two W−H stretching absorptions are slightly lower than this mode for WH_6 in argon but higher than the remaining tungsten hydrides. With ND_3 the two upper bands shifted into one band at 1377.0 cm^{-1}, which defines a 1.392 H/D ratio that is expected for tungsten hydrides. The W–N stretching frequency at 1092 cm^{-1} is 33 cm^{-1} higher than the diatomic W≡N molecule fundamental in solid argon suggesting triple-bond character. With $^{15}NH_3$ the two upper W–H stretching modes showed no displacement, but the W–N stretching mode shifted to 1058.4 cm^{-1}, 3.1%, which is similar to that for the W≡N diatomic molecule. These three bands are due to the W–H and W–N stretching modes of the $N{\equiv}WH_3$ molecule.

Since the *d* electrons of tungsten are strongly involved in σ bonding, reactions of W with NH_3 gave the major stable high-oxidation state product, $N{\equiv}WH_3$, which is in agreement with the DFT-calculated energy profile.[43] With the same rationale tungsten hexahydride, WH_6, has been identified in W atom reactions with H_2 in a neon matrix.[21] Natural orbital occupations have been calculated for the $N{\equiv}WH_3$ molecule.[43] The effective bond order is 2.9, depending on the theoretical method employed, which is a fully developed triple bond.

11. Conclusions

After more than five decades of matrix isolation spectroscopy, it is still possible to produce many additional novel chemical species for spectroscopic investigation and characterization. The unique combination of physical techniques such as laser ablation with low temperature trapping methods will sustain this evolution of new chemistry.

Acknowledgments

The author gratefully acknowledges support for this research from the National Science Foundation (US) and the contributions of the many coauthors whose names are given in the references below. The author also wishes to acknowledge numerous productive and pleasant professional and personal interactions with many scientists in the matrix isolation field over more than four decades.

References

1. Examples of early reviews include books (a) M. Moskovits and G. A. Ozin (Eds.), *Cryochemistry*, Wiley-Interscience, New York (1976). (b) L. Andrews and M. Moskovits (Eds.), *Chemistry and Physics of Matrix Isolated Species*, North-Holland, Amsterdam (1989).
2. (a) M. Zhou, L. Andrews and C. W. Bauschlicher, Jr., *Chem. Rev.* **101**, 1931 (2001) and references therein. (b) L. Andrews and A. Citra, *Chem. Rev.* **102**, 885 (2002) and references therein. (c) L. Andrews, *Chem. Soc. Rev.* **33**, 123 (2004) and references therein. (d) L. Andrews and H.-G. Cho, *Organometallics* **25**, 4040 (2006) and references therein.
3. V. E. Bondybey, A. M. Smith and J. Agreiter, *Chem. Rev.* **96**, 2113 (1996).
4. M. E. Jacox, *Chem. Phys.* **189**, 149 (1994).
5. W. L. S. Andrews and G. C. Pimentel, *J. Chem. Phys.* **44**, 2361 (1966).
6. W. E. Billups, S.-C. Chang, R. H. Hauge and J. L. Margrave, *J. Am. Chem. Soc.* **117**, 1387 (1995).
7. L. B. Knight, Jr., B. W. Gregory, S. T. Corbranchi, D. Feller and E. R. Davidson, *J. Am. Chem. Soc.* **109**, 3521 (1987).
8. T. R. Burkholder and L. Andrews, *J. Chem. Phys.* **95**, 8697 (1991).
9. S. Li, H. A. Weimer, R. J. Van Zee and W. Weltner, *J. Chem. Phys.* **106**, 2583 (1997).
10. S. Tam, M. Macler, M. E. DeRose and M. E. Fajardo, *J. Chem. Phys.* **113**, 9067 (2000).
11. L. Andrews, *J. Chem. Phys.* **50**, 4288 (1969).
12. D. A. Hatzenbuhler and L. Andrews, *J. Chem. Phys.* **56**, 3398 (1972). (b) L. Andrews and R. R. Smardzewski, *J. Chem. Phys.* **58**, 2258 (1973).
13. W. D. Allen, D. A. Horner, R. L. Dekock, R. B. Remington and H. F. Schaefer, III, *Chem. Phys.* **133**, 11 (1989) and references therein.
14. X. Wang and L. Andrews, *Mol. Phys.* **107**, 739 (2009).
15. F. A. Cotton, G. Wilkinson, C. A. Murillo and M. Bochmann, *Advanced Inorganic Chemistry*, 6th ed., Wiley, New York (1999).
16. J. W. Turley and H. W. Rinn, *Inorg. Chem.* **8**, 18 (1969).
17. L. Andrews and X. Wang, *Science* **299**, 2049 (2003).
18. X. Wang, L. Andrews, M. E. DeRose, S. Tam and M. E. Fajardo, *J. Am. Chem. Soc.* **125**, 9218 (2003).
19. A. J. Downs and C. R. Pulham, *Chem. Soc. Rev.* **23**, 175 (1994).
20. (a) G. V. Chertihin and L. Andrews, *J. Phys. Chem.* **97**, 10295 (1993). (b) F. A. Kurth, R. A. Eberlein, H. Schnöckel, A. J. Downs and C. R. Pulham, *J. Chem. Soc. Chem. Commun.* 1302 (1993). (c) P. Pullumbi, Y. Bouteiller, L. Manceron and C. Mijoule, *Chem. Phys.* **185**, 25 (1994).
21. X. Wang and L. Andrews, *J. Phys. Chem.* **A106**, 6720 (2002).
22. M. Kaupp, *J. Am. Chem. Soc.* **118**, 3018 (1996).

23. X. Wang, L. Andrews, I. Infante and L. Gagliardi, *J. Am. Chem. Soc.* **130**, 1972 (2008).
24. G. J. Kubas, R. R. Ryan, B. I. Swanson, P. J. Vergamini and H. J. Wasserman, *J. Am. Chem. Soc.* **106**, 451 (1984).
25. X. Wang and L. Andrews, *Organometallics* **27**, 4273 (2008).
26. (a) H.-G. Cho and L. Andrews, *J. Phys. Chem.* **A108**, 6294 (2004). (b) H.-G. Cho and L. Andrews, *Inorg. Chem.* **43**, 5253 (2004).
27. G. S. McGrady, A. J. Downs, N. C. Bednall, D. C. McKean, W. Thiel, V. Jonas, G. Frenking and W. Scherer, *J. Phys. Chem.* **A101**, 1951 (1997).
28. G. von Frantzius, R. Streubel, K. Brandhorst and J. Grunenberg, *Organometallics* **25**, 118 (2006).
29. H.-G. Cho and L. Andrews, *Chem. Eur. J.* **11**, 5017 (2005).
30. H.-G. Cho and L. Andrews, *Organometallics* **24**, 5678 (2005).
31. H.-G. Cho and L. Andrews, *J. Am. Chem. Soc.* **127**, 8226 (2005).
32. H.-G. Cho L. Andrews and C. Marsden, *Inorg. Chem.* **44**, 7634 (2005).
33. R. R. Schrock, *Chem. Rev.* **102**, 145 (2002).
34. J. W. Herndon, *Coord. Chem. Rev.* **251**, 1158 (2007), and earlier review articles in this series.
35. J. T. Lyon, H.-G. Cho and L. Andrews, *Organometallics* **26**, 6373 (2007).
36. H.-G. Cho and L. Andrews, *J. Am. Chem. Soc.* **130**, 15836 (2008).
37. L. Andrews and H.-G. Cho, *J. Phys. Chem.* **A109**, 6796 (2005).
38. J. T. Lyon and L. Andrews, *Inorg. Chem.* **44**, 8610 (2005)
39. J. T. Lyon and L. Andrews, *Inorg. Chem.* **45**, 1847 (2006).
40. J. T. Lyon, L. Andrews, P.-Å. Malmqvist, B. O. Roos, T. Yang and B. E. Bursten, *Inorg. Chem.* **46**, 4917 (2007).
41. J. T. Lyon and L. Andrews, *Eur. J. Inorg. Chem.* **1047** (2008).
42. J. T. Lyon, H.-S. Hu, L. Andrews and J. Li, *Proc. Natl. Acad. Sci.* **104**, 18919 (2007).
43. X. Wang and L. Andrews, *Organometallics* **27**, 4885 (2008).
44. C. Liang, R. D. Davy and H. F. Schaefer, III, *Chem. Phys. Lett.* **159**, 393 (1989).
45. M. Shen and H. F. Schaefer, III, *J. Chem. Phys.* **96**, 2868 (1992).

CHAPTER 3

CONFORMATIONAL CHANGES IN CRYOGENIC MATRICES

Rui Fausto

Department of Chemistry, University of Coimbra
P-3004-535 Coimbra, Portugal
E-mail: rfausto@ci.uc.pt

Leonid Khriachtchev

Laboratory of Physical Chemistry, University of Helsinki
P. O. Box 55, FIN-00014 Helsinki, Finland
E-mail: leonid.khriachtchev@helsinki.fi

Peter Hamm

Institute of Physical Chemistry, University of Zurich
Winterthurerstrasse 190, CH-8057 Zurich, Switzerland
E-mail: phamm@pci.uzh.ch

This chapter describes thermal and light-induced conformational changes experienced by small compounds in cryogenic matrices. For thermally induced processes, changing the temperature of the cold substrate during matrix deposition and annealing of the deposited matrix are useful methods to change the relative populations of the conformers. In the case of light-induced conformational changes, the use of both UV–visible and IR excitations is described, for the latter with focus on processes promoted by narrowband light. Time-dependent experiments and computations on the conformational change of HONO are described providing a prototype system for dynamical studies of conformational changes in matrix-isolated species.

1. Introduction

Molecular conformation is a central concept in molecular physics, chemistry, and biochemistry. Conformations of molecules determine the physical and chemical properties of substances and their biological functions. Many studies on conformers of small molecules have been performed, and these studies aim to constitute a basis for our understanding of the conformational behavior of more complex chemical systems including biological macromolecules.

The molecular conformations of small molecules determine not only the physical properties directly connected with the molecular structure (dipole moment or spectroscopic characteristics) but also properties influenced by intermolecular interactions. Different conformers of a given molecule may show distinct aggregation abilities, with some of them showing greater tendency to interact with other species[1] or promoting specific arrangements of aggregates, which in turn may lead, for example, to polymorphism.[2,3] Conformations are also important in determining the photochemical reactivity.[4]

The importance of conformations in biochemistry is well accepted. Conformations determine, for instance, catalytic mechanisms in enzymes, antifreeze cryoprotectants efficiency or membrane properties, and are the basis of the general phenomenon of molecular recognition. Conformational changes are key events in the vision process,[5] in the control of membrane permeability and different mechanisms of molecular transport in cells[6] and adenosine-5′-triphosphate (ATP) synthesis.[7] Protein folding is also controlled by conformation of the constituting amino acids. A misfolded protein may be not only just unable to perform its usual beneficial function, but also act as a powerful infectious agent, as in the case of the bovine spongiform encephalopathy in cattle and the Creutzfeldt–Jakob disease in humans.[8] However, it is often difficult to apply mechanisms known for small species to biological molecules, and this is not the purpose of this chapter. Building the bridge between small conformers in cryogenic matrices and real biochemical species is a great challenge for future research.

In this chapter, conformational changes occurring upon thermal or photoexcitation of small compounds in cryogenic matrices are described. Section 2 focuses on thermally induced processes, while Sections 3 and 4 describe UV–visible- and IR-induced conformational changes, respectively. In the latter case, we concentrate on the processes promoted by narrowband excitation. The last section of the chapter considers what can be learned from time-dependent dynamical studies of conformational changes in matrix-isolated species.

2. Thermally Induced Conformational Changes

2.1. *Conformational cooling*

For matrix-isolated compounds, accessible thermal processes are in principle restricted to low-energy barrier reactions (a few kJ mol^{-1}) since the work temperature is very low. However, conformational changes are frequently observed thermal processes in matrices. In the simplest case, a molecule has only two conformational states. Three different basic situations are possible:

(i) The conformational barrier is high enough to prevent any reaction, the gas-phase conformational distribution is preserved in a deposited matrix, and no reaction is observed upon annealing the matrix at higher temperatures.

(ii) The energy barrier is very low (below 1 kJ mol^{-1}) and can be easily surmounted even at the matrix temperature. In this case, a true equilibrium between the conformers can be reached in the matrix and the trapped conformational distribution strongly differs from the gas phase. Annealing the matrix produces reversible conversion, and higher temperatures increase the population of the higher-energy conformer.

(iii) The energy barrier is a few kJ mol^{-1} and partial conformational conversion can take place during deposition; hence, the matrix population of the conformers differ from the gas phase with the increased population of the more stable conformer. Annealing the matrix to a higher temperature leads to conversion of the higher-energy conformer to the lower-energy form.

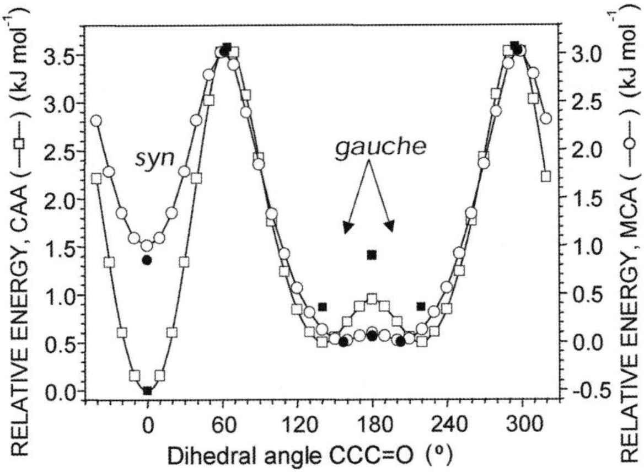

Fig. 1. Potential energy profiles for internal rotation about the C_α–C bond in cyanoacetic acid (CAA) and methyl cyanoacetate (MCA). The calculations are performed at the MP2/aug-cc-pVDZ level of theory; in all input geometries, the O=C–O–H(CH_3) dihedral was 0°. Solid symbols show positions of the critical points in the potential energy surface with inclusion of zero-point vibrational energy.[9]

The complexity is even greater because the matrix environment can perturb the potential energy landscape, change the relative energy of the conformers and the barriers, or even create or destroy specific conformational states, so that the knowledge of the gas-phase potential energy surface may not be enough to predict the experimental results in a solid matrix. Nevertheless, the gas-phase data is always an adequate starting point to analyze experiment and it can indicate the significance of matrix–guest interactions. In general, a more polarizable matrix environment tends to favor more polar conformations and the ability of a matrix to dissipate thermal energy, providing efficient conformational cooling, increases along the series Ne < Ar < Kr < Xe. In addition, the requirement for a guest to fit the matrix structure seems to favor more planar conformations.

Cyanoacetic acid ($NCCH_2COOH$, CAA) and methyl cyanoacetate ($NCCH_2COOCH_3$, MCA) are two closely related molecules which can provide a good illustrative example.[9] These molecules differ only by replacement of the acid hydrogen atom by a methyl group and they have similar potential energy surfaces; in particular, the estimated barriers for

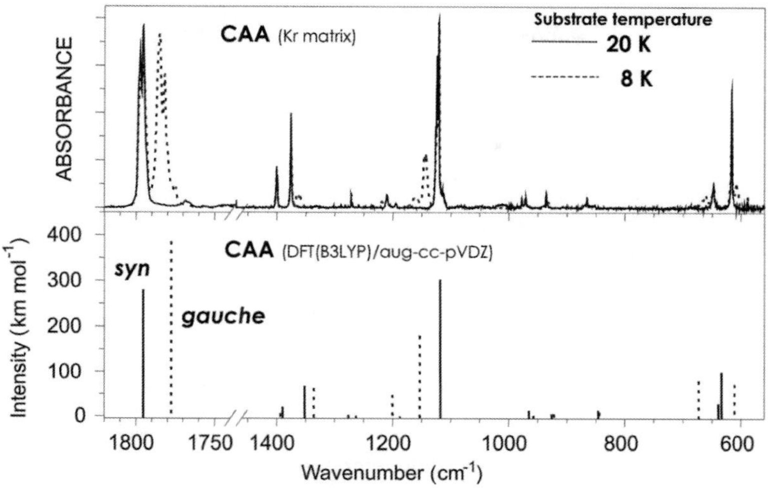

Fig. 2. IR spectra of cyanoacetic acid in a krypton matrix deposited at 8 and 20 K (top) and calculated spectra for *syn* and *gauche* CAA conformers (bottom).[9]

the *gauche* to *syn* conversion are equal (Fig. 1). In CAA, the *syn* conformer is theoretically the lower-energy form, whereas in MCA the relative stability of the conformers is opposite.[9]

The experimental results for CAA in krypton matrices deposited at 8 and 20 K (Fig. 2) clearly revealed conformational cooling during deposition. The less stable *gauche* conformer is totally converted into the *syn* form when the deposition is performed with the substrate at 20 K. Identical experiments carried out for the compound in xenon matrices led to the same conclusion. In addition, when deposition was performed at 8 K, the amount of the *gauche* conformer observed in a xenon matrix is smaller than that obtained in a krypton matrix. Hence, a xenon matrix is more efficient in promoting the conformational cooling than a krypton matrix, which is in agreement with general expectations. The next studies were done with MCA, and unexpected results were found. Deposition of the compound in a xenon matrix at 20 K traps both *syn* and *gauche* conformers, but contrarily to what we could expect, the population of the most stable *gauche* form was reduced compared to the gas phase.

The estimates of the relative populations for the two conformers in CAA and MCA at room temperature yielded the *gauche*/*syn* population

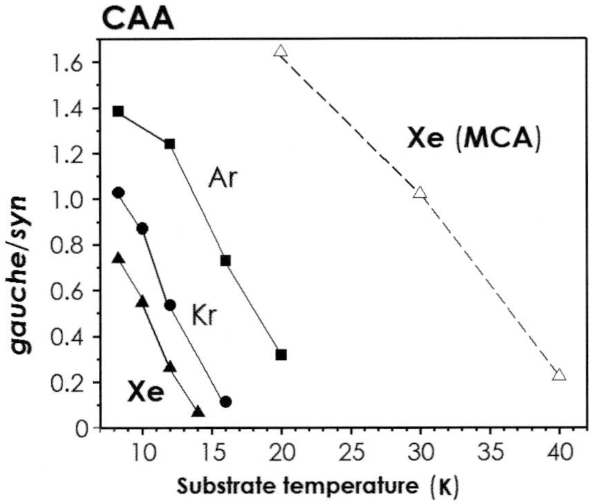

Fig. 3. Experimental *gauche/syn* population ratios for CAA and MCA in different matrices, deposited at various temperatures. The population ratios were obtained from the intensities of the C=O stretching bands, normalized by the calculated intensities (MP2/aug-cc-pVDZ). Each point in the graphic corresponds to a different experiment.[9]

ratios of 1.40 and 2.83, respectively.[9] For CAA, the calculated ratio is equal to that extracted from the IR intensities in an argon matrix deposited at 8 K. This shows that the substrate temperature of 8 K is low enough to prevent any conformational change during argon matrix deposition. Depositions at higher temperatures or using heavier matrix atoms yielded smaller *gauche/syn* populations, evidencing conformational cooling in an extension compatible with the expected relaxant properties of the matrix host, Ar < Kr < Xe (see Fig. 3).

On the other hand, the *gauche/syn* population ratio estimated for MCA in the gas phase (2.83) differs substantially from the values obtained in xenon matrices. The experimentally obtained population ratios were below 1.60 and decreased with the deposition temperature (Fig. 3). These results are explained assuming extensive conformational cooling and inversion of the relative stability of the two conformers in the matrix, i.e., the less stable in gas-phase *syn* conformer becomes the most stable form in the solid matrix. This change in the stability order of the two MCA conformers was explained in terms of additional stabilization in the matrix of the more polar and more planar *syn*

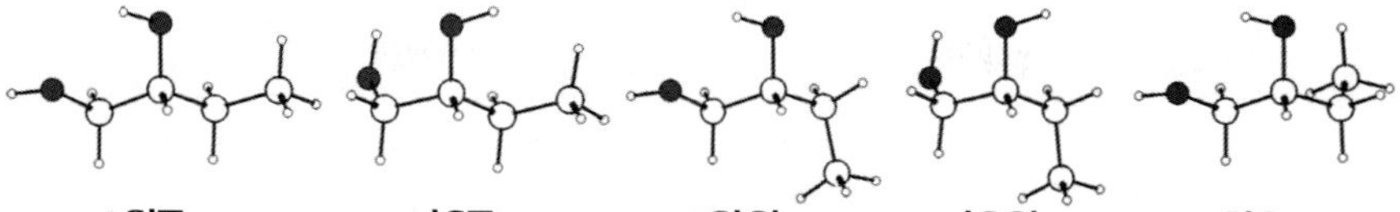

Fig. 4. Observed conformers of 1,2-butanediol in an as-deposited argon matrix.[11]

conformer.[9] The greater stability of this conformer was confirmed by annealing experiments, where the increase of the matrix temperature promoted additional *gauche* to *syn* conversion.

The next remarkable result on CAA was the observation of a matrix-induced conformational structure. In addition to the *syn* and *gauche* forms existing in the gas phase, the *anti* conformation is stabilized by a matrix and becomes the minimum energy structure. The *anti* form in the gas phase is the transition state between the two equivalent *gauche* forms. According to matrix annealing experiments, the *anti* conformer is more stable in matrices than the *gauche* form, which has a less planar heavy atom skeleton.[9]

A strong matrix effect was observed for the methyl ester of *N,N*-dimethylglycine.[10] For this molecule, the stability order of the two most stable conformers was inverted in xenon matrices compared to the gas phase and the more polar and more planar Lp-N-C-C *gauche* conformer was stabilized in matrices. Thus, the potential energy profile for conformer interconversion in a xenon matrix should be considerably different from the gas phase, where the conformer energy difference is *ca.* 3 kJ mol^{-1} and the conformational conversion barrier is over 20 kJ mol^{-1}.[10]

The conformational cooling in 1,2-butanediol is another remarkable case. In this molecule, a massive conformational cooling was observed, which converted 81 gas-phase conformers to a single conformer in a xenon matrix.[11] The conformational cooling occurred in steps accordingly to the conformation barriers. 1,2-Butanediol is a four internal rotor system and the conformational barriers involving rotation of the two hydroxyl groups are much lower than the barriers for rotation around the two central C–C bonds. The first barriers are easily surpassed

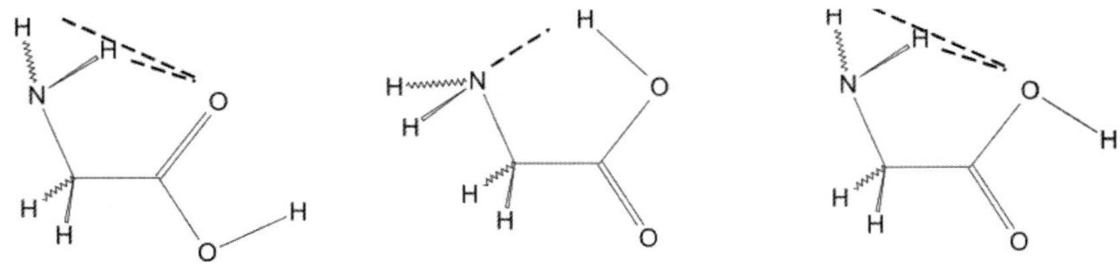

Fig. 5. Conformers of the non-ionic form of glycine. Dashed lines represent intramolecular hydrogen bonds. The numbers in parenthesis are zero-point corrected relative energies in kJ mol^{-1}, as predicted by the MP2/aug-cc-pVDZ *ab initio* calculations.[15]

during deposition of the matrix so that in the as-deposited matrices only the most stable member of each family of conformers sharing a common conformation about the two C–C bonds is observed (Fig. 4). Upon annealing the matrix up to 30 K, one of the conformers converted to the most stable form tG′Tg since the corresponding barrier is relatively small (9.1 kJ mol^{-1}). Further annealing at 50 K in a xenon matrix converted all higher-energy conformers to the most stable form featuring internal rotation barriers of 12–16 kJ mol^{-1}.

Similar studies were performed on other isomers of butanediol (1,3-, 2,3-, and 1, 4-butanediol).[12–14] In all these cases, stepwise conformational cooling determined by the relative rotational barriers and massive conformational cooling to the lowest-energy structures were observed. In the studies of butanediol isomers, the use of different matrix media and support from high-level theoretical calculations were essential to interpret the experimental data.

2.2. *"Elusive" conformers*

The change of temperature in a matrix isolation experiment provides an efficient way to observe conformers whose experimental detection is otherwise difficult by other methods. The 1,2-butanediol example given in Section 2.1 can be considered within this category since it demonstrates detailed structural and spectroscopic characterization of 1 among 81 possible conformers.[11] We provide here two additional examples. In the first study,[15] the third conformer of glycine was

Table 1. Calculated and observed characteristic frequencies (cm^{-1}) for three glycine conformers.[15]

Mode	Conformer	Calculated (MP2)	Experimental (Ar matrix)
νC=O	I	1787	1779
	II	1813	1790
	III	1779	1773/1767
νC–O	I	1113	1101/1099
	II	1210	1210
	III	1161	1147
τOH	I	642	619/617
	II	852	867
	III	675	644

observed experimentally for the first time. The second example describes the experimental observation of the elusive C_s symmetry conformer of trimethyl phosphate (TMP).[16]

Glycine possesses three internal rotation axes and the three most stable conformers shown in Fig. 5 should have significant populations in the gas phase. At the time of the matrix-isolation study by Stepanian *et al.*,[15] the question of the conformational structure of glycine in the gas phase remained open. An electron diffraction study showed that the most stable conformer I constituted *ca.* 75% of the total population at 219 °C,[17] which was in agreement with calculations.[18,19] In contrast, microwave spectroscopy studies suggested that conformer II is the lowest-energy form.[20–24] The difficulty in the observation of conformer I in the microwave spectra is caused by its significantly smaller dipole moment compared to conformer II. Since conformer III has also a small dipole moment, its identification in microwave spectra is difficult as well. On the other hand, conformers II and III differ from each other only by the position of the hydrogen atoms and they are indistinguishable by the electron-scattering intensities, thereby complicating their characterization with electron diffraction technique.

The matrix isolation studies on several isotopic analogs of glycine in different matrices could finally identify three low-energy conformers of glycine.[15] The calculated and experimental frequencies in an argon matrix are presented in Table 1. Figure 6 shows the carbonyl stretching

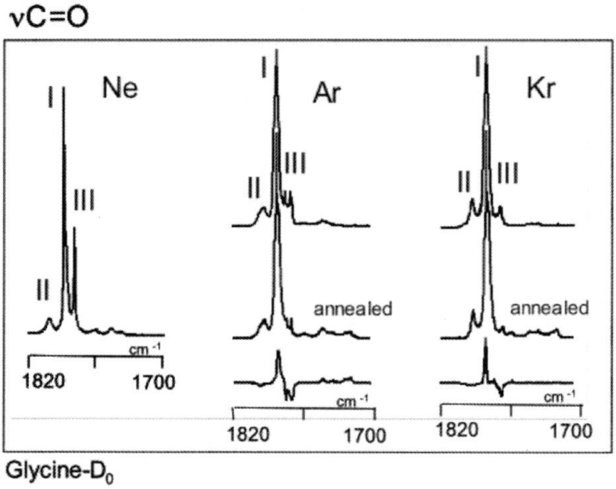

Fig. 6. Carbonyl stretching region of the IR spectra of glycine-D_0 in neon, argon, and krypton matrices showing bands of the three low-energy conformers.[15]

region of the IR spectra of glycine-D_0 in neon, argon, and krypton matrices and the results of annealing the latter two matrices at 20 K. Upon annealing the matrices above 13 K, conformer III converts to the most stable form I. The relative intensities of the bands assigned to each conformer show that the population of conformer I in the gas phase at 170 °C is *ca.* 70%, in good agreement with the previous electron diffraction study of Iijima *et al.*[17] and with the theoretical predictions.[15,18,19]

The study on TMP[16] provided the C_s symmetry conformer (Fig. 7), which had escaped experimental detection for many years. Two TMP stable conformations with C_1 and C_3 symmetries were demonstrated previously. The most polar C_1 form was more stable in polar environments, while the C_3 conformer predominates in the gas-phase and nonpolar environments.[25–27] However, vague indications for the third conformer with the C_s symmetry were reported.[28,29] In the experiments by Reva *et al.*,[16] xenon matrices were deposited at different temperatures. At the lowest deposition temperature (10 K), xenon solidifies very quickly and forms a disordered glass-like structure rather than a crystal. In this case, the TMP molecules are trapped in many different matrix morphologies (sites) and the absorption bands are broad. For deposition

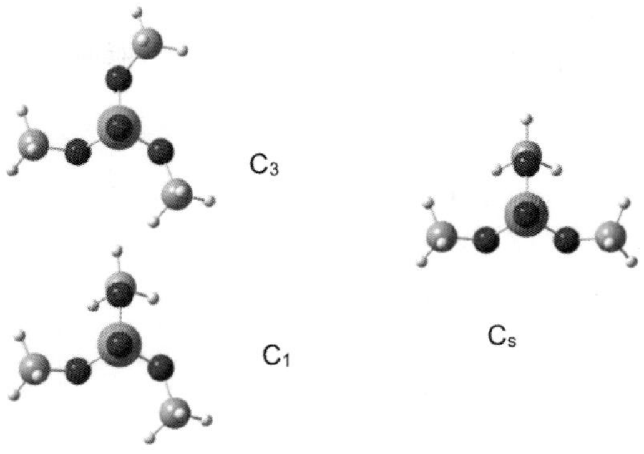

Fig. 7. Conformers of TMP by the DFT(B3LYP)/6-311++G(2d,2p) calculations. The relative zero-point corrected energies for the C_3, C_1, and C_S forms are 0.0, 4.3, and 8.2 kJ mol^{-1}, respectively.

at 10 K, conformational cooling is efficiently suppressed. Comparison of the experimental and theoretical spectra allow to identify the bands of the C_3 and C_1 conformers and also two bands assigned to νP–O–C of the C_s form (at 847 and 844 cm^{-1}). For deposition at 20–30 K, all conformers are still present in the matrix, but the bands become narrower due to partial matrix relaxation, which eliminates the less stable matrix sites. For deposition at 40 K, the C_s conformer is absent, indicating that it relaxes to the C_1 form, from which it is separated by an energy barrier of <1 kJ mol^{-1}. At 50 K, the bands corresponding to conformer C_1 become considerably less intense, whereas the bands of the C_3 form increased. These results indicate that at 50 K enough energy is available to overcome energy barriers of ~10 kJ mol^{-1} for the C_1 to C_3 conversion. At 60 K, extensive conformational cooling takes place and only the most stable conformer is trapped in the matrix. Besides the first experimental observation of the C_s conformer, this study is also a strong case of sequential conformational cooling of a matrix-isolated compound.

3. UV-Induced Conformational Changes

Irradiation of matrix-isolated compounds by UV light can induce a variety of chemical processes, from conformational changes to bond-breaking and bond-forming reactions. All these processes involve the

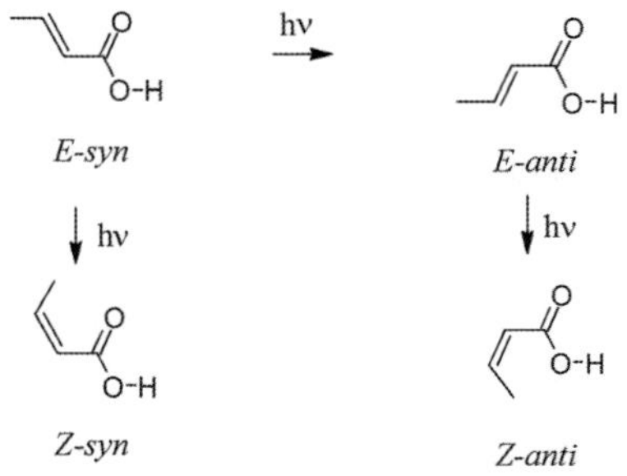

Fig. 8. Observed reactions upon UV irradiation (243 nm) of matrix-isolated *E*-crotonic acid.[31]

electronic excited states and they occur due to substantially different potential energy surfaces of the ground and excited electronic states.

The studies of matrix-isolated α,β-unsaturated carboxylic acids provide an interesting example.[30–32] In these compounds, the C_α–C_β linkage is a double bond in the ground electronic state and the minimum energy forms have the groups connected to the two carbon atoms in the same plane. In the excited states, this bond becomes essentially a single bond and the groups connected to the two carbon atoms are approximately perpendicular to each other.[33,34] This makes possible the internal rotation about this bond in the excited state and a different isomer can be obtained after relaxation to the ground state. This was exactly what was observed for *E*-crotonic acid,[31] where 243-nm light gives rise to two types of reactions: (i) conversion of the most stable *syn* form of *E*-crotonic acid to the *anti* conformer, and (ii) production of *Z*-crotonic acid *syn* and *anti* conformers (see Fig. 8). The latter processes involve internal rotation about the C_α–C_β bond in the excited state. In addition, the 266-nm light efficiently promotes conversion of fumaric acid to maleic acid,[32] and fumaryl chloride is changed to maleoyl chloride by irradiation at 340 nm.[35]

Another study showed conformational changes occurring upon photolysis of α-pyrone.[36] α-Pyrone general photochemistry has been known for many years[37–39]; however, only recently the relative importance in mechanistic terms of the different isomeric forms of the opening photoproduct has been addressed in detail.[36] The most efficient process was the Norrish type I ring opening reaction leading to a

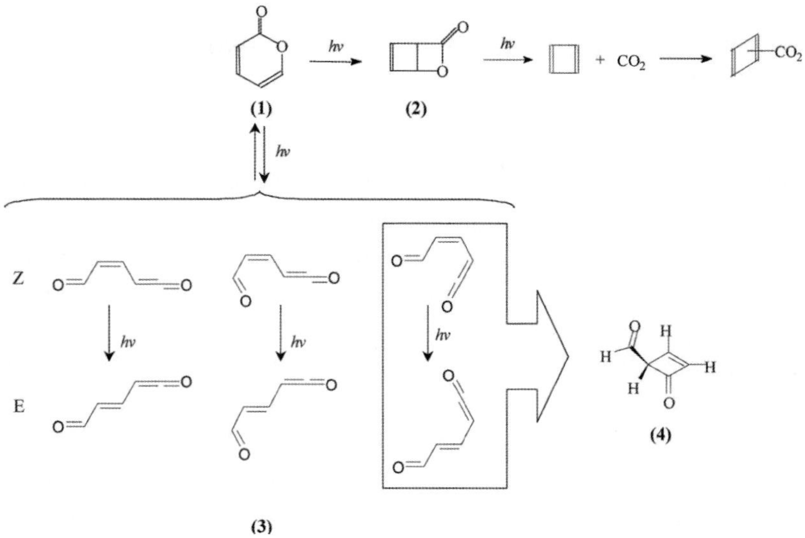

Fig. 9. Observed α-pyrone (1) photochemical reactions: fast photoequilibration with the conjugated aldehyde–ketene (3) and slow ring-closure to the Dewar isomer (2). In the first process, several conformers are produced, with some of them further reacting to yield 4-formyl-2-cyclobutene-1-one (4). The Dewar isomer of a pyrone subsequently expels CO_2 producing cyclobutadiene. The two latter species form a complex with the stacked structure and intermolecular distance of *ca*. 3.45 Å.[36]

conjugated aldehyde–ketene [(3) in Fig. 9]. Upon broadband irradiation with UV light (λ > 285 nm), rapid formation of the *Z* isomers of the aldehyde–ketene occurred. After a few minutes of irradiation, the IR bands assigned to these photoproducts stopped growing. Further irradiation generated new aldehyde–ketene forms, which were identified as conformers of the *E* isomer produced by excited state internal rotation around the central C=C bond. Upon subsequent UV irradiation (>337 nm), the *Z* aldehyde–ketene isomers reverted back to α-pyrone, while the *E* forms did not react further. For both *Z* and *E* ketene isomers, only the most stable conformers shown in Fig. 9 were observed experimentally. This result was interpreted in terms of thermal deactivation in the excited state.[36]

Another remarkable fact concerns the absence of specific conformers of the aldehyde–ketene photoproduct in the photolyzed matrix.[36] In the excited state, the aldehyde–ketene should have conformations where the C–C=C–C dihedral angle is *ca.* 90°, as it is usually the case for the α,β-unsaturated carbonyl compounds. Starting from such excited state geometry, the system can adopt during dissipation of the energy excess either *Z* or *E* configuration around the C=C double bond. Because the barriers for rotations around the single C–C bonds are much lower than the electronic excitation energy, the rotations around these bonds should be possible for a molecule carrying large excess energy. Hence, all the possible isomers of the conjugated aldehyde–ketene are in principle accessible. However, no spectral signatures of conformers with the ketene group and the central C=C bond in the *s-cis* geometry were found experimentally. It seems that these conformers, once formed, undergo a fast secondary reaction upon irradiation. Indeed, the presence in the spectra of the irradiated matrices of the cyclobutenone formyl derivative (4) suggests the pericyclic ring-closure reaction where the missing conformers of the aldehyde–ketene act as reactant species. The ground state of (4) was predicted to be higher by *ca.* 50 kJ mol^{-1} than the *s-cis* aldehyde–ketenes, but once (4) is photochemically created it should be stable because the barrier for the ground state opening of its ring is high (calculated value 55 kJ mol^{-1}).[36]

4. Conformational Changes Induced by IR Light and Tunneling

4.1. *IR-induced conformational changes*

The idea of IR-induced conformational change is the following. The initial conformer is transferred by IR light to a vibrationally excited state; however, in contrast to the previous section, the molecule stays in the electronic ground state. The vibrational excitation is followed by intramolecular vibrational energy redistribution (IVR), the excited molecule can relax with some probability to various possible conformers, and the conformational change may occur. In order to promote conformational changes, the excitation energy should be high enough, at least not much smaller than the reaction barrier. Many studies of conformational changes have been done using broadband light sources in

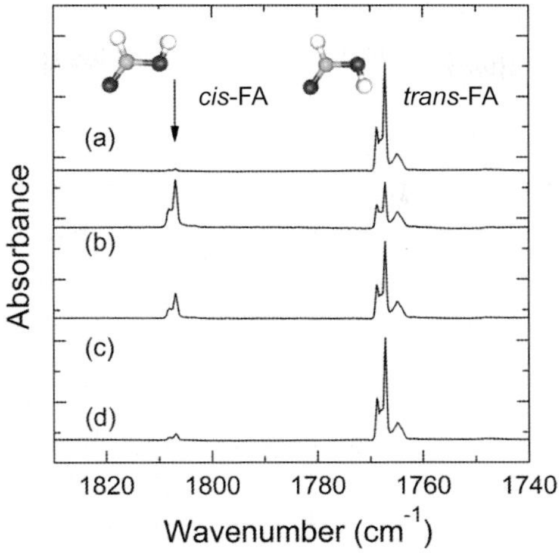

Fig. 10. FTIR spectra of formic acid in the C=O stretching region measured in solid argon at 8 K. Shown are (a) spectrum after deposition, where the *trans* form dominates, (b) spectrum after vibrational excitation of *trans*-FA increasing the *cis*-FA concentration, and spectra after (c) 5 and (d) 15 min without excitation. Structures of the *cis*-FA and *trans*-FA are also shown following the notations of Ref. 53.

low-temperature matrices.[40–51] However, in this section we concentrate on the processes promoted by *narrowband IR light*. In this situation, selective and efficient excitation of a chosen conformer is possible, which allows detailed studies of light-induced conformational changes.[52]

The first example of such study was done with formic acid (FA) in solid argon. The experimental *cis–trans* energy difference is 1365 cm^{-1} (16.3 kJ mol^{-1}),[53] and the *trans* to *cis* torsional barrier is computationally *ca.* 4000 cm^{-1} (48 kJ mol^{-1}).[54] The lower-energy form (*trans*-FA) was excited at the OH stretching overtone and new bands appear as shown in Fig. 10 for the C=O stretching mode.[55] The light-induced absorptions were assigned to the higher-energy *cis*-FA form based on the theoretical spectrum reported by Goddard *et al.*[56] It was the first experimental IR spectrum of the higher-energy FA conformer, and more spectroscopic information on this species in an argon matrix was obtained later.[57] In addition to FA, the higher-energy *cis* forms were later prepared by IR pumping for acetic acid (CH_3COOH) and propionic acid (CH_3CH_2COOH).[58–61] The vibrational frequencies for the OH and C=O stretching modes for various isotopic analogs of these species are shown

Table 2. Experimental OH and C=O stretching frequencies (in cm^{-1}) of the *trans* and *cis* forms of some carboxylic acids in an argon matrix[58].

	trans		***cis***		***cis–trans* shift**	
	νOH	νC=O	νOH	νC=O	$\Delta\nu$OH	$\Delta\nu$C=O
HCOOH	3550	1768	3617	1808	+67	+40
DCOOH	3550	1761[a]	3616	1777[a]	+66	+16
CH_3COOH	3564	1779	3623	1807	+59	+28
CD_3COOH	3564	1777	–	1798	–	+21
CH_3COOD	2629	1770	2674	1800	+45	+30
CD_3COOD	2630	1765	2673	1794	+43	+29
CH_3CH_2COOH	3567	1776	–	1802	–	+26
CH_3CH_2COOD	2630	1766	2674	1797	+44	+31

[a]Band position is influenced by Fermi resonance with the first overtone of ωCD.[57]

in Table 2. The *cis–trans* shifts are very characteristic for these modes, and they can be reproduced by *ab initio* computational methods.

Another important example is nitrous acid (HONO), which is the smallest molecule with rotational isomers. *trans*-HONO is 130 cm^{-1} (1.66 kJ mol^{-1}) lower in energy than *cis*-HONO, and the *cis–trans* barrier is ~3400 cm^{-1} (~41 kJ mol^{-1}). Hall and Pimentel reported conformational conversion in both directions induced by broadband IR light in a nitrogen matrix at 20 K.[40] The studies of the *trans* and *cis* forms of HONO were also performed using a narrowband IR excitation in a krypton matrix at 7.5 K, and all fundamentals were observed and assigned.[62] In particular, the weak HON and ONO bending absorptions of *cis*-HONO were found in agreement with the gas-phase data.[63] These weak bands were not observed in the previous matrix isolation studies.[40,64] The experimental IR spectra of the HONO conformers can be accurately obtained for both frequencies and intensities using the CC-VSCF anharmonic theory.[62]

The studies of HONO clearly demonstrated the resonant character of light-induced conformational changes.[62] The OH stretching band of *trans*-HONO is split in solid krypton into several components due to different matrix surroundings. It appears that molecules absorbing at different frequencies can be converted to another form selectively, which leads to hole burning in the inhomogeneously broadened absorption spectrum. This result evidences the localization of conformational

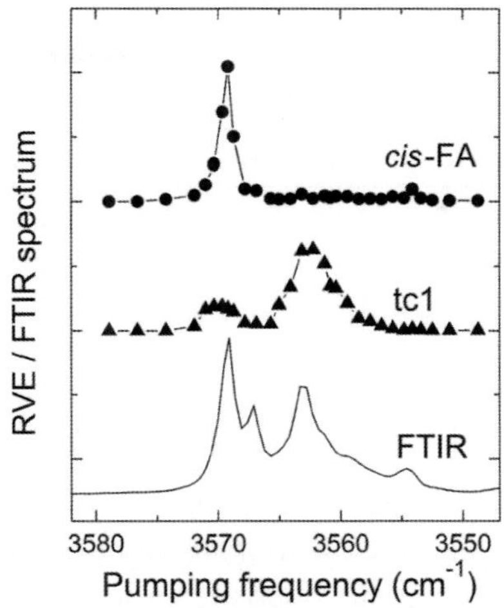

Fig. 11. RVE spectra of formation of *cis*-FA monomer and *trans–cis* FA dimer tc1 (see Ref. 67 for details) in a neon matrix and FTIR spectrum of a *trans*-FA/Ne matrix.[66]

process at the excited molecule and the negligible nonresonant energy flow in the matrix. Thus, the light-induced effect is presumably proportional to the excitation rate and the quantum yield.

The selectivity of IR-induced conformational change makes a basis for reactive vibrational excitation (RVE) spectroscopy. The RVE spectrum is defined as the dependence of conversion rate on the excitation frequency. The conformational change can be measured using strong absorption bands of the produced species, which makes the method very sensitive. The RVE spectra of FA were studied in argon and neon matrices, and matrix-site selectivity was shown.[65,66] The IR-induced *trans* to *cis* conversion rate follows the absorption profile of *trans*-FA. The high selectivity and sensitivity of RVE spectroscopy is demonstrated by measuring phonon bands.[65] The weak and broad phonon bands are invisible in the Fourier transform infrared (FTIR) spectra, but they clearly appear in the RVE spectra. In addition, RVE spectra are selective with respect to matrix sites and different species with overlapping spectral bands (Fig. 11).[65–67]

The efficiency of the light-induced process is numerically characterized by the quantum yield defined as η = (Process rate) / (Excitation rate). In the experiments on HONO in a krypton matrix, the

Table 3. Quantum yields of light-induced rotational isomerization of formic acid (FA), acetic acid (AA), and propionic acid (PnA) in solid argon.

Mode	**FA[a] (%)**	**AA[b] (%)**	**PnA[c] (%)**
2νOH, *trans* to *cis*	12	2.2	1.4
2νOD, *trans* to *cis*	–	0.36	0.41
2νOD, *cis* to *trans*	–	1.2	14

[a]FA, formic acid, from Ref. 65; [b]AA, acetic acid, from Ref. 60; [c] PnA, propionic acid, from Ref. 54.

ratio of quantum yields for the *cis* to *trans* and *trans* to *cis* processes was measured $\eta_c/\eta_t = 7.0$, i.e., the *cis* to *trans* light-induced process is seven times more efficient than the opposite process.[62] The experimentally estimated *cis* to *trans* quantum yield is very high, close to 100%. Similar results were obtained by Hall and Pimentel for HONO in solid nitrogen,[40] and the details of this process will be discussed in Section 5.

The quantum yields of light-induced conformational change have been measured for various species and selected results are presented in Table 3. The obtained values differ substantially between different species and the general trends have been noted as follows. Smaller molecules (HONO and formic acid) exhibit higher quantum yields compared to larger molecules (acetic and propionic acids). This observation looks reasonable because more channels exist in larger molecules that are in competition with the isomerization reaction. The *cis* to *trans* light-induced conversion studied in HONO, acetic, and propionic acids is more efficient than the *trans* to *cis* process indicating different coupling between involved states in the *cis* and *trans* geometries. Smaller quantum yields are found for OD species compared with OH species; hence, deuteration decreases coupling of the pumped 2νOH(D) mode with the reaction coordinate and possibly tunneling is involved into the conformational change. Acetic acid CD_3COOH shows a three times higher yield compared with CH_3COOH; however, the underlying mechanism is unclear.[60] No difference in isomerization quantum yields has been found for acetic acid in argon and krypton matrices and for formic acid in argon and neon matrices.[60,66] In contrast, the quantum yield of the *trans* to *cis* conversion of formic acid strongly decreases in solid hydrogen, which has been connected with coupling between molecular and host vibrations.[68]

Conformational changes have been achieved by selective vibrational excitation for a number of other species. For oxalyl fluoride $(FCO)_2$ the conformational change involves rotation of heavy atoms.[69] For oxalic ($C_2H_2O_4$)[70] and malonic ($C_3H_4O_4$) acids,[71] the operating overtone modes were found by scanning the IR light frequency and detecting the conformational change. The FTIR spectra of these species show no bands in this region demonstrating advantages of RVE spectroscopy.

4.2. *cis to trans tunneling reaction and matrix effect*

As described earlier, *cis*-FA can be prepared by vibrational excitation of *trans*-FA. It has been found that the *cis*-FA concentration decreases in the dark after the excitation is stopped (see Fig. 10). The observed conversion cannot occur over the *cis* to *trans* barrier (~2700 cm^{-1}) at such low temperatures (~10 K), and the effect was explained by quantum tunneling of hydrogen through the torsional barrier.[72] The *cis* to *trans* FA decay accelerates at higher temperatures, and a low-temperature limit of the *cis* to *trans* reaction is observed, which is a fingerprint of quantum tunneling.[73] Formic, acetic, and propionic acids were studied in argon matrices, and the *cis* to *trans* decay rate was the slowest for formic acid, in agreement with the highest computational barrier obtained for this molecule.[58] For the OD species, the tunneling rate decreases by a large factor (3×10^3 for formic, 3×10^4 for acetic, and 6×10^4 for propionic acids in solid argon), further supporting quantum tunneling as a mechanism for the dark decay of the *cis* conformers.

Formic acid can form a number of dimers studied experimentally and theoretically.[66,67,74–78] By using vibrational excitation of the free OH stretching mode of one of the *trans–trans* dimers in an argon matrix, the first *trans–cis* dimer (tc1) was prepared and identified.[67] The tc1 dimer differs from the initial *trans–trans* structure by rotation of the free OH group in one FA molecule. The tc1 dimer decays in the dark back to the corresponding *trans–trans* form by hydrogen tunneling; however, its lifetime is sufficiently longer than that of the *cis*-FA monomer. This difference can be explained by a complexation-induced change of the

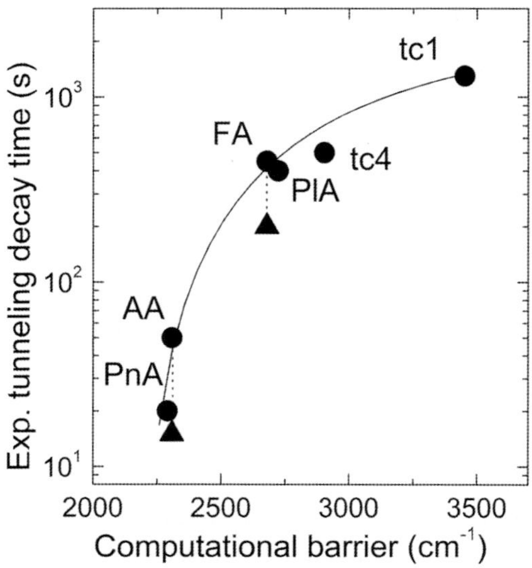

Fig. 12. Experimental tunneling decay time of the *cis* conformers versus computational reaction barrier for propionic acid (PnA), acetic acid (AA), formic acid (FA),[58] propiolic acid (PlA),[80] and two formic acid dimers (tc1 and tc4) shown by solid circles.[67,79] The dotted lines present the secondary deuteration effect for acetic and formic acids (triangles).[60,72] The solid line guides the eye.

torsional barrier. The tc1 dimer is computationally stabilized by a barrier of 3432 cm^{-1} (41.05 kJ mol^{-1}) [MP2/6-311++G(2*d*,2*p*)], which is substantially higher than the stabilization barrier for the *cis*-FA monomer (2676 cm^{-1}). The decay of *cis*-FA monomer is enhanced at elevated temperatures, by ~10 times at 30 K in solid argon. In contrast, the decay of the tc1 dimer does not change in this temperature range, and at 30 K the tc1 dimer is ~30 times more stable than the *cis*-FA monomer. Another *trans–cis* dimer structure is identified in an argon matrix, which is named tc4 following the nomenclature of the original work.[79] At the lowest temperatures, its stability is close to that of *cis*-FA monomer, which is consistent with the close reaction barrier (2900 cm^{-1}). At elevated temperatures, the tc4 dimer is more stable than the monomer. Many of these observations have no explanation at the moment.

Figure 12 presents the experimental decay time versus the computational barrier for the *cis* forms of various carboxylic acids including propiolic acid[80] and two *trans–cis* FA dimers. Slower decay generally occurs for higher barriers; thus, the reaction barrier is evidently an important factor in the *cis* to *trans* tunneling rate. On the other hand, a large secondary H/D isotope effect was observed for acetic and formic acids, which is applied to differences between the CH and CD species.[60,72] This observation shows that the reaction barrier is not the only determining factor for hydrogen tunneling rate and evidences the role of the vibrational manifold of the *trans* form behind the torsional barrier near the energy of the initial *cis* state. Due to the asymmetry of the torsional potential energy curve, the hydrogen tunneling involves excited vibrational states of the *trans* conformer as intermediates in the vibrational energy redistribution. The energy mismatch between the initial conformer and the excited vibrational levels of the final conformer must be dissipated by some mechanism, and this process presumably involves the matrix phonon bath.

The FA–water complex has been studied theoretically and experimentally, the experiments being previously limited to the low-energy *trans*-FA conformer.[81–85] In the recent studies in an argon matrix, the prolonged excitation at ~6200 cm^{-1} bleaches the *trans*-FA-water complex absorptions and promotes a number of new absorptions, in particular, a well-defined band at 1786 cm^{-1}, which is blue-shifted from the *trans*-FA–water band at 1737 cm^{-1}.[86] The new bands were assigned to the lowest-energy *cis*-FA–water complex. The assignment is based on good agreement of the experimental data with the computational results obtained by Zhou *et al.*[85]

The *cis*-FA–water complex was found to be stable at 9 K on a day scale. This stabilization effect was explained based on the computational and experimental energetics. The interaction energy of the *cis*-FA–water complex[85] is −29.6 kJ mol^{-1}, whereas the *cis*-FA monomer is higher in energy than the *trans*-FA monomer by 16.3 kJ mol^{-1}.[53] In this situation, hydrogen tunneling from the *cis*-FA–water complex would prepare a pair of *trans*-FA and water without a hydrogen bond, which is substantially higher in energy than the initial complex. This process is not allowed in the adiabatic approximation, i.e., without rearrangement of heavy atoms.

Pettersson *et al.* studied the *cis* to *trans* FA tunneling process in argon, krypton, and xenon matrices.[72] The tunneling rate constants were obtained in the order $k(\mathrm{Ar}) > k(\mathrm{Kr}) > k(\mathrm{Xe})$ so that the *cis*-FA decay is slowest in a xenon matrix. At the first glance, this trend can be easily explained by solvation of the molecule in the matrix material. *cis*-FA has a larger dipole moment than *trans*-FA and probably than the transition state structure. It follows that the *cis* to *trans* energy barrier should increase with polarizability of the matrix. Despite these arguments, a different order of tunneling rates occurs for acetic acid (AA) in various matrices: $k(\text{Xe}) > k(\text{Ar}) > k(\text{Kr})$, i.e., the *cis*-AA decay is fastest in a xenon matrix.[60] Thus, a number of solvation effects should be carefully analyzed, which can change barrier heights, energy gaps, nature of accepting levels, matrix reorganization energies, and coupling of the embedded molecule with lattice phonons.

It has been recently found that the *cis*-FA tunneling decay is very fast in solid neon.[66] The lifetime of *cis*-FA monomer in solid neon is ~5 s, i.e., about two orders of magnitude shorter than in solid argon. This huge matrix effect was explained by the lower *cis–trans* barrier in solid neon and the vibrational manifold of the *trans* conformer properly adjusted for the efficient transition from the virtual levels of tunneling *cis*-FA to the excited vibrational levels of *trans*-FA. The difference between lifetimes of *cis*-FA monomer and the tc1 dimer in solid neon is by a factor of ~300, i.e., the *cis* form is strongly stabilized by hydrogen bonding in this dimer. The *trans–cis* dimer tc1 in solid neon has the lifetime of ~20 min, which is close to the values obtained in argon, krypton, and hydrogen matrices at 4.2 K.[68,87] Similar stabilities of the *trans–cis* FA dimer in various matrices tentatively suggest the intrinsic nature of hydrogen tunneling in this system. It is possible that the *trans–cis* FA dimer decays similarly slowly in the gas phase, which makes promises for its experimental preparation and observation.

All previous conformational changes described in this section were obtained by resonant vibrational excitation of the molecule itself. In contrast, the *trans* to *cis* FA conversion in solid hydrogen was also achieved by exciting rovibrational transition of the solid hydrogen host, suggesting that the lattice vibrational energy is efficiently transferred to the *trans* to *cis* reaction coordinate.[68] The *trans* to *cis* reaction rate precisely follows the absorption profile of the hydrogen matrix as seen in Fig. 13. The conformational process is promoted with similar efficiencies at the FA-induced band at 4139 cm^{-1} and at "pure" hydrogen absorptions. This observation shows that the host exciton

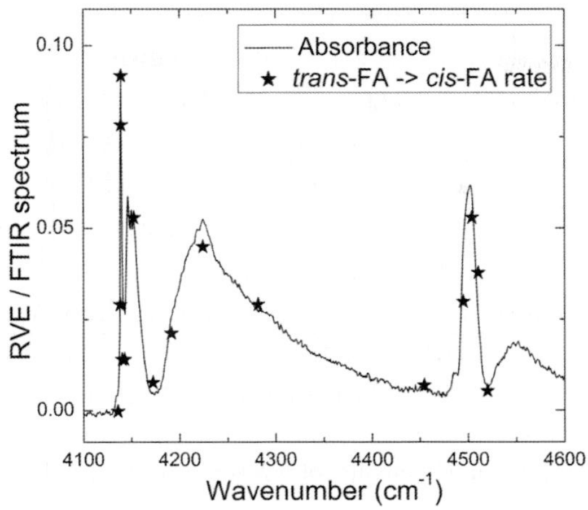

Fig. 13. RVE spectrum of the formation of *cis*-FA and FTIR spectrum of an FA/H_2 matrix measured at 4.2 K.[68]

diffusion length is at least not smaller than the mean distance between the FA molecules, which is ~10 nm for the FA/H_2 = 1/30000 matrix. The quantum yield of the *cis* to *trans* FA conversion at hydrogen absorptions is even higher than that at the OH stretching mode by a factor of 3–10. This difference can originate from the lower energy of the OH stretching vibration because the OH stretching vibration energy is slightly below the computational reaction barrier and tunneling is presumably involved into the process. The quantum yield of conformational change upon the OH stretching excitation is quite low compared to argon, which indicates the participation of host in vibrational energy redistribution.

5. Dynamics

So far, we had looked at molecular conformation with the help of stationary spectroscopy, which essentially characterizes the conformation in the initial (reactant) and final (product) state. Information about the reaction mechanism and the transition state can be deduced only indirectly from stationary spectroscopy. On the other hand, UV-induced conformational changes have been studied extensively with ultrashort

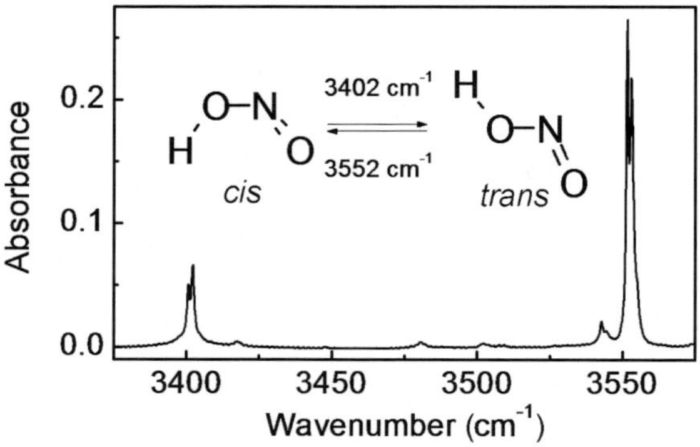

Fig. 14. Absorption spectrum of a mixture of *cis*- and *trans*-HONO in solid Kr, right after matrix deposition.

laser pulses at room temperature and in the liquid or gas phase. This led to a research field that Ahmed Zewail (and many others) coined femtochemistry, and for which he was awarded the Nobel Prize in Chemistry in 1999.[88] The information gain of time-resolved spectroscopy is that one can observe in real time how the reaction proceeds, and in particular how it crosses the transition state. Somewhat surprisingly, very little time-resolved spectroscopy has been performed in the solid matrix environment.[89]

The matrix environment is in fact quite attractive for time resolved-spectroscopy, in particular since many examples of IR-driven photochemical reactions are known (see Section 4.1). The same is not the case at room temperature in the liquid phase, probably since IVR is too efficient under these conditions (see discussion below). In contrast to the vast majority of UV-induced photochemical reactions studied by Ahmed Zewail and many others, the molecule stays on the electronic ground state surface during an IR-driven reaction, and hence mimics more closely thermally driven chemistry.

We will concentrate here on the time-resolved studies of the IR-induced *cis–trans* rotational isomerization of nitrous acid (HONO),[90–95] which has already been introduced in Section 4.1. HONO is the prototype example since it is the smallest possible system that actually can undergo the *cis–trans* isomerization. The smallness allows one to

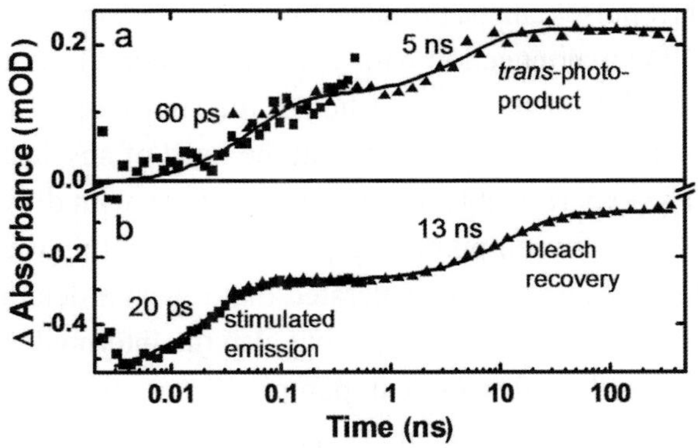

Fig. 15. (a) Appearance of *trans*-HONO after selective excitation of the OH_{cis} stretching vibration at 30 K in solid Kr. (b) Stimulated emission and bleach signal from the OH_{cis} stretch vibration, reporting on vibrational relaxation of the former and cooling of the molecule as a whole into the matrix environment, respectively.

study the reaction also theoretically on the highest feasible level.[92,96–98] Furthermore, the experimental *cis–trans* quantum yield after pumping the OH_{cis} stretching vibration is close to 100%.[40,62] This high quantum yield is indeed remarkable given the fact that it is not the reaction coordinate that is initially excited. Hence, energy must be funneled from the initially pumped state preferentially into reactive states (i.e., the tortional coordinate), whereas other degrees of freedom go away empty-handed. The high quantum yield is a clear indication that energy is not thermalized or randomized on the timescale of isomerization, and that the reaction is not statistical. Finally, it should be noted that the reaction has not been observed in the gas phase as of yet, and we actually will argue that the reaction is facilitated by the dissipating forces of the matrix environment.

The time-resolved pump-probe experiments are conceptually easy. Figure 14 shows an absorption spectrum of HONO in solid Kr in the region of the OH stretching vibration with two lines corresponding to the *cis* and *trans* conformers. Selective excitation of the OH_{cis} stretching vibration, e.g., will deplete *cis*-HONO and at the same time produce *trans*-HONO. The only difference from the experiments discussed in Section 4.1 is that we now use ultrashort femtosecond IR pulses for both pumping *cis*-HONO and for probing the progress of the reaction. Intense femtosecond IR pulses can be produced routinely with optical parametrical amplifiers (OPA) in connection with Ti:S femtosecond laser technology.[99] The time resolution of the method lies in the shortness of the IR pulses (which can be as short as 50 fs), without the need of any fast detectors or electronics. The experiment is repeated many times with different settings of the delay time between pump and probe pulse. In this way, a series of time-dependent spectra can be generated that reveal a detailed picture of not only the initial and the final state, but also the intermediate states.

In simple words, after optical pumping, e.g., the OH_{cis} stretching vibration, we will start to see additional intensity in the OH_{trans} spectral region once the molecule isomerizes (the result is a bit more complicated since the molecule will initially be hot, revealing pronounced hot bands for the *trans* photoproduct (see Refs. 90, 91, and 95 for details). Figure 15a shows the appearance of the *trans* photoproduct as a function of the delay time after pumping the OH_{cis} stretching vibration. Interestingly, the photoproduct appears in two steps, one on a 60 ps timescale and a second one on a significantly slower 5 ns timescale. Both steps are about equally high and amount to a total *cis* to *trans* quantum yield of 40 ± 10% at 30 K. The timescale of the two isomerization steps roughly correlates with two relaxation times of *cis*-HONO: the vibrational relaxation of the initially pumped OH_{cis} stretch vibration (20 ps, deduced from the disappearance of the stimulated emission from the OH_{cis} stretch vibration (see Fig. 15b) and the final cooling of the molecule into the matrix (13 ns, deduced from the partial recovery of the bleach of the OH_{cis} stretch vibration).

We explained the two-step conformational process with a tier model illustrated in Fig. 16. The initially pumped OH_{cis} stretching vibration relaxes preferentially into a subset of relatively few *reactive states*, which are closely resonant to the former. Mixed quantum-classical simulations suggested that these reactive modes are combination modes such as $2\nu_{tors} + 2\nu_{bend}$ and $4\nu_{tors} + \nu_{bend}$.[92] Because of their large number of quanta in the torsional coordinate, they can tunnel through the isomerization barrier relatively efficiently, giving rise to the first isomerization step with rate k_{trans}. However, the molecule contains many more degrees of freedom (NO stretching and ONO bending modes)

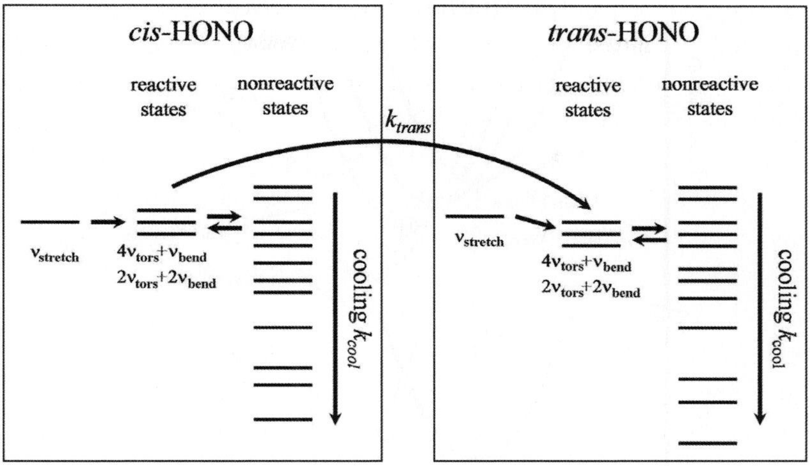

Fig. 16. Tier model explaining the two-step isomerization of *cis*-HONO.

which are *nonreactive* and eventually depopulate the reactive states. Nevertheless, as long as energy is still in the molecule, some sort of equilibrium is established between reactive and nonreactive states. The small fraction of population that is still in the reactive states in this equilibrium will lead to further isomerization at a reduced effective rate (second isomerization step). Hence, although the majority of states are nonreactive, they act as energy storage for some time. Finally, once energy drops below a certain threshold due to overall cooling (rate k_{cool}), conformational change stops.

The temperature dependence of the *cis* to *trans* isomerization quantum yield, which increases from 40 ± 10% at 30 K to 60 ± 10% at 15 K, gave the final clue about the reaction mechanism.[94] First, this increasing quantum yield fills the gap between what has been observed in the time-resolved experiments at 30 K (Fig. 15a), and the close to 100% quantum yield previously reported for 7 K.[62] Second and more importantly, it provided evidence for the funnel that guides energy from the *cis* to the *trans* side. We interpret the results in analogy to Marcus theory of electron transfer (Fig. 17). A rise in isomerization rate at lower temperature is obtained only in the so-called barrierless case, where the product free-energy surface cuts the reactant free-energy surface in more

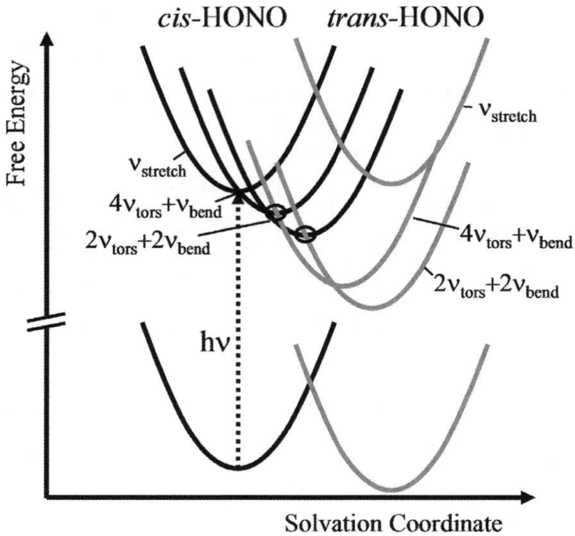

Fig. 17. Set of free-energy surfaces of the relevant reactive states in *cis*- and *trans*-HONO as a function of a generalized solvation coordinate.

or less exactly its minimum. The reaction can occur only at these cutting points, i.e., when the two states happen to be resonant due to an accidental matrix configuration. In other words, once one of the reactive states in *cis*-HONO is populated, it is funneled down in free energy by adjusting matrix degrees of freedom such that the crossing point with the corresponding reactive state of *trans*-HONO is readily found.

Putting these considerations together, we can discuss to what extent these results, in particular the high quantum yield, are of general nature. First, the high quantum yield is the result of a relatively special arrangement of energy levels, leading to the barrierless transitions depicted in Fig. 17, which act as effective energy funnels. In other words, not all molecules of equal size necessarily have equal quantum yields. The second important ingredient is the cooling rate that is of roughly the same order of magnitude as the isomerization rate. This is seen from the fact that the sum of forward and backward yields is close to 100%, but does not quite reach this value at temperature 15 K. If cooling was significantly slower, then the molecule would have time to fully thermalize on either side, in which case the sum of forward and backward yields would be one and their ratio would reflect the densities of states in both conformations. However, the densities of states are almost identical in the relevant energy range[96]; hence, we could expect a quantum yield of maximal 50%. On the other hand, if the isomerization was significantly slower than cooling, the molecule would not have time to isomerize and the sum of forward and backward yields would be significantly below 100%. At higher temperatures, in particular at room temperature, IVR is so efficient that IR-induced isomerization will hardly be observed. The same happens when increasing the size of the molecule and thereby the number of relaxation channels. Macoas *et al.* have recently studied the vibrational relaxation and cooling of formic acid and acetic acid in matrix environment.[100] In the latter case, the large separation of timescales between vibrational relaxation of the initially pumped state and cooling of the molecule as a whole into the solvent we observe for HONO (Fig. 15b) is essentially gone. It is evident from Fig. 16 that a long cooling time is prerequisite for a high quantum yield.

Consequently, the high quantum yield of the IR-driven *cis–trans* isomerization of HONO probably is very special feature of this particular molecule. Nevertheless, it helped significantly to study the reaction mechanism in unprecedented detail. Much of the principles of Fig. 17 survive in larger molecules at higher temperatures in real-world thermally driven chemistry.

6. Concluding Remarks

As shown in the various examples described in this chapter, the low-temperature matrix isolation method provides optimal conditions for studying therma and light-induced conformational changes experienced by small compounds.

Thermally induced processes may be conveniently investigated by taking advantage of the versatility of a standard matrix isolation set up regarding the points where temperature can be efficiently controlled (cold substrate, input valve nozzle, sample chamber, premixing container). In these processes, conformational cooling represents a key event and, e.g., changing the temperature of the cryostat substrate during matrix deposition or annealing of the deposited matrix allows to change the relative populations of the conformers. In some favorable cases, stepwise or selective conformational cooling can be promoted, making possible the detailed investigation of specific isomerization reactions as well as the structural and spectroscopic characterization of a particular conformer.

UV–visible and IR excitations can induce conformational changes and constitute very powerful and elegant approaches to study conformational processes. As discussed in this chapter, broadband excitation in the UV–visible region works well in many cases, in particular for unsaturated molecules where the geometries of the ground and excited electronic states usually differ considerably. On the other hand, narrowband IR vibrational excitation provides a precise control of the activated process. When IR irradiation is performed, it is possible in principle to selectively activate a specific vibrational coordinate of a given conformer in a particular matrix site, i.e., to achieve a triply selective experiment. This approach allows studies of quantum tunneling of hydrogen in carboxylic acids showing remarkable temperature, host, and isotope effects. An interesting observation concerns the conformational change obtained by vibrational excitation of the host observed for formic acid in solid hydrogen.

Finally, as demonstrated in this chapter for the simplest prototype system HONO, the recent developments in time-dependent experiments on matrix-isolated species opened another window to explore conformational processes. Indeed, the time-resolved spectroscopy experiments allow one to observe in real time how the reaction proceeds, and in particular, it is now possible to extract direct information on how it crosses the transition state. The consideration of HONO shows that similar conformational processes in large molecules are probably extremely difficult to describe in detail.

Acknowledgments

The authors thank the colleagues involved in these studies, especially L. Adamowicz, V. Botan, S. Breda, A. Domanskaya, M. L. Duarte,

E. Eusébio, A. Gómez-Zavaglia, J. Juselius, A. Kulbida, L. Lapinski, S. Lopes, A. Lopes de Jesus, J. Lundell, E. M. S. Maçôas, K. Marushkevich, M. Pettersson, E. Radchenko, M. Räsänen, J. S. Redinha, I. Reva, M. Rosado, R. Schanz, O. Schrems, A. Simão, S. Stepanian, and H. Willner. We also acknowledge support from the Academy of Finland, partially through the Finnish Center of Excellence "Computational Molecular Science," the Portuguese Science Foundation (Project PTDC/QUI/71203/2006), and the Swiss National Science Foundation.

References

1. A. Gómez-Zavaglia, I. Reva and R. Fausto, *Phys. Chem. Chem. Phys.* **5**, 52 (2003).
2. X. Mei and C. Wolf, *CrystEngComm.* **8**, 377 (2006).
3. A. Nangia, *Acc. Chem. Res.* **41**, 595 (2008).
4. L. Khriachtchev, E. M. S. Maçôas, M. Pettersson and M. Räsänen, *J. Am. Chem. Soc.* **124**, 10994 (2002).
5. G. F. X. Schertler, *Nature* **453**, 292 (2008).
6. J. Saunders, *Nat. Rev. Microbiol.* **1**, 6 (2003).
7. D. Sabbert, S. Engelbrecht and W. Junge, *Nature* **381**, 623 (1996).
8. A. Aguzzi, *Proc. Natl. Acad. Sci. USA* **105**, 11 (2008).
9. I. D. Reva, S. G. Stepanian, L. Adamowicz and R. Fausto, *Chem. Phys. Lett.* **374**, 631 (2003).
10. A. Gómez-Zavaglia and R. Fausto, *Phys. Chem. Chem. Phys.* **5**, 52 (2003).
11. I. D. Reva, A. J. Lopes. de Jesus, M. T. S. Rosado, R. Fausto, M. E. S. Eusébio and J. S. Redinha, *Phys. Chem. Chem. Phys.* **8**, 5339 (2006).
12. A. J. Lopes de Jesus, M. T. S. Rosado, I. D. Reva, R. Fausto, M. E. S. Eusébio and J. S. Redinha, *J. Phys. Chem.* **A110**, 4169 (2006).
13. A. J. Lopes de Jesus, M. T. S. Rosado, I. D. Reva, R. Fausto, M. E. S. Eusébio and J. S. Redinha, *J. Phys. Chem.* **A113**, 7499 (2008).
14. M. T. S. Rosado, A. J. Lopes de Jesus, I. D. Reva, R. Fausto and J. S. Redinha, *J. Phys. Chem.* **A113**, 7499 (2009).
15. S. G. Stepanian, I. D. Reva, E. D. Radchenko, M. T. S. Rosado, M. L. T. S. Duarte, R. Fausto and L. Adamowicz, *J. Phys. Chem.* **A102**, 1041 (1998).
16. I. D. Reva, A. Simão and R. Fausto, *Chem. Phys. Lett.* **406**, 126 (2005).
17. K. Iijima, K. Tanaka and S. Onuma, *J. Mol. Struct.* **246**, 257 (1991).
18. S. Vishveshwara and J. A. Pople, *J. Am. Chem. Soc.* **99**, 2422 (1977).
19. H. L. Sellers and L. Schäfer, *J. Am. Chem. Soc.* **100**, 7728 (1978).
20. R. D. Brown, P. D. Godfrey, J. W. V. Storey and M. P. Bassez, *J. Chem. Soc. Chem. Commun.* 547 (1978).
21. R. D. Suenram and F. J. Lovas, *J. Mol. Spectrosc.* **72**, 372 (1978).

22. R. D. Suenram and F. J. Lovas, *J. Am. Chem. Soc.* **102**, 7180 (1980).
23. P. D. Godfrey and R. D. Brown, *J. Am. Chem. Soc.* **117**, 2019 (1995).
24. F. J. Lovas, Y. Kawashima, J.-U. Grabow, R. D. Suenram, G. T. Freser and E. Hirota, *Astrophys. J.* **455**, 201 (1995).
25. O. A. Raevskii, A. N. Vereshchagin and F. G. Khalitov, *Bull. Acad. Sci. USSR Div. Chem. Sci.* **21**, 305 (1972).
26. A. V. Yarkov and O. A. Raevskii, *J. Gen. Chem. USSR* **55**, 42 (1985).
27. R. Streck, A. J. Barnes, W. H. Herrebout and B. J. van der Veken, *J. Mol. Struct.* **376**, 277 (1996).
28. G. Mastrantonio and C. O. Della Védova, *J. Mol. Struct.* **561**, 161 (2001).
29. K. Taga, N. Hirabayashi, T. Yoshida and H. Okabayashi, *J. Mol. Struct.* **212**, 157 (1989).
30. A. Kulbida, M. N. Ramos, M. Räsänen, J. Nieminen, O. Schrems and R. Fausto, *J. Chem. Soc. Faraday Trans.* **91**, 1571 (1995).
31. R. Fausto, A. Kulbida and O. Schrems, *J. Chem. Soc. Faraday Trans.* **91**, 3755 (1995).
32. E. M. S. Maçôas, R. Fausto, J. Lundell, M. Pettersson, L. Khriachtchev and M. Räsänen, *J. Phys. Chem.* **A105**, 3922 (2001).
33. A. Y. Hirakawa and M. Tsuboi, In: J. R. Durig (editor), *Vibrational Spectra and Structure.* Vol. 12(3), pp. 174–175. Elsevier, Amsterdam (1983).
34. P. D. Foo and K. K. Innes. *J. Chem. Phys.* **60**, 4582 (1974).
35. N. Piétri, B. Jurca, M. Monnier, M. Hillebrand and J. P. Aycard. *Spectrochim. Acta* **A56**, 157 (1999).
36. S. Breda, I. Reva, L. Lapinski and R. Fausto, *Phys. Chem. Chem. Phys.* **6**, 929 (2004).
37. R. G. S. Pong, B. S. Huang, J. Laureni and A. Krantz, *J. Am. Chem. Soc.* **99**, 4153 (1977).
38. R. G. S. Pong and J. S. Shirk, *J. Am. Chem. Soc.* **95**, 248 (1973).
39. O. L. Chapman, C. L. McIntosh and J. Pacansky, *J. Am. Chem. Soc.* **95**, 244 (1973).
40. R. T. Hall and J. C. Pimentel, *J. Chem. Phys.* **38**, 1889 (1963).
41. L. Homanen and J. Murto, *Chem. Phys. Lett.* **85**, 322 (1982).
42. H. Takeuchi and M. Tasumi, *Chem. Phys.* **77**, 21 (1983).
43. T. Lotta, J. Murto, M. Räsänen and A. Aspiala, *Chem. Phys.* **86**, 105 (1984).
44. M. Räsänen, G. P. Schwartz and V. E. Bondybey, *J. Chem. Phys.* **84**, 59 (1986).
45. M. Räsänen and V. E. Bondybey, *J. Phys. Chem.* **90**, 5038 (1986).
46. M. Räsänen and V. E. Bondybey, *Chem. Phys. Lett.* **111**, 515 (1984).
47. J. Nieminen, M. Räsänen and J. Murto, *J. Phys. Chem.* **96**, 5303 (1992).
48. A. Kulbida and R. Fausto, *J Chem. Soc. Faraday Trans.* **89**, 4257 (1993).
49. A. Olbert-Majkut, I. D. Reva and R. Fausto, *Chem. Phys. Lett.* **456**, 127 (2008).
50. L. Lapinski, R. Ramaekers, B. Kierdaszuk, G. Maes and M. J. Nowak, *J. Photochem. Photobiol.* **A163**, 489 (2004).
51. I. A. Reva, S. Jarmelo, L. Lapinski and R. Fausto, *J. Phys. Chem.* **A108**, 6982 (2004).

52. L. Khriachtchev, *J. Mol. Struct.* **880**, 14 (2008).
53. W. H. Hocking, *Z. Naturforsch.* **A31**, 1113 (1976).
54. M. Pettersson, E. M. S. Maçôas, L. Khriachtchev, R. Fausto and M. Räsänen, *J. Am. Chem. Soc.* **125**, 4058 (2003).
55. M. Pettersson, J. Lundell, L. Khriachtchev and M. Rasanen, *J. Am. Chem. Soc.* **119** 11715 (1997).
56. J. D. Goddard, Y. Yamaguchi and H. F. Schaefer, *J. Chem. Phys.* **96**, 1158 (1992).
57. E. M. S. Maçôas, J. Lundell, M. Pettersson, L. Khriachtchev, R. Fausto and M. Räsänen, *J. Mol. Spectrosc.* **219**, 70 (2003).
58. E. M. S. Maçôas, L. Khriachtchev, M. Pettersson, R. Fausto and M. Räsänen, *Phys. Chem. Chem. Phys.* **7**, 743 (2005).
59. E. M. S. Maçôas, L. Khriachtchev, M. Pettersson, R. Fausto and M. Räsänen, *J. Am. Chem. Soc.* **125**, 16188 (2003).
60. E. M. S. Maçôas, L. Khriachtchev, M. Pettersson, R. Fausto and M. Räsänen, *J. Chem. Phys.* **121**, 1331 (2004).
61. E. M. S. Maçôas, L. Khriachtchev, M. Pettersson, R. Fausto and M. Räsänen, *J. Phys. Chem.* **A109**, 3617 (2005).
62. L. Khriachtchev, J. Lundell, E. Isoniemi and M. Räsänen, *J. Chem. Phys.* **113**, 4265 (2000).
63. C. M. Deeley and I. M. Mills, *J. Mol. Struct.* **100**, 199 (1983).
64. Z. Mielke, Z. Latajka, J. Kolodziej and K. G. Tokhadze, *J. Phys. Chem.* **100**, 11610 (1996).
65. E. M. S. Maçôas, L. Khriachtchev, M. Pettersson, J. Juselius, R. Fausto and M. Räsänen, *J. Chem. Phys.* **119**, 11765 (2003).
66. K. Marushkevich, L. Khriachtchev and M. Räsänen, *J. Chem. Phys.* **126**, 241102 (2007).
67. K. Marushkevich, L. Khriachtchev, J. Lundell and M. Räsänen, *J. Am. Chem. Soc.* **128**, 12060 (2006).
68. K. Marushkevich, L. Khriachtchev and M. Räsänen, *Phys. Chem. Chem. Phys.* **9**, 5729 (2007).
69. S. Sander, H. Willner, L. Khriachtchev, M. Pettersson, M. Räsänen and E. L. Varetti, *J. Mol. Spectrosc.* **203**, 145 (2000).
70. E. M. S. Maçôas, R. Fausto, M. Pettersson, L. Khriachtchev and M. Räsänen, *J. Phys. Chem.* **A104**, 6956 (2000).
71. E. M. S. Maçôas, R. Fausto, J. Lundell, M. Pettersson, L. Khriachtchev and M. Räsänen, *J. Phys. Chem.* **A104**, 11725 (2000).
72. M. Pettersson, E. M. S. Maçôas, L. Khriachtchev, J. Lundell, R. Fausto and M. Räsänen, *J. Chem. Phys.* **117**, 9095 (2002).
73. V. I. Goldanskii, M. D. Frank-Kamenetskii and I. M. Barkalov, *Science* **182**, 1344 (1973).
74. J. Chocholousova, J. Vacek and P. Hobza, *Phys. Chem. Chem. Phys.* **4** 2119 (2002).
75. W. Qian and S. Krimm, *J. Phys. Chem.* **A105**, 5046 (2001).

76. S. Roszak, R. H. Gee, K. Balasubramanian and L. E. Fried, *J. Chem. Phys.* **123**, 144702 (2005).
77. F. Madeja, M. Havenith, K. Nauta, R. E. Miller, J. Chocholousova and P. Hobza, *J. Chem. Phys.* **120**, 10554 (2004).
78. M. Gantenberg, M. Halupka and W. Sander, *Chem. Eur. J.* **6**, 1865 (2000).
79. K. Marushkevich, L. Khriachtchev, J. Lundell, A. Domanskaya and M. Räsänen, *J. Phys. Chem.* **A114**, 3495 (2010).
80. E. Isoniemi, L. Khriachtchev, M. Makkonen and M. Räsänen, *J. Phys. Chem.* **A110**, 11479 (2006).
81. P.-O. Åstrand, G. Karlström, A. Engdahl and B. Nelander, *J. Chem. Phys.* **102**, 3534 (1995).
82. D. Priem, T.-K. Ha and A. Bauder, *J. Chem. Phys.* **113**, 169 (2000).
83. S. Alosio, P. E. Hintze and V. Vaida, *J. Phys. Chem.* **A106**, 363 (2002).
84. L. George and W. Sander, *Specrtochim. Acta* **A60**, 3225 (2004).
85. Z. Zhou, Y. Shi and X. Zhou, *J. Phys. Chem.* **A108**, 813 (2004).
86. K. Marushkevich, L. Khriachtchev and M. Räsänen, *J. Phys. Chem.* **A111**, 2040 (2007).
87. K. Marushkevich and L. Khriachtchev (unpublished).
88. A. H. Zewail, *J. Phys. Chem.* **A104**, 5660, (2000).
89. V. A. Apkarian and N. Schwentner, *Chem. Rev.* **99**, 1481 (1999).
90. R. Schanz, V. Botan and P. Hamm, *J. Chem. Phys.* **122**, 044509 (2005).
91. V. Botan, R. Schanz and P. Hamm, *J. Chem. Phys.* **124**, 234511 (2006).
92. P. Hamm, *Chem. Phys.* **347**, 503 (2008).
93. V. Botan and P. Hamm, *J. Chem. Phys.* **129**, 044507 (2008).
94. V. Botan and P. Hamm, *J. Chem. Phys.* **129**, 114510 (2008).
95. V. Botan and P. Hamm, *J. Chem. Phys.* **129**, 164506 (2008).
96. F. Richter, M. Hochlaf, P. Rosmus, F. Gatti and H-D. Meyer, *J. Chem. Phys.* **120**, 1306 (2004).
97. F. Richter, P. Rosmus, F. Gatti and H-D. Meyer, *J. Chem. Phys.* **120**, 6072 (2004).
98. D. Luckhaus, *Chem. Phys.* **304**, 79 (2004).
99. P. Hamm and R. A. Kaindl and J. Stenger, *Opt. Lett.* **25**, 1798 (2000).
100. E. M. S. Maçôas, P. Myllyperkiö, H. Kunttu and M. Pettersson, *J. Phys. Chem.* **A113**, 7227 (2009).

CHAPTER 4

PHOTODYNAMICS AT LOW TEMPERATURES, IN TIME DOMAIN

V. Ara Apkarian

Department of Chemistry, University of California
2125 Natural Sciences II, Irvine CA 92697-2025, USA
E-mail: aapkaria@uci.edu

Mika Pettersson

Nanoscience Center, Department of Chemistry
P.O. Box 35, FI-40014 University of Jyväskylä, Finland
E-mail: mika.j.pettersson@jyu.fi

We trace a path of synergistic evolution in concepts and tools of time-domain photodynamics in cryogenic rare-gas hosts — experiment and theory. From kinetic studies, to controllable classical dynamics, to the interrogation of quantum coherences and decoherence, the tools that have enabled quantitative atomistic understanding of observables is traced along one logical path. We connect the foundations built on the minimalist systems — a diatomic in an atomic host — to the current activities in the more complex systems of polyatomics.

1. Introduction

Without a doubt, the principle aim of studies of photophysics and molecular dynamics in cryogenic rare-gas solids is the elaboration of models — prototypes of many-body dynamics that can be investigated at a commensurate level of rigor through experiment and theory. This has been particularly true for the time-domain studies. Investigations of nominally simple systems, such as molecular halogens isolated in rare-gas solids, have elucidated the effects of the environment on elementary processes. As importantly, they have allowed deep insights into the experimental tools of interrogation, i.e., the information content in various nonlinear optical processes that are essential to probe events in time. It has been possible to make the connection between system observables and various methods of interrogation at an atomistic level of detail through explicit simulations. To this end, there has been significant development of theoretical methods driven by these model studies. In the first part of this chapter, we give a trend line of the concerted evolution of foundational concepts through investigations of the "simplest" of systems. We will mainly trace the nascence of concepts connected to activities at Irvine, leaving the coverage of the field that has burgeoned to the comprehensive reviews that have already appeared.[1,2] In the second part, we highlight bridges toward complexity, from molecular complexes to polyatomics in Jyväskylä.

2. Toward Quantitative Photodynamics: A Bimolecular Reaction in Real Time

In the Irvine group, the drive toward quantitative photodynamics in rare-gas solids (RGS) was initiated under a program devoted to condensed phase photoinduced harpoon reactions.[3] In contrast with isolated molecules, in condensed media intermolecular transitions are accessible. Of these, intermolecular charge transfer transitions are the brightest. Where the transition is between guest and host, rich photophysics that connects localized molecular states to delocalized collective states of the solid becomes accessible. The original work on RGS doped with molecular halogens, hydrogen halides, and atomic halogens, stimulated diverse developments.[4] Briefly, photoexcitation of charge transfer transitions between guest = Cl_2, HCl and host = Xe, leads to efficient dissociation. The product Cl and H atoms can be directly probed via their charge transfer excitations. The former leads to the fluorescing exciplex, $Xe_2^+Cl^-$, while the latter can be seen through strong, deep UV absorption progressions of $Xe_n^+H^-$.[5] In contrast to the valence states of diatomic halogens, which are subject to a strong cage effect, harpooning to $Xe^+Cl_2^-$ is followed by facile dissociation of the molecule and cage exit of the product atoms. Now the cage atoms actively catalyze dissociation. The branching of ensuing developments is interesting to trace.

(i) To understand the underlying dynamics, the many-body forces on excited electronic states must be described. To this end, Last and George offered the diatomics in ionic systems (DIIS) modification of the diatomics in molecules (DIM) theory.[6] The required modifications were the careful construction of ionic states solvated by the dielectric of the host, and the connection of electrostatics with valence interactions. Variants of DIM and DIIS were to become a staple in simulations of open shell systems and of dynamics on electronically excited states.

(ii) Perhaps the more significant step toward quantitative photophysics was the development of molecular dynamics (MD) simulations as an interpretive tool. The first simulations in RGS, photodissociation of HI in solid Xe, was carried out by Alimi and Gerber.[7] They were inspired by Lawrence's ongoing experiments in single crystals.[8] Due to fracture, single-crystal measurements proved less useful than matrices, or polycrystalline solids, which were to displace them. These simulations already defined motifs, such as direct versus delayed cage exit, that were to become organizing principles in future studies of photodissociation dynamics. Beyond their value as interpretive and quantitative analysis tools of photophysics, the MD simulations inspired the quest to capture the highlighted processes experimentally, as vividly, in real time. This was also the beginning of a long collaboration between Irvine and Jerusalem.

(iii) DIIS potentials, as constructed, did not describe the delocalized nature of vertically accessed charge transfer states of halogen atoms in Xe. Their identification as a sequence of Rydberg holes orbiting the localized negative ion was made in collaboration with Schwentner,[9] who then extended it to trapped H/D atoms.[5] The self-trapping of such states allows for charge separation, mobility, and storage — models for the ubiquitous processes of consequence from biological systems to solar cells.

(iv) The facile photogeneration of atoms in the solid state allowed for a host of studies on photomobility, thermomobility, bimolecular reactions with hot and cold atoms, the observation of transition states; and applications, such as the construction of solid-state exciplex lasers.

This was the subject of the synergy between Berlin and Irvine.[1] The elucidation of atom versus charge mobility was a target in IR studies of the products of photodissociation of HI in a Xe matrix,[10] which subsequent to identification of $XeHXe^+$, blossomed into the exciting field of studies on rare-gas hydride molecules.[11]

The above points, based on kinetic studies, set the stage for real-time measurements. The technical difficulty of time-resolved chemistry in solids is the accumulation of products in the same volume as the starting material. This was bypassed by Zadoyan, who carried out the real-time measurements of photoinduced harpoon reaction on Cl_2 dissolved in liquid xenon.[12] Among other features, the study is notable for time resolving a bimolecular reaction in the liquid phase, by staging the solute–solvent reaction:

$$Xe^+Cl^-(\ell) + Xe(\ell) \rightarrow Xe_2^+Cl^-(\ell). \tag{1}$$

The Xe_2^+ bond formation, from the equilibrium distance of the neutral $R_i = 4.7$ Å to that in the exciplex, $R_f = 3.2$ Å occurs in $t = 330$ fs, as an over-damped motion:

$$t = \left[\frac{M(R_i^8 - R_f^8)}{12\alpha\mu^2 2}\right]^{1/2} \tag{2}$$

with the Xe^+Cl^- dipole (μ) reeling in the polarizable (α) Xe atom. Since the motion is that of an over-damped population, the observables could be reproduced using rate equations with previously known spectroscopic constants.

3. Pump–Probe Measurements: Classical Coherence and Control

An alternative approach to time-dependent chemistry in the solid state is the choice of reversible reactions; hence, the selection of I_2 isolated in an Ar matrix for the first real-time studies in matrices.[13] The breaking and remaking of the molecular bond could be repeatedly measured due to the nearly perfect cage effect. Moreover, the extensive prior time-resolved studies on I_2 provided the background for analysis of condensed phase peculiarities. The first pump–probe (pu–pr) measurements in matrices were carried out on the A-state of I_2 in solid Ar. Preparation above the gas phase dissociation limit, when probed via laser-induced fluorescence (LIF) from the ion-pair states, show oscillatory signals with non-trivial timing. While clearly coherence in the dynamics of cage-induced recombination could be identified, the details of the time structure of the signal could not have been convincingly interpreted absent simulations. The classical trajectories of Li and Martens provided the key to the details. Adopting the reflection approximation, the action of the probe laser could be reduced to interception of the I–I bond in a Gaussian window centered at a distance q^*. The classical trajectory ensemble, $\{\delta(q,q_0,p,p_0;t)\}$, obtained from the many-atom simulation could then be inverted to reproduce the observed signal with surprisingly good fidelity. Ultrafast pu–pr measurements combined with the direct inversion of classical simulations are now inseparable tools where quantitative analysis is sought. Implicit in this strategy is that pu–pr measurements using LIF as probe interrogate the classical coherence of a system.

The distinction between classical and quantum coherence was, and remains, the source of considerable confusion and deliberation. The essence deserves formal definition. Thus, for an impulsively prepared wavepacket, prepared in short enough time to preclude environmental response, the global wavefunction may be factorized as the product of system (s) and bath (b):

$$|\Psi(0)\rangle = |\varphi\rangle^{(s)}|\Phi\rangle^{(b)}. \tag{3}$$

Coupled quantum evolution in the tensor space $s\otimes b$ produces the density:

$$\rho = |\Psi\rangle\langle\Psi| = \rho_c + \sum_{n\neq m}\sqrt{P_n P_m}|\varphi_n\rangle\langle\varphi_m|\otimes|\phi_n\rangle\langle\phi_m|, \tag{4}$$

in which the classical density ρ_c consists of the ensemble of correlated states:

$$\rho_c = \sum_n P_n|\varphi_n\rangle\langle\varphi_n|\otimes|\phi_n\rangle\langle\phi_n|. \tag{5}$$

The interferences in Eq. (4) represent the quantum coherence. Since projective measurements on the system trace out the bath:

$$<P> = Tr_b\,[\hat{P}\rho] = \hat{P}\rho_c \tag{6}$$

the observable density collapses into Eq. (5). In pu–pr measurements where the signal is given by the rate of population transfer from the evolving excited state, $d|\phi_n^e\rangle\langle\phi_n^e|/dt$, to the fluorescing state, it is the diagonal system density that is interrogated.

What is required for a classical coherence to be launched is that the preparation time be short relative to the characteristic frequencies of motion. Conversely, from the extent of observable coherence, $\rho(q,q_0,p,p_0,t)$ the distribution of initial conditions may be extracted. Deconvolution of the signal to extract $\rho_c(q_0,p_0)$ was recently demonstrated by Goldschleger *et al.* in a study of bromine clathrate hydrate crystals and bromine dissolved in water.[14] In these resonant transient grating experiments, the A/A′ state population is observed after rebound from the cage. The extracted rebound time distribution allows the characterization of size, structure, and an elasticity of the cage.

That control over a classical coherence may be accomplished by time-ordering initial conditions of trajectories $\rho_c(q_0;t_i, p_0;t_i)$ through chirped pulses was not so obvious prior to the first condensed phase control experiments carried out in collaboration with the Wilson group.[15] This followed in the heels of the gas phase control experiments on I_2, which were formulated according to paradigms of optimal quantum coherence control, absent dissipation and dephasing.[16] The effect of chirp on both pump and probe pulses was observed in the first measurements done on the A/A′ states of I_2/Kr, and the observed effect was explicitly reproduced by Sterling *et al.* through classical MD simulations.[17] The essential assumption is the reflection approximation to map the time-frequency profile of the laser pulse on the time ordering of the initial conditions of system trajectories. The effect is small because of the significant scrambling of trajectories on the cage wall prior to recombination. That the experimental results were withheld in these reports reflects the resistance at the time to notions of classical coherence control. The A-state experiments were later to be repeated and reproduced by the same methods.[18] The second set of experiments carried out on the deeply bound B state demonstrated more effective control than their gas phase counterpart.[19] The scrutiny of the measurements, cloaked under "semiclassical" and "nearly classical" qualifications, showed that classical MD with sampling from an initial quantum distribution reproduces the essence of the observable effects. A population prepared deep in the B state evolves as a damped Morse oscillator. In effect, chirped pump preparation reduces the problem to that of runners on a race-track — time-ordering the start line from inner to outer tracks controls the distribution of arrivals at the finish line. The true quantum effects are overlooked due to limitations inherent in measurements that only probe diagonal densities.

4. Nonadiabatic Vibronic Dynamics

The discussion up to this point pertains to vibrational many-body dynamics, presumed to proceed on a single electronic state. In molecular halogens, 23 electronic states correlate with the $^2P + ^2P$ asymptote of ground state atoms. Born–Oppenheimer breakdown in these systems, and in particular solvent-induced nonadiabatic dynamics, serves as a prototype for electronic curve crossings that play a central role in photochemistry. Starting with Eisenthal and co-workers,[20] the predissociation of I_2(B) has served as the model for time-domain studies in condensed media. In the ultrafast era, Scherer had demonstrated that I_2(B) predissociates in 230 fs in liquid CCl_4.[21] Yet, in the first time-resolved study of I_2(B), it was learned that the predissociation time is an order of magnitude longer in a Kr matrix.[22] This electronic caging effect is explained as the cancellation of signed coupling matrix elements (as in dipolar coupling) by the local symmetry of the solvent.[23] As significantly, expectations of well-defined crossing point(s) in energy or coordinate space prove to be naive. Systematic experiments and analysis reconstruct an energy-dependent population leakage as a rate process, determined by the competition between vibrational relaxation and predissociation.[24] Rather than the one-dimensional Landau–Zener picture, the multi-dimensional intersections between surfaces are distributed and dynamical, which combined with the spread of vibrational packets leads to nonlocality. In contrast, sharply localized spin-flipping curve crossings occur on two-photon accessed, cage-bound I*I* and I*I potentials on which the molecular bond undergoes large amplitude motion, as demonstrated by Bihary *et al.*[25] The full MD machinery — using semiempirical DIM potentials in the complete electronic basis set (valence and ion-pair states), including nonadiabatic processes among all 23 valence states, and computing all optical resonances in full dimensionality — was assembled in the most complete atomistic simulations of the vibronic dynamics of I_2 in cryogenic rare-gas solids (and liquids) by Coker *et al.*[26] The principles developed from these simulations, and their explicit juxtaposition with experimental observables, defined the standard in the field. Nevertheless, there was significant discrepancy with the experiments. The surface hopping trajectories predicted faster and more localized curve crossing events than observed experimentally. It can be suspected that an ensemble of independent trajectories cannot reproduce effects of quantum nonlocality. There are significant efforts to redress this shortcoming,[27] a fully quantum system embedded in a classical bath is one such approach,[28] which while limited by the number of degrees of freedom that can be treated exactly and the approximations necessary to marry the quantum and classical subsystems, is useful to test the importance of effects in controlling the observed behavior. The subject of nonadiabatic dynamics of dihalogens and halogen halides has been extensively developed in Berlin and Jerusalem.[2]

5. Quantum Coherence and SemiClassical Dynamics

Quantum coherent dynamics is interrogated through nonlinear spectroscopies. Resonant Raman (RR) spectra relay timescales of vibronic dephasing, which occur on ultrafast scales, as pointed out in the analysis of the I_2 spectra in condensed rare gases by Chergui *et al.*[29] Electronic dephasing determines intensities, while vibrational dephasing on the ground electronic state determines linewidths of RR progressions. In addition to the phase coherent radiation that generates a sharp vibrational progression, a broad pedestal due to dephased hot luminescence occurs.[30] The two processes can be distinguished by their time circuit diagrams, which are shown in Fig. 1: (a) absorption followed by emission, which can proceed even if the system is fully dephased during the t_2 time interval; and (b) the nonclassical RR path, which involves electronic coherence during t_1 and t_3, and vibrational coherence during t_2. During t_1 the system (I_2) is in the two-electronic state superposition, $\rho_{sys} = |\varphi_B\rangle\langle\varphi_X|$; accordingly through the system–bath

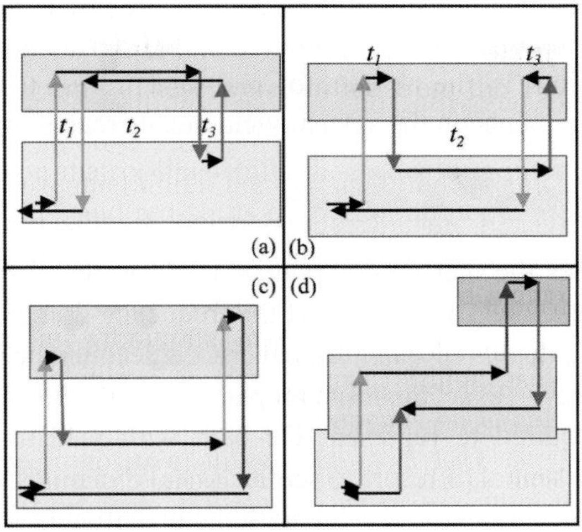

Fig. 1. Time-circuit diagrams define the time correlations responsible for various observables: (a) absorption followed by emission, which is part of the RR response if there is no dephasing during t_2; (b) the nonclassical RR response contribution in which the fields of the excitation and scattering photons are intertwined; (c) CARS diagram designed to interrogate the ground state vibrational coherence, (d) four-color four-wave mixing designed to interrogate quantum coherences in the excited electronic state in real time. Note, in (a) and (b) a photon is represented by its two fields, E and E^*. In (c) and (d) the complex conjugate diagrams are not shown, now each up/down arrow corresponds to interaction with a different photon. Time progresses along the horizontal, forward on ket states and backward on bra. The background colors designate different electronic states to highlight the resonant processes. For the case of the I_2 experiments discussed, the X, B, and E states are represented by the colored bands. See also Colour Insert.

entanglement (4), the bath is also in the two-state superposition, $\rho_{\text{bath}} = |\phi_B\rangle\langle\phi_X|$. The bath-bath correlation amplitude:

$$c_{\text{bath}}(t) = \langle\phi_B(t)|\,\phi_X(t)\rangle = \langle\phi(t)e^{iH(\varphi_B)t}|e^{-iH(\varphi_X)t}\,\phi(t)\rangle \qquad (7)$$

decays due to the difference in evolution in response to the system being in the X/B state.

The correlation in Eq. (7) describes the forward evolution of the bath when the system is in the X state and backward evolution when in the B state. It measures the time-reversibility of dynamics. As such, its decay signifies the generation of quantum entropy, or equivalently, the onset of decoherence. An explicit semiclassical treatment of this process was given by Bihary *et al.*,[31] to which we return below. Here we note that the decay of the correlation amplitude [Eq. (7)] terminates the coherent Raman scattering. In the absence of real-time tracking, the frequency-domain measurements on the system yield decoherence time scales:

$$c_{\text{bath}} = e^{-\gamma_2 t^2 - \gamma_1 t} \tag{8}$$

where γ_2 defines the inertial component while γ_1 defines the dephasing rate. This leads to the Voigt profile of the frequency-domain spectra.

Mechanistic details of quantum coherence/decoherence are contained in Eq. (7). However, exact quantum propagation in the hundreds of degrees of freedom required to represent the bath with realism is simply not feasible. It is essential to resort to semiclassical dynamics (SC). This was developed by Ovchinnikov, starting with the invention of the mixed-order SC method.[32] Quantum coherences can now be treated atomistically, and the required correlations for various nonlinear spectra can be evaluated in hundreds of degrees of freedom, using trajectories that carry phase. The full implementation using first-order (in $\hbar$) propagators, such as van Vleck[33] or Herman and Kluk,[34] is intensive because of the overhead in evaluating the monodromy matrices that maintain flux normalization. The mixed-order treatment allows for the majority of the degrees of freedom to be treated in zeroth order, as an ensemble of trajectories $\{\delta[q - q(q_0, p_0;t)]e^{-S(t)/\hbar}\}$ with phase $S(t)$ given by the classical action. The evaluation of the latter requires no more overhead than classical MD. The method was introduced with applications to linear and RR spectra containing quantum interferences,[32] and used to dissect spectral features, such as phonon sidebands in terms of the motions of atoms (Cartesian or normal coordinates).[35] Variants of the method have since been advanced in diverse applications. In condensed media, for which purpose the method was developed, the relatively short decoherence times allow for a nearly exact and explicit treatment of quantum MD.

6. Quantum Coherences in Real Time Through Four-Wave Mixing

Quantum coherences in real time are most directly interrogated through four-wave mixing (FWM) spectroscopies, $\chi^{(3)}$ processes that appear under many different acronyms. Of these, coherent anti-Stokes Raman scattering (CARS) was implemented in time domain to interrogate vibrational coherences of I_2 on its ground electronic state. The process is illustrated by the time-circuit diagram (Fig. 1c). The similarity of the diagrammatic representation with RR is not accidental — both experiments probe the ground state vibrational coherence. Note though, in RR the four arrows represent two photons, while in CARS the four arrows represent four different photons — three input pulses and one coherently radiated polarization. First implemented by Karavitis *et al.*[36,37] then extended by Pettersson and co-workers,[38] a rather complete analysis of the ground state vibrations, $v = 1$–19 at $T = 2$–45 K, of iodine in solid Kr was generated. The decay of the coherences prepared is characteristic of dephasing in the weak coupling limit, where each vibrational state is characterized by the single dephasing constant, $\gamma_1 = \gamma_{v0}$ — namely, the decay of phase correlation with respect to $v = 0$. Spectrally resolved CARS measurements were conducted to search for the possibility of coherence transfer.[37] None could be observed — vibrational dephasing rates are independent of the superposition prepared, decoherence is due to environmental noise in this weak coupling limit. These measurements serve the very purpose of prototypes, as testing grounds for novel theory and understanding. To characterize the environmental noise mechanistically, SC tools should be sharpened. Such analyses have already been provided for dephasing in various vibrational basis sets by Chapman and Cina,[39] the linearized density matrix dynamics by Ma and Coker,[40] and the SC Liouville method by Riga *et al.* as a trajectory-based method that reproduces the observed dephasing rates and their T dependence.[41]

The situation is dramatically different in the case of vibrational coherences in the excited B state of the molecule. This was interrogated with the FWM scheme of Fig. 1d. In its most general four-color implementation, where three different input colors are used, the evolving quantum coherences can be followed in real time in great detail. Electronic dephasing processes are measured along the t_1 and t_3 intervals where the system is in the $(|\varphi_B\rangle + |\varphi_X\rangle)$ and $(|\varphi_E\rangle + |\varphi_B\rangle)$ superpositions, respectively. During t_2 the vibrational superposition $(|\varphi_B(v)\rangle + |\varphi_B'(v)\rangle) = |s_1\rangle + |s_2\rangle$ that prepares the system as two distinct packets with timing and composition controlled by the two different color lasers is interrogated on the B state. The quantum interference term, $|s_1\rangle\langle s_2|$ is measured. Because of the strong coupling between the driven Morse oscillator and the lattice, the system can drive the bath and can sense the bath feedback — the bath–bath correlation in Eq. (7) is no longer a simple decay, but contains oscillatory behavior characteristic of phase coherent evolution in the environment. Segale's published work gives a preview of the insights that can be extracted from such measurements,[42] and a rather complete set of multicolor measurements are contained in his thesis.[43] The cross-correlation between the two prepared wavepackets of two different energies is recorded via resonant coherent Raman scattering as spatially distinct Stokes and anti-Stokes radiation $\langle P_{12}\rangle$ and $\langle P_{21}\rangle$. This direct measure of the evolving dissipative off-diagonal density of the system has direct implications about the coherences in the entangled bath [Eq. (4)]. Coherent dissipation, event-driven punctuated decoherence, bath-induced system coherence, echoes in the bath, decoherence-free einstates of system–bath, are among the direct observables. Pu–pr measurements, now using phase-locked and shaped pulses were advanced in Berlin to probe these very issues,[2] with most direct and elegant demonstrations of controlled system–bath coherences by selectively preparing the admixture of zero-phonon lines and phonon sidebands.[44,45] The implied coherent superposition states of the bath that last for ~2 ps was qualified as a nontrivial Schrödinger Cat.[42] With the collapse of such states emerges classicality — reduction to the diagonal density [Eq. (3)] — and this process can be directly monitored. These measurements, on the model system of a diatomic in a cryogenic rare-gas host, provide a fundamental window on *the quantum in the mechanics of molecular matter.*

7. Beyond Diatomic Molecules

In going to polyatomic species new processes arise, the most important ones being chemical reactions and intramolecular energy relaxation. It is of fundamental importance to understand these processes and how they are affected by the condensed phase environment and the nature of interactions with the surrounding atoms or molecules. However, relatively little is known about the ultrafast dynamics of polyatomic species in low-temperature solids. The reason for this is partially technical. Laser spectroscopic experiments with short pulses usually require repetitive measurements and averaging of signals over millions of pulses. This means that any photoprocesses occurring during irradiation must be reversible or they should be compensated somehow. For diatomic halogens isolated in low-temperature rare-gas matrices the quantum yield for permanent dissociation is very low allowing such easurements to be performed even with photons energetic enough to induce bond breaking. The low quantum yield is due to the strong cage effect, which forces geminate recombination. For polyatomic species many reaction channels become open at high energies and even in the case of strong cage effect geminate recombination of the produced radical pair leads to irreversible chemical reactions. Thus, most of the low temperature dynamical investigations of polyatomic species have concentrated on vibrational dynamics on the ground electronic state by using time resolved infrared or Raman techniques. In Sections 8 and 9, representative examples of dynamical studies of molecular complexes and polyatomic molecules in low-temperature rare-gas matrices are discussed.

8. Molecular Complexes

Complexes of diatomic molecules bridge the gap between isolated molecules and bulk condensed phase and their studies thus provide a way to understand how dynamics is affected by intermolecular interactions. Effects of hydrogen bonding and van der Waals interactions on vibrational energy relaxation and dephasing dynamics were studied by Crépin *et al.* by using H(D)Cl as a model system.[46–52] By using free electron laser (FEL) and optical parametric oscillator (OPO) radiation sources they were able to perform accumulated photon echo measurements with a picosecond time resolution on various complexes of H(D)Cl. Starting from the monomeric DCl in solid N_2, the dephasing time (T_2) was found to be 175 ps at 5.6 K decreasing to 8 ps at 20 K.[48] The temperature dependence indicated coupling to local phonons having a wavenumber of 19 cm^{-1}. Dimerization reduces the dephasing time so that at 5–7 K the T_2 is 55 ± 5 ps and 20 ± 3 ps for the proton acceptor (ν_1) and donor (ν_2), respectively.[49,51] It is interesting that hydrogen bonding affects significantly the free mode (ν_1) including modifying the temperature dependence in such a way that at 20 K the ν_1 of $(DCl)_2$ has longer T_2 (20 ps) than the monomer (8 ps). This indicates that hydrogen bonding causes decoupling of the free oscillator in the dimer from the nitrogen lattice. For corresponding HCl species the dephasing times were shorter than for the DCl species which can be understood based on the smaller amplitude of the vibration in the heavy isotopic species and thus reduced interaction with the lattice molecules.

In addition to hydrogen bonding the effect of van der Waals type of interaction with N_2 molecules was studied.[51,52] Even such a weak interaction has a substantial effect on the dephasing dynamics. For the series DCl-N_2, DCl-$(N_2)_2$, and DCl/(N_2 solid), the T_2 values are 1.3, 0.3, and 0.18 ns, respectively.[52] The effect of adding a second nitrogen molecule is substantial compared to the change upon filling the shell of 12 nearest neighbors in the *fcc* lattice. Finally, the effect of the trapping environment is important as evidenced by the observation of very fast dephasing of few ps for monomeric DCl in solid argon as compared to 175 ps in solid nitrogen.[52] This difference can be explained by the quasi-free rotation of DCl in solid argon, whereas in solid nitrogen rotation is quenched. Low-frequency rotational modes and their strong coupling to lattice phonons explain the difference in dynamics. Altogether, the examples above point out the importance of delicate interactions on the vibrational dynamics of molecules. For small complexes the effects on dynamics seem to be specific for each structure and no general conclusions can be made based on limited amount of available data.

Kiviniemi *et al.* investigated vibrational dynamics of I_2–Xe complex isolated in solid krypton by using time-domain CARS.[53] The experimental method was similar to the CARS studies of iodine molecules in low-temperature matrices. The sample contained complexes of iodine with xenon in addition to monomeric species and the two species had different vibrational frequencies. When both species are coherently excited with broad-band femtosecond pulses, the CARS radiation $P^3(t)$ from the two species interfere with the effect of producing modulations in the time-domain signal. This is the origin of polarization

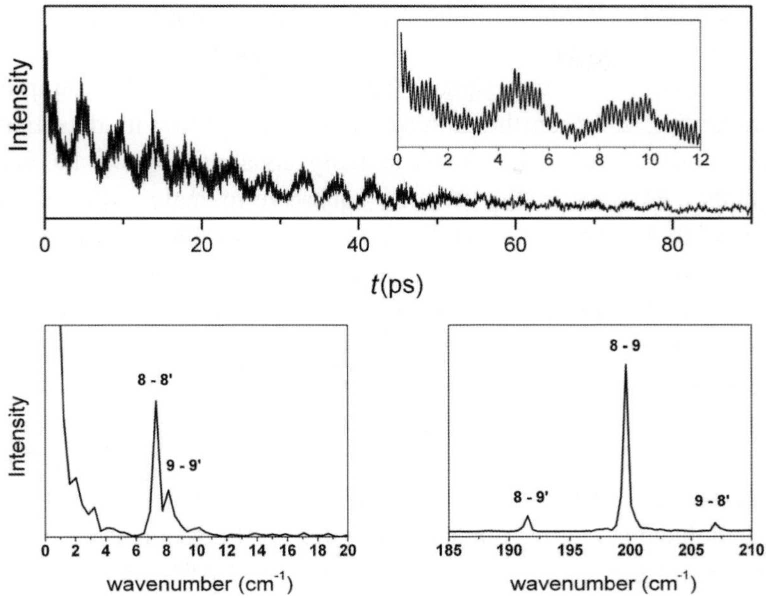

Fig. 2. TR-CARS data of a I_2-Xe in solid Kr at 4 K. Upper trace: Time-domain CARS signal showing modulation due to polarization beating between the polarizations from I_2–Xe and I_2. The inset shows a magnified view showing fast oscillations due to motion of the vibrational wavepacket of iodine. Lower left trace: FFT of the time-domain signal showing the wavenumbers of the low-frequency beats. The figures indicate the quantum states that participate in beating. Bare figures indicate I_2 states and primed figures indicate I_2–Xe states. Lower right trace: The same data as on the left trace at different spectral region.

beating, or polarization interference, as distinct from quantum beating.[54] The effect of multiple species on the CARS signal can be seen from Eq. (9):

$$S(t) = \int \left| P_A^{(3)}(t') + P_B(t') \right|^2 dt' \tag{9}$$

where A and B refer to different species, e.g., uncomplexed and complexed iodine. From Eq. (9) it can be seen that in addition to signals from individual species the signal contains a contribution from a cross-term between the third-order polarizations of A and B. In frequency-domain polarization beats appear as sidebands to the quantum beats and they appear at the difference frequencies between the quantum states of both excited species. As such their information content in a spectroscopic sense is exactly the same as in quantum beats.

An example of the signals appearing in I_2–Xe/Kr samples is shown in Fig. 2. When the amount of complex is low its detection via polarization beat phenomenon is actually enhanced by the heterodyne advantage. As seen in Eq. (9), the square term makes the individual signal weak but if the signal from the other species is strong it also enhances the observation of the species with a smaller contribution. With this technique the spectroscopic constants of the I_2–Xe complex were accurately determined.[53] The harmonic wavenumber in the complex is redshifted by 0.9 cm^{-1} relative to “free” I_2 in solid krypton. The dephasing times for different vibrational states were determined for both I_2 and I_2–Xe and they were found to be on average about 2–3 times shorter for the complex than for I_2. This indicates enhanced coupling of iodine molecule to lattice phonons mediated by the xenon atom.

An interesting aspect in the time-resolved CARS technique is the possibility not only to measure vibrational dynamics but also to control it. Two first interactions with the field involve sudden preparation of vibrational wavepackets on the ground electronic state via stimulated Raman scattering. If a real electronic state is involved in the preparation there is an additional possibility to prepare wavepackets with high vibrational quantum numbers.[55] This was demonstrated for I_2 isolated in RGS by preparing vibrational wavepackets with excitation up to $v =$ 19.[37,38] This type of excitation can be used to control the I–I coordinate in a complex involving iodine molecule. Classically thinking the motion of a wavepacket corresponds to periodic modulation of the bond distance. If the complex has an excited state which leads to photoreaction between the complex partners, this modulation can be used to initiate the reaction from different geometries of the complex by using a third excitation pulse. Although the outcome of a reaction is determined by the shape of the potential energy surface on which the reaction occurs, it also depends on where on the potential surface the system is prepared. Thus, a photoreaction in the complex could, in principle, be guided by controlling a vibrational coordinate in the complex via stimulated Raman scattering and using a third pulse after suitable delay time to transfer the wavepacket on the

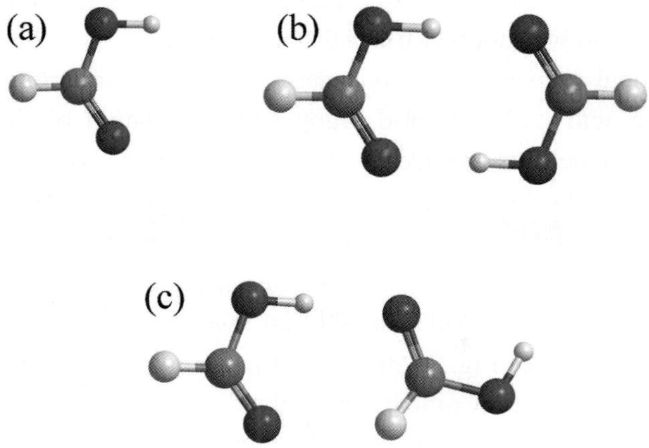

Fig. 3. Structure of formic acid and its dimers: (a) monomer, (b) cyclic dimer, (c) open chain dimer.

excited state potential energy surface. This is essentially the Tannor–Kosloff–Rice control scheme suggested theoretically in the 1980s.[56] However, it has never been applied to control a bimolecular reaction via control of vibrational ground state dynamics. Recently, Kiviniemi *et al.* investigated the complex between iodine molecule and benzene in order to evaluate its suitability for such an experiment.[57] However, from the resonance Raman spectrum it was deduced that, due to fast electronic dephasing, highly excited vibrational wavepackets cannot be reached in this complex and more lately the interest has been turned to the I_2–Xe complex again.

9. Polyatomic Molecules

There are relatively few studies of time-resolved dynamics of polyatomic molecules in low-temperature solids. The most thorough investigation is a series of studies on HONO in solid argon by Hamm and co-workers.[58] They used transient femtosecond mid-IR absorption techniques to reveal the dynamics related to the IR-induced isomerization reaction of HONO. Despite the apparent simplicity of tetratomic HONO, its isomerization reaction is rich, with process that involve dynamics on picoseconds to hundreds of nanoseconds. These investigations are described in Chapter 3 of this book and will not be further discussed here.

Vibrational relaxation of monomers and dimers of formic acid (FA) and acetic acid (AA) in solid argon upon excitation of the carbonyl stretching mode was studied by femtosecond transient IR absorption techniques by Maçôas *et al.*[59] The use of the matrix isolation technique allowed to compare the dynamics of monomers with that of dimers, which has not been possible in room temperature liquids since the equilibrium is strongly on the dimer side. In addition, two different dimer structures (cyclic and open, see Fig. 3) are trapped in the matrix allowing selective studies of both species. Comparison of the dynamics in monomeric acids allows revealing the effect of methyl group on the relaxation dynamics of carboxylic group. A qualitative picture of the relaxation chain is that the initial excitation of the stretching mode of the carbonyl group relaxes to intermediate hot vibrational states that cool by energy transfer to the lattice. For FA, the depopulation of the $v = 1$ state of νC=O occurs with a time constant of 0.5 ns. The recovery of the ground state and the decay of the intermediate states could not be observed within the time window of the experiment but it was estimated to be longer than 5 ns. This long recovery time is consistent with the sparse state density where energy gaps with several hundred cm^{-1} exist requiring high-order phonon processes for relaxation. As expected, the dynamics becomes much faster in AA where the recovery of the ground state occurs with a time constant of 84 ps. The apparent depopulation of the $v = 1$ state of νC=O has the same time constant as the decay of the intermediate hot states, which in turn is the same as the recovery time constant of the ground state (84 ps). This indicates that the vibrational manifold is at internal equilibrium and the cooling rate is determined by the energy transfer to the lattice. The role of methyl group in the relaxation is very significant enhancing the ground state recovery time at least two orders of magnitude compared to FA. This is due to increased state density and efficient coupling of the states allowing one-phonon processes to facilitate the relaxation chain. This direct observation of the much faster vibrational relaxation in AA compared to FA is consistent with the earlier observation of one order of magnitude lower quantum yield for *trans*-to-*cis* isomerization in AA compared to FA.[60,61] Increased state density and efficient coupling of vibrational states increase competing relaxation pathways, which do not lead to isomerization.

The relaxation dynamics of FA dimers show some interesting features. As expected for a strongly hydrogen bonded dimer, the depopulation of the $v = 1$ state of νC=O and population of the intermediate states is much faster (~40 ps) than in the monomer. However, the intermediate states decay relatively slowly with a time constant of 0.5 ns, which also coincides with the ground state recovery time. It seems that the strong hydrogen bonding creates large density of strongly coupled states within the dimer, but the coupling to the surrounding lattice remains relatively weak. This interpretation is consistent with the picture that emerged from studies of H(D)Cl dimers where it was observed that hydrogen bonding leads to decoupling of the H(D)Cl oscillator from the lattice. Note also that the overall state density in the FA dimer should be much higher than in the AA monomer, but still the ground state recovery time in AA is much shorter, pointing out the importance of the methyl group as the species providing strong coupling to the lattice. Nevertheless, the ground state recovery time of the FA dimer is at least an order of magnitude shorter than in the monomer, which probably is due to the coupling of intermolecular low-frequency modes to the lattice.

For AA dimer the ground state recovery time is essentially the same (88 ps) as in the monomer. This again shows that the determining factor in the cooling rate is the coupling of the methyl group to the lattice and increased state density and introduction of intermolecular modes provided by dimer formation do not enhance the overall relaxation rate. These studies indicate that it is possible to tune the relaxation time in molecules and molecular complexes in condensed phase by using suitable chemical groups as facilitators of vibrational energy transfer to the surrounding medium. In this respect, methyl group seems to be a good candidate. On the other hand, by avoiding such groups it would be possible to hinder the cooling rate and this effect might have some use in applications targeted for carrying out photoreactions requiring thermal activation in molecular centers isolated in condensed phase.

Acknowledgments

We would like to apologize to all of our colleagues whose seminal contributions to the field were not included. Our aim was to follow one trend line in the evolution of concepts and methods. In that spirit, in most cases we have cited the first of a series of papers on a subject, rather than the most recent. We have noted the first-generation synergistic connections between centers, Irvine, Jerusalem, Berlin, Helsinki, and Jyväskylä, but the community is much larger, and growing even faster.

References

1. V. A. Apkarian and N. Schwentner, *Chem. Rev.* **99**, 1481 (1999).
2. M. Bargheer, A. Borowski, A. Cohen, M. Fushitani, R. B. Gerber, M. Gühr, P. Hamm, H. Ibrahim, T. Kiljunen, M. V. Korolkov, O. Kühn, J. Manz, B. Schmidt, M. Schröder and N. Schwentner, In: O. Kühn and L. Wöste (editors), *Analysis and Control of Ultrafast Photoinduced Reactions*, Vol. 87, Ch. 4, Springer Series in Chemical Physics, Springer, Heidelberg (2007).
3. M. E. Fajardo, R. Withnall, J. Feld, F. Okada, W. Lawrence, L. Wiedeman and V. A. Apkarian, *Laser Chem.* **9,** 1 (1988).
4. M. E. Fajardo and V. A. Apkarian, *J. Chem. Phys.* **85**, 5660 (1986); *J. Chem. Phys.* **89**, 4102 (1988); *J. Chem. Phys.* **89**, 4124 (1988).
5. H. Kunz, J. McCaffrey, M. Chergui, R. Schriever, Ö. Ünal, V. Stepanenko and N. Schwentner, *J. Chem. Phys.* **95**, 1466 (1991).
6. I. Last and T. F. George, *J. Chem. Phys.* **86**, 3787 (1987); I. Last, T. F. George, M. E. Fajardo and V. A. Apkarian, *J. Chem. Phys.* **87**, 5917 (1987).
7. R. Alimi, R. B. Gerber and V. A. Apkarian, *J. Chem. Phys.* **89**, 174 (1988).
8. W. Lawrence, F. Okada and V. A. Apkarian, *Chem. Phys. Lett.* **150**, 339 (1988).
9. N. Schwentner, M. E. Fajardo and V. A. Apkarian, *Chem. Phys. Lett.* **154**, 237 (1989).
10. H. Kunttu, J. Seetula, M. Räsänen and V. A. Apkarian, *J. Chem. Phys.* **96**, 5630 (1992).
11. L. Khriachtchev, M. Räsänen and R. B. Gerber, *Acc. Chem. Res.* **42**, 183 (2009).
12. R. Zadoyan and V. A. Apkarian, *Chem. Phys. Lett.* **206**, 475 (1993).
13. R. Zadoyan, Z. Li, C. Martens and V. A. Apkarian, *J. Chem. Phys.* **101**, 6648 (1994).
14. I. U. Goldschleger, G. Kerenskaya, V. Senekerimyan, K. C. Janda and V. A. Apkarian, *Phys. Chem. Chem. Phys.* **10**, 7226 (2008).
15. J. W. Che, M. Messina, K. R. Wilson, V. A. Apkarian, Z. Li, C. C. Martens, R Zadoyan and Y. J. Yan, *J. Phys. Chem.* **100**, 7873 (1996).

16. B. Kohler, V. V. Yakovlev, J. Che, J. L. Krause, M. Messina, K. R. Wilson, N. Schwentner, R. M. Whitnell and Y. J. Yan, *Phys. Rev. Lett.* **74**, 3360 (1995).
17. M. Sterling, R. Zadoyan and V. A. Apkarian, *J. Chem. Phys.* **104**, 6497 (1996).
18. R. Zadoyan, N. Schwentner and V. A. Apkarian, *Chem. Phys.* **233**, 353 (1998).
19. C. J. Bardeen, J. W. Che, K. R. Wilson, V. V. Yakovlev, V. A. Apkarian, C. C. Martens, R. Zadoyan, B. Kohler and M. Messina, *J. Chem. Phys.* **106**, 8486 (1997).
20. T. J. Chuang, G. W. Hoffman and K. B. Eisenthal, *Chem. Phys. Lett.* **25**, 201 (1974).
21. N. F. Scherer, D. M. Jonas and G. R. Fleming, *J. Chem. Phys.* **99**, 153 (1993).
22. R. Zadoyan, M. Sterling and V. A. Apkarian, *J. Chem. Soc. Faraday Trans.* **92**, 1821 (1996).
23. R. Zadoyan, M. Sterling, M. Ovchinnikov and V. A. Apkarian, *J. Chem. Phys.* **107**, 8446 (1997).
24. M. Gühr, M. Bargheer, P. Dietrich and N. Schwentner, *J. Phys. Chem.* **A106**, 12002 (2002).
25. Z. Bihary, R. Zadoyan, M. Karavitis and V. A. Apkarian, *J. Chem. Phys.* **120**, 7576 (2004).
26. V. S. Batista and D. F. Coker, *J. Chem. Phys.* **105**, 4033 (1996); *J. Chem. Phys.* **106**, 6923 (1997).
27. N. Yu, C. J. Margulis and D. F. Coker, *J. Phys. Chem.* **B105**, 6728 (2001).
28. A. Borowski and O. Kuhn *J. Photochem. Photobiol.* **A190**, 169 (2007).
29. J. Xu, N. Schwentner and M. Chergui, *J. Chem. Phys.* **101**, 7381 (1994); J. Xu, N Schwentner, S. Hennig and M. Chergui, *J. Chim. Phys.* **92**, 541 (1995).
30. J. Almy, K. Kizer, R. Zadoyan and V. A. Apkarian, *J. Phys. Chem.* **A104**, 3508 (2000).
31. Z. Bihary, M. Karavitis, and V. A. Apkarian, *J. Chem. Phys.* **120**, 8144 (2004).
32. M. Ovchinnikov, and V. A. Apkarian, *J. Chem. Phys.* **105**, 10312 (1996); *J. Chem. Phys.* **106**, 5775 (1997).
33. J. H. Van Vleck, *Proc. Nat. Acad. Sci.* **14**,178 (1928).
34. M. F. Herman and E. Kluk, *Chem. Phys.* **91**, 27 (1984).
35. M. Ovchinnikov and V. A. Apkarian, *J. Chem. Phys.* **108**, 2277 (1998).
36. M. Karavitis, R. Zadoyan and V. A. Apkarian, *J. Chem. Phys.* **114,** 4131 (2001).
37. M. Karavitis, T. Kumada, I. Goldschleger and V. A. Apkarian, *Phys. Chem. Chem. Phys.* **7**, 791 (2005).
38. T. Kiviniemi, J. Aumanen, P. Myllyperkiö, V. A. Apkarian and M. Pettersson, *J. Chem. Phys.* **123,** 064509 (2005).
39. C. T. Chapman and J. Cina, *J. Chem. Phys.* **127**, 114502 (2006).
40. Z. Ma and D. F. Coker, *J. Chem. Phys.* **128**, 244108 (2008).
41. J. M. Riga, E. Fredj and C. C. Martens, *J. Chem. Phys.* **124,** 064506 (2006).
42. D. Segale, M. Karavitis, E. Fredj and V. A. Apkarian, *J. Chem. Phys.* **122,** 111104 (2005).
43. D. Segale, *Ph. D. Thesis*, Irvine (2007).

44. M. Fushitani, M. Bargheer, M. Guhr, H. Ibrahim and N. Schwentner, *J. Phys.* **B41**, 074013 (2008).
45. M. Fushitani, M. Bargheer, M. Guhr and N. Schwentner, *Phys. Chem. Chem. Phys.* **7**, 3143 (2008).
46. M. Broquier, A. Cuisset, C. Crépin, H. Dubost and J. P. Galaup, *J. Lumin.* **94-95**, 575 (2001).
47. C. Crépin, *Phys. Rev.* **A67**, 013401 (2003).
48. M. Broquier, C. Crépin, A. Cuisset, H. Dubost, J. P. Galaup and P. Roubin, *J. Chem. Phys.* **118**, 9582 (2003).
49. C. Crépin, M. Broquier, A. Cuisset, H. Dubost, J. P. Galaup and J. M. Ortega, *Nucl. Instr. Meth* **A528**, 636 (2004).
50. M. Broquier, C. Crépin, A. Cuisset, H. Dubost and J. P. Galaup, *J. Phys. Chem.* **A109**, 4873 (2005).
51. M. Broquier, B. Lebech and C. Crépin, *Chem. Phys. Lett.* **416**, 121 (2005).
52. M. Broquier, C. Crépin, A. Cuisset, H. Dubost and J. P. Galaup, *Eur. Phys. J.* **D36**, 41 (2005).
53. T. Kiviniemi, T. Kiljunen and M. Pettersson, *J. Chem. Phys.* **125**, 164302 (2006).
54. J. Faeder, I. Pinkas, G. Knopp, Y. Prior and D. J. Tannor, *J. Chem. Phys.* **115**, 440 (2001).
55. I. Pinkas, G. Knopp and Y. Prior, *J. Chem. Phys.* **115**, 236 (2001).
56. D. J. Tannor, R. Kosloff and S. A. Rice, *J. Chem. Phys.* **85**, 5805 (1986).
57. T. Kiviniemi, E. Hulkko, T. Kiljunen and M. Pettersson, *J. Phys. Chem.* **A113**, 6326 (2009).
58. R. Schanz, V. Botan and P. Hamm, *J. Chem. Phys.* **122**, 044509 (2005); V. Botan,R. Schanz and P. Hamm, *J. Chem. Phys.* **124**, 234511 (2006); V. Botan and P. Hamm, *J. Chem. Phys.* **129**, 044507 (2008); P. Hamm, *Chem. Phys.* **347**, 503 (2008); V. Botan and P. Hamm , *J. Chem. Phys.* **129**, 114510 (2008); V. Botan and P. Hamm, *J. Chem. Phys.* **129**, 164506 (2008).
59. E. M. S. Maçôas, P. Myllyperkiö, H. Kunttu and M. Pettersson, *J. Phys. Chem.* **A113,** 7227 (2009).
60. E. M. S. Maçôas, L. Khriachtchev, M. Pettersson, J. Juselius, R. Fausto and M. Räsänen, *J. Chem. Phys.* **119**, 11765 (2003).
61. E. M. S. Maçôas, L. Khriachtchev, M. Pettersson, R. Fausto and M. Räsänen, *J. Chem. Phys.* **121**, 1331 (2004).

CHAPTER 5

MATRIX ISOLATION OF H AND D ATOMS: PHYSICS AND CHEMISTRY FROM 1.5 TO 0.1 K

Vladimir V. Khmelenko and David M. Lee

Laboratory of Atomic and Solid State Physics, Cornell University, Ithaca, NY 14853 USA, and Department of Physics and Astronomy, Texas A&M University, College Station, TX 77843 USA
E-mails: khmel@ccmr.cornell.edu, dml20@cornell.edu

Sergey Vasiliev

Department of Physics, Wihuri Physical Laboratory, University of Turku, 20014 Turku, Finland
E-mail: servas@utu.fi

Recent work on matrix isolated hydrogen and deuterium atoms in impurity helium condensates (IHC), which consist of nanoclusters of the matrix atoms or molecules (H_2, D_2, HD, and Kr), is discussed. Continuous wave (CW) and pulsed electron spin resonance (ESR) techniques at 9 GHz have both been employed. The kinetics of exchange tunneling reactions of hydrogen isotopes have been studied at T = 1.35 K. Pulsed ESR has shown that a large fraction of the stabilized atomic free radicals resides on the surfaces of the clusters. High concentrations of hydrogen atoms in IHCs containing Kr atoms and H_2 molecules have been attained. Studies of hydrogen atoms embedded in thin H_2 films have been conducted at temperatures down to ~0.1 K via ESR at 128 GHz in magnetic fields of 4.6 T. Dissociation of H_2 molecules at very low temperatures leads to the accumulation of films containing H atom concentrations ~10^{19} atoms cm^{-3}. Some remarkable effects have been observed in these samples, including a large departure from the Boltzmann factor for the populations of the two lowest H atom hyperfine states and a complete absence of recombination at the lowest temperatures.

1. Introduction

Some of the most exciting recent developments in the field of matrix isolation have involved atomic hydrogen free radicals embedded in bulk solid molecular hydrogen, in nanoclusters of molecular hydrogen and in thin films of solid molecular hydrogen. Because of the small mass of hydrogen, solid H_2 belongs to a unique class of substances known as quantum solids, for which zero point energy plays an important role in determining their properties. For the case of quantum solids, also including solid ^3He, ^4He, and D_2, the large zero-point energy results in large oscillation amplitudes about the equilibrium positions, so that the atoms are not completely localized on specific lattice positions. This requires that the correlations of nearby atoms be taken into account in determining the overall wave functions characterizing quantum solids. The molar volumes of quantum solids are considerably larger than would be the case for classical solids as a result of these large zero-point energies. In fact, to stabilize solid helium, an external pressure must be applied, even at temperatures close to absolute zero. Otherwise, helium will remain liquid all the way down to absolute zero. This is not the case for solid H_2 which exhibits a triple point at 13.95 K. When an atomic hydrogen free radical is confined in a crystal of solid H_2, its zero-point energy will tend to distort the crystal lattice in its immediate vicinity.

In common with other free radicals, atomic hydrogen possesses an unpaired electron spin. Powerful magnetic resonance technique can thus be employed to investigate the behavior of H and D atoms embedded in various inert matrices. This chapter is mainly concerned with electron spin resonance (ESR) studies of H and/or D atoms embedded in solid matrices of H_2, D_2, and/or HD. Early studies of these systems were conducted in the mid-1950s by Jen *et al.*,[1,2] Wall *et al.*,[3] and Pietter *et al.*[4] The studies were motivated by the desire to store large amounts of chemical energy in atomic free radicals, which, for H atoms, can be released by recombination into H_2 molecules with an energy of 4.5 eV produced per reaction. It was hoped that sufficiently high concentrations of H atoms could be stabilized, leading to an attractive light-weight rocket fuel. Concentrations of H atoms of 7.5% or greater are required for performances comparable to those of the best existing chemical rocket fuels.[5] In spite of their strong efforts, researchers in the 1950s were never able to obtain concentrations exceeding about 0.01%, far below the criterion for rocket fuels, or, for that matter, any practical energy storage applications. Long-term energy storage in mixtures of H atoms in solid H_2 was also limited by the recombination reaction $H + H \rightarrow H_2$.

Following the efforts of the 1950s, this research area languished for almost three decades, until it was realized that samples of H atoms embedded in solid molecular hydrogen could exhibit interesting quantum diffusion phenomena and quantum tunneling chemical reactions. The large mass difference between H and D atoms raised the possibility that isotopic effects could be clearly discerned in these systems.

Two diffusion mechanisms for H atoms embedded in molecular H_2 crystals must be considered. The first process is physical diffusion in which the H atoms, with their large zero-point motions, probe the cage formed by surrounding H_2 neighbors for the minimum potential barriers. This mechanism allows the atoms to move from one lattice position to the next. Andreev and Lifshitz have proposed a model that predicts that at sufficiently low temperatures, an impurity within a periodic crystal can become delocalized, allowing it to move in energy bands analogous to the conduction band for electrons in metals.[6] Kagan and co-workers performed calculations in which a phonon-assisted mechanism contributed strongly to the physical diffusion of H atoms in solid H_2.[7] Density fluctuations associated with the passage of a phonon alter the spacing between the H_2 molecules of the matrix, allowing a freer channel for H atoms to pass through. On the basis of these calculations, they predicted that the recombination rate for H atoms should be a linear function of temperature according to the relation $K_H(T) = AT$.

Ivliev *et al.*[8] performed experiments to study the decay due to recombination in the H atom population in solid H_2 samples. They observed the linear dependence of $K_H(T)$ predicted by Kagan *et al.*,[7] where $A = (2 \pm 1) \times 10^{-24}$ cm^3 s^{-1} K^{-1} in the temperature range 1.3 K – 4.2 K. This result indicated that physical diffusion might be playing an important role in the decay of the H atoms in H_2 samples. Above 4.3 K, the rate of recombination obeyed an Arrhenius law, $K_H(T) \sim \exp(-E_a/k\mathrm{T})$ with $E_a \sim 103$ K. This result indicated that the barrier height for physical diffusion was ~100 K.

Chemical diffusion is a second possible mechanism contributing to recombinational decay of the H atom population. This process involves the exchange tunneling chemical reaction $H + H_2 \rightarrow H_2 + H$, in which a hydrogen atom and a hydrogen molecule change places in the crystal lattice. After multiple repetitions of this reaction, a hydrogen atom will encounter a second hydrogen atom, which can lead to recombination into a hydrogen molecule. This mechanism was suggested by Tsuruta *et al.*[9] and Gordon *et al.*[10] to explain the recombination rates observed in their experiments. The potential barrier for this tunneling reaction is shown to be 4600 K,[11] far larger than that associated with the physical diffusion process for hydrogen atoms in solid H_2 (~100 K). For the case of chemical diffusion, however, the barrier is considerably narrower than the barrier for physical diffusion. Theoretical calculations were made by Takayanagi *et al.*[12] for the chemical diffusion process associated with tunneling reactions. A very fast rate constant was obtained for the reaction $H + H_2 \rightarrow H_2 + H$. To help choose between physical and chemical diffusion, new information was provided by subjecting the hydrogen samples to pressures up to 13 MPa.[13] In these experiments, the rate constant K_H was found to be independent of pressure. This result could not be explained by physical diffusion, which would lead to a decrease in K_H with increasing pressure. On the other hand, the result was consistent with diffusion by chemical tunneling. Further work is needed to determine the relative importance of physical and chemical diffusion.

For recombination of two H atoms to take place in a solid H_2 matrix, they must approach each other to within a lattice constant. A hydrogen atom tends to distort the lattice potential in its immediate vicinity. As a hydrogen atom moves through the crystal it will encounter various lattice sites. For tunneling to take place efficiently, the energy levels of these sites must match. The lattice distortion by a nearby H atom will result in a misalignment between these energy levels, thus reducing the tunneling probability. Hence the recombination rate associated with diffusion will also be greatly reduced.[14] In fact, this effect has been observed experimentally and can lead to extremely long H atom storage times at very low temperatures.[15]

The recombinational decay has also been studied for deuterium atoms in a D_2 matrix. Since deuterium atoms have double the mass of hydrogen atoms, their larger mass leads to a much slower rate of recombination. The D atoms can move through a D_2 crystal by physical diffusion, and/or by chemical diffusion in analogy with the case for atomic hydrogen moving through solid H_2. For the case of chemical diffusion the exchange tunneling reaction $D + D_2 \rightarrow D_2 + D$ must be involved. A number of groups performed experiments on the decay of D atom populations contained in molecular D_2 environments. Sharnoff and Pound,[16] Iskovskikh *et al.*,[17] and Miyazaki *et al.*[18–20] studied the decay of D atoms in bulk solid D_2 via ESR. Gordon *et al.*[10,21] and Bernard *et al.*[22,23] also used ESR to study the decay of D atom populations in samples composed of D_2 nanoclusters. Thermally activated diffusion was found to be the dominant recombination mechanism above 7 K, whereas below this temperature two phonon quantum physical diffusion became more important. The rate constant obtained by Iskovskikh *et al.*[17] below 7 K is consistent with the migration of D atoms through a D_2 crystal via physical diffusion. On the other hand, Miyazaki *et al.*[18,19] and Gordon *et al.*[10] favored chemical diffusion via exchange tunneling reactions as an explanation of the observed recombination rates in their experiments on D atoms in D_2 matrices, especially at the lowest temperatures.

The importance of exchange tunneling reactions is strongly supported by studies of systems involving both hydrogen and deuterium atomic free radicals moving through matrices containing mixtures of D_2, H_2, and HD. The exchange tunneling reactions

$$D + H_2 \rightarrow HD + H \tag{1}$$

$$D + HD \rightarrow D_2 + H \tag{2}$$

lead to a depletion of the atomic deuterium population and an enhancement of the atomic hydrogen population. It can be shown from zero-point ground state energy considerations for H_2, HD, and D_2 molecules that the reverse reactions do not occur at low temperatures because they are endothermic. The effect of exchange tunneling reactions was first dramatically demonstrated in experiments by Gordon *et al.*[10] in their study of nanoclusters containing H and D atoms in a matrix of H_2, D_2, and HD molecules. After their samples were formed at T = 1.8 K, the populations of H atoms were found to increase substantially and that of the D atoms were found to be greatly reduced. The H and D atoms in these experiments were easily identified via the continuous wave (CW) ESR technique since H and D atoms have different signatures due to their very different hyperfine interactions.[10] In these experiments, the H atoms traveled through the matrix until they encountered another H atom and recombined or until they were surrounded by a cage of D_2 molecules. Migration through the crystal was also facilitated by the tunneling reactions H + HD → HD + H and D + HD → HD + D.

Soon after the experiments by Gordon *et al.*[10] results were obtained by Ivliev *et al.*[24] in Moscow and Tsuruta *et al.*[9] in Nagoya verifying the occurrence of tunneling exchange reactions for the H and D atoms embedded in solid molecular H_2–D_2 systems. A wide ranging set of experiments was performed in Japan by the Nagoya group on tunneling reactions in bulk solid hydrogen–deuterium crystals in which the dissociation of the molecules was produced by gamma irradiation. They determined the rates of various tunneling reactions[9,18–20,25] and also studied the way in which the presence of ortho-H_2 influenced the reaction rates in solid H_2.[26–28] In addition they determined the effect of temperature on the reaction rates.[29] Information on the trapping sites of atoms in the molecular matrices was also obtained.[30–34] A comprehensive discussion of these results is contained in the monograph edited by Miyazaki.[14]

More recently, experiments were performed at Cornell University to study the kinetics of the exchange tunneling reaction D + HD → H + D_2 in nanoclusters prepared by the technique of Gordon *et al.*[35,36] The gel-like solids, consisting of aggregates of nanoclusters, are now known as impurity-helium condensates (IHCs) or alternatively as impurity-helium solids. They have a structure similar to that of the light aerogels with a wide range of pore sizes, ranging from a few nanometers in diameter to macroscopic channels. Early ultrasound measurements by Kiselev *et al.* were instrumental in determining the pore sizes.[37] The sizes of the clusters (5–9-nm diameters) were obtained by X-ray powder pattern studies performed at the National Synchrotron Light source at Brookhaven National Laboratory by Kiryukhin *et al.*[38] The solid helium layers, which formed on the cluster surfaces prevented them from forming a single large mass. It also tended to suppress recombination of atomic free radicals by separating atoms on the surfaces of adjacent clusters from coming in contact with one another. With typical cluster sizes of 5–9 nm, each cluster will contain $\sim 10^4$ atoms and molecules. The average density of atoms and molecules contained in the clusters over the entire sample volume can be as low as 10^{20} cm^{-3}, leading to a concentration of clusters $\sim 10^{16}$ cm^{-3}. Tunneling exchange reaction rates in the interior of the clusters is expected to be similar to those in the bulk solid. It is expected that tunneling rates will be enhanced at the cluster surfaces where the mobility of the reacting atoms should be larger.[39] This effect is expected to be especially dramatic in samples consisting of clusters containing H and/or D atoms and host matrices composed of much heavier inert atoms, such as Kr. The tunneling exchange reactions and the molecular recombination both should be faster at the cluster surfaces in these systems. The exciting possibility of observing quantum correlations for H and D atoms requires temperatures below 1 K, for which the thermal de Broglie wavelength becomes comparable to the spacing between H atoms or D atoms. For *free* H atoms, the thermal de Broglie wavelength, $\lambda = h/(2\pi mkT)^{1/2}$, and for concentrations of 1.57×10^{19} cm^{-3} is found to be ~100 mK.[40] The geometry associated with the matrix confinement will lead to an effective mass m^* in place of the H atom bare mass. This will have the effect of significantly lowering the onset temperature of any quantum fluid behavior for an assembly of hydrogen atoms embedded in a matrix. So far very few experiments have been performed in the temperature range below 1 K for matrix isolated hydrogen and deuterium atoms. In a 1976 publication, Webeler described studies of the recombinational heating of H atoms in a sample of solid H_2 containing 0.03% radioactive tritium.[41] The energetic electrons emitted during the beta decay of tritium dissociated some of H_2 molecules, thereby giving rise to a population of atomic hydrogen. The H atom concentrations increased until spontaneous decay took place when the H atom population reached a critical value at which recombination occurred. For lower temperatures, larger amounts of energy were released during spontaneous decay, indicating, possibly, that larger concentrations of H atoms were stored at lower temperatures. The sample evolution has been analyzed in a later work of Rosen,[42] which accounted for the recombination and dissociation processes and made an estimate of the maximum concentration 4.6×10^{17} cm^{-3} of trapped H atoms in the experiments of Webeler.[41] Rosen[42] predicted that concentrations of atoms approaching 10^{20} cm^{-3} could be achieved by a proper optimization of the experimental conditions. However, no further attempts to reach higher concentrations with this method were reported.

More recently, Ahokas *et al.*[15,43] studied samples consisting of H atoms stored in thin films of solid H_2 at temperatures between 0.15 and 1 K. The samples were studied via high-field ESR.[44] This work utilized most of the experimental advances and techniques developed in the long-term research on spin-polarized atomic hydrogen in the gas phase (H↓)[15] and is discussed in detail later in this chapter.

2. Studies of Exchange Tunneling Reactions of Hydrogen Isotopes in Clusters Forming IHC

2.1. *Setup for studies of tunneling reactions in nanoclusters*

IHCs, composed of nanoclusters of H_2 and D_2 molecules containing H and D atomic free radicals, were studied at Cornell University. ESR on the unpaired spins of the H and D atoms was the main technique employed in this work.[45] Experiments were performed in a Janis metal cryostat containing a 4 K reservoir for the main liquid helium bath and a variable temperature insert (VTI) thermally insulated from the main bath but connected to it by a thin tube. A needle valve installed in this tube allowed the VTI to be filled with liquid helium from the main bath when the needle valve was open. When the needle valve was closed, the VTI and the main bath were isolated from one another. Temperatures down to 1.2 K were achieved by pumping on the VTI with a roots blower backed by a mechanical pump. The temperature could be regulated by adjusting the needle valve setting and/or pumping line impedance. Careful adjustment of the needle valve and the pumping speed permitted a constant bath level to be maintained in the VTI, facilitating long-term

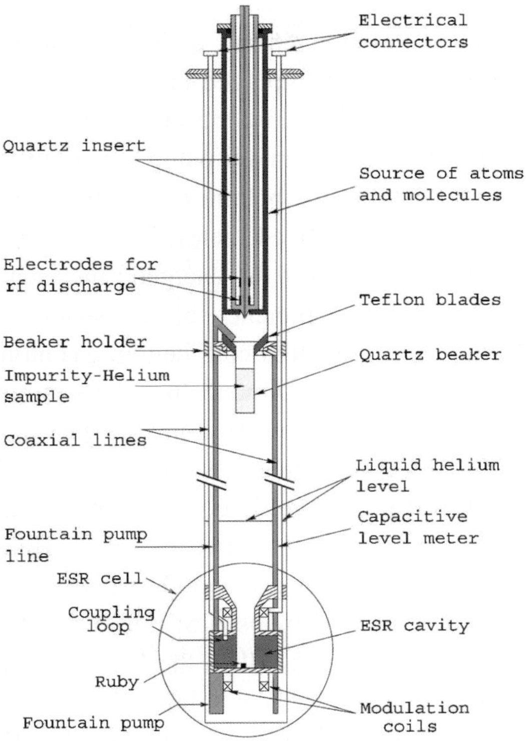

Fig. 1. Low-temperature insert used in the ESR investigation of impurity-helium condensates. The quartz beaker is lowered into ESR resonant cavity for measurements after sample preparation.[45]

experiments below the lambda temperature (2.17 K) of liquid helium. The diameter of the lower portion of the dewar was small enough to fit between the pole pieces of a homogeneous Varian electromagnet, which was required for magnetic resonance investigations. Both CW and pulsed ESR studies on H and D atoms were performed.

Figure 1 shows the cryostat insert specially designed for the preparation and study of the IHC samples. Gas mixtures containing D_2 and/or H_2 molecules along with helium gas were transported from a gas handling system through a quartz capillary surrounding by liquid N_2 into the low-temperature region. Electrodes were placed around the lower portion of the capillary to create a radiofrequency (rf) discharge ($f \sim 50$ MHz, power ~70 W), for the purpose of dissociating the hydrogen and/or deuterium molecules into their constituent atoms. The mixed gases formed a jet as they passed through a small diameter orifice at the bottom of the quartz capillary. The jet impinged on the surface of superfluid helium placed in a beaker 2 cm below the orifice. The level of ^{4}He in the beaker was maintained by a fountain pump placed in the VTI helium bath (maintained at 1.5 K) at the bottom of the insert. The helium gas in the jet forms relatively long-lived metastable states as it passes through the discharge. The presence of the energetic metastable helium atoms is thought to increase the efficiency of dissociation of the H_2 and D_2 molecules. As the jet encountered cold helium gas atoms evaporating from the liquid helium, formation of nanoclusters containing hydrogen and/or deuterium molecules and D and H atoms trapped in or on the clusters took place. This process continued as the jet penetrated through the superfluid He surface to form a gel-like sample of clusters, each thought to be separated by a monolayer or two of solid helium. The structures of the IHC samples so formed are highly porous and closely resemble light aerogels. For IHCs composed of pure deuterium or hydrogen–deuterium mixtures, the samples fell to the bottom of the sample beaker, where they appeared as translucent solids. This was not true if only pure H_2 and He gases were used to prepare the samples. For this latter case, the sample remained at the liquid helium surface because H_2 floats on liquid helium. Therefore, it was not possible to prepare IHC samples containing only pure hydrogen nanoclusters.

Note in Fig. 1 that the top of the beaker is funnel shaped to maximize the collection of the sample. Teflon blades were used to scrape the sample from the walls of the funnel while the beaker was rotated. This procedure assured that all of the sample falling onto the funnel surface was collected in the lower (cylindrical) part of the sample beaker. The jet typically transported $\sim 5 \times 10^{19}$ atoms and molecules per second into the sample cell, leading to the accumulation of ~0.3–0.4 cm^3 of sample in 10 min.

X-ray scattering studies of D_2–He samples played a key role in determining that these samples did indeed consist of nanoclusters, each surrounded by thin layers of solid helium. The X-ray studies showed typical deuterium clusters to be roughly 5–9 nm in diameter.[22,23] The IHC porous gel structure was essentially composed of aggregates of these tiny clusters. By comparing the intensity of the D_2 cluster peak in the X-ray powder pattern with the intensity of scattering from the surrounding liquid helium, it was possible to estimate that the average molecular deuterium densities were ~10^{20} cm^{-3}.

X-ray studies of D_2–Ne–He IHCs indicated that the Ne and D_2 molecules were mainly localized in separate clusters rather than in mixed D_2–Ne clusters. This suggested that it might be possible to form IHCs containing pure H_2 nanoclusters from mixtures of H_2 and Ne. The extra mass of the Ne clusters would give the samples a high enough density to sink to the bottom of the cell. In fact, this technique made it possible to study pure H_2 nanoclusters.

Once the sample was collected, the sample beaker was lowered into the TE_{011} cylindrical ESR resonant cavity. For this mode, the performance was not degraded by an axial hole in the top of the cavity. This hole provided for the entry of the lower (cylindrical) part of the sample beaker. The sample occupied a volume of 0.35 cm^3. The cavity was situated in the region of maximum magnetic field homogeneity between the pole pieces of the Varian electromagnet. Aside from the sample region, the cavity was filled with quartz whose high dielectric constant permitted a more compact geometry to be used. CW and pulsed ESR studies at ~9-GHz (X band) frequencies were conducted in this work. The cavity was operated at a Q of 3000 for the CW measurements, to obtain maximum sensitivity. The cavity Q was lowered to 700 for the pulsed ESR studies to prevent long cavity ring-down times from obscuring the free induction decay signals. For this purpose a thin chromium strip was placed in the cavity to absorb some of the microwave energy.

Most of the ESR signals were obtained from IHC samples immersed in superfluid helium and held at temperatures ~1.35 K. The CW ESR spectra were measured with a traditional homodyne ESR reflection spectrometer. A klystron operating at ~9.1 GHz provided the microwave power for the spectrometer. A small 100-kHz sinusoidal magnetic field was applied via special modulating coils surrounding the cavity, whose cylindrical sidewalls consisted of a stack of copper washers, insulated from one another, to admit the modulation field into the interior of the

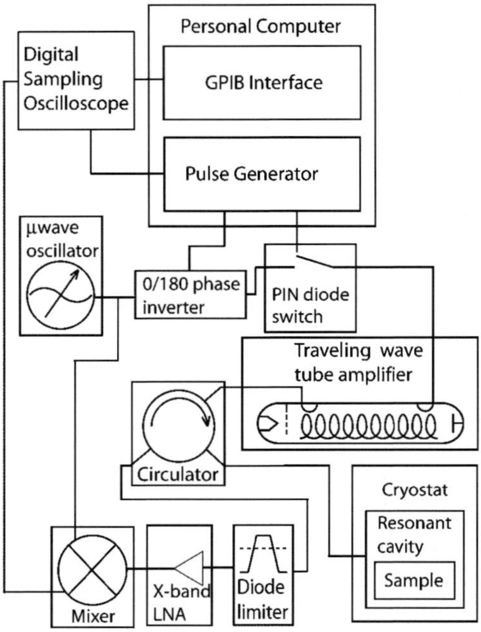

Fig. 2. Block diagram of pulse ESR spectrometer.[47]

cavity. The modulation magnetic field was employed so that the derivatives of the ESR signal could be detected by means of a lock-in detector following the amplifier and detector chain. Signals were obtained by slowly sweeping the main magnetic field and recording the data on a computer and a strip chart recorder. The magnetic field was measured by a nuclear magnetic resonance (NMR) magnetometer and the microwave frequencies were monitored by a frequency counter. Concentrations of the H and D atoms were obtained by comparison of their signal intensities with the signal intensity from a small ruby crystal attached to the bottom of the cavity. The ruby crystal was in turn calibrated by comparison with signals from a diphenyl-picrylhydrazyl (DPPH) sample containing ~2.4×10^{17} spins. All of the intensities were obtained by doubly integrating the derivative signals.

Pulsed ESR measurements required the assembly of a special X-band homodyne detection spectrometer operating at 8930 MHz.[46,47] A block diagram of this spectrometer is shown in Fig. 2. Logic signals from a

12-channel programmable pulse generator operating at a clock frequency of 125 MHz drive a PIN diode switch, biphase modulator, oscilloscope trigger, and traveling wave tube (TWT) grid modulator. The output from the tunable microwave oscillator shown in Fig. 2 is amplified and split into pulse and reference branches. Amplification of the short, intense pulses required for pulsed ESR experiments is accomplished with the 1 kW, 50-dB TWT amplifier operating in the saturated regime. The amplified pulses are transported to a circulator which sends the microwaves into the cryostat via a 20-cm length of stainless steel coax. The low thermal conductivity stainless steel minimized the heat flow into the cryostat. The lower end of the stainless steel coax is connected to a low-loss low-temperature coax 107 cm in length which transfers the microwaves into the TE_{011} resonant cavity placed between the Varian magnet pole pieces. The echo signals generated in the cavity by the sample in response to the applied pulses return through the same coaxes to the circulator where they are routed to a low-noise amplifier (LNA) which is protected from the high power applied microwave pulses by a diode limiter. Following amplification, a quadrature mixer provides homodyne detecting of the echoes by referencing them to the microwave oscillator. A digital oscilloscope is employed to sample the quadrature signals which are then delivered by a General Purpose Interface Bus (GPIB) to a computer. As mentioned previously, Q suppression of the resonant cavity was required to reduce ring-down time of the cavity to provide detection of the free induction decay in its earliest stages. Furthermore, the low Q (~700) facilitated broadband excitation by the short pulses. The value of Q in this arrangement was adjusted by changing the position of a chromium strip (discussed earlier) evaporated onto a quartz ring.

Pulse ESR measurements provided the opportunity to explore the interaction between H and D atoms and their surrounding environments. In addition to the standard $\pi/2$–τ–π pulse spin echo sequences and free induction decay studies employed in T_1 and T_2 measurements, a technique known as electron spin echo envelope modulation (ESEEM)[48] was used to specifically probe the molecular environment in the neighborhood of D or H atoms. In this work, a typical $\pi/2$ pulse had a duration of 40 ns. According to the ESEEM method, a series of

θ–π pulses was applied to a sample. The succession of echo amplitudes so obtained was found to be modulated due to the interaction of the H or D atoms spins with the nuclear spins of surrounding molecules. The technique was particularly sensitive to the nuclear moments of D atoms contained in D_2 or HD molecules. The technique was not useful for studies of ESEEM associated with protons in ortho-H_2 or HD molecules due to the small echo modulations. The spectrometer output was sampled by a digital oscilloscope rather than a boxcar integrator. A narrow section of the echo peak could be integrated to recover the standard boxcar ESEEM data. Alternatively, for two pulse ESEEM studies, the echo can be sectioned in time and the phase correction of Atashkin *et al.*[49] can then be employed. This allows the entire echo to contribute to the signal, thereby improving the signal-to-noise ratio.

2.2. *CW ESR studies of tunneling exchange reactions of hydrogen isotopes in clusters forming IHCs*

Experiments on the kinetics of tunneling exchange reactions between atoms and molecules of hydrogen isotopes in the nanoclusters comprising IHCs have been performed at Cornell University.[45,50–53] CW ESR techniques at X-band frequencies (~9 GHz) were employed in this work. With this method, the concentrations of H and D atoms could be monitored as a function of time as the tunneling exchange reactions proceeded. The method also made possible studies of the changes of the immediate surroundings of H and D atoms during the course of the reactions. The Cornell work was primarily devoted to the tunneling exchange reaction D + HD → H + D_2. The large difference in the ESR spectra of H and D atoms made it very straightforward to account for the evolution of the population of H and D atoms as a function of time. The CW derivative ESR spectra of H and D atoms are shown in Fig. 3. The nuclear spin ½ of the proton gives two allowed lines as shown in Fig. 3a, whereas deuterons have a nuclear spin of 1 leading to three allowed lines as shown in Fig. 3b.

Studies were first performed in pure hydrogen and pure deuterium clusters. It was found that diluting the H_2 gas with neon gave the samples a large enough density to sink into the superfluid helium in the collection

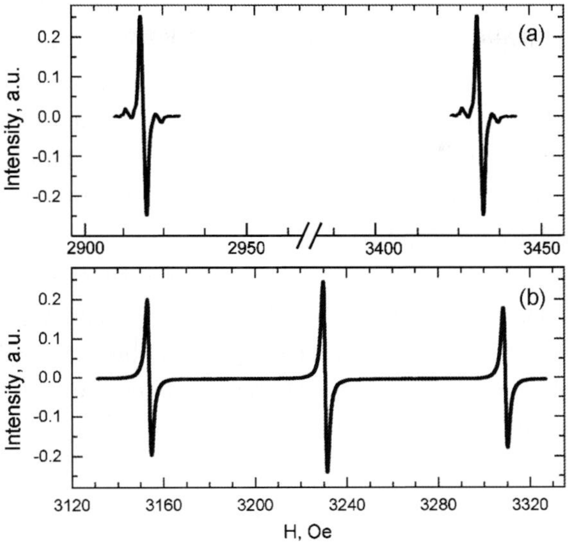

Fig. 3. CW ESR derivative signals of atomic hydrogen in IHCs prepared from H_2:Ne:He = 1:4:100 gas mixture (a) and of atomic deuterium in IHCs prepared from a D_2:He = 1:20 gas mixture (b). ESR spectra were obtained at $T = 1.35$ K.[47]

beaker. As mentioned earlier, the hydrogen and neon tended to form in separate clusters during the formation of the IHC samples so that it was actually possible to study the ESR spectrum of H atoms in relatively pure solid H_2 nanoclusters. Since solid D_2 sinks in superfluid helium, it was not necessary to employ this strategy to form IHC samples from pure deuterium.

The spectra of H and D atoms shown in Fig. 3 correspond to data obtained for samples formed from initial gas mixtures in the ratios H_2:Ne:He = 1:4:100 and D_2:He = 1:20, respectively. The hydrogen ESR lines are accompanied by two satellite lines, one on either side of the main line. These satellite lines correspond to forbidden transitions involving an electron spin flip and a simultaneous spin flip of a proton on a neighboring ortho-H_2 molecule. In parahydrogen, the proton spins are antiparallel, so parahydrogen is not magnetically active. The ESR signals of the H atoms were fitted with a sum of three Gaussians (one central line and two symmetric satellites). The ratio of the intensities of the satellite lines to the main line was found to be 0.2. The observed splitting between the satellite lines and the main line was 4.5 ± 0.1 G, which corresponds to a proton spin flip in the applied field of 3 kG. As shown in Fig. 3b the satellites are unobservable in the deuterium spectrum because of the small size of the nuclear magnetic moment of the deuteron, which gives a calculated splitting in the applied field of only 0.7 G. This value is much smaller than the 2-G width of the main D atom ESR lines, so the D atom satellites could not be resolved.

Let us now turn our attention to mixed samples that contain both H and D atoms. The observation of satellites in these samples indicates that a large fraction of the H and D atoms are contained within or on the surfaces of clusters of molecular H_2 (containing some ortho-hydrogen) and/or HD molecules. If the H or D atoms were individually isolated from the molecular H_2 or HD by layers of solid He, then satellite lines would not have been observed. The presence of neighboring ortho-H_2 and/or HD molecules makes it possible to carry out quantitative studies of the environments of the atoms in the course of tunneling reactions of hydrogen isotopes in nanoclusters.

Trammel *et al.* have derived an expression for the ratio of the ESR satellite line intensities to the respective main line intensities:[54]

$$\frac{I_{\text{sat}}}{I_{\text{main}}} = \left(\frac{3}{20}\right)\left(\frac{g_e\beta_e}{H_0}\right)^2 \langle r_n^{-6}\rangle n. \tag{3}$$

Equation (3) shows that the intensity ratio is directly proportional to the number n of neighboring protons interacting magnetically with a stabilized atom. It also is strongly dependent on the separation r_n between a typical stabilized atom and protons in surrounding ortho-H_2 or HD molecules, as indicated by the $<r_n^{-6}>$ term. In Eq. (3), g_e is the electronic g-factor and β_e is the Bohr magneton. The average distance between the atoms and the surrounding molecules can be estimated via the Trammel formula if a reasonable estimate of the number of protons interacting magnetically with the atom can be made. By monitoring the time dependence of this ratio during the course of the tunneling exchange chemical reactions, the time variation of the population of ortho-H_2 and HD molecules in the immediate neighborhood of the atoms can be obtained.

Let us now consider tunneling reactions taking place in clusters prepared from a mixture of H_2, Ne, and He gases. The satellite data showed that the hydrogen molecules and neon atoms tended to cluster separately in these samples. The clusters of interest in this study were therefore assumed to be mainly composed of H_2 molecules with embedded H atoms. The H atoms moved through these clusters rather quickly as a result of the tunneling exchange reaction $H + H_2 \rightarrow H_2 + H$, allowing the H atoms to approach one another closely enough to result in recombination. Similar considerations involving the reaction $D + D_2 \rightarrow D_2 + D$ apply to D atoms embedded in molecular deuterium clusters, but because of the larger mass of deuterium, the recombination is expected to take place more slowly. Recombination data for H atoms contained in IHC samples prepared from a H_2–Ne–He gas mixture and IHC samples prepared from a D_2–He gas mixture are shown in Fig. 4. The measurements for both samples were made at a temperature of 1.35 K. It is very clear from the graph in Fig. 4 that the deuterium atom population decay takes place at a much slower rate than the hydrogen atom population decay. The curves in Fig. 4 represent the time dependence of the average concentrations of H and D atoms, respectively, as obtained from the ESR intensities.

For the case of H atoms, the dependence of the concentration n on time is given by the equation for two-body recombination,

$$\frac{\mathrm{d}n}{\mathrm{d}t} = -2K_{\mathrm{H}}(T)n^2, \tag{4}$$

when the sample temperature T is held constant. With this equation, the decay constant can be obtained from the data shown in Fig. 4. The concentration drops to half its initial value n_0 in a time $\tau_{1/2}$, defined as the half-life. An expression for $\tau_{1/2}$ is easily derived from Eq. 4 by noting that

$$\frac{\mathrm{d}(1/n)}{\mathrm{d}t} = -\left(\frac{1}{n^2}\right)\left(\frac{\mathrm{d}n}{\mathrm{d}t}\right) = 2K_{\mathrm{H}}(T) \tag{5}$$

and then integrating. This leads to $1/n = 2K_{\mathrm{H}}(T)t + 1/n_0$ corresponding to a linear time dependence. For $t = \tau_{1/2}$, we have $2/n_0 = 2K_{\mathrm{H}}(T)\tau_{1/2} + 1/n_0$, so that

$$\tau_{1/2} = [2K_{\mathrm{H}}(T)n_0]^{-1} \tag{6}$$

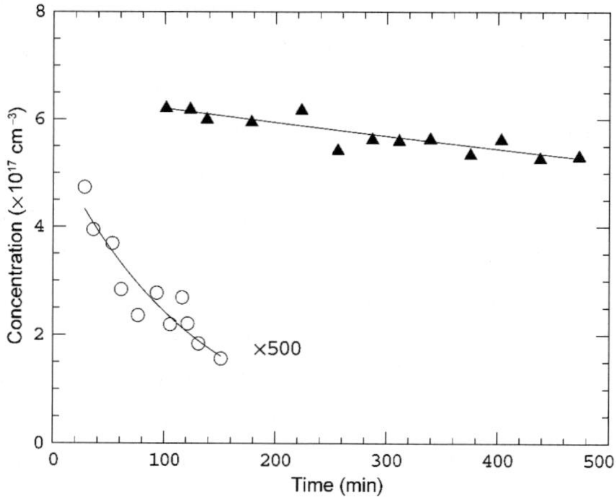

Fig. 4. Time dependence of concentrations of atomic deuterium [prepared from a D_2:He = 1:20 mixture (solid triangles)] and atomic hydrogen [prepared from H_2:Ne:He =1:4:100 (open circles)] in impurity-helium condensates stored at T = 1.35 K.[51] Note that the H atom concentrations were multiplied by 500 so that D and H recombination data could be displayed on the same figure.

where n_0 is the initial *local* concentration of hydrogen atoms. Note that Fig. 4 depicts the average concentration of atomic free radicals, whereas the above discussion corresponds to the local concentrations. For the case of mechanically stable samples for which the clusters maintain the same spatial configuration, it is reasonable to assume that the local and average concentrations have the same time dependence.

For samples prepared from H_2–Ne–He gases, we noted previously that H_2 and Ne formed separate clusters. We shall assume that the density of H_2 molecules in the H–H_2–Ne–He samples is about the same order of magnitude as that of D_2 molecules in D_2–He samples, which is $\sim 10^{20}$ cm^{-3}. This is more than 100 times smaller than the density of solid molecular hydrogen (2.4×10^{22} cm^{-3}), i.e., the H_2 molecules in the sample occupy less than 1% of the available space in total sample volumes due to the voids associated with the space between strands. We note that the average concentration of H atoms at the beginning of the measurements shown in the lower curve of Fig. 4 is $\sim 10^{15}$ cm^{-3}. The local concentration must therefore be 100 times larger corresponding to a value $n_0 = 10^{17}$ cm^{-3}. The half-life $\tau_{1/2}$ for H atoms calculated from Fig. 4

is 150 ± 20 min. From the expression for half-life (Eq. 6), the decay or rate constant for H atoms in solid H_2 at $T = 1.35$ K was found to be $K_H(1.35) = 5.5 \times 10^{-22}$ cm^3 s^{-1}. This result is in marked contrast to the results found for H atoms entrapped in bulk solid H_2 by the Nagoya group,[20] $K_H(1.9–4.2) = (4.4–5.9) \times 10^{-23}$ cm^3 s^{-1}, and the Moscow group,[8,17] $K_H(1.35) = 2.7 \times 10^{-24}$ cm^3 s^{-1}. This discrepancy can possibly be explained by the higher mobility of H atoms on the cluster surfaces.

The decay rate of the D atom populations shown in Fig. 4 is dramatically slower than that for H atoms. As stated earlier, the larger mass of D atoms is expected to significantly retard their motion through a molecular deuterium sample via the reaction $D + D_2 \rightarrow D_2 + D$.

Equations (4), (5), and (6) also will apply to the D–D_2 sample in describing the time dependence of the magnitude of the D atom population for the sample shown in Fig. 4, with an initial local concentration $n_0 = 6.25 \times 10^{19}$ cm^{-3} and a half-life of 2500 ± 500 min, which is more than an order of magnitude longer than that for hydrogen. From Eq. (6) as applied to deuterium, one finds the $K_D(1.35\ \text{K}) = 5 \times 10^{-26}$ cm^3 s^{-1}.This value can be compared with those obtained in bulk solid D_2 by Iskovskikh *et al.*[17] $K_D(4.2) = 4 \times 10^{-26}$ cm^3 s^{-1} and Miyazaki *et al.*[19] $K_D = 3.3 \times 10^{-27}$ cm^3 s^{-1}. There is good qualitative agreement with Iskovskikh *et al.*,[17] but the reaction rate is considerably faster than that found by Miyazaki *et al.*[19]

Quite different behavior was observed for IHCs containing H and D atoms as well as H_2, D_2 and HD molecules. Figure 5a displays an overall view of the atomic H and D spectra obtained from a sample with an initial gas mixture in the ratio H_2:D_2:He = 1:4:100. The small features symmetrically positioned on either side of both the D and H main lines are the satellites corresponding to the spin flips of the protons on neighboring HD or ortho-H_2 molecules in the applied magnetic fields, which occur simultaneously with an electron spin flip on the atomic H or D free radicals. These are forbidden transitions and are therefore much smaller than the main lines. Figures 5b and c provide an expanded view of the main and satellite lines for the low-field lines of H and D atoms, respectively. In addition, there is a small feature denoted by F between two of the deuterium lines. This feature corresponds to a forbidden transition of atomic hydrogen between the lowest hyperfine state |a> and

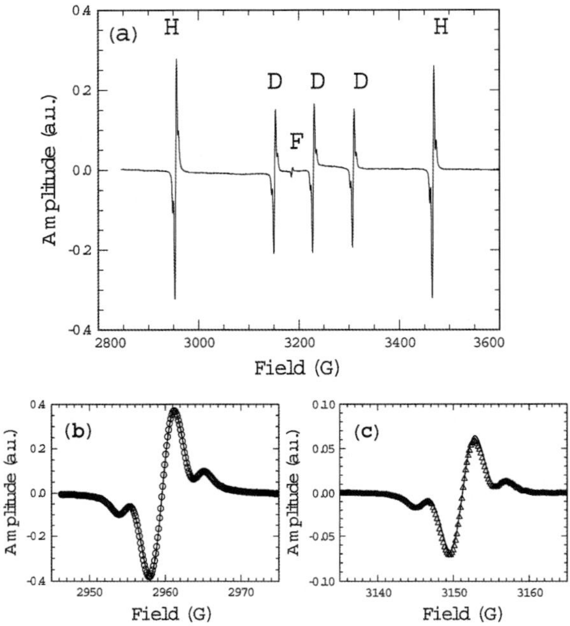

Fig. 5. (a) CW ESR derivative signal from IHC sample prepared from gas mixture H_2:D_2:He = 1:4:100. "H" marks the allowed ESR transitions between the hyperfine levels of atomic H, "D" indicates the hyperfine transitions of atomic D, and "F" indicates the forbidden ESR transition of H atoms. At the bottom are shown typical derivative signals (circles) for the low-field hyperfine lines of (b) H atoms and (c) D atoms demonstrating, at an expanded scale, both the allowed and the accompanying satellite lines. The lines in (b) and (c) represent simulations of the signals as the sum of three Gaussians.[51]

the third highest hyperfine state |c>. Figures 6a and b display energy of hyperfine states versus applied magnetic field plots for atomic hydrogen and atomic deuterium, respectively. The allowed transitions for the main lines and the forbidden line, F, for H atoms are also indicated in Fig. 6a, where the dashed line corresponds to the forbidden transition.

In experiments by Kiselev *et al.*[45,51] many different initial gas mixtures of H_2 and D_2 were employed for the preparation of IHCs containing H and D atoms. Immediately following the preparation of the IHCs a large excess of the H atoms was noted relative to the D atom concentration in comparison to the ratio for the amount of H_2 to D_2 in the initial gas mixtures. This enhancement of H atoms in initial stages of

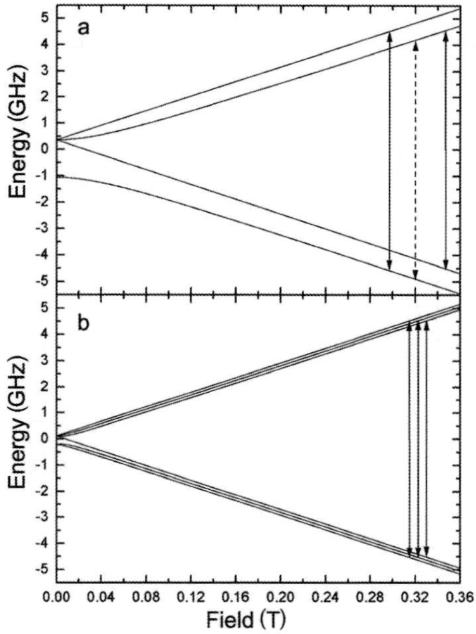

Fig. 6. Energy level diagram for the ground electronic states of (a) hydrogen and (b) deuterium atoms in a magnetic field.

sample preparation is associated with the tunneling reaction D + H_2 → HD + H, which is known to have a very short time constant (~1 min). The rate constant has been calculated theoretically to be 3.56×10^{-25} cm^3 s^{-1} by Takayanagi *et al.*[12] The rapid loss of D atoms and increase of the H atom population thus takes place over only about a minute or two. This process could not be monitored with the CW ESR technique of Kiselev *et al.*[51] because of the long time required to sweep through the ESR lines. In a very short time, a large fraction of the H_2 molecules having neighboring D atoms is converted into HD molecules with the consequent release of hydrogen atoms, which can then diffuse through the sample.

Kumada was actually able to study the fast tunneling reaction in bulk solid HD–D_2 and D_2–H_2.[55] He applied CW ESR technique to study the progress of this fast reaction in bulk HD–D_2 and D_2–H_2 solid mixtures at temperatures in the range 4–8 K. Deuterium atoms were produced by the ultraviolet photolysis of deuterium iodide molecules embedded in the

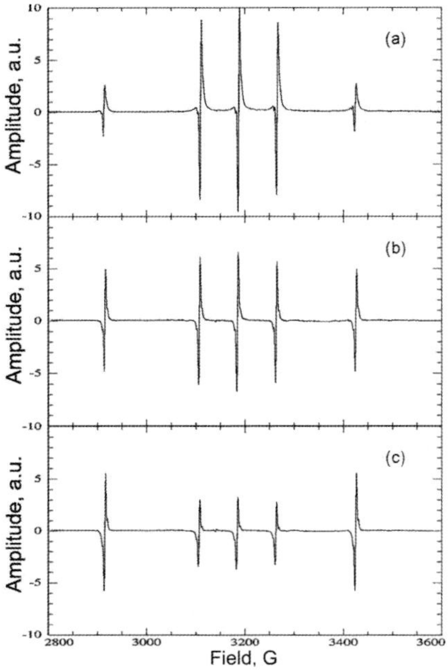

Fig. 7. ESR spectra from H atoms (doublet) and D atoms (triplet) in a H_2–D_2–He condensate taken at $T = 1.35$ K at (a) $t = 23$ min, (b) $t = 205$ min, and (c) $t = 429$ min after sample collection. The sample was prepared from gas mixture H_2:D_2:He = 1:8:180.[52]

samples. The photolysis radiation was applied for 30 s, after which the time variation of the ESR signals of D atoms and H atoms was measured. Observations over a duration of 300 s allowed the decay rate constants to be estimated for this fast reaction which resulted in a rapid buildup of the H atoms and a loss of D atom population over this period. After this time, the first reaction was essentially complete. The rate constants found by Kumada were $(2.9 \pm 0.3) \times 10^{-25}$ cm^3 s^{-1} in the solid HD–H_2 mixture and $(5.0 \pm 1.0) \times 10^{-25}$ cm^3 s^{-1} in the solid D_2–H_2 mixture at 4.1 K. These values are consistent with the theoretical results of Takayanagi *et al.*[12] for the tunneling exchange reaction $D + H_2 \rightarrow HD + H$. The agreement between experiment and theory provides further evidence for tunneling reactions in this system.

Following the initial fast reaction, the decay of the D atom population and the buildup of the H atom population continues over a much longer time period (2–40 h). This slower process is associated with the tunneling exchange reaction D + HD $\rightarrow$ D_2 + H, which further increases the H atom population in the sample. This slow reaction has been studied extensively at Cornell University in IHCs.[51,52] In these experiments, performed at 1.35 K, mixed gases of D_2 and H_2 were used to prepare the IHC samples. As the gas mixture passed the dissociator and entered into collection beaker, IHC samples were formed which contained H atoms, D atoms, HD molecules, H_2 molecules, and D_2 molecules. The time evolution of the ESR spectrum for the gas mixture of H_2:D_2:He = 1:8:180 is shown in Fig. 7. The decrease in the D atom concentration as the H atom concentration increases is dramatically illustrated in this figure.

Figure 8 provides a comparison of the time evolution of ESR spectra for samples prepared from three different initial gas mixtures. Figure 8a shows the time evolution of the H atom and D atom populations for a sample prepared from an initial gas mixture H_2:D_2:He = 1:2:60. Despite the much larger D_2 concentration in the initial gas mixture the D atom concentration became two orders of magnitude smaller than that of the H atoms over the duration of the experiment. In this particular sample, the H atom population decayed slowly as a result of the recombination reaction H + H $\rightarrow$ H_2, which overcame the buildup of the H atom population. After corrections were made for the effects of the buildup, the actual hydrogen recombination rate derived from this result was much slower than that obtained from Cornell studies of IHCs prepared from H_2–Ne–He gas mixtures with no deuterium present. The reason for the slower decay of the H atom population in Fig. 8a is that the H atoms must diffuse through a medium containing many HD and D_2 molecules which impedes their motion through the sample. The results of an experiment on a sample prepared from the gas mixture H_2:D_2:He = 1:4:100 are shown in Fig. 8b. The H atom population is nearly constant throughout the course of the experiment because the H atoms buildup is almost completely balanced out by recombinational decay. Figure 8c shows data from a sample prepared from a gas mixture with ratios H_2:D_2:He = 1:8:180. The large excess of D_2 over H_2 in this mixture provides an ample supply of D atoms in the sample, which leads to a higher production of H atoms. Therefore, the H atom population increases throughout the entire experiment for this sample.

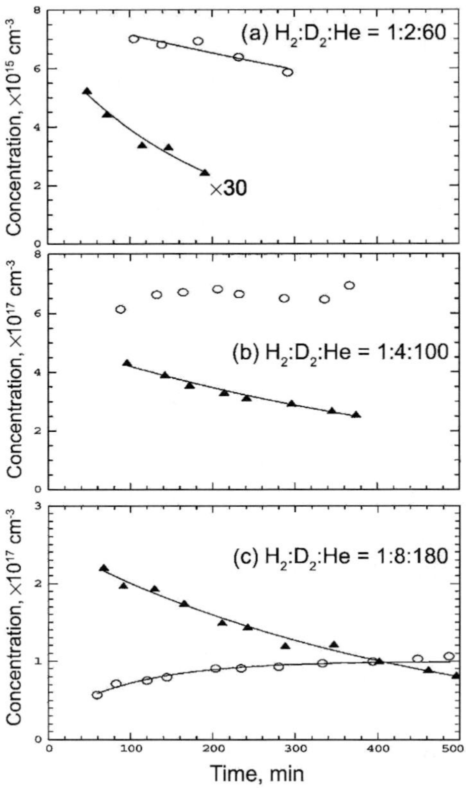

Fig. 8. Time dependence of concentrations of atomic deuterium (closed triangles) and atomic hydrogen (open circles) in different IHCs at $T = 1.35$ K.[51]

Very large concentrations of atomic hydrogen can be produced by the exchange tunneling reactions. In the Cornell experiments a variety of initial gas mixtures were studied in order to find the mixture leading to the largest H atom population.[45] The gas mixture H_2:D_2:He = 1:4:100 was found to produce the maximum population $(7.5 \pm 3.0) \times 10^{17}$ cm^{-3} (see Fig. 8b).

The very long-term stability of the sample obtained from the gas mixture H_2:D_2:He = 1:4:100 was investigated for a time period of 40 h at a temperature of 1.35 K.[53] Figure 9 shows the time dependence of the logarithm of the average concentration of H and D atoms in the sample. (Note that the time scale for Fig. 8 is in minutes while that for

Fig. 9 is in hours.) The concentration of H atoms was nearly constant throughout the 40-h time period, whereas the D atom concentration decreased from its initial value of 10^{17} cm^{-3} by 2.5 orders of magnitude. This was a consequence of the D + HD $\rightarrow$ D_2 + H reaction, which removed D atoms and produced H atoms over the entire time period. The rate constant for the above reaction is estimated from the formula $d[D]/dt = -K_{D,HD}[HD][D]$. At late stages of the experiment, the D atoms are indeed severely depleted, so D_2 and HD concentrations are always almost constant and the production of H atoms is negligible. Therefore, the main change in the H atom population will be via the recombination reaction. This is extremely slow because if we take the HD concentration to be fixed at its initial value (estimated from the concentration of D_2 molecules obtained from the X-ray experiments, 2.5×10^{21} cm^{-3}) we then find $d[D]/[D] = -K_{D,HD}[HD]dt$ leading to $\ln[D] = -K_{D,HD}[HD]t +$ *constant*. As a result, ln[D] will decrease linearly with time. (There will be an error here due to the fact that [HD] is not perfectly constant in time, but it becomes nearly so as the experiment progresses.) From the slope of the log versus time plot of [D] shown in Fig. 9, the rate constant $K_{D,HD}$ was found to be equal to 3.6×10^{-27} cm^3 s^{-1}, which agrees with the value obtained by Kumada *et al.* in bulk samples.[29]

The local environments of the H and D atoms during the course of the exchange tunneling reactions can be monitored by observing the ratios of the satellite lines to the main ESR lines. Since most of the ortho-H_2 molecules have been converted to the para state at an early stage by the atomic free radicals, the main contributors to the satellite lines are the HD molecules. The spin flips of the protons in the HD molecules give rise to the satellite signals for both the H and D atom ESR lines. Trammel *et al.*[54] have shown that the relative amplitude of the satellite line to the main line is proportional to the number of HD nearest neighbors. The time dependence of the ratio of the satellite line intensities to those of the main lines for H and D atoms was studied for a HD–D_2–He sample while the tunneling reaction D + HD $\rightarrow$ H + D_2 progressed.[53] For the case of the H atom satellite lines, the ratio was found to be equal to 0.11. This value remained constant for the duration of the experiment. The tunneling reaction H + HD $\rightarrow$ HD + H should permit the H atoms to move through the sample and possibly

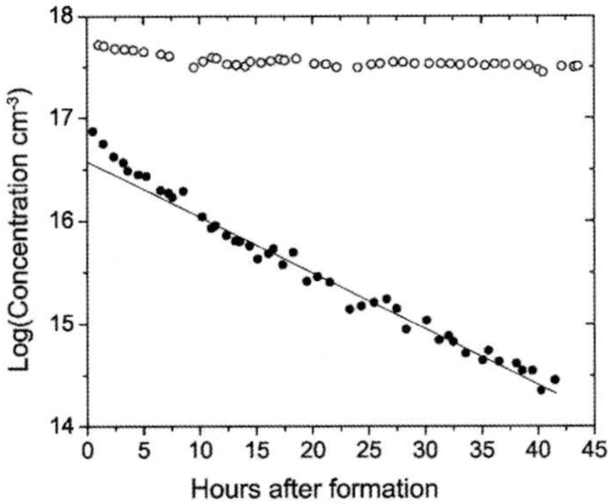

Fig. 9. Time dependence of the logarithms of the average concentration of H atoms (open circles) and D atoms (closed circles) in HD–D_2–He condensates immersed in superfluid helium at $T = 1.35$ K. Sample was prepared from H_2:D_2:He =1:4:100 gas mixture.[53]

meet another hydrogen atom, which could lead to a recombination event in which the two H atoms formed a hydrogen molecule. This would in turn lead to a reduction in the number of H atoms. In this experiment, however, the hydrogen atom content remained remarkably constant. While it is true that the reaction H + HD → HD + H is expected to be ~200 times slower than the diffusion of H through H_2 via tunneling,[12] nevertheless, a significant effect on the hydrogen atom population should be observed over a long time period (14 h). The absence of the effect may be explained as follows: the percolation of HD molecules further slows the diffusion because the process is obstructed by the presence of D_2 molecules.

On the other hand, studies of the satellites of the D atoms associated with the proton spin flips on the HD neighbors showed a reduction in the satellite to main line ratio from 0.2 to 0.14 over a 14-h period,[53] indicating that the number of HD molecules surrounding the D atoms decreased by 30%. In considering this behavior, we must realize that initially the D atoms will be in sites containing various numbers of HD molecules as nearest neighbors. A D atom can react with the HD molecule to yield a deuterium molecule and a hydrogen atom. Over time this process will deplete the number of D atoms having numerous close HD neighbors, leading to a reduction of the D atom satellite signal from the proton in HD molecules. In other words, the surviving D atoms will have fewer HD nearest neighbors, which leads to a reduction of the satellite signal.

2.3. *Pulse ESR studies of H and D atoms in clusters forming IHCs*

As shown in the previous section, the satellite signals associated with the spin flips of neighboring protons provide a useful technique for studying the environment of hydrogen or deuterium atomic free radicals in the IHC samples. The protons associated with HD molecules are mainly responsible for the satellites since a rather large fraction of the ortho-H_2 molecules will be converted to para-H_2 soon after sample creation by the unpaired electron spins of H or D atoms. As stated earlier, the satellite technique cannot be used for studying the D_2 molecule neighbors because the small magnetic moment of the deuterons gives only a very small satellite shift. This shift is smaller than the ESR line width associated with inhomogeneous broadening, which precludes observation of the satellites associated with D_2 molecular neighbors. Therefore, another technique must be employed. Pulsed ESR has been particularly helpful in this regard. The time-resolved spectroscopy provided by pulse ESR allows the effects of the interactions between H or D atoms and molecular D_2 to be investigated.

The first observation of spin echoes by Erwin Hahn[56] in 1950 changed the world of magnetic resonance. First seen in NMR, the effect has later been observed in ESR and optical studies. A comprehensive discussion of spin echoes is given by Slichter.[57] In the discussion of the spin echo technique, one works in the frame of reference rotating with the Larmor frequency of the spins to be studied. Here we consider a spin ½ system. The axis is along the applied field, $\boldsymbol{H}_0$. Suppose an rf pulse is applied along the y-axis in the plane perpendicular to $\boldsymbol{H}_0$ (the z direction). This plane is defined to be the equatorial plane. The spins will be rotated through an angle θ around the y-axis by an amount $\theta = \gamma \boldsymbol{H}_1 \Delta t$ where $\boldsymbol{H}_1$ is the applied rf field and Δt is the pulse duration. The values of $\boldsymbol{H}_1$ and Δt can be adjusted to give $\theta = \pi/2$, which corresponds to a rotation of the spins into the equatorial plane along the x-axis. As the time progresses, the individual spins will precess away from x-axis at different rates in the equatorial plane due to inhomogeneities in local field. They, therefore,

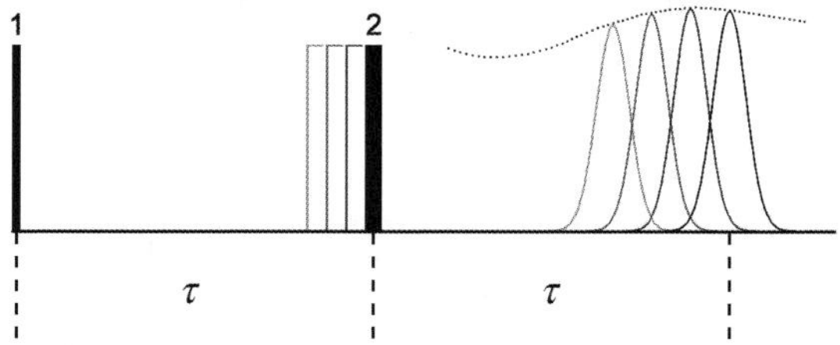

Fig. 10. The two pulse electron spin echo experiment. Microwave pulses 1 and 2 are applied by the spectrometer and separated by a time τ. The "spin echo" is a microwave pulse produced by the sample after another period τ has elapsed. Incrementing τ reveals the echo envelope modulation pattern.

fan out in the plane and eventually will not produce any macroscopic magnetization. After a time τ, a π pulse with twice the duration Δt of the $\pi/2$ pulse is applied along the y-axis. This will rotate each of the spins around the y-axis by an angle π back into the equatorial plane, but with their x components reversed in sign. After the π pulse, each of the spins will precess in the equatorial plane in the same manner they did before the π pulse. The local spin directions will all converge after a time τ to produce an echo. Studies of the amplitude and shape of this refocused signal provide a method of studying relaxation processes and other properties of spin systems.

The spin echo method as applied to ESR typically requires specially designed spectrometers due to the short relaxation times for most samples. Thus short pulses, ~10–100 ns in duration, must be applied with sufficient power to tilt the spins through angles up to 180°. The spectrometer used in the Cornell experiments was described in an earlier section.

To determine the effect of the magnetic moments of the deuterons in the HD/D_2 matrix surrounding the H and D atoms, an ESEEM technique is employed. According to this method, a succession of $\pi/2$–τ–π pulse pairs is applied to the sample with sufficient time intervals between successive pulse pairs to allow complete recovery of the magnetization to its initial (Boltzmann) value to take place. After each $\pi/2$–τ–π pulse pair, the corresponding echo is recorded. The amplitude of each echo is carefully measured. The echo train thus obtained for samples containing deuterium atoms and molecules, or hydrogen and deuterium atoms in a matrix composed of HD and D_2 molecules is found to be modulated in time. A schematic plot of the echo amplitudes versus τ is shown in Fig. 10. The modulation pattern is traced by the amplitudes of successive echoes. The depth of modulation is dependent on the average number of deuterons surrounding the H or D atoms. This corresponds to the same type of information obtained from the proton-induced satellites in the CW measurements. The ESEEM technique, which is sensitive to small magnetic couplings, thus provides a means of overcoming the problem associated with resolving deuterium satellites.

How does this echo modulation arise and why is it effective in canceling out the effects of inhomogeneous broadening? In this discussion, we more or less follow the treatments of Dikanov and Tsvetkov,[48] and Slichter.[57] The explanation can be traced to the fact that the process of echo formation is a reversible process. After the first ($\pi/2$) pulse, inhomogeneous broadening is manifested by the spreading out of the various spins in the rotating frame. The second (π) pulse refocuses the spins to produce the echo, and the effect of the inhomogeneous broadening is basically eliminated. Effects hidden by broadening due to the inhomogeneous applied field in CW magnetic resonance can now be studied.

Let us now consider the simple situation where we have the unpaired electron of an H or D atom with electron spin $\boldsymbol{S} = ½$ interacting with a nucleus of spin $\boldsymbol{I}$ on an adjacent molecule. Initially, let the nucleus (spin $\boldsymbol{I}$) and the electron (spin $\boldsymbol{S}$) be pointing along the constant applied field, $\boldsymbol{H}_0$. If the electron and nucleus are separated by a distance r, the total field acting on the nucleus will be $\boldsymbol{H}_I = \boldsymbol{H}_0 + \boldsymbol{H}_{SI}$, where $\boldsymbol{H}_{SI}$ is the field at the nucleus due to the electron. Similarly $\boldsymbol{H}_S = \boldsymbol{H}_0 + \boldsymbol{H}_{IS}$ where $\boldsymbol{H}_{IS}$ is the field at the electron due to the nucleus. For typical separations between the electrons and nuclei of a few angstroms, $\boldsymbol{H}_{SI}$ will be a few hundred gauss. Since nuclear moments are $\sim 10^3$ times smaller, $\boldsymbol{H}_{IS}$ will be correspondingly smaller. Thus, the nuclear field will have very little influence on the electron spin direction in an external field for an applied external field appropriate for X-band ESR (~3000 G). On the other hand, the nuclear spin direction will be strongly influenced by the magnetic field of the electron, $\boldsymbol{H}_{SI}$. More specifically, the quantization axis for the electron will nearly coincide with $\boldsymbol{H}_0$, whereas the nuclear quantization axis may be tilted away from the external field during the course of an experiment. For example, when a $\pi/2$ microwave pulse tilts the electron spin, the nucleus will find itself in a new field which has been tilted away from the direction parallel to $\boldsymbol{H}_0$. The nucleus will then precess about the combination of the electron field and the applied field. The combined field serves as the quantization axis for the precession of the nuclear spins. In this discussion, we assume that the microwave pulse is short compared with the nuclear precession period so that there is no time for the nuclei to adiabatically adjust to the suddenly changed local field. The precession of the nuclei leads to a time varying field acting back on the electrons. The electrons will precess in the time varying field that is associated with the nuclear precession in the combined field of the electron and $\boldsymbol{H}_0$. Suppose after a time interval τ, a π pulse is applied. The nuclear precession will respond to the changed field direction, produced by the electrons. This changed field direction will, in general, prevent perfect refocusing at a time τ after the π pulse, since the electron precession rate will be disrupted by the precessing nuclei. Only if τ corresponds to an integral number of complete nuclear precessions will the refocusing be almost perfect, which leads to a maximum in the echo. As τ is varied for successive $\pi/2$–τ–π pulses, the echo amplitude will be modulated at the Larmor period of the nucleus in the applied field along with the contribution from the electron magnetic field, which can be very small for the experiments discussed here. The quantum mechanical treatments given by Slichter[57] and by Dikanov and Tsvetkov[48] should be consulted in order to gain a more complete understanding of the ESEEM phenomenon.

We now discuss ESEEM experiments performed at Cornell[58-60] on IHC samples containing deuterium atoms and molecules, using the X-band pulsed spectrometer described earlier. A series of $\pi/2$–τ–π pulse sequences were applied with pulse lengths of 40 and 80 ns, respectively for the $\pi/2$ and π pulses. The value of τ was increased for subsequent applied pulse pairs from 460 ns to 12 μs. A recovery time t of 400 ms was used to separate successive pulse pairs.

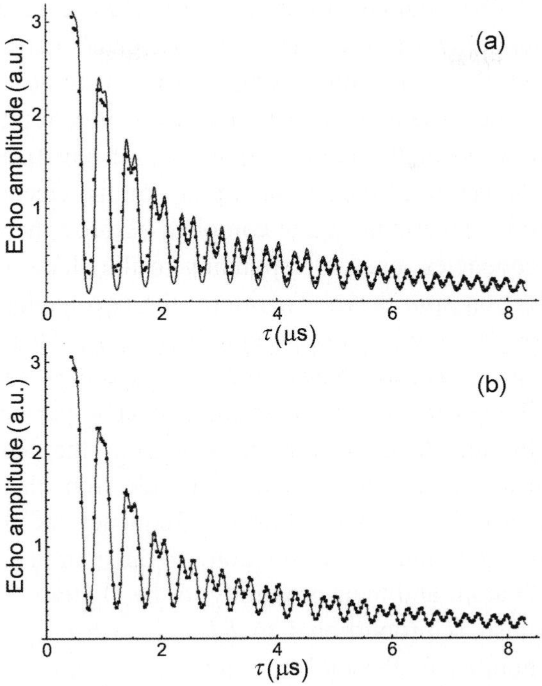

Fig. 11. Two-pulse ESEEM trace of the low-field hyperfine line of D atoms in a D_2–He sample. (a) The dots show every fourth data point. The line shows the best fit of the bulk population model. (b) The dots show every fourth data point. The line shows the best fit of the two population (bulk and surface) model $E(\tau)$.[58] Clearly, the data in panel b is a better fit to the model.

Figure 11 shows two plots of the ESEEM echo amplitude versus τ for a D–D_2–He sample prepared from a gas mixture with a ratio D_2:He = 1:20 for the low-field hyperfine line of atomic deuterium. The upper frame (Fig. 11a) shows the data fitted to a simulation (solid line) for D atoms embedded on bulk solid D_2. The lower frame (Fig. 11b) displays the best fit for the same data to a model (solid line) in which some D atoms are embedded in the solid D_2 clusters of the IHC and some D atoms reside on the cluster surfaces. The overall envelope pattern decreased steadily with time, but the intensity was strongly modulated by the first and second harmonics of the Larmor frequencies corresponding to the deuterons on the surrounding D_2 molecules. The large dips and peaks are the ESEEM signals of the first harmonic and the small dips on each of the main peaks are associated with the second harmonics.

As discussed above, the refocusing action giving rise to the peaks takes place for the ensemble of electron spins only for the case of a complete 2π rotation of the nuclear spins whose quantization axis has been tilted by the action of the $\pi/2$ or π pulse on the electrons. Even for this case, the refocusing is not quite complete because there are actually two nuclear frequencies, one corresponding to the electron spin up and the other to electron spin down. Therefore, the two nuclear frequencies do not go through complete cycles simultaneously. This leads to two separate refocusing events. Since the two nuclear frequencies are nearly equal, however, the refocusing is nearly perfect. Since the applied magnetic field $\boldsymbol{H}_0$ couples more strongly to the deuterons on the molecules than does the dipolar field of the electron, the applied field mainly determines the ~2 MHz Larmor frequency of the deuterons. The modulation amplitude is a function of the number of deuterons surrounding a D atom and their proximity to the D atom. The amplitude of modulation also is proportional to $I(I + 1)$, where I is the nuclear spin quantum number of the molecule interacting with the D atom. D_2 molecules in the $I = 1$ and $I = 2$ states are responsible for the ESEEM signals shown in Fig. 11. The ESEEM signals should therefore be greatly enhanced in comparison with the modulation produced by protons ($I = ½$) on HD molecules surrounding H or D atoms in samples containing HD. This difference is even greater at low temperatures where the $I = 1$ state is depopulated by conversion to the $I = 2$ state, which is catalyzed by the powerful field gradients of the electronic spins within a sample. In the IHC experiments described herein, the proton moment ESEEM signal has never been observed.

The model that Kumada *et al.*[33] developed for studies of ESEEM signals for D atoms in bulk solid D_2 has been used to analyze the IHC data discussed here. Their ESEEM signals were simulated from the standard theoretical results for orientationally disordered crystalline systems developed by Mims *et al.*[61] Kumada *et al.* make the assumption that all orientations of the face-centered-cubic (*fcc*) deuterium crystal axis with respect to the applied magnetic field are equally represented within the sample. The Hamiltonian contains a term corresponding to the dipolar interactions between the D atom electrons and the D_2 nuclear moments and also a term corresponding to the Fermi contact interaction of the D atom electronic wave function with the nuclear wave functions of the nearest neighbor D_2 molecules. This latter term corresponds to the superhyperfine interaction. The internuclear couplings are weak and so have been neglected. Therefore, it is reasonable to take the echo modulation associated with the nuclei surrounding an electron spin to be a product of the modulations due to each individual nucleus.

Hence the echo signal as a function of the interpulse time τ can be modeled as

$$E(\tau) = E_{\text{dec}}(\tau) \times E_b(\tau);\; E_b(\tau) = \prod_{i=1}^{5} < E_i(\tau) >_{\theta}^{N_i};\; E_{\text{dec}}(\tau) = Ae^{-\tau/T_2}.$$

The expression for $E_{\text{dec}}(\tau)$ represents the exponential decay processes associated with the relaxation time T_2; $E_b(\tau)$ corresponds to the modulation of the echo due to the N_i molecules populating the closest five coordination shells (i = 1–5) surrounding the electron spin. The modulation functions $E_b(\tau)$ due to each individual molecule are averaged over all orientations before they are multiplied in the expression for $E_b(\tau)$. This latter procedure is an unphysical feature of the model, but it facilitates the computations and causes only small changes in the predicted echo signals for systems with weak electron–nuclear couplings such as those found for H or D atoms in substitutional positions in a molecular D_2 matrix.[62] (These interactions are weak because of the large spacing between a D or H atom in a substitutional site and molecules in the surrounding lattice.)

Only the lowest molecular rotational states, J = 0 (ortho) and J = 1 (para) will be present in an ensemble of deuterium molecules at liquid helium temperatures. Although the para state is 85 K more energetic than the ortho state of D_2 molecules, it is quite stable against conversion to the ortho state except in the presence of paramagnetic H or D atoms, which catalyze the conversion process. Since deuterons are bosons, the product of the spin and spatial wave functions is symmetric under particle exchange. The J = 1 (odd) orbital angular momentum of the para state must be therefore have an I = 1 (odd) nuclear spin state in order to satisfy this symmetry requirement. On the other hand, the J = 0 even ortho state must be accompanied by either an $I = 0$ or $I = 2$ spin state. The $I = 2$ state is $2I + 1 = 5$ fivefold degenerate, whereas the $I = 0$ state is nondegenerate. The energy splitting between the $I = 0$ and $I = 2$ states will be negligible at liquid helium temperatures, meaning that the five $I = 2$ states and the single $I = 0$ state will be almost equally populated. In modeling the ESEEM experiments, we first assume that in the vicinity of the D atoms the entire population of $I = 1$, $J = 1$ para states has been converted to $J = 0$, $I = 0$, or $I = 2$ ortho states. We therefore neglect any effects of the $I = 1$ state on the ESEEM signals. Since the $I = 0$ molecules are not magnetically active, they will not contribute to the ESEEM signal, so only five-sixth of the deuterium molecules will be involved. Therefore the number of effective molecules in each coordination shell must be reduced by this factor. As an example, we consider an atom on a substitutional site in *fcc* D_2. The first coordination shell of 12 nearest neighboring molecules is thus modeled with 10 spin $I = 2$ particles, each having the Larmor frequency of the deuteron.

The simple analysis of a system with a spin ½ electron on the atomic free radical and a spin ½ nucleus is not applicable to higher nuclear angular momentum states. A scheme developed by Ponti[63] allows one to use Chebyshev polynomials of the second kind to relate the ESEEM modulation function due to a particle of any spin to the function that would result if the particle were spin ½. Equation (7) gives the modulation function $E_i(\tau)$ for any spin I in terms of modulation function $V_i(\tau)$ for the spin ½ case:

$$E_i(\tau) = \frac{1}{2I+1} U_{2I}\,[V_i(\tau)] \qquad (7)$$

where U_n is the nth Chebyshev polynomial of the second kind and $V_i(\tau)$ is given explicitly by the following equations:

$$V_i(\tau) = \frac{K}{2}[1 - \cos(\omega_\alpha \tau)][(1 - \cos(\omega_\beta \tau)],$$

$$K = \left(\frac{B\omega_L}{\omega_\alpha \omega_\beta}\right)^2,$$

$$\omega_\alpha = \left[\left(\frac{A}{2} + \omega_L\right)^2 + \left(\frac{B}{2}\right)^2\right]^{\frac{1}{2}},$$

$$\omega_\beta = \left[\left(\frac{A}{2} - \omega_L\right)^2 + \left(\frac{B}{2}\right)^2\right]^{\frac{1}{2}},$$

$$\omega_L = g_n \beta_n H_0,$$

$$A = g_n \beta_n g_e \beta_e (3\cos^2\theta - 1) / \hbar(r_i)^3 + 2\pi A_{sup},$$

$$B = 3 g_n \beta_n g_e \beta_e \cos\theta \sin\theta / \hbar(r_i)^3,$$

where r_i is the distance from the electron to coordination shell i, θ is the angle between the applied field $\boldsymbol{H}_0$ and the vector pointing from the atomic free radical to the molecule, g_e and g_n are the g-factors of the electron and nuclear spins, β_e and β_n are the Bohr and nuclear magnetons, and A_{sup} is the isotropic superhyperfine coupling associated with the Fermi contact interaction between the electronic wave function of the atom and the nuclear wave function of the molecule.

The model presented above was used by Kumada *et al.*[33] to fit A_{sup} and r_i to data they obtained from pure irradiated solid D_2 for two sets of N_i and r_i. They studied two cases, one in which the D atoms were situated in substitutional sites in the *fcc* solid D_2 lattice and the other in which the D atoms were located in interstitial sites. The results as obtained from the model for the two cases were quite different. As expected, the case for the D atoms located at interstitial sites gave much larger modulation depths as compared with the analysis for substitutional states because the interstitial atoms were much closer to their nearest neighbors. The data were found to favor the case for which the D atoms occupied substitutional sites. The results from numerical fittings indicated that for the D atoms in the samples, r_i = 3.6 Å, which is the nearest neighbor spacing in the unstrained molecular lattice. The value of A_{sup}, the superhyperfine parameter, was 70 kHz.

The above model was applied to the D–D_2–H_2 samples studied at Cornell. The agreement between the model and the ESEEM results was rather poor, as indicated in Fig. 11a. The poor agreement was attributed to the fact that the IHC samples were composed of clusters and that the

D atoms could either reside on the cluster surfaces or in the interior of the clusters. This motivated the procedure of fitting the ESEEM data to two decaying exponentials multiplied with two oscillatory modulation terms, as indicated below.

$$E(\tau) = A\left(Qe^{-\tau/T_{2Q}} + (1-Q)e^{-\tau/T_{2L}}\right)\left(bE_b(\tau) + (1-b)\,E_s(\tau)\right). \quad (8)$$

In this expression, A, Q, T_{2Q}, and T_{2L} correspond to the echo decay and b weights the modulation functions E_b and E_s.[58,60] The function E_b simulates the ESEEM signal of deuterium atoms in the interior of the solid D_2 clusters under the assumption that they are in substitutional sites of an *fcc* lattice composed of deuterium molecules. The associated modulation signal E_b was identical to that discussed above for D atoms substituted in bulk solid D_2. The function E_s is obtained from the same model, but the population of each coordination shell around the atom is reduced by a factor of 2. E_s is thus a simulation for D atoms in contact with helium at the cluster surfaces. The values of r_i and A_{sup} are fixed to those found by Kumada *et al.*[33] to describe bulk solid deuterium. The modulation function $E(\tau)$ will then have only a single free parameter, the ratio of atoms embedded in the solid D_2 clusters to atoms residing on the cluster surfaces at the cluster–helium interface. The functional form of the echo decay, $E(\tau)$, was determined to fit the data rather well with only a small number of parameters. The function $E(\tau)$ corresponding to the modified model described by Eq. (8) was applied to the ESEEM measurements shown in Fig. 11b. A good fit for the data was found when 50%–60% of the atoms occupied positions at the surfaces of the molecular deuterium nanoclusters.[58] The remainder of the atoms was embedded in substitutional sites inside the clusters. These results are in accord with the idea of Gordon,[64] who first suggested that a large fraction of the atomic free radicals in IHC samples would occupy positions at the surface of the clusters at the cluster–helium interfaces. If the deuterium atoms simply occupied positions randomly spread through the cluster interiors and surfaces, only 10%–20% of the atoms would be found at the cluster surfaces. (This distribution depends on the sizes of the clusters.)

The ESEEM technique was also applied to IHC samples containing hydrogen and deuterium atomic free radicals along with D_2, H_2, and HD

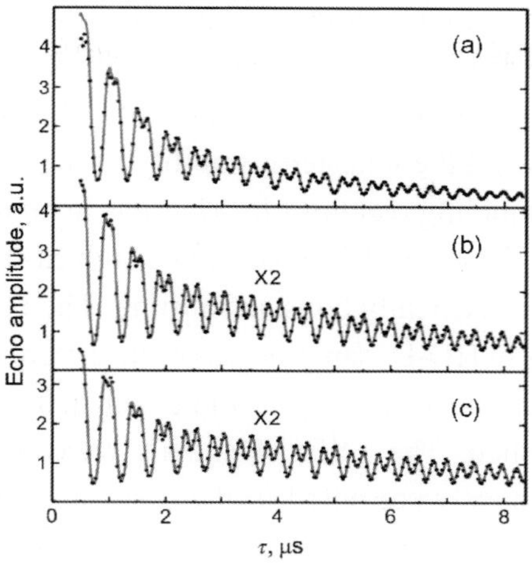

Fig. 12. Two-pulse ESEEM signals (dots) for the H and D atoms in HD–D_2–He condensates at 1.35 K taken at 8.93 GHz: (a) low-field hyperfine (hf) component of H atoms at 2910.1 G; (b and c) low-field hf components for D atoms at 3105.5 G at the beginning of the experiment and 6 h later, respectively. Solid lines are the best fits of model calculations to the experimental data (see text).

molecules. CW and ESEEM techniques are complementary. CW can only probe the coupling between H or D atoms and protons contained in ortho-H_2 and HD molecules. On the other hand, ESEEM probes the interactions between the H and D atoms and deuterons contained in D_2 and HD molecules. (The small magnetic moment of the deuteron precludes CW observations since the satellite lines are obscured by the line width of the deuterium atom ESR signals.)

Figure 12a shows the ESEEM signals obtained from H atoms in a HD–D_2–He IHC sample formed from the initial gas mixture H_2:D_2:He = 1:4:100. The analysis of the ESEEM data required the presence of both D_2 and HD molecules to be taken into account. The model used for studies of the pure deuterium samples was therefore modified. The ratio between HD and D_2 molecules was taken to be a fitting parameter. It was assumed that the fraction of atoms at the cluster surfaces was 55%, in accordance with the results for the pure D_2 IHC samples.[58,60] The solid line in Fig. 12a represents the best fit for the model calculation with a configuration in which 10 D_2 and 2 HD molecules surrounded the H atoms. The behavior of ESEEM signals did not change for the H atoms over a period of 6 h indicating that the number of HD and D_2 molecules surrounding the H atoms did not change throughout the course of the reactions. Earlier CW experiments[53] showing the long-term stability of the satellite signals over the course of the tunneling reactions for H atoms in mixed HD-D_2 samples are consistent with these ESEEM results. For both cases the environments of the H atoms were basically constant in time. The behavior of ESEEM signals for D atoms in HD-D_2 samples was quite different. As the tunneling reaction proceeded, the ESR signals indicated that the overall D atom population diminished. The ESEEM signals were recorded over a 6-h period. Subtle changes were noted in these signals for data taken at the beginning and at the end of the run. Fits to these D atom ESEEM signals demonstrated substantial changes in the D atom environments as shown in Figs. 12b and c. By modeling ESEEM signals at the beginning and ending of the 6-h runs, it was found from the fits for the IHC samples prepared from H_2:D_2:He = 1:4:100 samples that at the beginning of the experiment, a typical D atom within a cluster was surrounded by nine D_2 molecules and three HD molecules. At the end of the 6-h period, a typical D atom was surrounded by 10 D_2 molecules and 2 HD molecules. The loss of HD molecules and the gain of D_2 molecules are to be expected when the $D + HD \rightarrow D_2 + H$ reaction is taking place. The ESEEM result is also consistent with CW ESR observations over 6 h on samples prepared from H_2:D_2:He = 1:4:100 gas mixtures. The satellite lines on the D atom signals due to the protons on neighboring HD molecules decreased over time by approximately 30% over a 6-h period. The agreement between the two different studies gives some confidence in the correctness of the results.

3. Stabilization of High Concentrations of Hydrogen Atoms in Aggregates of Krypton Nanoclusters

Studies of IHCs containing krypton and atomic and molecular hydrogen have also been performed at Cornell.[65,66] It has been possible to stabilize very high concentrations of atomic hydrogen in these samples, with most of the atomic free radicals residing on the nanocluster surfaces. Figures

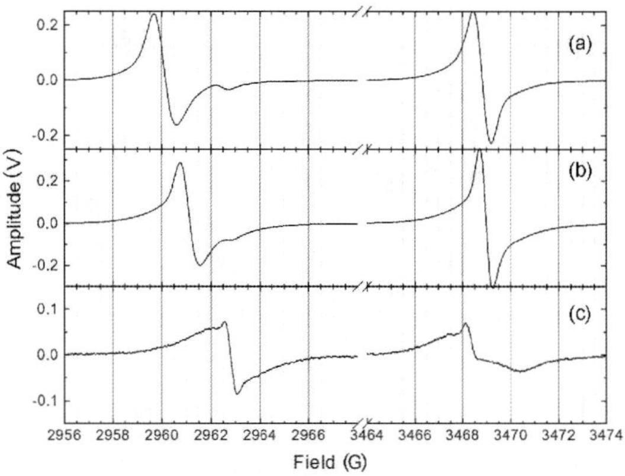

Fig. 13. ESR spectra of hydrogen atoms in as-prepared H–Kr–He samples at 1.35 K. Samples were prepared from gas mixtures (a) with the ratios H_2:Kr:He = 1:1:200 and (b) with the ratios H_2:Kr:He = 1:50:10 000. (c) ESR spectra of the sample shown in Fig. 1a after annealing to 14.5 K and cooling to 1.35 K. All spectra were obtained at 1.35 K. The amplification for measuring spectra of the annealed sample was increased by 15 times.[66]

13a and b display the results of a set of runs taken at 1.35 K. The CW ESR spectra for IHC samples prepared from gas mixtures with ratios H_2:Kr:He = 1:1:200 and 1:50:10000, respectively, are shown in the figures. Results for the high-field (HFL) and low-field hydrogen lines (LFL) are given in Figs. 13a and b for samples prepared directly from the jet and cooled to 1.35 K. The centers of the HFL and LFL lines are separated by about 508 G in accordance with the Breit–Rabi equation[67] for hydrogen atoms in a ~0.3-Tesla applied magnetic field. Both Figs. 13a and b show a small feature on the high-field wing of the LFL. No such feature is found on the high-field lines. In comparing Figs. 13a and b, we find that both the small LFL feature and the high-field line occur at the same value of the magnetic field for both figures. In contrast, the main LFL in Fig. 13b is found to be shifted by 0.8 G to the right as compared to the main LFL line in Fig. 13a. This fact is already hinting that the main part of the low-field line is responding to changes in the local environment of the hydrogen atoms.

Following the studies at 1.35 K, shown in Figs. 13a and b, the samples were slowly warmed to 14.5 K and then cooled back to 1.35 K. This

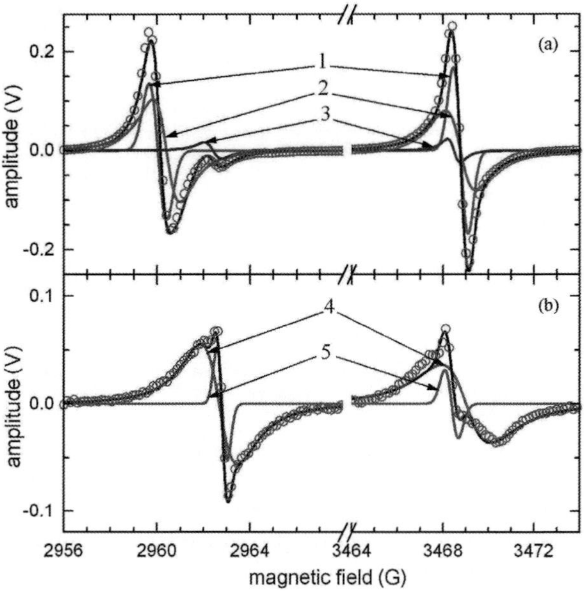

Fig. 14. (a) Experimental ESR spectra of H atoms for as-prepared H–Kr–He sample (open red circles) shown in Fig. 13a and three pairs of fitting lines (1, magenta are Gaussian lines; 2, green and 3, blue are Lorentzian lines). (b) Experimental ESR spectra of H atoms for annealed sample (open red circles) shown in Fig. 13c and two pairs of fitting lines (4, green are Lorentzian lines and 5, magenta are Gaussian lines). The sums of fitting lines are shown as black lines for both (a) and (b).[66] See also Colour Insert.

procedure resulted in annealing of the sample. During this process, the population of the H atoms dropped by a large factor as a result of the atoms recombining to form H_2 molecules. The result of annealing of the sample portrayed in Fig. 13a is shown in Fig. 13c. The position of the small feature in Figs. 13a and b and the resulting spectra shown in the left side of Fig. 13c after annealing closely coincide (at 2962.5 G). Following the annealing process, the HFL and the LFL are shown. Each consists of superposition of a narrow line and a broad line as shown in Fig. 13c. Previous work by Dmitriev[68] showed similar spectra for samples prepared by fast deposition of H and Kr atoms on cold substrates.

Spectra shown in Figs. 13a and b for the as-prepared samples were each fitted by one Gaussian and two Lorentzians lines. The fitting

Table 1. Hyperfine structure constants, A, and g-factors for H atoms in H–Kr–He condensates, in H_2 and in Kr matrices.[66]

Matrix	*A* (MHz)	Δ*A*/*A* (%)	*g*-factor
H–Kr–He, curve 1	1416.33(20)	−0.29	2.00228(5)
H–Kr–He, curve 2	1415.49(62)	−0.35	2.00220(16)
H–Kr–He, curve 3	1408.74(43)	−0.82	2.00176(11)
H–Kr–He, curve 4	1409.60(20)	−0.76	2.00160(5)
H–Kr–He, curve 5	1407.88(43)	−0.88	2.00178(11)
H–Kr–He, curve 1′	1414.71	−0.40	2.00199
H–Kr–He, curve 2′	1414.105	−0.45	2.00192
H–Kr–He, curve 3′	1409.935	−0.74	2.00158
Free state[73,74]	1420.40573(5)	0	2.002256(24)
H_2^1	1417.13(45)	−0.23	2.00243(8)
Kr (substitutional)[70]	1411.799(30)	−0.61	2.00179(8)
Kr (substitutional)[71]	1409	−0.80	2.0013
Kr (interstitial)[70]	1427.06(280)	+0.47	1.99967(31)

curves in Fig. 14a are shown only for the data in Fig. 13a. The curves in Fig. 13a show only two well-resolved peaks before annealing. Nevertheless, it was found that three curves were needed in order to fit the data in this figure. While the fitting curves for the LFL were displaced significantly from one another, this was not the case for the HFL. The data for the annealed sample was fitted by only two curves, one Lorentzian and one Gaussian as shown in Fig. 14b. The number of unpaired spins was determined from the amplitudes and widths of the fitting curves. By adding the number of spins of the LFL and the HFL, and dividing the result by the volumes of the sample, the average concentrations for the sample were found. Since dipolar magnetic coupling is the predominant cause of ESR line broadening, it was possible to estimate the local concentration in atoms per cm^3 of the H atom free radicals by the formula $n_l = 2.7 \times 10^{19} \Delta H_{pp}$, where ΔH_{pp} is the peak-to-peak width in Gauss of the ESR lines.[69] The *g*-factors and the hyperfine constants, *A*, for the H atoms were obtained by applying the Breit–Rabi equation to the data.[67]

Table 1 gives values of *A* in MHz and the *g*-factors for the five pairs of fitting curves shown in Fig. 14. The three curves in the top panel (curves 1–3) are for as-prepared samples. The curves 4–5 shown in the lower panel are for the annealed samples. The data for the sample shown in Fig. 13b was analyzed by the same method. Analysis of the data shown in Fig. 13b are labeled curves 1′, 2′, and 3′ in Table 1. A figure showing curves 1′, 2′, and 3′ has not been installed in this chapter, since the fitting procedures and general conclusions are the same as those for curves 1, 2, and 3.

There have been several studies of atomic hydrogen trapped in Kr matrices in the past.[68,70,71] Values of g and A from these investigations are also included in the Table 1 for the purpose of comparison with the IHC results discussed here. It was found in some of this earlier work that the hydrogen atoms could occupy either substitutional or interstitial sites. The hyperfine constant A is expected to be larger than the free atom value for the H atoms occupying interstitial sites since the unpaired electron of a hydrogen atom will be repelled by the close Kr neighbors via the Pauli principle, leaving a larger probability for this electron to be at the proton position. For the substitutional sites the attractive van der Waals forces of the surrounding Kr neighbors tends to enlarge the electron orbit, leading to a smaller electron wave function at the proton and a smaller hyperfine constant. The effect of these competing effects is discussed more thoroughly in theoretical studies by Adrian.[72]

In some of the earlier investigations by Foner *et al.*[70] and Dmitriev,[68] H and Kr atoms were deposited onto a cold substrate. This procedure always led to the H atoms being located in substitutional sites of the krypton matrix. Hydrogen atom samples contained in solid krypton have also been obtained by photolysis. Studies of Foner *et al.*[70] and Vaskonen *et al.*[71] of samples prepared in this manner found ESR signals corresponding to substitutional sites and interstitial sites. In the work of Foner *et al.*,[70] hydrogen atoms mainly occupied substitutional sites at 4.2 K. On the other hand, the experiments by Vaskonen *et al.*[71] at higher temperatures revealed that the H atoms first occupied interstitial (octahedral) sites. Following annealing at 24 K, however, the H atoms were found to occupy substitutional sites.

The lowest three entries in Table 1 provide values of g and A found experimentally by Foner *et al.*[70] and Vaskonen *et al.*[71] for both substitutional and interstitial sites. Results of the above investigations showed reasonable quantitative agreement with the theoretical studies of Adrian[72] for interstitial and substitutional sites of H atoms in solid Kr matrices. Table 1 also includes entries corresponding to g and A values for free H atoms as measured by Beringer and Heald,[73] and by Kusch.[74] In addition, Table 1 also gives values of g and A for H atoms embedded in solid hydrogen.[1]

The top three entries in Table 1 correspond to the Cornell studies of IHC samples represented by the curves in Fig. 14a. The values of g and A obtained from curves 1 and 2 in this figure agree reasonably well with the values of g and A for H atoms embedded in solid H_2. This is interpreted as evidence that the hydrogen atoms are embedded in thick layers of solid H_2 adsorbed on the surfaces of krypton nanoclusters. Curve 3 is identified with the small feature seen on the LFL line in the ESR spectrum. The g and A values obtained from curve 3 are in reasonable agreement with the Table 1 entries corresponding to substitutional sites of H atoms in a solid krypton matrix. Curves 4 and 5, which are obtained from spectra taken after the samples were annealed, also correspond to results expected for H atoms in substitutional sites in solid Kr. The g and A values have also been calculated for the ESR spectra shown in Fig. 13b, where the main LFL has clearly shifted to the right. Thus, calculations of g and A from curves 1′ and 2′ give somewhat smaller values of A and g, as shown in the Table 1. The influence of the Kr substrate is thought to be manifested by the observed reduction in the g and A values. This would imply that the H atoms have a greater opportunity to interact with the Kr, which would in turn correspond to a much thinner H_2 coating on the Kr nanoclusters for the sample in Fig. 13b. The small feature on the low-field lines in Fig. 13b, represented by curve 3′ also has g and A values that are quite close in value to H atoms occupying substitutional sites in solid Kr again indicating that only a small population of hydrogen atoms is embedded in the Kr clusters.

We have seen that the effect of annealing on the samples is to greatly reduce the H atom populations and that the main LFL has vanished completely leaving only a signal associated with the small feature to the right of the main line in Fig. 13a. The data after annealing, shown in Fig. 13c, are fitted by the sum of curves 4 and 5. The values of g and A for curves 4 and 5 agree quite well with the results obtained by Foner *et al.*[70]

and Vaskonen *et al.*,[71] for hydrogen atoms occupying substitutional sites in solid krypton.

The observed changes in the ESR spectra of the H-Kr IHCs during the annealing process are summarized here, as follows: the large broad main ESR lines observed before annealing and shown in Figs. 13a and b correspond to signals from hydrogen atoms embedded in layers of solid molecular hydrogen at the surfaces of krypton nanoclusters. This interpretation is in agreement with Gordon's model,[64] which shows that a large proportion of the atomic free radicals should be trapped at the nanocluster surfaces. The interpretation is based on the fact that during the early high-temperature stages of sample formation, the strong van der Waals forces between colliding Kr atoms lead to the binding of these atoms into solid krypton nanoclusters.[75] At lower temperatures, the more weakly attracted H atoms, H_2 molecules, and He atoms bind to the surfaces of the Kr nanoclusters. We recall that the pulsed ESR experiments discussed previously on IHCs containing H and/or D atoms and HD and/or D_2 molecules also showed an analogous effect, with the tendency of the atomic free radicals to populate the cluster surfaces.[60]

In Figs. 13a and b, we have identified the main ESR lines as being associated with H atoms trapped in thick H_2 films and thin H_2 films, respectively, on the nanocluster surfaces, corresponding (for the case of Fig. 13a) to curves 1 and 2 in Fig. 14a. Furthermore, the small shoulder seen both in Figs. 13a and b is identified with a small population of H atoms embedded within the Kr nanoclusters. When the samples are heated and annealing occurs, the large population of H atoms in the surface layers will diffuse more rapidly and recombine. The Kr clusters will aggregate into larger crystallites during this process. At the end of annealing, basically the only hydrogen atoms remaining will be the small population inside the krypton nanocluster surfaces. In the sample corresponding to Fig. 13a, 95% of the H atoms were on surfaces and 5% in the interior of the clusters.

We now discuss the high concentrations of H atoms achieved in IHC samples prepared from gas mixtures containing molecular hydrogen, krypton atoms, and helium atoms. We first consider a sample prepared from the gas mixture H_2:Kr:He = 1:1:200. The average concentration of H atoms in this IHC sample was found from the ESR spectral line intensities (see Fig. 13a) by double integration of the derivative signals to be 1.4×10^{18} hydrogen atoms per cm^3. The local concentration, as calculated from the dipole–dipole interaction broadened line width ($\Delta H = 2.2$ G), was 6×10^{19} hydrogen atoms per cm^3. The local concentration is expected to be larger than the average concentration since the porous IHC samples will possess voids in which no hydrogen atoms are present. The decay of the hydrogen atom population in the thick solid hydrogen film of this sample was measured at 1.35 K. The result showed a linear dependence of the reciprocal of the concentration on time as expected for two-body recombination, according to Eq. (6), shown earlier, $1/n = 2K_H(T)t$. The recombination rate constant $K_H(T) = (2n_0\tau_{1/2})^{-1}$ was obtained by measuring $\tau_{1/2}$, the time required for half the number of atoms in the sample to decay. For the sample with an initial concentration of $n_0 = 1.4 \times 10^{18}$ cm^{-3} (corresponding to the thick H_2 film), $\tau_{1/2}$ was found to be 161 min so that the rate constant at $T = 1.35$ K for this sample was 3.7×10^{-23} cm^3 s^{-1}. The tunneling exchange reaction $H + H_2 \rightarrow H_2 + H$ allows a hydrogen atom to tunnel through the thick film until it can find another hydrogen atom, which leads to recombination.

Only 5% of the initial population of H atoms survives annealing. These atoms are embedded in the Kr nanoclusters. The ESR signal from these atoms consists of a broad and a narrow line, superposed, as shown in Fig. 14b. This structure can possibly be explained by phase separation of atoms in the nanocrystals into regions with larger concentrations having closer spacing and large dipole–dipole interactions, leading to the broader component. Conversely, there will also be regions of smaller concentrations with less dipolar broadening leading to the narrower component. An alternative explanation is that the narrow lines are associated with atoms in substitutional *fcc* and *hcp* sites and the broad lines correspond to disordered regions.[68]

Other H–Kr IHCs prepared with different proportions of H_2, Kr, and He in the make-up gas were studied. The largest H atom concentration was achieved in the sample with an initial mixture ratio H_2:Kr:He = 1:50:10000 shown in Fig. 13b. As stated earlier, the Kr clusters in this sample were coated by a thin film of H_2 molecules, and the initial H atom concentration was 3.2×10^{18} cm^{-3}. Ninety-seven per cent of the atoms were on the Kr nanocluster surfaces and 3% were contained in cluster interiors. After initial measurements, the sample was compressed to attain an average concentration of 1.2×10^{19} cm^{-3}. A calculation showed that for this sample, with typical nanocluster diameters of 5 nm, the average spacing between H atoms on the cluster surfaces was 14 Å. In contrast to the earlier sample discussed (see Fig. 13a), this latter sample was almost completely stable against recombination at 1.35 K, making long-term storage of the H atoms possible. The low recombination rate can perhaps be explained by the fact that the H_2 film was very thin for this sample, leading to a stronger bonding of H atoms to the Kr substrate. The displacement of the main LFLs between Figs. 13a and b indicate that the H atom spectra are influenced by the underlying Kr nanocluster surfaces. We can conclude that for thin H_2 films the H atoms are more closely bound to the Kr surface and may therefore be partially immobilized, leading to a much lower recombination rate at 1.35 K.

4. Studies of H Atoms in Solid H_2 Films at Ultralow Temperatures

An entirely new approach toward the study of matrix isolated hydrogen atoms has been taken by groups at Turku University and Cornell University. They have created films of molecular hydrogen with embedded H atoms obtained from a dissociator in the low-temperature region of the cryostat. This has made possible studies of H atoms and their interactions at temperatures below 100 mK, achieved with dilution refrigerators. In a typical experiment the gas of spin polarized hydrogen atoms (H↓) is stabilized at $T < 1$ K in a strong magnetic field of 4.8 T. The atoms are generated in an rf discharge in a cryogenic dissociator and the high field seeking fraction is pulled into in the sample cell whose walls are covered by a superfluid helium film. The latter is required to suppress the strong adsorption of the gas on the walls and inhibit the processes of surface recombination and relaxation. However, these cannot be completely avoided and some of the atoms recombine back to molecules, which eventually stick to the sample cell walls beneath the helium film and form a solid H_2 layer. The rate of such a coating process is very slow, typically of the order of one molecular layer per hour, and therefore one has to keep the sample cell continuously filled with atoms for a few days or a week. It has been found that H atoms are also

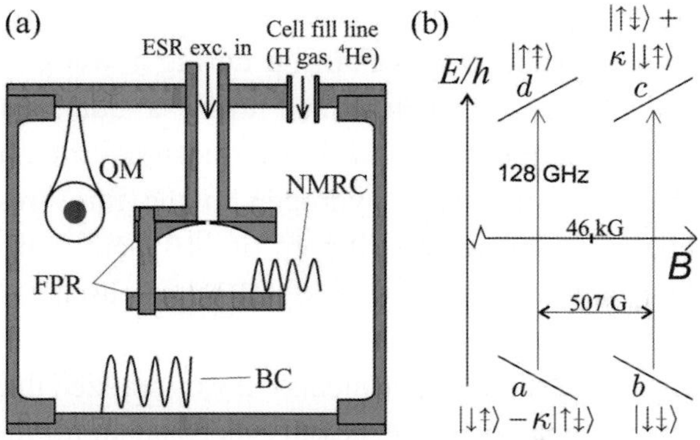

Fig. 15. (a) A schematic illustration of the sample cell used in experiments of Ahokas *et al.*[43] FPR, mirrors of the Fabry–Perot resonator; BC, bottom rf coil; NMRC, NMR coil; QM, quartz microbalance. (b) Hyperfine level diagram of hydrogen atom in a strong magnetic field. Spin wave functions are labeled by projections of electron and nuclear spins, and corresponding energy levels by letters *a, b, c*, and *d*.

captured inside the molecular film if the main recombination channel is a process of three-body recombination: H + H + H = H_2 + H.[15] In this case the third atom receives part of the recombination energy (~10–50 K), large enough to punch through the helium film and stick to the solid H_2 beneath. This novel method of creating the samples of H in H_2 has several important advantages: the samples are created at $T < 1$ K, implying a very small heat release during the sample formation, and the films of H_2 have the best possible purity since no other substance can penetrate into the sample cell at such low temperatures. An obvious disadvantage of this method is the very slow speed of growing the films.

The sample cell used in experiments with H atoms stabilized in the films of solid H_2 at ultralow temperatures[43] is schematically presented in Fig. 15. The H_2 films containing H atoms were situated on the surfaces of the flat and spherical mirrors of a Fabry–Perot resonator (FPR). In preliminary experiments,[15] a small (~0.1 cm^2) part of the sample cell wall was thermally insulated and actively cooled to a temperature much lower than the rest of the sample cell. Such a cold spot provided a local compression of adsorbed H atoms and made the three-body recombination a dominant loss mechanism.[76] The films of H in H_2 were accumulated on the sample cell walls by pushing atoms from the H source with a flux of $\sim 10^{13}$ s^{-1} into the sample cell where they recombined back into molecules and were deposited on the walls. A film of ~50-nm thickness was built up during 1 week of continuous coating. The thickness was determined using a quartz microbalance located inside the sample cell to within an accuracy of one molecular layer. Concentrations of atomic hydrogen in the films resulting from three-body recombination were ~50 ppm or $\sim 10^{18}$ cm^{-3}. Both ESR and electron-nuclear double resonance (ENDOR) were used to characterize the films of H in H_2.

In the second stage of the experiments[43] it was realized that the process of three-body recombination is a necessary mechanism for the H atoms to be captured into the matrix during H_2 coating process. H in H_2 ESR lines were no longer detected after removing the cold spot (the source of the three-body recombination) from the sample cell, though the molecular film growth had been observed to occur at the same rate. This motivated a search for another method of implanting H atoms into the H_2 films. A low-power rf discharge running inside the sample cell itself at temperatures ~0.5 K was utilized. The discharge was initiated by applying rf pulses to one of the helical resonators as indicated in Fig. 15a. The rf pulses 5–20-μs long with repetition period of several ms and peak power of ~10 mW were strong enough to produce a gas of low-energy (≤100 eV) electrons in the sample cell. The strong cyclotron resonance line of the electrons was detected by the ESR spectrometer during operation of the discharge. The electrons collided with the H_2 film on the sample cell walls and dissociated molecules into atoms inside the solid. Although dissociating by the electron impact is an old and well-known method, this was the first time it had been used at such low energies and temperatures. It provided a rapid ($t \sim$ 10–100 min) and controllable build up of very high concentrations of H atoms in the H_2 films and facilitated more detailed studies of their properties. The H densities of 2×10^{19} cm^{-3} reached in this work were a factor of 3 larger than were ever observed in this system.

Typical ESR spectra of atomic hydrogen in the gas phase and inside the solid H_2 films are presented in Fig. 16.[43] The H ESR spectrum consists of two lines separated by ~507 G due to the hyperfine interaction. The two lines correspond to the allowed *a–d* (*a*-line) and *b–c*

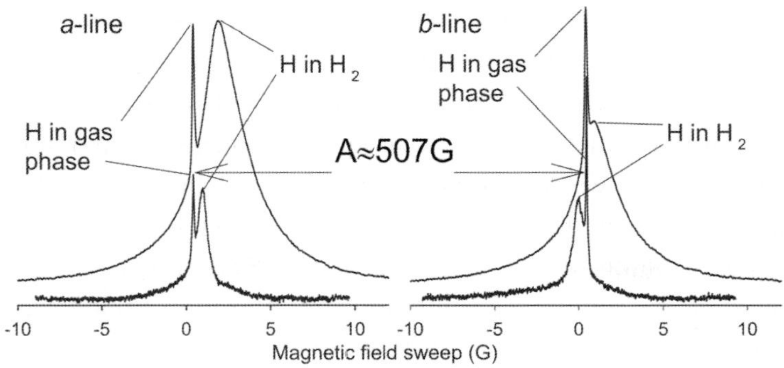

Fig. 16. ESR spectra of atomic hydrogen in the gas phase and in solid H_2 observed in experiments with thin films of H in H_2 at ultralow temperatures $T < 0.5$ K. Upper traces are recorded at $n \sim 2 \times 10^{19}$ cm^{-3}, lower traces are recorded at $n \sim 6 \times 10^{17}$ cm^{-3} and multiplied by 3.5.[43]

(*b*-line) transitions. The possibility of detecting ESR lines of free H and of the atoms inside the H_2 matrix in the same experiment provided a nice opportunity to measure the effects of the host matrix and make an accurate measurement of the dipolar fields inside samples. Resonance peaks of the hydrogen atoms in the solid H_2 turned out to be shifted from those of the gas phase. The overall change of the spectrum can be characterized by two parameters: the shift of the center of the spectrum and the change of the separation between the lines. The latter is defined by the hyperfine constant and was more accurately measured using the ENDOR method, which will be considered later. The shift of the spectrum center was found to be proportional to the density of H atoms and is caused by the magnetization of the sample, a noticeable effect at the high densities studied.

As one can see in Fig. 16, the peaks of the H in H_2 lines at high density are much broader than the low-density peaks. The density-dependent broadening is caused by the dipolar interactions and was observed in Ref. 43 for the first time because of the high densities reached in this work. The shape and the shift of the ESR line in this case are sensitive to the distribution of the atoms inside the film and the film thickness. These effects were incorporated in a model developed by the authors of Ref. 43 to calculate the distributions of the local dipolar fields inside the samples of different thickness and density.

Comparing the synthetic line shapes with the experimental ones these authors confirmed the even distribution of the atoms inside the film and provided a correct determination of the absolute values of the H density and the film thickness. Notice that these experiments did not use a modulation technique to record derivative signals. Instead the absorption signals were measured directly.

In their studies of the ESR transitions at high-excitation powers Ahokas *et al.*[15,43] studied the saturation and relaxation behavior of the H atoms in solid H_2 films. An interesting effect was observed by saturating the *b–c* transition. This has lead to a permanent destruction of the *b*-line, while the *a*-line has grown by the same amount. This was explained by the relaxation of the excited hyperfine state *c* to the *a*-state via the forbidden *a–c* transition, and is an analog of the Overhauser effect, a well-known method of dynamic nuclear polarization.[57,77] The *a–c* transition involves a double electron and nuclear flip and its probability is strongly ($\sim 10^5$ times) reduced in a 5-T magnetic field. The measured *a–c* relaxation times were found to be much faster than estimated using this factor and the measurement of the relaxation rates of the allowed ESR transitions. This discrepancy indicates the presence of some other fast relaxation mechanism, remaining as yet unexplained but most likely non radiative. Saturating the *b–c* transition the authors were able to transfer the entire *b*-state population into the *a*-state within a few minutes. The second method of manipulating the hyperfine state populations was to equalize them by saturating the *a–b* transition using the NMR coil. Both methods lead to a substantial deviation of the system from equilibrium. Following these processes, the *a* and *b* ESR lines were recorded as a function of time at low-excitation power and the relaxation of the samples was studied at different sample cell temperatures.

A typical relaxation of the populations of the *a*- and *b*-hyperfine states measured at a temperature of 150 mK is presented in Fig. 17. In this figure, the nuclear polarization of the sample $p = (n_a - n_b)/(n_a + n_b)$ is plotted against time. The recovery of p to its equilibrium value is controlled by the nuclear relaxation process, which turns out to be very slow, $T_{ba} \sim 60$–80 h at 150 mK. A most striking result of these experiments is that the nuclear polarization relaxes to the value $p(t \rightarrow \infty) = 0.5(1)$ at $T = 150$ mK, which is much larger than $p \approx 0.14$ dictated by the Boltzmann

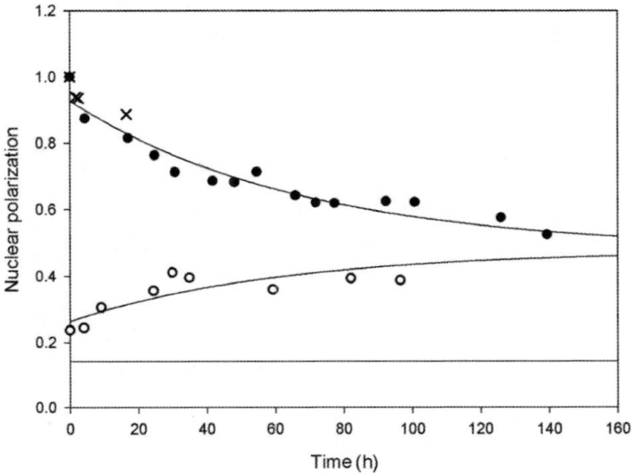

Fig. 17. Evolution of the nuclear polarization $p = (n_a - n_b)/(n_a + n_b)$ after saturation of the b–c (●) and a–b (○) transitions at $T = 150$ mK. Solid horizontal line gives the value of the polarization defined by the Boltzmann factor for this temperature.[15]

statistics with $n_a/n_b = \exp(E_{ab}/T)$. Here $E_{ab} \approx 43$ mK is the energy difference between the b- and a- states.[15] The authors considered the formation of a Bose–Einstein condensate in the ensemble of atoms as a possible explanation of the non-Boltzmann population ratio observed in these experiments.

Having the possibility of exciting b–c and a–b transitions simultaneously, Ahokas *et al.*[15] performed a study of the nuclear spin transition of the H atoms inside solid H_2 films. Conventional NMR cannot be detected directly in this system because of the low sensitivity of this method. However, one can detect the NMR transition indirectly, using ENDOR. High-power ESR excitation is applied at the b–c transition frequency. This transfers all the b-state population into the a-state via the Overhauser effect. In the steady state, when the transfer is complete, the power absorption detected by the ESR spectrometer goes to zero. Then, the rf excitation at the a–b resonance frequency is applied to the NMR coil. This transfers atoms back to the b-state and thus leads to the increase of the ESR absorption signal. Sweeping the frequency of the rf excitation applied to the NMR coil, one can record the shape of the

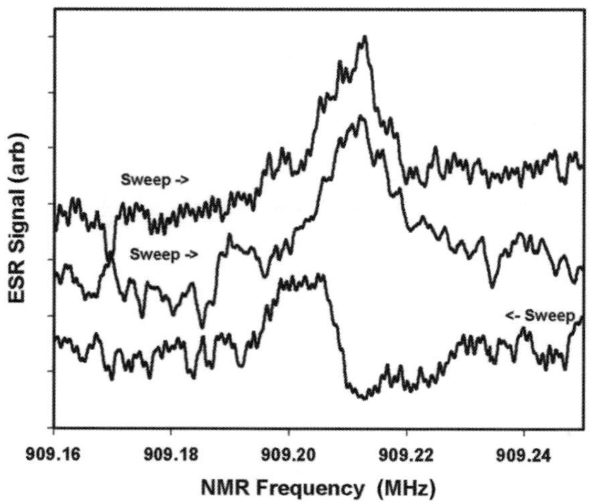

Fig. 18. Typical ENDOR spectra of atomic hydrogen observed in thin films of H_2 at ultralow temperatures.[15] Traces were recorded for different directions of the sweep, as marked on the picture.

a–b transition. Typical spectra recorded in such an experiment are presented in Fig. 18.

Even in a strong magnetic field, one of the main factors determining the nuclear *a–b* transition frequency is the hyperfine interaction. Its strength depends on the density of the electron at the location of the proton, and is sensitive to locations of the other electrons surrounding the impurity atom in the host matrix. If the atom turns out to be in an interstitional site of the lattice, the effect of the neighboring molecules is to push the impurity atom electron towards the proton, which leads to an increase of the hyperfine interaction. The effect has the opposite sign for the atoms in substitutional sites. The ENDOR lines detected in the experiments with H_2 films[15] turned out to be shifted to the red from the free atom transition by ~1.3 MHz, indicating that the atoms are located in the substitutional sites of the H_2 crystal. This result is in agreement with the conclusion of Kumada *et al.*[32] who have also performed ENDOR studies on γ-irradiated H_2 samples. Another important finding of the ENDOR experiments of Ahokas *et al.*[15] is the very small width of the NMR transition ~20 kHz. This indicates a very homogeneous crystalline field and the high quality of the H_2 films grown by the above-mentioned slow-deposition method.

The stability of the hydrogen atoms inside solid H_2 matrices has been extensively studied for samples prepared by different methods. Most of these studies were done at temperatures above 1 K. It has been found that above ~4 K the recombination rate constant follows the exponential Arrhenius law, while at lower temperatures it tends to decrease linearly with temperature. To recombine, the atoms need to approach closely enough to one another to be within a distance on the order of the lattice constant. The diffusion of hydrogen atoms towards each other turned out to be responsible for recombination. At low temperatures $T < 4$ K this motion proceeds via quantum mechanical tunneling.[14] The rate of hopping of the atoms from one lattice site to the other depends on the energy level mismatch between the neighboring sites. Each atom distorts the lattice and shifts the energy levels. If these shifts substantially exceed the matrix temperature, a close approach of the atoms to one another and recombination are strongly suppressed. The theory of quantum diffusion of light impurities in cryocrystals has been developed by Kagan *et al.*,[7,78] and has been verified in solid mixtures of ^{3}He in ^{4}He. However, in the case of H in H_2, the mechanism of diffusion may be somewhat different. It has been demonstrated by Kumada[13] and discussed earlier in this chapter that the atoms can move via the quantum tunneling exchange chemical reaction $H + H_2 = H_2 + H$. In this mechanism, the presence of some other impurity nearby may also suppress the reaction if the new position of the atom after tunneling requires gaining some extra energy.

Recombination of H has also been a subject of studies in films of H_2 at temperatures above and below 1 K. Already in preliminary experiments, it has been found that recombination of atoms is imperceptible at temperatures below 150 mK.[15] Some slow decays in H atom densities were observed at temperatures of 300 and 480 mK. In the second stage of the experiments[43] studies of recombination were extended to higher concentrations of atoms. Having the possibility of easily regenerating the atomic populations, the authors performed detailed measurements of the recombination rate constant for different temperatures and densities of atoms. A summary of this study together with the results obtained in earlier experiments is presented in Fig. 19. At

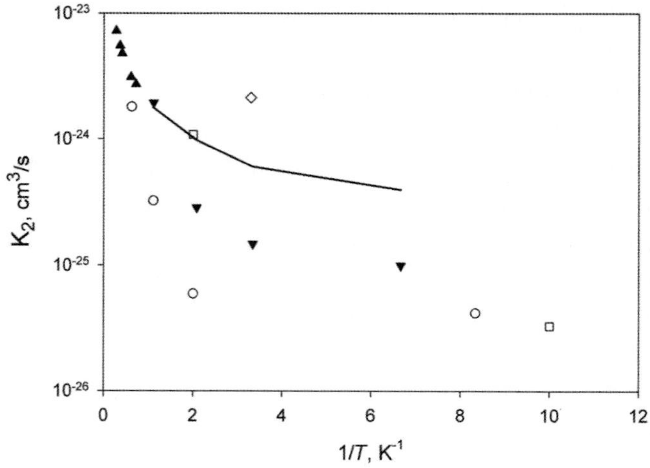

Fig. 19. Summary of the measurements of the recombination rate constant of H atoms in H_2 films. (▲) Experiments at $1.3 < T < 4.2$ K, (▼) low-H-density sample below 1 K,[15] (○) high-density samples below 1 K,[43] (◊) high-density samples under H_2 coating,[43] (□) low-density samples created by bottom coil,[43] (solid line) theory.7

the highest temperatures in the experiments with H_2 films, the data tend to coincide with the previous work and with the theory. However, below 1 K the recombination rate constant drops down by an order of magnitude. These observations confirm the suggestion that the energy level difference between the neighboring lattice sites may substantially slow down the motion of atoms towards each other. At low enough temperatures this may stop recombination entirely. The authors also tried to measure the loss rate of atoms via the process of H_2 coating described above, i.e., pushing continuously a high flux of gas phase atoms into the sample cell. In this case the density of H in the H_2 film decreased much more rapidly (see Fig. 19). During the coating process a flux of high-energy phonons is transferred into the film by the energetic H_2 molecules resulting from the recombination of H in the gas phase. This increases the probability for the atoms to hop to the next site and stimulate recombination. The same effect also explains why the method of sample preparation involving a low-energy discharge allows one to obtain much higher densities than other methods, where much higher energy fluxes were used to dissociate H_2 and some atoms recombined during sample preparation.

High-stability atomic hydrogen in the experiments with H_2 films at temperatures below 1 K opens up the possibility of creating samples with even higher densities and cooling them to much lower temperatures. This makes this system a good candidate for observing collective quantum phenomena, such as Bose–Einstein condensation and supersolid behavior.

5. Conclusions

Considerable progress has been made in studies of matrix isolated hydrogen atoms and deuterium atoms during the past decade. Matrices studied have included IHCs made up of nanoclusters (≤10 nm in diameter) of solid H_2, solid D_2, and mixed solids consisting of solid H_2, D_2, and HD. More recently, IHCs with Kr matrices have also been studied. Pulse and CW ESR techniques have been employed to study the behavior and location of hydrogen and deuterium atomic free radicals contained in the samples. In these studies, information about exchange tunneling reactions has been obtained. Rate constants for the reactions involving tunneling exchange reactions for D atoms interacting with D_2 and HD molecules have been measured. The highest average concentration of H atoms achieved in these studies was 7×10^{17} atoms per cm^3. Recombination rates of hydrogen atoms into H_2 molecules and deuterium atoms into D_2 molecules have also been determined. Evidence seems to favor a strong role for the quantum tunneling reaction $H + H_2 \rightarrow H_2 + H$ in the diffusion of H atoms through solid H_2 crystals, which brings two H atoms into contact with one another, leading to the recombination reaction $H + H \rightarrow H_2$. Pulse ESR experiments on H or D atoms in nanoclusters of D_2, or HD and D_2, have shown that a large fraction (~55%) of the H or D atoms reside on the cluster surfaces, in accordance with Gordon's prediction.[64] Hydrogen atoms contained in IHC samples prepared from mixtures of molecular hydrogen and krypton have also been studied by CW ESR at 1.35 K. Exceptionally high concentrations, up to 10^{19} atoms per cm^3 in compressed samples, have been achieved. It was shown in these experiments that 95% or more of the hydrogen atoms resided on the cluster surfaces. The average distance between hydrogen atoms on the cluster surfaces was as small as 14 Å. Recent studies of IHC samples prepared from molecular deuterium and Kr gases showed that concentrations of atomic deuterium even higher than those for atomic hydrogen could be achieved. These high concentrations approach the regime where the thermal de Broglie wavelength is comparable to the average interatomic separation of H atoms or D atoms. This may, in the future, lead to interesting quantum overlap effects possibly associated with Bose–Einstein correlations in atomic hydrogen at lower temperatures.

Investigations have been performed on H atoms trapped in thin films of molecular hydrogen at the University of Turku and Cornell. H_2 molecules were either dissociated in a separate dissociator and the resulting atoms were transferred to the sample cell, or alternatively the molecules were dissociated directly in the sample cell with an rf coil. A dilution refrigerator was used to cool the thin film samples down to temperature below 100 mK. Recombination rates were vanishingly small at the lowest temperatures so that the samples could be studied for many days. The ensembles of H atoms in the films were polarized by a 4.6 T magnetic field. CW millimeter wave ESR experiments were performed at a frequency of 128 GHz in a Fabry–Perot resonator fabricated with a curved mirror and a flat mirror. The thin films were formed on the interior walls of the sample chambers and the Fabry-Perot mirrors. In some cases one of the walls of the cell consisted of a thin mylar membrane separating the liquid ^3He–^4He mixture in the dilution refrigerator mixing chamber from the sample in the sample cell. The antinodes in the cavity were adjusted to give peak signals at the sample locations on the mirrors or the mylar membrane. Very high concentrations of hydrogen atoms were obtained in the solid H_2 films in these experiments. A number of unusual effects were seen. The ratio of the populations for $T < 150$ mK of the two lowest hyperfine states deviated strongly from the Boltzmann factor with an unexpectedly large population in the lowest hyperfine state, suggesting the possibility of Bose–Einstein correlations. The studies of the population of the two lowest hyperfine states give seemingly contradictory results at high rf powers. It was not possible to completely saturate the transition between the lowest hyperfine states, even though the spin lattice relaxation time was many hours. ENDOR experiments have provided a powerful tool for investigating this transition. More experimental work on these films is clearly in order. Although some progress has been made, more effort will be required before effects of quantum statistical correlations or exotic low-temperature magnetic transitions can be unambiguously observed.

Physics is still full of surprises, but these surprises can happen only if there is sufficient dedication and effort.

Acknowledgments

This work was supported by National Science Foundation Grant DMR 0504683, the Academy of Finland Grants 122595 and 124998, the Wihuri Foundation and the Texas Norman Hackerman Advanced Research Program under Grant No. 010366-0137-2009. The authors of this chapter thank Janne Ahokas, Neil Ashcroft, Ethan Bernard, Roman Boltnev, Peter Borbat, Jack Freed, Jarno Jarvinen, and Erich Mueller for helpful discussions and many suggestions.

References

1. C. K. Jen, S. N. Foner, E. L. Cochran and V. A. Bowers, *Phys. Rev.* **104**, 846 (1956).
2. C. K. Jen, S. N. Foner, E. L. Cochran and V. A. Bowers, *Phys. Rev.* **112**, 1169 (1958).
3. L. A. Wall, D. W. Brown and R. E. Florin, *J. Phys. Chem.* **63**, 1762 (1959).
4. L. H. Piette, R. C. Rempel, H. E. Weaver and J. M. Flournoy, *J. Chem. Phys.* **30**, 1623 (1959).
5. A. M. Bass and H. P. Broida, *Formation and Trapping of Free Radicals*. Academic Press, New York and London (1960).
6. A. F. Andreev and I. M. Lifshits, *Sov. Phys. JETP.* **29**, 1107 (1969).
7. Yu. Kagan, L. A. Maksimov and N. V. Prokof'ev, *JETP Lett.* **36**, 253 (1982).
8. A. V. Ivliev, A. Ya. Katunin, I. I. Lukashevich, V. V. Sklyarevskii, V. V. Suraev, V. V. Filippov, N. I. Filippov and V. A. Shvetsov, *JETP Lett.* **36**, 472 (1982).
9. H. Tsuruta, T. Miyazaki, K. Fueki and N. Azuma, *J. Phys. Chem.* **87**, 5422 (1983).
10. E. B. Gordon, A. A. Pelmenev, O. F. Pugachev and V. V. Khmelenko, *JETP Lett.* **37**, 282 (1983).
11. R. N. Porter and M. Karplus, *J. Chem. Phys.* **40**, 11059 (1964).
12. T. Takayanagi, K. Nakamura and S. Sato, *J. Chem. Phys.* **90**, 1641 (1989).

13. T. Kumada, *Phys. Rev.* **B68**, 052301 (2003).
14. T. Miyazaki (editor), *Atom Tunneling Phenomena in Physics, Chemistry and Biology*. Springer, Berlin (2004).
15. J. Ahokas, J. Jarvinen, V. V. Khmelenko, D. M. Lee and S. Vasiliev, *Phys. Rev. Lett.* **97**, 095301 (2006).
16. M. Sharnoff and R. V. Pound, *Phys. Rev*. **132**, 1003 (1963).
17. A. S. Iskovskih, A. Ya. Katunin, I. I. Lukashevich, V. V. Sklyarevskii, V. V. Suraev, V. V. Filippov, N. I. Filippov and V. A. Shvetsov, *Sov. Phys. JETP*. **64**, 1085 (1986).
18. T. Miyazaki, K.-P. Lee, K. Fueki and A. Takeuchi, *J. Phys. Chem*. **88**, 4959 (1984).
19. K.-P. Lee, T. Miyazaki, K. Fueki and K. Gotoh, *J. Phys. Chem.* **91**, 180 (1987).
20. T. Miyazaki, N. Iwata, K.-P. Lee, and K. Fueki, *J. Phys. Chem.* **93**, 3352 (1989).
21. E. B. Gordon, A. A. Pelmenev, O. F. Pugachev and V. V. Khmelenko, *Sov. J. Low Temp. Phys*. **11**, 307 (1985).
22. E. P. Bernard, R. E. Boltnev, V. V. Khmelenko, V. Kiryukhin, S. I. Kiselev and D. M. Lee, *J. Low Temp. Phys*. **134**, 169 (2004).
23. E. P. Bernard, R. E. Boltnev, V. V. Khmelenko, V. Kiryukhin, S. I. Kiselev and D.M. Lee, *Phys. Rev.* **B69**, 104201 (2004).
24. A. V. Ivliev, A. S. Iskovskih, A. Ya. Katunin, I. I. Lukashevich, V. V. Sklyarevskii, V. V. Suraev, V. V. Filippov, N. I. Filippov and V. A. Shvetsov, *JETP Lett*. **38**, 379 (1983).
25. T. Miyazaki and K.-P. Lee, *J. Chem. Phys.* **90**, 400 (1986).
26. T. Miyazaki, S. Mori, T. Nagasaka, J. Kumagai, Y. Aratono and T. Kumada, *J. Phys. Chem.* **A104**, 9403 (2000).
27. T. Kumada, S. Mori, T. Nagasaka, J. Kumagai and T. Miyazaki, *J. Low Temp. Phys.* **122**, 265 (2001).
28. T. Kumada, M. Sakakibara, T. Nagasaka, H. Fukuta, J. Kumagai and T. Miyazaki, *J. Chem. Phys.* **116**, 1109 (2002).
29. T. Kumada, K. Komagauchi, Y. Aratono and T. Miyazaki, *Chem. Phys. Lett.* **261**, 463 (1996).
30. T. Miyazaki, N. Iwata, K. Fueki and H. Hase, *J. Phys. Chem*. **94**, 1702 (1990)
31. T. Miyazaki, *Chem. Phys. Lett.* **176**, 99 (1991).
32. T. Kumada, N. Kitagawa, T. Noda, J. Kumagai, Y. Aratono and T. Miyazaki, *Chem. Phys. Lett.* **288**, 755 (1998).
33. T. Kumada, T. Noda, J. Kumagai, Y. Aratono and T. Miyazaki, *J. Chem. Phys.* **111**, 10974 (1999).
34. J. Kumagai, T. Noda and T. Miyazaki, *Chem. Phys. Lett.* **321**, 8 (2000).
35. E. B. Gordon, L. P. Mezhov-Deglin and O. F. Pugachev, *JETP Lett.* **19**, 63 (1974).
36. E. B. Gordon, V. V. Khmelenko, E. A. Popov, A. A. Pelmenev and O. F. Pugachev, *Chem. Phys. Lett.* **155**, 301 (1989).
37. S. I. Kiselev, V. V. Khmelenko, D. M. Lee, V. Kiryukhin, R. E. Boltnev, E. B. Gordon and B. Keimer, *Phys. Rev.* **B65**, 024517 (2002).

38. V. Kiryukhin, B. Keimer, R. E. Boltnev, V. V. Khmelenko, and E. B. Gordon, *Phys. Rev. Lett.* **79**, 1774 (1997).
39. E. B. Gordon, *Dokl. Phys. Chem.* **378**, 156 (2001).
40. T. J. Greytak and D. Kleppner, In: G. Grynberg and R. Stora (editors), *New Trends in Atomic Physics,* p. 1125, Elsevier, Amsterdam, (1984).
41. R. W. H. Webeler, *J. Chem. Phys*. **64**, 2253 (1976).
42. G. Rosen, *J. Chem. Phys*. **65**, 1735 (1976).
43. J. Ahokas, O. Vainio, J. Järvinen, V. V. Khmelenko, D. M. Lee and S. Vasiliev, *Phys. Rev*. **B79**, 220505 (2009).
44. S. Vasiliev, J. Jarvinen, E. Tjukanov, A. Kharitonov and S. Jaakola, *Rev. Sci. Instrum.* **75**, 94 (2004).
45. S. I. Kiselev, V. V. Khmelenko, E. P. Bernard and D. M. Lee, *Low Temp. Phys.* **29**, 505 (2003).
46. E. P. Bernard, V. V. Khmelenko, E. Vehmanen, P. P. Borbat, J. H. Freed, and D. M. Lee, *AIP Conference Proceedings 850: 24th International Conference on Low Temperature Physics*, p.1659, Orlando, Florida (10–17 August 2005).
47. V. V. Khmelenko, E. B. Bernard, S. Vasiliev and D. M. Lee. *Russ. Chem. Rev.* **76**, 1107 (2007).
48. S. A. Dikanov and Y. D. Tsvetkov, *Electron Spin Echo Envelope Spectroscopy (ESEEM spectroscopy)*, CRC Press, New York (1992).
49. A. V. Astashkin, V. V. Kozlyuk and A. M. Raitsimring, *J. Mag. Res.* **145**, 357 (2000).
50. S. I. Kiselev, V. V. Khmelenko, C. Y. Lee and D. M. Lee, *J. Low Temp. Phys.* **128**, 37 (2002).
51. S. I. Kiselev, V. V. Khmelenko and D. M. Lee, *Phys. Rev. Lett.* **89**, 175301 (2002).
52. V. V. Khmelenko, S. I. Kiselev, D. M. Lee and C. Y. Lee, *Physica Scripta.* **T102**, 118 (2002).
53. E .P. Bernard, R. E. Boltnev, V. V. Khmelenko and D. M. Lee, *J. Low Temp. Phys.* **138**, 829 (2005).
54. G. T. Trammell, H. Zeldes and R. Livingston, *Phys. Rev.* **110**, 630 (1958).
55. T. Kumada, *J. Chem. Phys.* **124**, 094504 (2006).
56. E. L. Hahn, *Phys. Rev.* **80**, 580 (1950).
57. C. P. Slichter, *Principles of Magnetic Resonance*, Springer, New York (1996).
58. E. P. Bernard, V. V. Khmelenko, E. Vehmanen, P. P. Borbat, J. H. Freed and D. M. Lee, *AIP Conference Proceedings 850: 24th International Conference on Low Temperature Physics,* p. 372, Orlando, Florida (10–17 August 2005).
59. V. V. Khmelenko, E. P. Bernard, E. Vehmanen and D. M. Lee, *AIP Conference Proceedings 850: 24th International Conference on Low Temperature Physics,* p. 376, Orlando, Florida (10–17 August 2005).
60. E. P. Bernard, V. V. Khmelenko and D. M. Lee, *J. Low Temp. Phys.* **150**, 516 (2008).
61. W. B. Mims, J. Peisach and J. L. Davis, *J. Chem. Phys*. **66**, 5536 (1977).

62. M. Iwasaki, K. Toriyama and K. Nunome, *J. Chem. Phys.* **86**, 5971 (1987).
63. A. Ponti, *J. Mag. Res.* **127**, 87 (1997).
64. E. B. Gordon, *Low Temp. Phys.* **30**, 756 (2004).
65. J. Järvinen, E. P. Bernard, R. E. Boltnev, V. V. Khmelenko and D. M. Lee, *J. Phys.: Conf. Ser.* **150**, 03235 (2009).
66. R. E. Boltnev, E. P. Bernard, J. Järvinen, V. V. Khmelenko and D. M. Lee, *Phys. Rev.* **B79**, 180506 (2009).
67. G. Breit and I. I. Rabi, *Phys. Rev.* **38**, 2082 (1931).
68. Yu. A. Dmitriev, *J. Low Temp. Phys.* **33**, 493 (2007).
69. C. Kittel and E. Abrahams, *Phys. Rev.* **90**, 238 (1953).
70. S. N. Foner, E. L. Cochran, V. A. Bower and C. K. Jen, *J. Chem. Phys.* **32**, 963 (1960).
71. K. Vaskonen, J. Eloranta, T. Kiljunen and H. Kunttu, *J. Chem. Phys.* **110**, 2122 (1999).
72. F.J. Adrian, *J. Chem. Phys.* **32**, 972 (1960).
73. R. Beringer and M. A. Heald, *Phys. Rev.* **95**, 1474 (1954).
74. P. Kusch, *Phys. Rev.* **110**, 1188 (1955).
75. V. Kiryukhin, E. P. Bernard, V. V. Khmelenko, R. E. Boltnev, N. V. Krainyukova and D. M. Lee, *Phys. Rev. Lett.* **98**, 195506 (2007).
76. J. Järvinen, J. Ahokas, S. Jaakkola and S. Vasilyev, *Phys. Rev.* **A72**, 052713 (2005).
77. A. Abragam, *Principles of Nuclear Magnetism,* Oxford University Press, (1961), p. 367.
78. Yu. Kagan and A. J. Leggett, *Quantum Tunneling in Condensed Media*. North-Holland, Amsterdam (1992).

CHAPTER 6

MATRIX ISOLATION SPECTROSCOPY IN SOLID PARAHYDROGEN: A PRIMER

Mario E. Fajardo

Munitions Directorate, US Air Force Research Laboratory AFRL/RWME, 2306 Perimeter Rd., Eglin AFB, Florida, 32542-5910, USA

E-mail: mario.fajardo@eglin.af.mil

Solid parahydrogen (pH_2) offers several advantages as a host for matrix isolation spectroscopy over conventional rare-gas solids. These arise from the special "quantum solid" nature of the pH_2 host, and include facile preparation of thick, optically transparent samples, enabling true isolation of dopants at vanishingly low concentrations; minimal inhomogeneous broadening of dopant rovibrational transitions allowing for high-resolution infrared spectroscopy; a greatly reduced cage effect opening up new dopant photochemical processes; and the participation of the delocalized excitations of the pH_2 host in a number of unique phenomena. The price of these advantages is additional complexity in both sample preparation and spectral data interpretation. This chapter is intended as a short introduction to the field, with an emphasis on practical lessons learned over the past two decades of research.

1. Introduction

This chapter is intended as a convenient starting point for anyone interested in learning more about matrix isolation spectroscopy (MIS) in solid parahydrogen (pH_2, see Ref. 1 annotation). In MIS experiments chemical impurities, or "dopants," are deliberately incorporated into a nominally inert cryogenic solid matrix host, and the resulting sample is characterized by spectroscopic means. The principal target audience includes students and researchers who are already familiar with MIS in rare-gas solid (RGS) hosts and are considering working with solid pH_2.

Clearly, a wide variety of spectroscopic techniques are applicable to MIS, for an equally broad range of scientific motivations. However, this chapter will focus almost exclusively on infrared (IR) absorption spectroscopy,[2] with which the author is most familiar. This will permit at least a brief *mention* of everything thought to influence IR absorption spectra in solid pH_2, and an emphasis on those cases having no familiar analog in RGS hosts. Readers with different scientific agenda are invited to consider if, and how, the phenomena described below may influence their own investigations. Space limitations preclude detailed discussions for every topic, engendering a heavy reliance on references to the appropriate literature. A handful of tables and plots of original data are included to illustrate specific points, along with a few personal anecdotes describing instances of transformational learning experiences.

Section 2 is primarily a listing and description of "essential references," recommended for every personal library. These include compendia of the extensive studies on the *pure* solid molecular hydrogens (SMH; H_2 and its isotopologes: D_2, HD, etc.) — as well as prior reviews of MIS in solid pH_2. The single most important fact to keep in mind about the SMH is the survival in the solid phase of the (nearly) free rotation of the individual molecules. In particular, the ground state $J = 0$ hydrogen molecules exist as spherical objects, and arrange themselves into close-packed solids, just as rare-gas atoms do.

Much of the early excitement surrounding the use of solid pH_2 as a matrix host arose from the observation of very sharp IR absorption lines for certain rovibrational transitions. The great precision of such measurements promises improved insights into a system's intermolecular interactions, structures, and dynamics. The reviews of this field capture its origins and scientific development, and discuss the special "quantum solid" properties of solid pH_2, which afford many advantages as a matrix host beyond just host compatibility with high-resolution IR spectroscopy.

Section 3 includes a survey of the extant experimental approaches for producing doped pH_2 solids, with comments on their relative advantages and disadvantages. Thereafter, the focus shifts primarily to results from H_2 solids produced by the rapid vapor deposition (RVD) technique, by which an unusually fast flow of cold pH_2 host gas (and an independent flow of vaporized dopant species) are cocondensed onto a conventional

MIS cryogenic substrate-in-vacuum. This provides the first of several opportunities to highlight the differences encountered between MIS in solid H_2 versus RGS hosts, e.g., the ubiquitous presence of highly mobile metastable $J = 1$ orthohydrogen (oH_2) impurities.[1] Unlike the spherically symmetrical $J = 0$ pH_2 molecules, the $J = 1$ oH_2 molecules have a nonzero rotationally averaged electrostatic quadrupole moment. The resulting long-range intermolecular interactions lead them to cluster together, and with dopant molecules, in low-temperature solids — with profound effects on the observed spectra.

One important practical advantage obtained in RVD solid pH_2 versus RGS matrix hosts is the superior sample optical clarity, which permits the use of millimeters- or even centimeters-thick matrices. This effect is attributed to the orders-of-magnitude larger crystallite grain sizes achieved in vapor depositions; i.e., ~100 μm in solid pH_2 versus ~100 nm in RGS. The attendant very low utilizable dopant concentrations, ~1 part-per-million (ppm), are in the true matrix isolation limit, allowing for the unambiguous identification of dopant monomer spectroscopic features. Trading increasing dopant concentration against decreasing optical pathlength, while maintaining roughly constant column density (i.e., concentration × pathlength), provides a controlled approach for the deliberate study of dopant–dopant cluster formation.

Section 4 begins with a mid-IR absorption survey spectrum of a pure pH_2 solid, showing the condensed-phase-induced H_2 features, as well as the quiet spectral regions available for studying direct dopant IR absorptions. Explaining the observed H_2 transitions and lineshapes requires concepts such as: nearly free molecular rotation, the Cancellation Principle, single- and double-transitions, phonon sidebands, and delocalized rotational and vibrational excitations. These ideas are also important to understanding the direct dopant, and dopant-induced, absorptions in solid pH_2; the latter being unknown in RGS matrices, where the hosts remain spectrally silent in the IR. The mobility of the energy carried by the delocalized "rotons" and "vibrons" leads to novel IR photoprocesses in doped SMH.

These intrinsic solid pH_2 absorptions also provide a means of determining an individual sample's thickness — a necessary first step towards measuring absolute dopant concentrations *via* the application of Beer's law.[2] This raises a personal pet theme: the notion that since dopant concentration is directly related to dopant clustering and to the mean separation between dopants, such quantification is valuable to the rigorous study of condensed-phase chemical reaction kinetics and dynamics, as well as energy relaxation and transfer phenomena.

Section 5 presents a discussion of IR spectroscopy in doped pH_2 solids. In MIS, the rotational dynamics of the dopant molecules has a profound influence on the appearance of a given vibrational absorption band. As this topic alone has already filled volumes, only a *very brief* mention will be made of the expected differences between the IR spectra of rotating versus non-rotating dopants. The reader is directed to references describing the models used to connect these spectra to the anisotropic dopant–host interactions, and to the interactions between dopant rotation–translation and host trapping-cage motions. This literature also includes excellent discussions of homogeneous versus inhomogeneous line-broadening effects in solid pH_2.

New spectra of CO_2/pH_2 solids containing varying amounts of residual oH_2 impurities are presented to illustrate the highly nonstatistical oH_2-dopant clustering accompanying the large mobility of the $J = 1$ excitations via "quantum rotational diffusion." Hopefully, this example of facile $(oH_2)_n$–CO_2 clustering will drive home the importance of working with the lowest achievable residual oH_2 concentrations, at least when spectra of truly isolated dopant molecules are desired. Even at oH_2 concentrations of only ~1%, the multitude of overlapping $(oH_2)_n$-CO_2 absorption features blend together into broad peaks, raising the interesting notion of inhomogeneous broadening as a discrete, countable entity.

Annealing of these same $CO_2/oH_2/pH_2$ solids reveals the existence of multiple trapping site structures for the CO_2 molecules. As-deposited pH_2 solids produced by RVD are polycrystalline, containing metastable face-centered-cubic (*fcc*) regions, as well as thermodynamically stable hexagonal-close-packed (*hcp*) regions. Even within *hcp* regions, the CO_2 molecules appear to occupy two distinct trapping sites.

Annealing also results in improved alignment of the *hcp* crystallites' unique *c*-axes with the substrate surface normal. One exciting (but unproven) possibility is that this signals the consolidation of the crystallites into much larger oriented single crystals. In any case, laboratory-frame alignment of the *hcp* crystallites can sometimes result in a pronounced polarization dependence of the IR absorptions. This effect can be strong enough that even nominally "unpolarized" IR spectra may depend noticeably on such usually mundane details as the angle between the IR beam propagation vector and the deposition substrate. This can complicate the comparison of spectra obtained in different laboratories, and is a good reason for encouraging the reporting of optical setup details in accounts of MIS studies in pH_2 solids.

The various dopant-induced IR absorptions known to occur in solid pH_2 are included as a final example of novel phenomena that may be unfamiliar to practitioners of MIS in RGS. These examples show that, in addition to probing the influences of the matrix environment on the dopant "solute" species, IR spectroscopy in solid pH_2 permits direct observation of the response of the host "solvent" as well.

2. Essential Literature References

2.1. *Pure solid molecular hydrogens*

It is only a small exaggeration to state that the pure SMH have been investigated using *every* known experimental and theoretical technique. This information can be a valuable resource to practitioners of MIS in solid pH_2, but the enormity of the literature demands efficient and selective access. Fortunately, a handful of well-written books and review articles provide a viable gateway to the data and knowledge most relevant to rovibrational spectroscopy in solid pH_2.

A good starting point is the 1980 review article by Silvera,[3] which covers a wide range of topics concerning the pure SMH. These include the properties of single hydrogen atoms and molecules, intermolecular interactions, free molecular rotations, ortho/para species, ortho/para nuclear spin conversion (NSC), quantum rotational diffusion of $J = 1$ impurities, crystal structures and phase transitions at low and high pressures, thermodynamic properties, and "renormalized" interactions in quantum solids.

The 1983 book by Van Kranendonk[4] is a theoretical *tour de force* which expands upon many of the same topics, and summarizes/analyzes the available nuclear magnetic resonance (NMR), microwave, IR, Raman, and neutron scattering spectroscopic data. He explicitly addresses the delocalized nature of rotational and vibrational excitations in SMH; the resulting rotons and vibrons are touted in the Preface as "…the simplest and most perfect examples of Bloch states in all of solid-state physics."[4] One chapter is devoted to an introduction to quantum solids, and the next to "renormalization" phenomena associated with the quantized lattice vibrations (phonons). This nomenclature arises by analogy to quantum electrodynamics, and includes both the calculation of "effective" intermolecular interactions as expectation values of the instantaneous interactions over the distribution of molecular positions, as well as dynamical "self-energy" effects associated with the absorption and emission of virtual phonons.

The 1986 book by Souers was written to provide future fusion engineers a comprehensive collection of physical properties data for the various hydrogen isotopologes in the gas, liquid, and solid phases. Incidentally, this information is also extremely valuable to anyone designing a SMH MIS experiment.[5] Because of the importance of the radioactive tritium isotope to nuclear fusion, nearly half of the book deals with the properties of, and processes in, irradiated SMH. This is critical background information for those interested in the extensive literature on MIS studies of ionic and radical species in irradiated SMH.

Part I of the 1997 book by Manzhelii and Freiman,[6] i.e., Chapters 1–9 titled "Quantum Molecular Crystals" authored by Manzhelii and Strzhemechny, deals almost exclusively with the SMH. While the subject matter overlaps strongly with the sources listed above, the text is exceptionally well written and translated. It also captures the benefits of more than a decade of continuing research. I have found it to be a valuable complementary presentation of this material, and have experienced several "Aha!" moments of improved comprehension by comparing these discussions with the earlier sources.

The extensive body of experimental and theoretical work dedicated to understanding the *absolute* intensities of IR absorptions in the SMH was reviewed in 2004 by Mishra, Balasubramanian, Tipping, and Ma.[7] By comparing multiple theories and experiments, those authors are able to comment on the consistency of the experimental data, and on those theoretical aspects in need of improvement. This manuscript also constitutes the most up-to-date compilation of IR spectroscopy of pure SMH available at this time.

As for the primary literature, the original full report by Gush, Hare, Allin, and Welsh on the IR absorptions of pure condensed-phase H_2 includes an excellent discussion of the "Cancellation Principle," and how it applies to single- but not to double-transitions.[8] A difficult to find 1984 manuscript by Rao, Gaines, Balasubramanian, and D'Cunha is well worth any interlibrary cooperation required for acquisition.[9] It amounts to a greatly simplified, and thus more easily digestible, description of the core subject matter in van Kranendonk's book. The 1998 report by Mengel, Winnewisser, and Winnewisser documents their comprehensive high-resolution,[10] high-sensitivity solid pH_2 spectral survey covering the 600–14000 cm^{-1} region, as well as their detection of a *triple* transition in solid pH_2, i.e., the simultaneous excitation of three neighboring H_2 molecules by a single photon. Finally, the 2003 communication by Hinde, on modeling of substitutional impurities in solid pH_2, distills the description of the IR absorptions in pure solid pH_2 down to two succinct paragraphs.[11] He provides an updated depiction of the Cancellation Principle, showing how symmetry breaking by dopants leads to additional IR activity, and relating the shapes and widths of the resulting IR absorptions to the strength of the dopant–pH_2 interactions.

2.2. *Solid pH_2 MIS reviews*

While the SMH have seen use as matrix hosts dating back to the dawn of MIS,[12] the modern era of MIS in solid pH_2 began in the late 1980s with the efforts of Oka and co-workers at the University of Chicago. The first half of Oka's 1993 review article deals with the application of high-resolution laser spectroscopy to pure pH_2 solids; providing a bridge to, and illuminating, the earlier IR and microwave studies.[13] The second half presents results from similar experiments on pH_2 solids containing isotopic substitutional dopants, i.e., D_2 and HD, as well as preliminary results on chemically distinct dopants such as CH_4 and CO, and ionic species produced by electron bombardment. Oka's masterful discussion of the various sources of line broadening in pure and doped pH_2 solids remains an excellent introduction to this material. This manuscript is also the first to connect the "quantum solid" nature of solid pH_2 to a list of advantages as a matrix host. The weak intermolecular interactions due to the spherical nature of the $J = 0$ pH_2 molecules, in combination with their low molecular mass, results in large quantum zero-point energy (ZPE) and zero-point motion (ZPM) effects. These quantum effects inflate the solid, producing spacious trapping sites within which small dopant molecules can "… vibrate and rotate nearly freely." The easy preparation of "… transparent and fairly stress-free crystals…" by gas condensation in an enclosed cell at $T \approx 7$ K is credited to "… delicate diffusion and tunneling…" during solidification. Oka ascribes the robustness of the pH_2 solids against intense laser irradiation to their high thermal conductivity, setting the stage for future nonlinear optical experiments.

Contemporary with this effort, our group at Edwards Air Force Base (AFB) was working on trapping metal atoms in normal-H_2 and normal D_2 solids (nH_2 and nD_2; room temperature equilibrium compositions: 3:1 oH_2:pH_2, and 2:1 oD_2:pD_2). Our 1994 review article on atom-doped SMH can be commended for its extensive list of references, and perhaps for the novel ultraviolet/visible (UV/Vis) spectra of Al and B atom-doped SMH matrices.[14] At that time, I had just started thinking about the desirability of working in solid pH_2 versus nH_2, and about the possibility of incorporating a modest resolution (~0.1 cm^{-1}) IR spectroscopic diagnostic into our experiment. We finally decided to design and build an ortho-H_2/para-H_2 converter ("o/p converter"), not to facilitate the IR work, but to allay our concerns that heat released by uncontrolled ortho/para NSC in the doped samples was limiting the achievable concentrations of isolated metal atoms.

The 1998 review by Momose and Shida at Kyoto University summarizes their high-resolution spectroscopic work on the CH_4/pH_2 system performed in collaboration with Oka's group, as well as their own home-grown successes at producing novel dopant species by in-situ photochemistry in precursor-doped pH_2 solids.[15] Photodissociation of dopants in RGS matrices is usually impeded by the so-called "cage effect," in which the rigid surrounding matrix host prevents the permanent separation of all but the smallest photolysis fragments (e.g., H and F atoms). Momose and Shida attribute the successful photolysis of

CH_3I and C_2H_5I in solid pH_2 to the large amplitude ZPM of the pH_2 host molecules, and the corresponding large compressibility of the pH_2 solid. They propose that the "... soft, compressible ..." pH_2 matrix permits separation of the photofragments, after which the solid's high thermal conductivity quickly freezes them in place.

The 2004 review by Momose, Hoshina, Fushitani, and Katsuki at the University of British Columbia summarizes a decade of experimental and theoretical work on the rovibrational spectroscopy of methane isotopologes (CH_xD_{4-x}) in solid pH_2.[16] This manuscript is an excellent entry point into the extensive literature on crystal field theory (CFT) treatments of the rotations of small molecules in anisotropic trapping environments in solids. It also includes a valuable discussion of the generalized dephasing processes which determine the homogeneous spectral linewidths for dopant rovibrational transitions in solid pH_2.

The 2006 review by Yoshioka, Raston, and Anderson at the University of Wyoming summarizes their work on the effects of oH_2–dopant clustering on the observed IR spectra, and on ortho/para H_2 NSC.[17] They show that these clusters are formed during sample deposition at $T \approx 2.5$ K, and again by thermal annealing at $T \approx 4.4$ K. This manuscript also includes data on rare-gas atom-induced IR activity, and IR-induced chemical reactions mediated by the H_2 vibron, demonstrating the importance of the delocalized vibrational excitations in doped pH_2 solids.

3. Production of Doped pH_2 Solids

3.1. *ortho/para hydrogen NSC*

A necessary step in any solid pH_2 MIS scheme is enrichment of the even-J component of the room temperature nH_2 gas. Figure 1 shows the odd-J fractions [$f(oH_2)$ and $f(pD_2)$] for *equilibrated* hydrogen and deuterium (eH_2 and eD_2) samples over the 10–120 K temperature range, calculated for a canonical ensemble of noninteracting rigid-rotors[1] with rotational constants[3] $B(H_2) = 59.3\ \text{cm}^{-1}$ and $B(D_2) = 29.7\ \text{cm}^{-1}$. While any H_2 or D_2 sample held at a fixed temperature must *eventually* come to equilibrium, the ortho/para NSC process is strongly forbidden for an isolated odd-J molecule, requiring interactions with an inhomogeneous magnetic field.

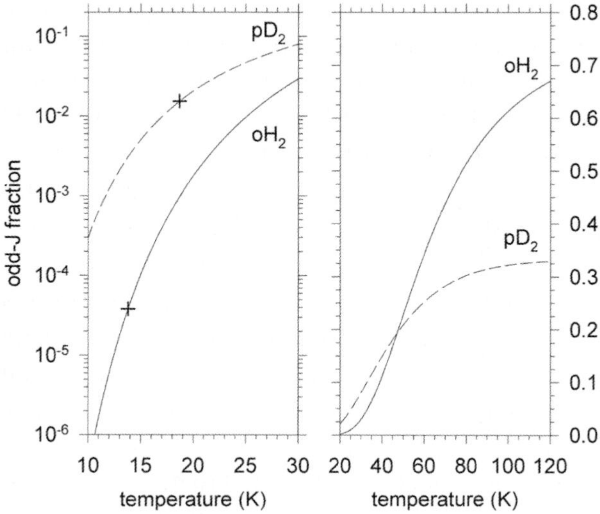

Fig. 1. Equilibrium odd-J fractions for eH_2 and eD_2 versus temperature.[1] The crosses indicate the compositions for ortho/para equilibrated hydrogens at their triple points: $T_t(eH_2) = 13.80$ K, $P_t(eH_2) = 7.03$ kPa, $f(oH_2) = 38$ ppm, and $T_t(eD_2) = 18.69$ K, $P_t(eH_2) = 7.13$ kPa, $f(pD_2) = 1.54\%$.[5]

When a sample of pure nH_2 is rapidly frozen and held at liquid helium (ℓHe) temperatures, the pH_2 molecules will quickly relax to the $J = 0$ ground state, and the oH_2 to the lowest energy odd-J level, i.e., $J = 1$. However, the intrinsic rate of ortho/para NSC in *pure* solid H_2 is extremely slow, and a significant fraction of the $J = 1$ oH_2 molecules can survive for remarkably long times.[3,4,6]

In pure solid H_2, the NSC reaction is autocatalytic, proceeding through interactions between $J = 1$ molecules. These $J = 1$ molecules, or more correctly the $J = 1$ excitations, are mobile primarily via the "quantum rotational diffusion" or "resonant conversion" mechanism, by which an oH_2–pH_2 pair converts into a pH_2–oH_2 pair. For oH_2 fractions $x > 0.05$, and for $T > 4$ K, the NSC kinetics follow the spatially homogeneous "well-stirred reactor" bimolecular rate law:[18] $dx/dt = -kx^2$; with rate constant $k \approx 0.019\ h^{-1}$. Integration of this law yields: $1/x = 1/x_0 + kt$, where x_0 is the initial ortho fraction. At lower temperatures and ortho fractions, deviations from this law appear due to the increased clustering of the oH_2 molecules. The net result is that it takes several weeks for a flash-frozen pure nH_2 solid to reach an oH_2 fraction ~1%; conversion times in solid nD_2 under similar conditions are considerably longer. Thus, practical preparation of $pH_2(oD_2)$ solids requires the action of a catalyst on a low-temperature $H_2(D_2)$ sample, which can be accomplished prior to solidification, or *in situ* in the condensed solid.

3.2. *Condensation of solid pH_2 in an enclosed cell*

The pioneering Chicago[13] and Kyoto[15] solid pH_2 MIS experiments were performed using pH_2 gas produced in "immersion wand" style o/p converters. In this approach,[3] a long tube containing a granular paramagnetic solid catalyst is cooled by insertion into a storage dewar containing liquid hydrogen (ℓH_2) or ℓHe. Several variations on this basic design, employing different NSC catalysts and temperature control schemes, have been described recently.[19–21] The lowest practical residual $J = 1$ concentrations are limited to ~10–100-ppm oH_2 (~1% pD_2) by the steep drop off in vapor pressure, and hence in mass throughput, with decreasing temperature below the triple point.

In the Chicago and Kyoto work, the emerging pH_2 enriched gas is warmed to room temperature and mixed with the desired dopant gas. The doped pH_2 solids are produced by condensation of this gas mixture inside an enclosed cylindrical copper cell cooled to $T \approx 7$–8 K. The growth of the solid is monitored visually through the cell's transparent end windows, and proceeds radially inward from the copper surface. Spectroscopic measurements indicate that the resulting solid consists of fairly strain- and defect-free *hcp* crystallites, with their *c*-axes oriented in the radial direction.

The enclosed-cell sample preparation method yields a number of important advantages. It produces beautifully transparent samples, with any voids or other optical imperfections confined to the cell's centerline. Cell lengths of ≈1–12 cm have been demonstrated; these long, known optical pathlengths provide good sensitivity for low-concentration or weakly absorbing dopants. These low-defect solids yield the sharpest recorded dopant IR absorption lines, demonstrating minimal inhomogeneous broadening as compared to other sample preparation methods. Doped samples can be studied at temperatures up to $T \approx 8$ K, limited only by dopant–dopant aggregation and phase separation. The laboratory-frame spatial orientation of the *hcp* crystallites facilitates the application of polarization-dependent spectroscopic techniques. In short, these samples constitute a well-defined benchmark against which the results of all other sample preparation methods must be judged.

These advantages must be weighed against two potentially significant disadvantages. First, dopant selection is limited almost exclusively to gaseous species that can be mixed in a room temperature container with the pH_2 gas, or to those that can be produced by photolysis or radiolysis in the pH_2 solid. Some limited successes at trapping small C_n clusters[22] and $(C_{60})_n$ clusters[23] have been achieved by intra-cell laser ablation of graphite and solid C_{60}, respectively. However, better results on both systems have been reported using other sample preparation methods.[23–25] Second, except for the very highest vapor pressure dopants (e.g., HD, D_2, CH_4) the dopant monomer isolation efficiency is very poor. This makes it difficult to know *a priori* if the observed spectral features are due to isolated dopant molecules or to dopant clusters.

3.3. *Rapid vapor deposition of solid pH_2*

At Edwards AFB, we designed and fabricated a different type of o/p converter; a flow-through device cooled by a dedicated 10-K closed-cycle He cryostat.[26,27] This apparatus delivers cold pH_2 gas directly to a deposition substrate-in-vacuum, which is cooled by an independent ℓHe bath cryostat. In our design, the granular solid NSC catalyst is contained in a long, thin copper tube that is wound about, and thermally anchored to, a copper bobbin mounted on the 10-K cryostat's cold tip. We routinely operated this o/p converter at $T \approx 15$ K [$f(oH_2) \approx 100$ ppm]. Operation just above the triple point, say at $T \approx 14$ K [$f(oH_2) \approx 45$ ppm], is also practical, but requires careful temperature control to avoid the unsteady flow regime that accompanies the freezing and thawing of the H_2 within the converter. More recently at Eglin AFB, we have built a redesigned ultrahigh vacuum compatible o/p converter,[28] and operated this system below the triple point, $T \approx 13$ K, [$f(oH_2) \approx 18$ ppm], at the price of somewhat reduced maximum H_2 flow rates.

In our earliest experiments at Edwards AFB,[24] we operated the o/p converter at conventional MIS host gas-flow rates $q_{host} \approx$ 1–5 mmol h^{-1} (sample thickness growth rates $r \approx$ 0.3–1.5 μm min^{-1}). For this we were rewarded with the same miserably poor optical quality, highly scattering matrices we had come to expect from our previous nH_2 and nD_2 MIS experiments.[14] Optical scattering in matrices is maximized when the sample contains crystallites, or other physical structures, with dimensions comparable to the wavelengths in question. Indeed, vapor-deposited SMH have long been notorious for their strong optical scattering properties:[5] "*... a rapid freeze from the gas produces snow ... a 1-mm-thick layer of hydrogen crystallites can be a completely opaque brown-black.*" We persisted in operating at these slow pH_2 flow rates for quite some time; not only because we expected unacceptable scattering losses in thicker samples, but also to keep the uncondensed H_2 gas pressure in the deposition cryostat below $P_{cryo} \sim 10^{-5}$ torr, thereby minimizing gas-phase reactions in our laser ablation metal atom dopant source.

Our "big breakthrough" at producing optically transparent doped pH_2 solids came quite unintentionally, when we decided to attempt the deposition of ~1-mm-thick samples. This was my response to skeptical questions about the prospects for scaled-up production of any advanced "rocket fuel" we might develop. I remember my sense of confusion, and thinking "well, *something*'s broken" during our first RVD experiment, performed at $q_{host} \approx$ 50 mmol h^{-1}. Several minutes into the deposition, the IR diagnostic showed intense solid pH_2 absorptions, yet the sample avoided our best efforts at visual observation! Eventually, with the help of a strong hand-held flashlight, we were able to make out the faint Fresnel reflections off the outer growing matrix surface – thereby fixing my broken cognitive process, and allowing me to accept the possibility of simultaneously thick *and* transparent H_2 solids. Since then, we have demonstrated the deposition of almost 1-cm-thick optically transparent solid H_2 samples, at flow rates up to $q_{host} \approx$ 300 mmol h^{-1} ($r \approx$ 70 μm min^{-1}), and at oH_2 concentrations up to 70%. Thus, pure pH_2 gas is not a requirement for producing transparent solids. Maximum sample growth rates in our apparatus are limited by the need to maintain an effective thermal isolation vacuum in the deposition cryostat ($P_{cryo} < 10^{-3}$ torr).

We have speculated that the excellent optical quality of our RVD pH_2 solids is due to the formation of *larger* crystallites than normally formed during slower depositions.[26] We proposed that local heating near the accreting surface during the rapid gas condensation process results in "self-annealing" of the samples. All our efforts at visual observation of such large crystallites using optical polarizers have failed, possibly due to strain-induced birefringence in the cryostat windows. It should be noted that this idea of fast depositions producing larger crystallites runs counter to past experience with vapor-deposited SMH[29] and RGS[30] produced in enclosed cells held at constant temperatures. These experiments support the conventional wisdom that faster depositions should result in *smaller* crystallites, due to the larger number of initial nucleation sites. Indeed, solid pH_2 crystallites significantly smaller than ~100 nm would also explain the observed reduced optical scattering at visible wavelengths.

However, optical shadowgraphs[29] of ≈50-μm-thick solid pH_2 films deposited at $r \approx 40$ μm min^{-1}, and at $T = 3.6$ and 7.0 K, *do* show an increase in average crystallite size from <3 μm to ~100 μm with increasing temperature. The temperatures throughout these relatively thin samples are expected to remain fairly close to that of the deposition substrate. We have used the IR absorptions of CO molecules isolated in ~1-mm-thick pH_2 solids as a "molecular thermometer" to measure the bulk temperature rise during the RVD process.[31] Starting from a base substrate temperature of $T_{sub} = 1.9$ K, a pH_2 flow of $q_{host} \approx 240$ mmol h^{-1} results in an increase to $T_{sub} = 2.9$ K, and a further rise in the outermost ≈1-mm-thick layer of a total ≈3-mm-thick CO/pH_2 solid to $T \approx 4.6$ K. The magnitude of the temperature rise at the accreting matrix surface itself is still unknown, but it is plausible that the elevated temperatures encountered during the RVD process result in the initial formation, and/or the ripening growth, of larger crystallites. This explanation for the observed sample transparency is tied closely to the discussion of spectroscopic measurements of the solid pH_2 microscopic structure that follows in Section 5.3. However, absent some independent measurement of the as-deposited crystallite grain sizes, these speculations on the *mesoscopic* structure of our RVD pH_2 solids remain unproven.

The main advantages of the RVD approach over the enclosed-cell condensation method are its compatibility with most known dopant production schemes, and excellent dopant monomer isolation efficiency for concentrations up to ~1000 ppm. These make the RVD approach much more versatile, and thus applicable to a wider range of MIS investigations. Direct deposition of the cold pH_2 gas eliminates the possibility of back-conversion to form additional oH_2 during storage and handling. It also means that any condensable impurities in the pH_2 host gas (e.g., air, H_2O, CO_2) will be trapped in the o/p converter, and the in-vacuum deposition allows any He impurity to be pumped away. The RVD method also retains all the benefits of excellent sample optical transparency; however, with variable optical pathlengths up to ≈1 cm.

The main disadvantages of the RVD approach versus the enclosed-cell method are tied to the more complex microstructure of the resulting pH_2 solid (*vide infra*); i.e., a mixed *fcc/hcp* as-deposited structure and higher internal strains and defect concentrations, resulting in increased inhomogeneous spectral broadening. Our use of simple ℓHe bath and ℓHe transfer cryostats, configured for maximum cooling of the deposition substrate, limits our solid pH_2 sample temperatures to $T < 5$ K. Our attempts to exceed this limit have all resulted in runaway solid pH_2 sublimation, leading to catastrophic sample loss.

3.4. *Alternative conversion/condensation schemes*

Even a cursory online literature search reveals considerable activity on solid pH_2 MIS during the present century. The goal of this section is to provide a handful of entry points to this current literature, so only the most recent publications from the various laboratories are listed, and each citation is presented with an implicit "and references therein." Readers interested in pursuing MIS experiments in solid pH_2 are well advised to examine these latest developments carefully before proceeding.

Other research groups have implemented both the RVD and enclosed-cell sample productions schemes with minor modifications. As already noted,[17] the RVD technique has been put to very productive use at the University of Wyoming.[32] Recently, the enclosed-cell method has been adopted at research laboratories in Hefei,[33] Berlin,[34] and Hong Kong.[35]

A different technique, employing pulsed deposition of room temperature dopant/pH_2 gas mixtures onto a cold substrate-in-vacuum, was developed in 2004 in Taipei.[36] This is a modern adaptation of the "pseudo matrix isolation" or "pulsed matrix isolation" (PMI) method.[37,38] The very high instantaneous incident flux of pH_2 gas permits depositions at substrate temperatures $T_{sub} \approx 5.5$ K. The IR absorption spectra of RVD and pulse-deposited CH_2/pH_2 samples are very similar, with both showing a mixed *fcc/hcp* as-deposited structure. This is perplexing, since T_{sub} is already ≈1 K higher than the annealing temperature required to eliminate the *fcc* regions in RVD samples. Maintaining these samples at $T_{sub} = 5.0$ K results in slow sublimation, with the measured solid thickness decreasing at a rate of ≈40 μm h^{-1}, consistent with the solid pH_2 vapor pressure ($P_{vap} \sim 10^{-4}$ torr) at that temperature.

Of course, the range of possible adaptations to o/p converters and deposition schemes is limited only by the special requirements and ingenuity of the workers in each laboratory. Recent improvements in closed-cycle ℓHe refrigerators have brought their "no-load" cold-tip temperatures down well below 4 K. Thus, both closed-cycle and ℓHe cryostats can be used for slow or fast pH_2 depositions in more-or-less conventional MIS geometries, as demonstrated in laboratories in Hefei,[33] Kharkov,[39] Lausanne,[40] Nagoya,[41] Charlottesville,[42] Lund,[43] Vancouver,[44] Taipei,[45] and Urbana.[46]

3.5. *In-situ ortho/para hydrogen conversion*

In most solid pH_2 MIS schemes, the required oH_2 NSC is performed prior to sample preparation. However, it is also possible to incorporate a paramagnetic dopant into nH_2 (or nD_2) solids to catalyze the NSC process *in situ*. The most popular dopant for this purpose is molecular oxygen, O_2, which has two unpaired electrons in its $^3\Sigma_g^-$ ground state.[2]

This approach has been used to study the rotational dynamics of O_2 in solid nD_2 via visible laser-induced fluorescence (LIF)[47] and electron spin resonance (ESR)[48] spectroscopies. Both reports invoke the rapid NSC of nearby $J = 1$ pD_2 molecules, leaving the O_2 surrounded by $J = 0$ oD_2. Both conclude that the O_2 molecules exist in the strongly hindered rotor or "librational" limit in distorted trapping cages.

NMR experiments on ~0.1% O_2/nH_2 solids show that an O_2 molecule induces NSC of a nearest neighbor $J = 1$ oH_2 molecule on a timescale ~10 s.[49] Thus, the overall catalyzed conversion rate is controlled by oH_2 diffusion, and the time to reach an oH_2 fraction ~1% is reduced to several days, instead of many weeks as for pure nH_2.

Very recently, even higher O_2 dopant concentrations of ≈0.5% have been employed in IR experiments on the NSC process in nH_2 solids.[50–52] At these O_2 levels, the time to reach ~1% residual oH_2 at $T \approx 4$ K is reduced to about 3 days.[52] Double-doping such samples with a small amount of another species, X, permits the study of $(oH_2)_n$–X clusters; initially for very large *n*, but with the average *n* decreasing with time. The main advantage of this approach is the effort and complexity saved by avoiding constructing and incorporating a separate o/p converter. The disadvantages include the long waiting times required to reach very low oH_2 levels, and the added trouble of identifying spectral features associated with $(O_2)_n$–X complexes.

4. IR Absorptions in Pure pH_2 Solids

4.1. *Spectroscopic assignments and delocalized excitations*

Figure 2 shows an IR survey spectrum of solid pH_2 covering the 800–6000 cm^{-1} region. The spectrum is actually a composite of spectra from two samples, obtained using different Fourier Transform IR (FTIR) spectrometer source/beamsplitter/detector combinations. The data are presented without any correction for scattering losses, demonstrating the excellent transparency of the nearly 3-mm-thick samples. The wide, flat window from ≈1200 to 4200 cm^{-1} is open for dopant spectroscopy with only minor interference from weak absorptions;[10] the small peak seen near 3765.5 cm^{-1} is due to an unintentional ~0.2-ppm H_2O impurity.[53]

As a homonuclear diatomic molecule, H_2 lacks a permanent electric dipole moment, so an isolated molecule has no electric dipole allowed rotational or vibrational transitions.[2] The absorptions in Fig. 2 are *not* the ~10^{-6} weaker electric quadrupole allowed transitions, but arise due to intermolecular interactions in the condensed phase.[4,8,9,11] The occurrence

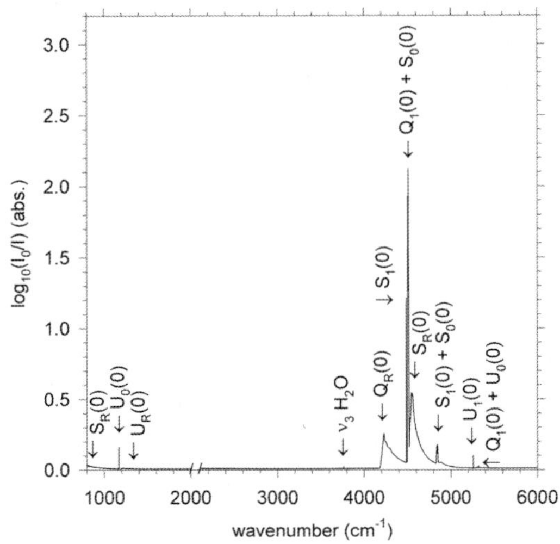

Fig. 2. Mid-IR survey spectrum of RVD pH_2 solids on a BaF_2 substrate at $T = 2.4$ K, presented at 0.1 cm^{-1} resolution. The low-wavenumber region is for an as-deposited 2.6-mm-thick 10 ppm $^{12}C^{18}O/pH_2$ sample, the rest for a nominally pure as-deposited 2.8-mm-thick pH_2 solid containing ~0.2-ppm H_2O as an unintentional impurity. The spectroscopic notation is explained in the annotation to Ref. 54.

of these absorptions at transition energies approximated by simple sums of the gas-phase H_2 vibration and rotation energies confirms the nearly free nature of the pH_2 molecules in the solid. The selection rules in the solid involve the combined symmetry of both the crystal and the pH_2 molecule.[4,13] Thus, the IR spectrum encodes information about the environments occupied by the pH_2 molecules.

The zero-phonon "single-transitions" are most sensitive to this effect, since the transition dipoles arising from the individual pairwise intermolecular interactions can sum to zero, as described by the Cancellation Principle.[8] For example, the observation of the $S_1(0)$ transition[2,54] near 4486 cm^{-1} requires that at least *some* of the pH_2 molecules reside in sites lacking true inversion symmetry; e.g., the D_{3h} symmetry *hcp* site. The nonobservation of a $Q_1(0)$ feature near 4150 cm^{-1} instead requires the sum of the vectors connecting each pH_2 molecule with its nearest neighbors to vanish; as happens for both the D_{3h} *hcp* and O_h *fcc* sites. Despite the delocalized nature of the $v = 1$ vibron and $J = 2$

roton excitations, the single-transition absorption features are typically sharp, due to the $\Delta\mathbf{k} = 0$ selection rule arising from exciton momentum conservation ($\mathbf{p} = \hbar\mathbf{k}$).[9,13]

The strongest features in the spectrum are the zero-phonon "double-transitions," in which two neighboring pH_2 molecules undergo simultaneous rovibrational transitions; e.g., $Q_1(0) + S_0(0)$ at 4495–4520 cm^{-1} and $S_1(0) + S_0(0)$ at 4828–4853 cm^{-1}. The double-transitions are not subject to cancellation, and so are relatively insensitive to the detailed structure of the solid. In a double-transition, two excitons with wavevectors $\mathbf{k}_1$ and $\mathbf{k}_2$ are created, and the full bandwidths are accessible, since momentum conservation is satisfied for all $\mathbf{k}_1 = -\mathbf{k}_2$. Thus, the widths of these features are very nearly the sum of the widths of the $v = 1$ vibron ($\approx$3 cm^{-1}) and $J = 2$ roton ($\approx$20 cm^{-1}) bands.[9] A given double-transition occurs at slightly higher energy than the corresponding single-transition [e.g., $\Delta v = +1$ and $\Delta J = +2$ for both $Q_1(0) + S_0(0)$ and $S_1(0)$] due to the *intramolecular* rotation–vibration interaction.[4]

Associated with each single- and double-transition is a $\approx$400-cm^{-1} wide phonon sideband, such as the $Q_R(0)$ and $S_R(0)$ features peaking near 4230 cm^{-1} and 4550 cm^{-1}, respectively. The $S_R(0)$ sideband seen growing in at the lowest wavenumbers in Fig. 2 is associated with the pure rotational $S_0(0)$ line[43] near 356 cm^{-1}. Since these features are many times broader than the $v_D \approx 80$ cm^{-1} Debye frequency of solid pH_2,[3] they correspond to the creation of one or more (apparently up to five)[4] phonons in concert with the rovibrational excitation.

The IR energy absorbed into the delocalized hydrogen vibron and roton excitations can lead to novel effects in *doped* SMH. For example, the reaction between ground state chlorine atoms and H_2 molecules, $Cl(^2P_{3/2}) + H_2(v = 0) \rightarrow HCl + H$, is endothermic by $\approx$360 cm^{-1}, so Cl atoms are stable in solid pH_2.[17] However, the same reaction with $H_2(v = 1)$ is exothermic by $\approx$3800 cm^{-1}, and IR excitation of the $v = 1$ vibron in Cl/pH_2 solids results in the formation of HCl.[17] A similar effect is observed in formic acid (FA, HCO_2H) doped nH_2 solids, where the action spectrum for generation of the higher energy *cis*-FA conformer tracks precisely the solid nH_2 IR absorption spectrum.[55] It is important to keep these photoprocesses in mind when considering the possible effects on a doped SMH sample of IR irradiation by a spectrometer, or a hot evaporative dopant source. This is especially important, since the closest analogy in RGS matrices, the delocalized electronic excitations,[56] are only accessible by much higher energy (e.g., vacuum ultraviolet, VUV) irradiation.

4.2. *Sample thickness determination*

We can exploit the insensitivity of the zero-phonon double-transitions to crystal microstructure to determine the thicknesses of both as-deposited and annealed pH_2 solids.[57] Our IR data are presented as decadic absorbance spectra: $A_{10}(\bar{v}) \equiv \log_{10}[I_0(\bar{v})/I(\bar{v})]$. Table 1 shows various functions of the $Q_1(0) + S_0(0)$ and $S_1(0) + S_0(0)$ absorption band integrated intensities, abbreviated as: $\int A_{Q+S} d\bar{v}$ and $\int A_{S+S} d\bar{v}$, respectively. Note that the integration baselines are not simply $A_{10} = 0$; rather, they are sloping line segments drawn between the values of $A_{10}(\bar{v})$ at the integration endpoints listed in the legend for Table 1.[57] These integrated intensities are calibrated against interferometrically determined values of the solid pH_2 sample thicknesses, d.

The values in Table 1 originally reported in Ref. 57 were derived from spectra recorded using an ℓN_2 cooled HgCdTe detector operated at IR intensities that were clearly in a nonlinear regime. In my efforts to account for the anticipated spectrometric errors,[57] I apparently missed a systematic error introduced in the phase correction step of the FTIR data processing procedure.[58] Thus, the sample thicknesses calculated by the approach presented in Ref. 57 (including all our results published prior to 2009) are too large by ≈10%. The new recommended values in Table 1 are derived from spectra recorded over the intervening years with a different FTIR spectrometer, employing an ℓN_2 cooled InSb detector with superior linearity characteristics.[59] So, I now recommend that the thickness of a pH_2 solid be calculated as: $d(\text{cm}) = \int A_{Q+S} d\bar{v}/90$, or $d(\text{cm}) = \int A_{S+S} d\bar{v}/7$, with estimated errors less than ±3% (95% confidence).

We have also recorded numerous spectra using a broadband antireflection-coated ≈2.1-μm long-wavelength-pass Ge filter. Such spectra show very poor signal-to-noise (S/N) ratios in the $Q_1(0) + S_0(0)$ and $S_1(0) + S_0(0)$ regions, so an alternative method for estimating those sample thicknesses is required. As it happens, the peak absorbance of the $Q_R(0)$ feature, $\Delta A_{10}[Q_R(0)] \equiv A_{10}(4228.5\ \text{cm}^{-1}) - A_{10}(4100\ \text{cm}^{-1})$, is an

Table 1. Solid pH_2 thickness determination.

Quantity	Reference 57	Recommended Values
$\int A_{Q+S}\, d\bar{\nu} / \int A_{S+S}\, d\bar{\nu}$	13.0 ± 0.4	12.9 ± 0.2
$\int A_{Q+S}\, d\bar{\nu} / d$	82 ± 3 cm^{-2}	90 ± 2 cm^{-2}
$\int A_{S+S}\, d\bar{\nu} / d$	6.3 ± 0.3 cm^{-2}	7.0 ± 0.2 cm^{-2}
$\Delta A_{10}[Q_R(0)]/d$	–	0.90 ± 0.03 cm^{-1}

"Q+S" = $Q_1(0) + S_0(0)$; 4495 – 4520 cm^{-1}.

"S+S" = $S_1(0) + S_0(0)$; 4825–4855 cm^{-1}.

$\Delta A_{10}[Q_R(0)] \equiv A_{10}(4228.5\ \text{cm}^{-1}) - A_{10}(4100\ \text{cm}^{-1})$.

The stated uncertainties indicate the 95% confidence limits.

excellent proxy for the sample thickness:[59] $d(\text{cm}) = 1.11\Delta A_{10}[Q_R(0)]$. This approach yields nearly the same level of accuracy as the integrated double-transitions, but is much simpler and quicker to use.

5. IR Absorptions in Doped pH_2 Solids

5.1. *Rotating versus non-rotating dopants*

The rotational dynamics of a dopant molecule in a matrix influences both the overall and detailed appearance of the IR absorption bands. The subject of molecular rotations in condensed phases is a long-standing and still active area of research.[60] In solids, two limiting cases are fairly well understood: (i) strongly hindered "libration" or torsional oscillation with respect to one or more orientational energy minima, and (ii) "nearly free" rotation with only minor host-induced perturbations to the gas-phase case. Most molecules in most solids do not rotate, but exist as librators. However, some small molecules with large rotational constants (typically hydrides) *can* rotate in weakly interacting environments — the SMH being a prime example. Intermediate cases show more complex dynamics, including quantum tunneling through rotational barriers, and thermally promoted reorientation between the librational minima.

In a nutshell, the IR spectrum of a nearly free rotor in a cryogenic matrix can be understood as a low-temperature gas-phase rovibrational spectrum,[2] with additional splittings and line broadening due to the dopant–host interactions. The low sample temperatures typically result in only one or two thermally populated low-lying rotational levels, allowing for the observation of at best a handful of transitions. Additional transitions are seen for dopants with equivalent H atoms, such as CH_4/pH_2,[15,16,44] and H_2O/pH_2,[53] in which metastable high-lying rotational levels remain populated due to incomplete NSC of the dopant.

Typically for librators, all of this spectral structure collapses down into a single line near each vibrational origin. This highlights an important difference between MIS of rotating dopants in SMH versus RGS matrices: the possibility of clustering with residual oH_2 molecules. At the ≈3.8 Å solid pH_2 nearest-neighbor separation, the magnitude of the quadrupolar electric field of a $J = 1$ oH_2 molecule is $|\mathbf{E}| \sim 10^6$ V cm^{-1}. This is sufficient to strongly perturb the rotational dynamics of a neighboring dopant, even converting some nearly free rotors into librators (e.g., H_2O/nH_2).[50,52]

5.2. *IR spectra of rotating dopants in solid pH_2*

Space limitations compel me to direct the interested reader's attention to the existing literature on this topic. We included a short review of this field in a recent publication describing our experiments and theoretical modeling of rovibrational spectra of CO molecules isolated in solid pH_2.[61] As mentioned above, Ref. 16 includes a more thorough review, and describes the state-of-the-art in the application of CFT to nearly freely rotating molecules in quantum solids. Intermediate cases, for which the IR spectra do not reflect either a clear gas-phase rotational dynamics or a simple librational collapse, are currently of great interest.

5.3. *IR Spectra of non-rotating dopants in solid pH_2*

Dopants in sufficiently anisotropic trapping environments will exist as librators. This can occur due to strong repulsive *or* attractive dopant–pH_2 interactions, related to the dopant's size or electrostatic properties, respectively. As noted above, interactions between dopants and residual

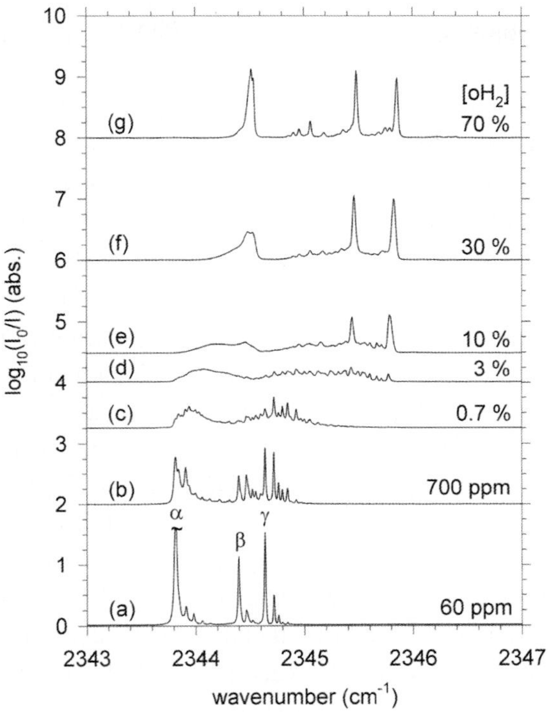

Fig. 3. oH_2 concentration dependence of the v_3-mode CO_2 absorption spectra in $^{12}C^{16}O_2/oH_2/pH_2$ samples, as-deposited on a BaF_2 substrate at $T = 2.4$ K. The spectra are presented at 0.005 cm^{-1} resolution. Trace (a) is for a 1.7-mm-thick 1.2-ppm CO_2 sample, trace (b) for a 1.2-mm-thick 1.4-ppm CO_2 sample; trace (c) for a 1.5-mm-thick 1.1-ppm CO_2 sample; trace (d) for a 1.1-mm-thick 1.5-ppm CO_2 sample; trace (e) for a 1.2-mm-thick 1.4-ppm CO_2 sample; trace (f) for a 1.4-mm-thick 1.2-ppm CO_2 sample; and trace (g) for a 1.1-mm-thick 1.3-ppm CO_2 sample.

oH_2 molecules can also result in libration. Figure 3 shows IR spectra of as-deposited $CO_2/oH_2/pH_2$ samples produced by the RVD method. The oH_2 fractions are calculated from the temperature of the o/p converter. We previously presented IR spectra of CO_2/pH_2 samples at low-oH_2 concentrations,[31] and a similar sequence of spectra of $^{12}C^{18}O_2/oH_2/pH_2$ samples has appeared very recently.[33]

The main α, β, and γ peaks in Fig. 3a are due to isolated non-rotating CO_2 molecules in three distinct trapping sites in the as-deposited solid.[31]

However, even at the lowest oH_2 concentrations, each of these peaks shows small, blue-degraded satellites, which were not definitively assigned in our original report. These features grow in both size and number with increasing oH_2 fraction, and hence are assigned here to $(oH_2)_n$–CO_2 clusters. The extensive clustering seen in Fig. 3b, for an oH_2 concentration of only 700 ppm, clearly demonstrates that the oH_2 molecules are not simply randomly distributed in the pH_2 solid. In that case, only ~1% of the CO_2 molecules should have one or more oH_2 nearest neighbors. Rather, the mobile $J = 1$ oH_2 excitations appear to be finding, and preferentially binding with, the nonrotating CO_2 molecules.

The complexity of the IR spectra of $(oH_2)_n$–dopant clusters depends on a number of factors, including the underlying dopant rotational dynamics, and the variety of distinct cluster structures. The cleanest examples involve a nonrotating dopant with a strong permanent electrostatic dipole moment, e.g., $CH_3F/oH_2/pH_2$.[17,62] Actually, CH_3F can rotate about its symmetry axis, but this axis can only librate; so most IR bands match those of a nonrotating molecule.[45] The strong electric dipole–quadrupole interaction between CH_3F and oH_2 dictates the cluster structure, and only one or two peaks are observed for each $(oH_2)_n$–CH_3F cluster. At the other extreme, the individually rotating oH_2 molecules in the oH_2–oH_2 pair are coupled primarily by the weaker electric quadrupole–quadrupole (EQQ) interaction,[3–6] and the solid pH_2 lattice accommodates several distinct cluster structures. The complexity of the high-resolution oH_2–oH_2 pair IR spectrum is remarkable, containing literally hundreds of well-resolved (and assigned!) lines.[63]

Since CO_2 possesses a nonzero electrostatic quadrupole moment (but no dipole), the binding in $(oH_2)_n$–CO_2 clusters is due mainly to the EQQ interaction. Admixtures of quadrupoles and “noninteracting” spherical species (e.g., oH_2 and pH_2 molecules), constrained to occupy close-packed three-dimensional lattices, are easily frustrated,[3,6,64] existing at low temperatures in many nearly iso-energetic configurations. At oH_2 concentrations of ~1%, the CO_2 IR spectrum is smeared out into two broad lumps; presumably composed of hundreds of overlapping sharp lines. With the further addition of oH_2, the spectrum becomes simpler, reverting to a pattern of three main peaks, but now with red-degraded satellites (perhaps signifying interloping pH_2 neighbours?).

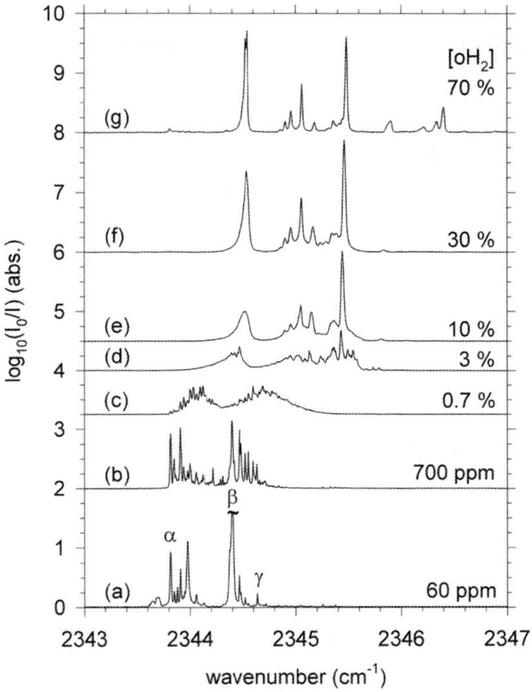

Fig. 4. Effects of annealing on the $CO_2/oH_2/pH_2$ samples depicted in Fig. 3. The spectra were obtained at $T = 2.4$ K after annealing at $T = 4.8$ K for 1 h.

Figure 4 shows the effects of annealing on the spectra of these same RVD $CO_2/oH_2/pH_2$ samples. At the lowest oH_2 concentration, the α line diminishes in intensity, the β line grows substantially, and the γ line nearly disappears. These changes can be understood by considering the annealing-induced structural changes in RVD pH_2 solids.[26,27,65]

A direct comparison of IR spectra of CH_4/pH_2 samples produced by the enclosed-cell and RVD methods[65] shows the as-deposited RVD pH_2 solids to be a mixture of *fcc* and *hcp* regions, with the *fcc* regions virtually disappearing upon annealing for several minutes at $T > 4$ K. These observations and assignments are confirmed independently by Raman spectroscopy of RVD pH_2 solids.[27,29] Additionally, we have detected only very weak polarization dependences in IR spectra of as-deposited samples, consistent with the absence of any significant laboratory-frame orientation of the nascent *hcp* regions.

Now, the *hcp* and *fcc* structures are just two of an infinite number of possible close-packed solids, all sharing the same maximum bulk density for packing of spherical objects, but differing in the stacking order of close-packed planes.[4,66] The *fcc* structure (…ABCABCABC…) has the highest symmetry, with only one pH_2 molecule in the unit cell, and each successive close-packed plane maintaining the same translation vector established by the previous two. The *hcp* structure (…ABABAB…) has two molecules in the unit cell, and each successive close-packed plane reverses its predecessor's lateral displacement. An isolated deviation from an *fcc* or *hcp*-stacking sequence is a "stacking fault." One can imagine close-packed solids with any combination of *fcc* or *hcp* stacking, in repeating or random patterns, with any arrangement of stacking faults.

Aside from polarization-dependent measurements, the IR and Raman spectral diagnostics are sensitive only to very short-range intermolecular interactions. If they are in fact limited to nearest neighbor interactions, then they could not distinguish between a mixture of true *fcc* and *hcp* crystallites versus a mixture of much smaller regions containing local *fcc* stacking (XABCX, "*fcc*-like") and local *hcp* stacking (XABAX, "*hcp*-like"). This leads to two extreme idealized mesoscopic structures for our as-deposited pH_2 solids: (a) an aggregate of distinct *fcc* and *hcp* crystallites in which ordered stacking persists over dimensions larger than optical wavelengths, or (b) a random-stacked close-packed solid within which *fcc*-like and *hcp*-like layers exist between stacking faults occurring randomly at the molecular level.

The strong polarization-dependent features observed in IR spectra of *annealed* RVD pH_2 solids indicate a strong alignment of the *hcp* crystallites' *c*-axes with the deposition substrate surface normal.[65] We verified these older Edwards AFB results in a new apparatus at Eglin AFB, in which the pH_2 samples are grown on a reflective metal substrate, and then characterized by polarized IR absorption spectroscopy (PIRAS).[28,61,67] The belated annealing-induced alignment in the RVD samples matches that known for enclosed-cell pH_2 samples, in which the *hcp* crystallites *grow* inward from the cell walls with *c*-axes oriented in the radial direction. The physical mechanism by which annealing of the RVD samples results in crystallite alignment has not been determined.

With this picture of the possible structures for RVD pH_2 solids in mind, we can assign the CO_2/pH_2 spectrum as follows. The γ peak in Fig. 3a is due to CO_2 molecules initially trapped in the metastable *fcc*-like sites. The α and β peaks are due to CO_2 molecules in two distinct trapping sites in *hcp*-like regions. The most likely candidates being the "in-plane" (ip) and "out-of-plane" (oop) double-substitutional vacancies; so named because the missing pair of pH_2 nearest neighbors are removed from the same, or adjacent, close-packed planes, respectively. The v_3 mode of CO_2 is the asymmetric stretch of the linear molecule, so its transition dipole moment lies along the main molecular axis. In these experiments, the focused IR beam passes through the pH_2 solid, and transparent BaF_2 substrate, with its Poynting vector[68] (**S**) roughly parallel to the substrate surface normal vector. Since light propagates as a transverse electromagnetic wave [i.e., $\mathbf{S} = (c/4\pi)\,\mathbf{E} \times \mathbf{B}$], the electric field vector **E** in our nominally unpolarized IR beam is always roughly perpendicular to the substrate surface normal, and hence to the *hcp* crystallites *c*-axes in the annealed sample. The increase in the β peak upon annealing identifies it with the perpendicularly polarized absorption expected for the ip trapping site, and the α peak is assigned to the remaining oop site.[31] A similar assignment to ip and oop double substitutional trapping sites holds for the N_2O/pH_2 system.[69]

I was surprised to finally realize that such "hidden" polarization dependencies could complicate the comparison of IR spectra obtained in different apparatus. For example, in our report from Edwards AFB on water molecules trapped in solid pH_2,[53] I mistakenly assigned several minor peaks in spectra of annealed samples to surviving *fcc*-like structures. Recent PIRAS data from Eglin AFB reveal that these features are actually due to crystal-field-split transitions of water molecules in *hcp*-like trapping sites.[67] Universal reporting of these experimental details necessary to uniquely specify the geometry of the IR beam Poynting vector with respect to the pH_2 sample would help to avoid similar confusion in the future.

5.4. *Dopant-induced IR absorptions in solid pH_2*

The mere presence of even an IR inactive, readily accommodated, substitutional impurity in crystalline solid pH_2 can influence the sample's IR absorption spectrum. The dopant breaks the local symmetry for

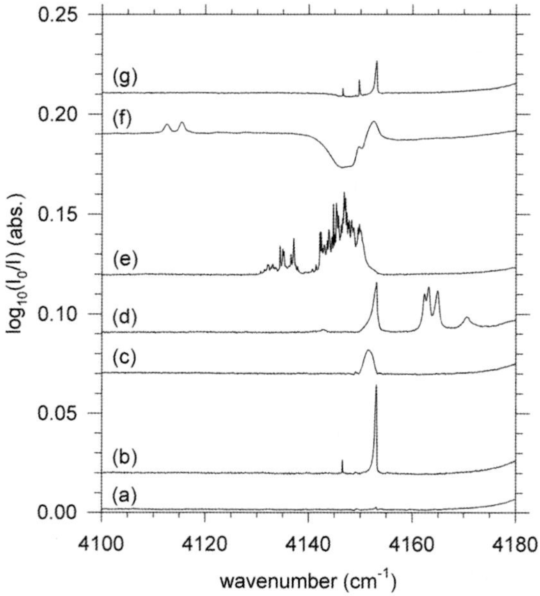

Fig. 5. Dopant-induced IR absorptions in pH_2 solids: (a) pure 2.8-mm-thick pH_2 sample depicted in Fig. 2; (b) 3.6-mm-thick 2500-ppm oH_2 sample; (c) 2.6-mm-thick 1000-ppm Ar sample; (d) 2.8-mm-thick 1000-ppm N_2 sample; (e) 1.6-mm-thick tropolone-doped sample; (f) 1.4-mm-thick sample doped by 308-nm laser ablation of solid aluminum; and (g) electron-bombarded 0.7-mm-thick pH_2 sample. The spectra are presented at 0.1 cm^{-1} resolution, except for trace (f), which was recorded at 1.0 cm^{-1} resolution. Traces (a)–(f) are single-pass absorptions of samples deposited on a transparent BaF_2 substrate, trace (g) is the 45°-incidence double-pass absorption of a sample deposited onto a polished metal substrate under bombardment by 5-keV electrons. All samples are shown as-deposited at $T = 2.4$ K, except for trace (d), which is undergoing annealing at $T = 4.8$ K.

nearby pH_2 molecules, unbalancing the delicate cancellation of short-range intermolecular interactions.[11,17] It also breaks the long-range translational symmetry of the lattice, modifying the exciton momentum conservation selection rules.[13] Some dopants can also participate in cooperative transitions with the pH_2 host,[27,53,70,71] reminiscent of the double-transitions in pure solid pH_2. Large,[23] complex,[27,62] or ionic[13,20,27,72,73] dopants can induce IR activity well outside the normal $v = 1$ vibron band spectral region. Figure 5 shows examples of dopant-induced IR absorptions for all of these cases; beginning with trace (a) highlighting the quiet 4100–4180 cm^{-1} region in pure solid pH_2 just to the red of the $Q_R(0)$ phonon sideband feature.

The most well-studied example of dopant-induced IR activity in solid pH_2 is the $Q_1(0)$ absorption associated with isolated oH_2 impurities.[8,9,13] As shown in Fig. 5b, at low-oH_2 concentrations this asymmetric, red-degraded feature peaks at 4153.1 cm^{-1}, then tails off to nil intensity near 4150 cm^{-1}. This absorption is best understood as a zero-phonon double-transition, in which the vibrational excitation of a pH_2 molecule is accompanied by an orientational transition of an oH_2 neighbor.[9] Symmetry breaking by the oH_2 relaxes the $\Delta\mathbf{k} = 0$ selection rule, and the entire ≈3 cm^{-1} wide $v = 1$ vibron band becomes allowed.[13] The asymmetric lineshape maps out the subset of the $v = 1$ vibrons enjoying nonzero spatial overlap with the oH_2 quadrupolar electrostatic field.[11,71] Also apparent in Fig. 5b is the sharp $Q_1(1)$ absorption at 4146.56 cm^{-1} with the $v = 1$ excitation localized on the oH_2 molecules themselves. At higher oH_2 fractions, interactions involving multiple oH_2 molecules cause additional features to appear, and the $Q_1(0)$ and $Q_1(1)$ absorptions broaden to encompass the 4135–4175 cm^{-1} region.[8,26]

The integrated intensity of the oH_2-induced $Q_1(0)$ absorption is related to the oH_2 and pH_2 fractions, f_o and f_p, through a variant of Beer's law:[8,26] $\int A_{10}[Q_1(0)]d\bar{\nu}/d = af_o^2 + bf_of_p = bf_o + (a-b)f_o^2$. The nonlinear dependence on f_o underscores the role of $(oH_2)_n$ clusters at higher oH_2 fractions. In our original o/p converter report,[26] we gave $b \approx 30$ cm^{-2}; still an acceptable effective value for $f_o > 0.01$. However, interest in lower oH_2 concentrations ($f_o < 0.005$), and our revised sample thickness determination method, justify a new value for b. Integration limits of 4151–4154 cm^{-1}, with a sloping integration baseline connecting the integration endpoints, yield $b = 35 \pm 3$ cm^{-2} (95% confidence).[59]

Other simple substitutional dopants can also induce IR activity within the $v = 1$ vibron band region. Spherical species, such as rare-gas atoms and $J = 0$ molecules, can only perturb nearby pH_2 molecules *via* extremely short-range overlap interactions.[11,17,53,70,71] The intensities and shapes of the resulting IR absorption features depend on the details of the dopant–pH_2 interactions. Figure 5c shows the absorption induced by Ar atoms, which spans the entire $v = 1$ vibron band. Stronger interactions (e.g., with Kr and Xe) lead to sharper lines, as the nearest neighbor pH_2

vibrations are "detuned" further from the main vibron band, localizing the $v = 1$ excitation in the vicinity of the dopant.[11,17]

Small rotating dopants, such as CO,[27] N_2,[27,71,73] H_2O,[53] and HCl,[70] can additionally participate in cooperative transitions, in which the $v = 1$ vibron excitation is accompanied by a pure rotational transition of the dopant. This has the net effect of up-shifting the (vibron-perturbed) dopant rotational spectrum from the far-IR spectral region into the mid-IR. While thermal population of low-lying dopant rotational levels can lead to IR activity at energies below the $v = 1$ vibron region, e.g., the weak $Q_1(0)H_2 + O_0(2)N_2$ transition[71,74] near 4142 cm^{-1} in Fig. 5d, most of the cooperative transitions appear at higher energies. In the case of H_2O/pH_2, the $Q_1(0)H_2 + (2_{02} \leftarrow 0_{00})H_2O$ transition occurs at 4217.5 cm^{-1}, nearly at the peak of the pH_2 $Q_R(0)$ phonon sideband.[53]

Larger, more complex dopants, such as methanol,[27] C_{60},[23] and CH_3F,[62] result in very elaborate induced-IR absorption spectra. Figure 5e shows the forest of peaks induced by an undetermined concentration of tropolone ($C_7H_6O_2$) doped in solid pH_2. The origin of the structure in these spectra is still unclear. I can only continue to speculate that the complicated line patterns somehow reflect the variety of different chemical environments (manifested as different Stark shifts for the different local electrostatic fields?) experienced by the nearby pH_2 molecules. This explanation is motivated by studies of ionic dopants in solid pH_2,[13,20,27,72,73] which include several Stark-shifted[75] absorptions well to the red of the $v = 1$ vibron region, as shown for likely Al-containing positive ions[20] near 4114 cm^{-1} in Fig. 5f.

The strangest spectrum I have encountered in solid pH_2 is the asymmetric derivative lineshape "interference" feature sometimes observed in samples doped by laser ablation of metal targets;[76] (e.g., Fig. 5f in this chapter, and Fig. 7 in Ref. 27). This spectrum at first appears to exhibit negative absorption in the 4140–4150 cm^{-1} region. As it turns out, the apparent zero-absorbance baseline is actually the top of a broad featureless absorption due to the lowest particle-in-a-box transition of excess electrons localized in cavity-like states. The decrease in absorption is due to interference between the $v = 1$ vibron transition, and the $1p \leftarrow 1s$ transition of the trapped electron. This interference feature is also just discernable in Fig. 5g, which shows the spectrum of a pH_2

sample deposited during simultaneous 5-keV electron bombardment of the metal deposition substrate. Figure 5g also shows the $Q_1(0)$ and $Q_1(1)$ peaks from oH_2 formed by back-conversion during the deposition, and the sharp $\Delta\mathbf{k} = 0$ feature peaking at 4149.68 cm^{-1} induced by the macroscopic electrostatic field produced by the trapped charges.[72]

6. Summary, Omissions, and Future Directions

A summary of this chapter was presented up front in the Introduction. Hopefully, after making it this far, the reader has gained an appreciation for the range of concepts employed in understanding IR spectra in doped pH_2 solids; borrowed from the fields of molecular spectroscopy and chemical- and solid-state-physics. More complete and definitive explanations of these phenomena can be found in the cited references.

My narrow focus on rovibrational spectroscopy has caused me to ignore large research areas relevant to MIS in solid pH_2, for which I apologize. Perhaps the worst omission is the extensive literature on ESR spectroscopic studies of quantum diffusion of H atoms, and tunneling chemical reactions, in solid pH_2. A 1991 review article and 2004 book by Miyazaki summarize this field.[77,78] A good entry point to more recent literature on applications of ESR spectroscopy in solid pH_2 is provided in Ref. 41. Readers interested more generally in tunneling phenomena in SMH can also examine the 1998 review by Meyer,[79] and the 2005 review by Momose, Fushitani, and Hoshina.[80]

The modern era of MIS in solid pH_2 began with a series of laser-based high-resolution spectroscopic studies in Oka's laboratory. I neglected the beautiful nonlinear optics (NLO) high-resolution spectroscopic studies of Hakuta and co-workers,[81,82] since they were performed exclusively on pure pH_2 single crystals grown carefully from the melt. This work has inspired other efforts to exploit the NLO properties of pH_2 solids produced by gas condensation in enclosed cells.[83,84] Refs. 34 and 40 describe the results of *ultrafast* laser-based spectroscopic studies of pure and doped pH_2 solids.

Many of the broad lines of research on the SHMs laid out in Refs. 3–6 have continued to the present. For example, efforts are still underway to produce *metallic* solid hydrogen at cryogenic temperatures and megabar pressures.[85,86] The various experimental and theoretical techniques described elsewhere in this book, to the extent that they have not already been applied to MIS in solid pH_2, point the way to the future of this field.

My own crystal ball gives me only generic answers to such questions. Most likely, established lines of research in solid pH_2 will continue alongside the newer efforts noted in Sections 3.4 and 3.5. Those mentioned in this chapter include: (i) high-resolution IR spectroscopy of $(oH_2)_n$–dopant and dopant–dopant van der Waals clusters, (ii) spectroscopic, photophysical, and photochemical processes involving the delocalized roton and vibron excitations, (iii) rotational dynamics of dopants intermediate between free rotors and librators, and (iv) NSC processes of both $J = 1$ hydrogens and dopants in SHMs. At Eglin AFB, we hope to exploit the aligned *hcp* pH_2 crystallites found in annealed RVD samples as high-efficiency moderators for converting fast nascent positrons (i.e., antielectrons) into more manageable slow positrons.[87] We also plan to apply positron-based gamma-ray spectroscopies[88] to illuminate the structures in our pH_2 solids. Ultimately, the range of possible MIS investigations in solid pH_2 is limited only by the imaginations and capabilities of the involved researchers.

Acknowledgments

I thank all my co-workers for their contributions to our various solid pH_2 projects over the years, especially, at Edwards AFB: S. Tam, M. E. DeRose (née Cordonnier), and M. Macler; at Eglin AFB: C. M. Lindsay and C. D. Molek, and our university collaborators: T. Shida, T. Momose, D. T. Anderson, R. J. Hinde, and L. Andrews. I thank J. L. Jordan for a critical reading of this manuscript.

References

1. D. A. McQuarrie, *Statistical Mechanics*. Harper & Row, New York (1976). The homonuclear H_2 molecule exists in two forms: "para" (pH_2) with paired nuclear spins ($I = 0$) and even rotational quantum numbers ($J = 0, 2, \ldots$), or "ortho" (oH_2) with $I = 1$ and odd-J. The two forms interconvert very slowly in the absence of a catalyst. The ortho label denotes the species with the highest nuclear spin multiplicity [$g_I = \Sigma_I(2I + 1)$]. Thus, oD_2 denotes molecules with $I = 0$ or 2, and even-J, and pD_2 denotes $I = 1$ and odd-J. This often confusing nomenclature persists, despite repeated recommendations to refer simply to "even-J" and "odd-J" species.
2. P. F. Bernath, *Spectra of Atoms and Molecules*, 2nd ed., Oxford (2005).
3. I. F. Silvera, *Rev. Mod. Phys.* **52**, 393 (1980).
4. J. Van Kranendonk, *Solid Hydrogen: Theory of the Properties of Solid H_2, HD, and D_2*, Plenum Press, New York (1983).
5. P. C. Souers, *Hydrogen Properties for Fusion Energy*, University of California Press, Berkeley (1986).
6. V. G. Manzhelii and Y. A. Freiman (editors), *Physics of Cryocrystals*, AIP Press, Woodbury, NY (1997).
7. A. P. Mishra, T. K. Balasubramanian, R. H. Tipping and Q. Ma, *J. Mol. Struct.* **695–696**, 103 (2004).
8. H. P. Gush, W. F. J. Hare, E. J. Allin and H. L. Welsh, *Can. J. Phys.* **38**, 176 (1960).
9. K. N. Rao, J. R. Gaines, T. K. Balasubramanian and R. D'Cunha, *Acta Physica Hungarica* **55**, 383 (1984).
10. M. Mengel, B. P. Winnewisser and M. Winnewisser, *J. Mol. Spectrosc.* **188**, 221 (1998).
11. R. J. Hinde, *J. Chem. Phys.* **119**, 6 (2003).
12. A. M. Bass and H. P. Broida (editors), *Formation and Trapping of Free Radicals*, Academic, New York (1960).
13. T. Oka, Annu. Rev. *Phys. Chem.* **44**, 299 (1993).
14. M. E. Fajardo, S. Tam, T. L. Thompson and M. E. Cordonnier, *Chem. Phys.* **189**, 351 (1994).
15. T. Momose and T. Shida, *Bull. Chem. Soc. Jpn.* **71**, 1 (1998).
16. T. Momose, H. Hoshina, M. Fushitani and H. Katsuki, *Vib. Spectrosc.* **34**, 95 (2004).
17. K. Yoshioka, P. L. Raston and D. T. Anderson, *Int. Rev. Phys. Chem.* **25**, 469 (2006).
18. F. Schmidt, *Phys. Rev.* **B10**, 4480 (1974).
19. A. M. Juarez, D. Cubric and G. C. King, *Meas. Sci. Technol.* **13**, N52 (2002).

20. L. Andrews and X. Wang, *Rev. Sci. Instrum.* **75**, 3039 (2004).
21. S. R. Lydzinski, D. Celik, A. Hemmati and S. W. Van Sciver, *AIP Conf. Proc.* **823**, 475 (2006).
22. M. Miki, T. Wakabayashi, T. Momose and T. Shida, *J. Phys. Chem.* **100**, 12135 (1996).
23. N. Sogoshi, Y. Kato, T. Wakabayashi, T. Momose, S. Tam, M. E. DeRose and M. E. Fajardo, *J. Phys. Chem.* **A104**, 3733 (2000).
24. S. Tam, M. Macler and M. E. Fajardo, *J. Chem. Phys.* **106**, 8955 (1997).
25. H. Hoshina, Y. Kato, Y. Mosisawa, T. Wakabayashi and T. Momose, *Chem. Phys.* **300**, 69 (2004).
26. S. Tam and M. E. Fajardo, *Rev. Sci. Instrum.* **70**, 1926 (1999).
27. M. E. Fajardo and S. Tam, *J. Chem. Phys.* **108**, 4237 (1998).
28. C. M. Lindsay and M. E. Fajardo, HCN/pH_2 PIRAS, manuscript in preparation.
29. G. W. Collins, W. G. Unites, E. R. Mapoles and T. P. Bernat, *Phys. Rev.* **B53**, 102 (1996).
30. J. A. Venables and B. L. Smith, *Crystal Growth and Crystal Defects*, In: M. L. Klein and J. A. Venables (editors), *Rare Gas Solids Vol. II,* Academic Press, London (1977).
31. S. Tam and M. E. Fajardo, *Fiz. Nizk. Temp.* **26**, 889 (2000) [*Low Temp. Phys.* **26**, 653 (2000)].
32. P. L. Raston and D. T. Anderson, *J. Chem. Phys.* **126**, 021106 (2007).
33. W. Deng, G. Xu, L. Wan, A. Liu, B. Gao, J. Du, S. Hu and Y. Chen, *Acta Phys. Chim. Sin.* **24**, 1329 (2008).
34. F. Königsmann, M. Fushitani, N. Owschimikow, D. T. Anderson and N. Schwentner, *Chem. Phys. Lett.* **458**, 303 (2008).
35. M. C. Chan, Y. Song and L. Yan, *Chem. Phys. Lett.* **468**, 166 (2009).
36. Y. J. Wu, X. Yang and Y. P. Lee, *J. Chem. Phys.* **120**, 1168 (2004).
37. M. M. Rochkind, *Anal. Chem.* **39**, 567 (1967).
38. R. N. Perutz and J. J. Turner, *J. Chem. Soc. Faraday Trans.* 2 **69**, 452 (1973).
39. N. N. Galtsov, A. I. Prokhvatilov, G. N. Shcherbakov and M. A. Strzhemechny, *J. Low Temp. Phys.* **29**, 784 (2003).
40. L. Bonacina, P. Larrégaray, F. Van Mourik and M. Chergui, *J. Chem. Phys.* **125**, 054507 (2006).
41. T. Kumada, Y. Shimizu, T. Ushida and J. Kumagai, *Rad. Phys. Chem.* **77**, 1318 (2008).
42. X. Wang, L. Andrews, I. Infante and L. Gagliardi, *J. Am. Chem. Soc.* **130**, 1972 (2008).
43. J. Ceponkus, P. Uvdal and B. Nelander, *J. Phys. Chem. A* **112**, 3921 (2008).
44. Y. Miyamoto, M. Fushitani, D. Ando and T. Momose, *J. Chem. Phys.* **128**, 114502 (2008).
45. Y. P. Lee, Y. J. Wu and J. T. Hougen, *J. Chem. Phys.* **129**, 104502 (2008).

46. B. A. Tom, S. Bhasker, Y. Miyamoto, T. Momose and B. J. McCall, *Rev. Sci. Instrum.* **80**, 016108 (2009).
47. A. V. Danilychev, V. E. Bondybey, V. A. Apkarian, S. Tanaka, H. Kajihara and S. Koda, *J. Chem. Phys.* **103**, 4292 (1995).
48. T. Kumada, *J. Chem. Phys.* **117**, 10133 (2002).
49. V. Shevtsov, E. Ylinen, P. Malmi and M. Punkkinen, *Phys. Rev.* **B62**, 12386 (2000).
50. L. Abouaf-Marguin, A. M. Vasserot, C. Pardanaud, J. Stienlet and X. Michaut, *Chem. Phys. Lett.* **454**, 61 (2008).
51. L. Abouaf-Marguin and A. M. Vasserot, *Chem. Phys. Lett.* **460**, 82 (2008).
52. L. Abouaf-Marguin, A. M. Vasserot and C. Pardanaud, *J. Chem. Phys.* **130**, 054503 (2009).
53. M. E. Fajardo, S. Tam and M. E. DeRose, *J. Mol. Struct.* **695-696**, 111 (2004).
54. R.S. Mulliken, *Phys. Rev.* **36**, 611 (1930); the standard notation "$L_v(J)$" labels rovibrational transitions to vibrational level v, originating from $v = 0$ and rotational level J, with L = O, P, Q, R, S, and U for $\Delta J = -2, -1, 0, +1, +2$, and $+4$, respectively.
55. K. Marushkevich, L. Khriachtchev and M. Räsänen, *Phys. Chem. Chem. Phys.* **9**, 5748 (2007).
56. N. Schwentner, E. E. Koch and J. Jortner, *Electronic Excitations in Condensed Rare Gases*, Springer-Verlag, Berlin (1985).
57. S. Tam and M. E. Fajardo, *Appl. Spectrosc.* **55**, 1634 (2001).
58. D. B. Chase, *Appl. Spectrosc.* **38**, 491 (1984).
59. M. E. Fajardo, Solid pH_2 thickness revisited, manuscript in preparation.
60. L. Pauling, *Phys. Rev.* **36**, 430 (1930).
61. M. E. Fajardo, C. M. Lindsay and T. Momose, *J. Chem. Phys.* **130**, 244508 (2009).
62. K. Yoshioka and D. T. Anderson, *J. Chem. Phys.* **119**, 4731 (2003).
63. Y. Zhang, T. J. Byers, M. C. Chan, T. Momose, K. E. Kerr, D. P. Weliky and T. Oka, *Phys. Rev.* **B58**, 218 (1998).
64. J. A. Hamida, N. S. Sullivan and M. D. Evans, *Phys. Rev. Lett.* **73**, 2720 (1994).
65. S. Tam, M. E. Fajardo, H. Katsuki, H. Hoshina, T. Wakabayashi and T. Momose, *J. Chem. Phys.* **111**, 4191 (1999).
66. C. Kittel, *Introduction to Solid State Physics,* 6th ed., Wiley, New York (1986).
67. M. E. Fajardo and C. M. Lindsay, *J. Chem. Phys.* **128**, 014505 (2008).
68. F. L. Pedrotti and L. S. Pedrotti, *Introduction to Optics*, Prentice-Hall, Englewood Cliffs, NJ (1987).
69. B. D. Lorenz and D. T. Anderson, *J. Chem. Phys.* **126**, 184506 (2007).
70. D. T. Anderson, R. J. Hinde, S. Tam and M. E. Fajardo, *J. Chem. Phys.* **116**, 594 (2002).
71. R. J. Hinde, D. T. Anderson, S. Tam and M. E. Fajardo, *Chem. Phys. Lett.* **356**, 355 (2002).
72. T. Momose, C. M. Lindsay, Y. Zhang and T. Oka, *Phys. Rev. Lett.* **86**, 4795 (2001).

73. L. Andrews and X. Wang, *J. Chem. Phys.* **121**, 4724 (2004).
74. A. P. Mishra and T. K. Balasubramanian, *J. Chem. Phys.* **129**, 194503 (2008).
75. J. D. Poll and J. L. Hunt, *Can. J. Phys.* **63**, 84 (1985).
76. S. K. Bose and J. D. Poll, *Can. J. Phys.* **65**, 58 (1987).
77. T. Miyazaki, *Rad. Phys. Chem.* **37**, 635 (1991).
78. T. Miyazaki (editor), *Atom Tunneling Phenomena in Physics, Chemistry and Biology,* Springer-Verlag, Berlin (2004).
79. H. Meyer, *J. Low Temp. Phys.* **24**, 381 (1998).
80. T. Momose, M. Fushitani and H. Hoshina, *Int. Rev. Phys. Chem.* **24**, 533 (2005).
81. K. Hakuta, M.Suzuki, M. Katsuragawa and J. Z. Li, *Phys. Rev. Lett.* **79**, 209 (1997).
82. K. Kuroda, A. Koreeda, S. Takayanagi, M. Suzuki and K. Hakuta,, *Phys. Rev.* **B67**, 184303 (2003).
83. M. Fushitani, S. Kuma, Y. Miyamoto, H. Katsuki, T. Wakabayashi, T. Momose and A. F. Vilesov, *Opt. Lett.* **28**, 37 (2003).
84. B. J. McCall, A. J. Huneycutt, R. J. Saykally, C. M. Lindsay, T. Oka, M. Fushitani,Y. Miyamoto and T. Momose, *Appl. Phys. Lett.* **82**, 1350 (2003).
85. M. I. Eremets, V. V. Struzhkin, H. Mao and R. J. Hemley, *Physica.* **B329–333**, 1312 (2003).
86. S. Deemyad and I. F. Silvera, *Phys. Rev. Lett.* **100**, 155701 (2008).
87. C. D. Molek, C. M. Lindsay and M. E. Fajardo (unpublished).
88. Y. C. Jean, P. E. Mallon and D. M. Schrader (editors), *Principles and Applications of Positron & Positronium Chemistry*, World Scientific, Singapore (2003).

CHAPTER 7

MATRIX ISOLATION SPECTROSCOPY IN HELIUM DROPLETS

Kirill Kuyanov-Prozument,[1] Dmitry Skvortsov,[2] Mikhail N. Slipchenko,[3] Boris G. Sartakov,[4] and Andrey Vilesov

Department of Chemistry, University of Southern California, Los Angeles, CA 90089, USA
E-mail: vilesov@usc.edu

Helium nanodroplets constitute an ideal medium for isolation and spectroscopy of molecules and clusters. Low-temperature (0.37 K) and superfluid state of the droplets as well as the weakness of the interaction of helium atoms with encapsulated species facilitate narrow infrared spectra, which often have resolved rotational structure. Some unique features of the helium matrix isolation technique are exemplified by our recent study of the interaction of molecular rotation with superfluid helium, measurements of infrared intensities in small water and ammonia clusters, study of the evolution of ammonia from single molecules to bulk as well as measurements of relative energy of conformers in 2-chloroethanol molecules.

1. Introduction

Extensive experiments conducted in the past decade have demonstrated helium droplets to be the *ultimate* matrix, characterized by extremely low temperatures, a small matrix broadening and unique possibilities to

---

[1] Present address: Department of Chemistry, Massachusetts Institute of Technology, Cambridge, MA 02139

[2] Present address: DCG Systems, Inc., 45900 Northport Loop East, Fremont, CA 94538

[3] Present address: Weldon School of Biomedical Engineering, Purdue University, West Lafayette, IN 47907

[4] Permanent address: A. M. Prokhorov General Physics Institute Russian Academy of Sciences, Vavilov str. 38, 119991 Moscow, Russia

synthesize new molecular complexes.[1–6] In this experimental technique, molecules are captured by superfluid helium droplets having temperature of 0.37 K and studied via laser spectroscopy. He droplets have some important complements to the more traditional cryogenic solid matrix. The latter serves to minimize the mobility of species preventing uncontrollable agglomeration of the studied molecules at low temperature. In liquid helium, this issue is addressed via encapsulation of molecules in myriads of isolated droplets in the beam. In addition, molecules remain highly mobile in the superfluid helium droplets even close to absolute zero temperature. Therefore, if several molecules are captured by the same droplet, they recombine in the interior of the droplet. The number of the captured species per droplet, and therefore the sizes of the formed clusters, is given by the Poisson distribution, i.e., at much better control as compared to solid matrices. However, helium droplet experiment is also more challenging. It involves beams of droplets inside a high-vacuum apparatus. Moreover, the optical density in the beam is far too low to apply directly absorption spectroscopic methods, calling for the use of laser light sources and special techniques for detection. The large portion of these techniques relies on evaporation of He atoms upon absorption of radiation.

In this chapter, the unique features of helium matrix isolation technique are exemplified by study of the interaction of molecular rotation with superfluid helium, measurements of infrared intensities in small water and ammonia clusters, study of the evolution of ammonia from single molecules to bulk, as well as measurements of relative energy of conformers in 2-chloroethanol (2-CLE) molecules. More details on He droplet technique as well as reviews of the experimental results can be found in Refs. 1–6.

Schematic of a helium droplet experiment is shown in Fig. 1. Helium droplets are produced in free jet expansion of gaseous He at temperature T_0 = 20–6 K and at source pressure of P_0 = 5–20 bar through a 5-μm diameter nozzle. The expanding He gas in the jet is cooled and condenses into droplets. The formed droplets enter the high-vacuum part of the apparatus, where they undergo further cooling by evaporation until they reach the measured temperature of 0.37 K.[7,8] After the droplets have captured one or more molecules in the pickup cell they are exposed to

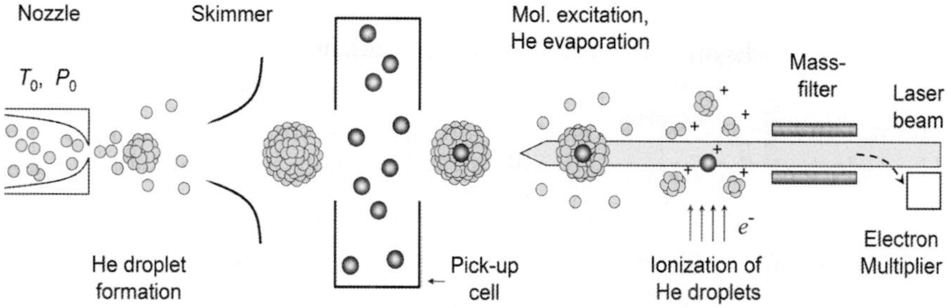

Fig.1. Schematic diagram of an apparatus for spectroscopy in helium droplets.

the laser beam, which enters the apparatus coaxially and counter-propagates to the beam of droplets, providing an optimal overlap of the two beams. Upon the absorption of a laser photon, the droplet size decreases due to evaporative loss of helium atoms as the photon energy, deposited in the chromophore, relaxes and temporarily heats the droplet. The absorption signal is detected as a decrease (depletion) of the mass spectrometer signal. Most of the infrared spectra shown in this chapter are obtained using a pulsed optical parametric oscillator–amplifier laser setup.

2. Quantum Matrix

Helium is an exceptional substance, which remains liquid down to absolute zero temperature. Moreover, at temperatures lower than about 2 K ^{4}He is superfluid, where it lacks viscosity and can be described by a macroscopic wavefunction.[9] Superfluidity has important implications for the spectra of the embedded molecules. Figure 2a shows the spectrum of the v_3 band of the CH_4 molecules in ^{4}He droplets.[10] The spectrum has well-resolved rotational structure, which indicates that molecules rotate coherently over many periods. For comparison, Fig. 2b shows the absorption spectrum of a dilute solution of CH_4 molecules in liquid argon at $T = 82$ K.[11] The v_3 band appears as a broad (~20 cm^{-1}) envelope, indicating that the rotation is quenched by interaction with the thermal bath. It would be desirable to compare molecular spectra in ^{4}He at different temperatures below and above the superfluid transition.

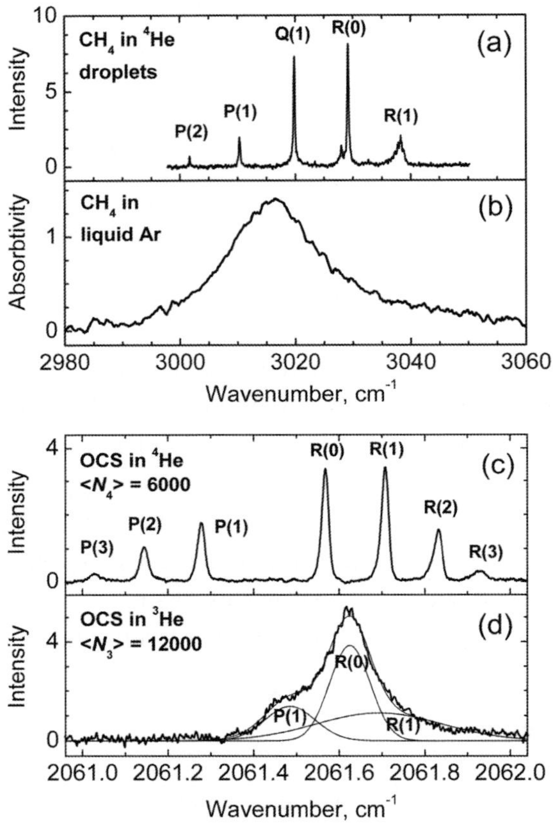

Fig. 2. Spectra of the v_3 mode of CH_4 in ^{4}He droplets (a) and in liquid argon at 82 K (b). Spectra of the v_3 mode of OCS in ^{4}He (c) and in ^{3}He droplets (d).

Unfortunately, the temperature of the droplets cannot be sufficiently changed due to extremely fast evaporative cooling.[12]

In order to gauge for the effect of superfluidity of the droplets, molecular spectra have been also measured in ^{3}He droplets[13]. ^{3}He droplets have lower temperature of 0.15 K,[13] but are still normal fluid.[9,14] Figures 2c and d show spectra of the OCS molecules in ^{4}He and ^{3}He droplets, respectively.[8,13] Once again, spectrum (c) in superfluid ^{4}He droplets shows rotational structure. The intensity distribution of the rotational lines in spectrum (c) gives the temperature of the ^{4}He droplets

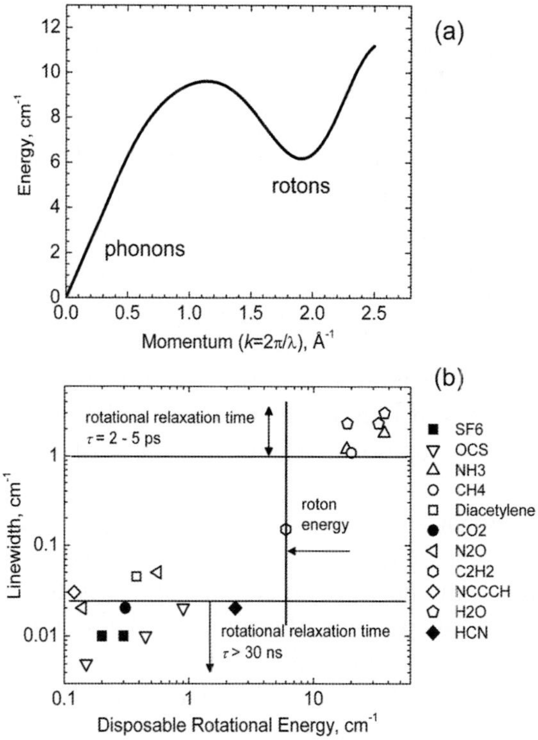

Fig. 3. (a) Spectrum of the elementary excitations in liquid ^{4}He. (b) Linewidth of the ro-vibrational lines of different molecules in ^{4}He droplets versus the disposable rotational energy.

of $T = 0.37$ K.[8] For comparison, the population of the rotational levels of the much lighter CH_4 in Fig. 2a is determined by the abundance of the corresponding nuclear spin states of CH_4 molecules at room temperature, which remain unrelaxed in He droplets. Spectrum (d) in ^{3}He droplets lacks any rotational structure and thus resembles that in liquid Ar rather than in ^{4}He. The drastic difference between spectra (c) and (d) confirms that the superfluidity indeed enables free molecular rotation in liquid ^{4}He.

The long lifetime of the molecular rotation in liquid helium is in agreement with the sharp spectrum of the elementary excitations in superfluid ^{4}He, which is shown in Fig. 3a. At low excitation energy it has a phonon branch. Coupling of molecular rotation with He environment originates from the short range van der Waals interaction between the molecules and He atoms of the droplet. Due to the long wavelength of the phonons they couple ineffectively with molecular rotation, which enables observation of the narrow rotational lines of heavy molecules in He droplets, as seen in Fig. 3b. In light rotors such as CH_4 the rotational energy may exceed the energy required for creation of the roton excitations in liquid helium. Rotons are elementary excitations with energy of about 6 cm^{-1} and a short wavelength ($\lambda \approx 3.3$ Å),[9] which enables efficient coupling to molecular rotation. Therefore, the rotational relaxation of light molecules possibly involves creation of rotons. This conjecture is in agreement with observation of broadening of the rotational lines of light molecules such as HF,[16] NH_3,[17] H_2O,[18] and CH_4,[10,19] which have rotational energy in excess of the roton gap (see Fig. 3b).

The ro-vibrational structure of the molecular spectra can be nicely fit using Hamiltonians characteristic for free molecules.[1,2,4,6] It suggests that the symmetry of the rotor in liquid helium is the same as in the free space. This is another manifestation of the softness of the liquid helium matrix, which assumes the shape of the embedded molecule, because the interaction between the He atoms is much weaker than that of He and embedded species. In contrast, in a solid matrix solute molecule is immersed in the "crystal field" which is a source for sizeable inhomogeneity. The latter causes broadening and frequently site splitting of the spectral lines of molecules in a solid matrix as well as significant shift of spectral lines. In spite of the weakness of the interaction of molecules with its ^{4}He solvent, it is not trivial. It was obtained that for heavy molecules, such as SF_6,[7,15] OCS,[8] and CO_2,[20] the effective moment of inertia of the molecules in liquid helium, I_{He}, is larger by about a factor of 3 as compared to those for free molecules. In contrast, very small increments of I_{He} of less than 5% have been observed for light molecules, such as CH_4,[10,19] NH_3,[17] H_2O,[18] and C_2H_2.[21]

The renormalization of the molecular moments of inertia (MOI) implies coupling of the molecular rotation with some states of the liquid helium surrounding. Recently observed satellite band in the infrared laser excitation spectra of CO_2 molecules in helium droplets was assigned to simultaneous ro-vibrational excitation of a molecule and its first helium solvation shell.[22] However, the mechanism of the coupling remains poorly understood. Studies of the effective MOI of molecules in small He clusters show that it levels out upon completion of the one or two solvation shells, i.e., in clusters having between 15 and 60 He atoms for molecules such as CO_2, N_2O, and OCS.[23–27] Nevertheless, this number is already far too large for the conventional basis set type molecular spectroscopic calculations. Therefore, spectroscopic constants of the above molecules in helium have been calculated by the projector operator imaginary time spectral evolution method,[24,26,28–31] in good agreement with experiments. Additional MOI for heavy rotors was also quantitatively accounted for by a rigid coupling of the molecular rotation to the nonsuperfluid helium density in the first solvation shell.[28–30] Theoretical calculations have shown that the helium density, which is inhomogeneous in the vicinity of molecule, cannot adiabatically follow the motion of the rapidly rotating light molecules.[28] In the case of fast rotating molecule the modulation of helium density is reduced. Apparently, the renormalization of MOI in liquid helium depends primarily on two factors. One is the strength of the interaction between a molecule and He atoms, which in turn determines the modulation of the local helium density close to the molecule. The other factor is the magnitude of the molecular rotational constant, which determines the speed of molecular rotation and thus adiabatic/nonadiabatic character of the coupling between the molecule and its helium environment.[10]

The rotational structure often remains unresolved in the spectra of larger molecules or clusters due to small spacing of the rotational lines. Such spectra consist of rather sharp vibrational bands. Large number of the studied systems indicate that the shift of the vibrational frequency in liquid helium is usually of the order of few wavenumbers.[1,2,4,6] Consequently, the frequencies in He, in the gas phase, and those obtained from quantum chemical calculations can be directly compared. Thus, He droplets indeed comprise an ultimate weakly interacting quantum matrix for spectroscopy of molecules and molecular clusters. Some experiments from our group illustrating this point will be discussed in the following sections.

3. Spectra and Infrared Intensities of Small Clusters

Here we demonstrate the utility of the He droplet technique for measurements of absolute infrared intensities in ammonia and water clusters as a function of size.[32–34] Water and ammonia clusters are textbook examples of hydrogen bonded (HB) systems, which have been extensively studied previously via spectroscopic experiments[35–46] and theoretical calculations.[42,47–51] The frequencies of OH and NH stretch bands in clusters have been measured in molecular beams[35,37–39,41] and in cryogenic matrices.[42,44] The absorption strength of vibrational bands, usually referred to as infrared intensity (IRI), is a complimentary spectroscopic observable, which can be used as a probe of the HB. At present, numerical calculations are the main source of our information on the IRIs in clusters.

There are few direct measurements of the IRIs in molecular clusters, largely because of the difficulties in determining the absolute number densities.[52,53] Strong enhancement of the infrared band intensities of different HB clusters suspended in a cryogenic matrix was discovered about 50 years ago.[54,55] However, the broadening of the spectra and complicated kinetics of the cluster formation in solid matrices did not allow for attainment of the IRI for specific cluster sizes.

Figures 4a and b show the spectra of ammonia and water in He droplets, respectively. Narrow lines in Fig. 4a are assigned to the ro-vibrational transitions of the $v_1(a_1)$ and $v_3(e)$ fundamental stretching bands as well as of the $2v_4(l = 2)$ overtone of the $v_4(e)$ bending band of single NH_3 molecules.[17] Ro-vibrational lines of the v_3 band of water molecules are assigned in Fig. 4b.[18] The strong broader bands in Figs. 4a and b are assigned to the donor bridge H stretch bands in the corresponding clusters. The spectra of water clusters in He droplets have been studied earlier.[38,41]

As a matter of fact He droplet is an isothermal nanocalorimeter, whereby the amount of the absorbed energy is related to the number of evaporated He atoms via binding energy of about 7.2 K (1 K = 0.69 cm^{-1} = 8.6 × 10^{-5} eV = 2 cal mol^{-1}). The integral of the depletion signal over the vibrational band of clusters comprising n molecules I_n is

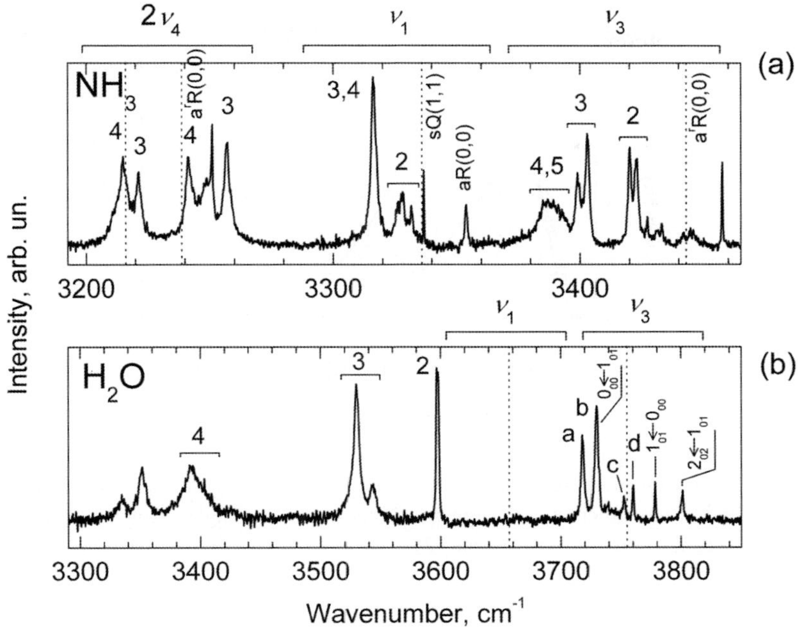

Fig. 4. Depletion spectra of NH_3 (panel a) and H_2O (panel b) single molecules and clusters in He droplets of about 3500 atoms with an average number of about 1.5 and 2 captured molecules per droplet, respectively. Vertical dashed lines give the frequencies of the vibrational band origins of monomers. Additional strong bands are assigned to the $(NH_3)_n$ and $(H_2O)_n$ clusters, which have been enhanced by hydrogen bonding. Cluster size, n, is indicated by numbers. Bands labeled a and b, c, and d in panel (b) are assigned to free OH stretches in water trimers and dimers, respectively.

proportional to the energy of the absorbed photons $h\nu_n$, the *IRI* of the cluster band A_n, and the abundance of the clusters $[M_n]$:

$$I_n = C \cdot h\nu_n \cdot A_n \cdot [M_n] \tag{1}$$

where C is an amplitude factor. The abundance of the clusters having n molecules in each cluster in the He droplet beam is given by the Poisson distribution[4,56]:

$$[M_n] = \frac{\langle n \rangle^n}{n!} \exp(-\langle n \rangle) \tag{2}$$

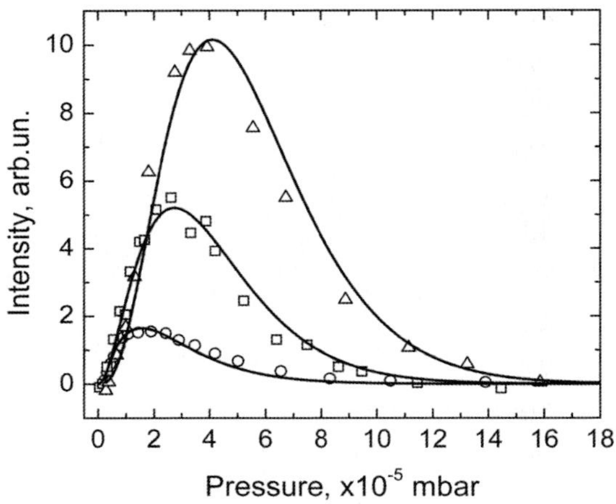

Fig. 5. The pickup pressure dependence of the total intensity of the studied bands of $(NH_3)_n$. Circles, monomer ($n = 1$); squares, dimer ($n = 2$); triangles, trimer ($n = 3$). Solid curves are fits according to Eqs. (1) and (2).

where

$$\langle n \rangle = \chi P_{\mathrm{M}} \tag{3}$$

is the average number of pickup events per droplet, P_{M} is the pressure of the molecular gas in the pickup cell, and χ is a numerical factor which is proportional to the cross section of the droplet and length of the pickup cell. Figure 5 shows the dependence of the total intensity, I_n, of the spectral bands of the $(NH_3)_n$ ($n = 1$–3) species versus P_{NH3}. The solid curves in Fig. 5 are fits of the experimental points by Eqs. (1) and (2). The analogous measurements for water clusters yielded similar results. The ratio of the IRI of the clusters of n and $(n–1)$ molecules, A_n/A_{n-1} could be obtained from the corresponding intensity ratio I_n/I_{n-1}. According to Eqs. (1) and (2):

$$\frac{I_n}{I_{n-1}} = \frac{\nu_n \cdot A_n}{\nu_{n-1} \cdot A_{n-1}} \cdot \frac{\chi \cdot P_{\mathrm{M}}}{n}. \tag{4}$$

Experimental results show that dependence of I_n/I_{n-1} versus P_{M} scales slower than linear at $P_{\mathrm{M}} > 3 \times 10^{-5}$ mbar. Similar effect has been observed previously[57] and was attributed to the effects of the droplet size distribution, scattering of the droplet beam at high-pickup pressure and to decrease of the droplet size upon multiple pickup events. Interplay of

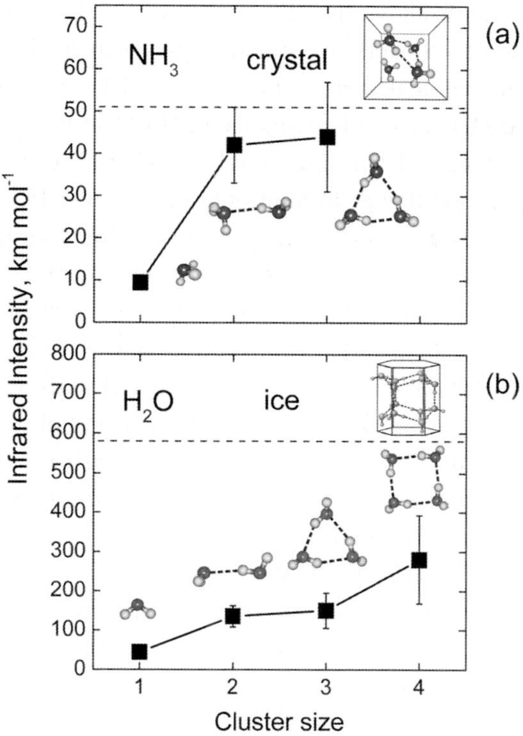

Fig. 6. Size dependence of the infrared intensity per hydrogen bond (IRIB) in ammonia (a) and in water clusters (b). Points represent the total intensity of the bands of a given cluster size labeled by numbers in the spectra of Figs. 4a and b. The crystal structures are also shown for each substance. Dashed horizontal lines give the IRIB in the corresponding crystal. The total IRI of the v_1, v_3, and $2v_4$ bands of NH_3 and of the v_3 band of H_2O are given as a reference for $n = 1$.

these effects, which could not be precisely quantified at present is the main source of the estimated error limits for the obtained values of IRI to be 20%, 30%, and 40% for dimers, trimers, and tetramers, respectively. In this work the values of A_n/A_{n-1} were obtained from the average slopes of the I_n/I_{n-1} versus P_M at $P_M \leq 2 \times 10^{-5}$ mbar. In order to obtain absolute values of the IRI, we used the known IRI of single molecules of $A_1 = (5.8 \pm 0.3)$ km mol^{-1} (Ref. 58) and $A_1 = (44.6 \pm 2)$ km mol^{-1} (Ref. 59) for the v_1 band of ammonia and v_3 band of water, respectively. The results for $(NH_3)_{n=2-3}$ and $(H_2O)_{n=2-4}$ bands are shown in Figs. 6a and b, respectively.

Figures 6a and b show that the IRIs of the H stretching bands in ammonia and water clusters are much larger than in the corresponding monomers, which indicates formation of the HB. In ammonia clusters, the formation of HB causes the enhancement of the ν_1, ν_3, and $2\nu_4$ bands, which have a strong anharmonic coupling. Therefore, the total IRIs of these bands were calculated. In ammonia dimers, the IRI amounts to 42 km mol^{-1}, which is about four times larger than the overall IRI of single ammonia molecules in the studied spectral range. In the case of clusters, it is useful to consider the infrared intensity per hydrogen bond (IRIB). The obtained values of the IRIB are plotted in Fig. 6. In $(NH_3)_3$ IRIB = 44 km mol^{-1} was obtained, i.e., very similar to that in dimers. The IRIB in the stretching region (from 3000 to 3600 cm^{-1}) in crystalline ammonia was obtained to be 51 km mol^{-1}.[60] It follows that the IRIB in ammonia dimers and trimers is the same as in the crystal within the experimental error. Similarity of the IRIB in small ammonia clusters and in crystal suggests that binding in ammonia is dominated by two-body interactions.

In water dimers, the IRI of the donor bridge H stretch band equals to 144 km mol^{-1},[32] which is an increase by a factor of about 3 with respect to the intensity of the ν_3 mode in single water molecules. In trimers, the IRIB was found to be 150 km mol^{-1}.[33] In tetramers of water we found IRIB equals to 280 km mol^{-1},[33] which shows the large nonadditive increase. The size dependence of the IRI in water clusters indicates that charge distribution continues to change beyond the tetramer, whereas both the intermolecular distance and the binding energy converge.[36,40,48,50]

The IRI of the donor H stretch in water dimers has previously been calculated to be in the range of 195–330 km mol^{-1}.[42,47–50] It is seen that the calculated values of IRI are about a factor of 2 larger as compared to the measurements in helium droplets. The same observation is valid for the other bands as well as for larger clusters. The detailed origin of this discrepancy is not clear at the moment. However, some error is certainly introduced by the use of the harmonic approximation and equilibrium cluster structure. Similar discrepancy has recently been confirmed in ammonia clusters $(NH_3)_{n=2,3}$ as well as in the H_2O–NH_3 and H_2O–N_2 complexes.[34,61,62]

Very recently IRI in water dimer have been evaluated using second-order vibrational perturbation theory (VPT), which involves an expansion of both the potential energy and the dipole moment functions about the molecular equilibrium[51] in good agreement with experiments in helium droplets.[32]

It follows that He droplet technique not only contributes to better resolution and assignment of the spectra to particular cluster sizes, but can also be used for measuring the infrared intensities in small clusters. The cluster size dependence of the IRI could give additional information on the charge redistribution in molecules upon development of the HB network. IRI has a number of applications, such as in atmospheric sciences, where the knowledge of the IRI is instrumental in calculations of the absorption of solar radiation by water clusters.[63] Previous estimates based on the calculated IRI can thus overestimate the effect by a factor of 2–3.

4. Formation and Study of Large Clusters

He droplets provide attractive hosts for assembling large molecular and atomic clusters. The pickup process leads to a decrease in the droplet size due to the kinetic energy and the internal energy of the captured molecules, as well as the cohesive energy of the formed cluster, all of which is transferred to the droplet and lead to evaporation of He atoms. The energy balance during the cluster growth has been discussed at some length in our previous paper.[64] As a rule of thumb, 1 eV of heat released corresponds to the evaporation of about 1600 ^{4}He atoms. For example, estimates show that addition of a molecule to large NH_3 cluster causes evaporation of about 600 He atoms from the droplet. Thus, the maximal size of the obtained clusters depends on the initial size of the He droplets. Readily accessible He droplets of about $\langle N_{He} \rangle = 10^3$ to 10^7 atoms allow to bridge a large range of molecular cluster sizes from few to about 10^4 particles. This section reviews the change of the spectra of ammonia all the way from single molecules to molecular solid.[65]

Figure 7 shows a series of depletion spectra, which were measured at increasing average number of ammonia molecules captured by He droplets, as shown in each panel. Traces (a) and (b), which were obtained

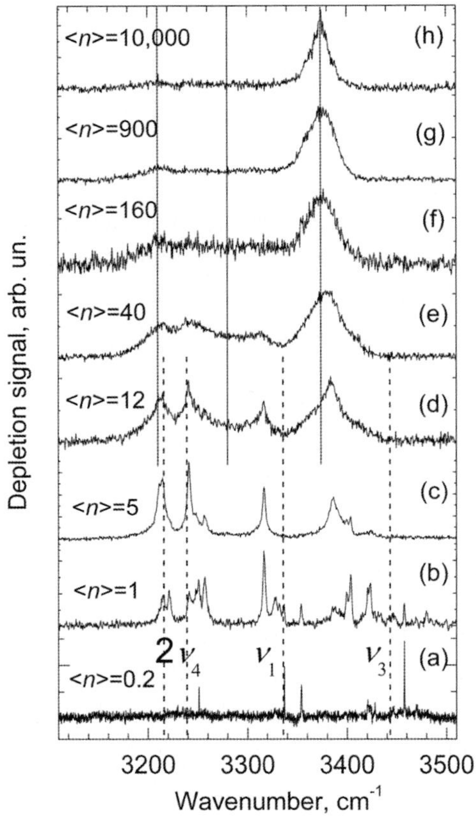

Fig. 7. Spectra of the $(NH_3)_n$ clusters of different average size in He droplets. Average number of ammonia molecules captured by the droplets is shown in each panel. Vertical solid lines show the positions of the maxima in the spectrum of solid NH_3.

in He droplets consisting of $\langle N_{He} \rangle = 3.5 \times 10^3$ atoms, show the v_1 and v_3 fundamental bands and the $2v_4$ overtone band of single NH_3 molecules as well as ammonia dimers through tetramers, similar to that in Fig. 4. No additional resolved spectral features appear at larger $\langle n \rangle$, which indicates that the vibrational shift converges and the bands of larger clusters pile up at approximately the same frequencies. In the case of larger clusters, the spectral traces represent a convolution of the spectra of different cluster sizes, whose abundances are given by the Poisson distribution [see Eq. (2)].

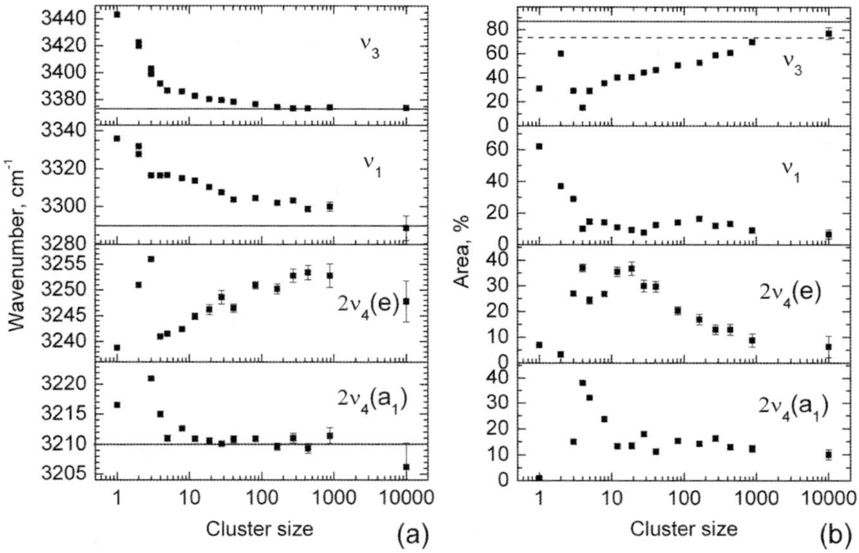

Fig. 8. (a) Band centers and (b) relative band intensities of the spectra versus cluster size.

Figure 7 shows the spectra of the $(NH_3)_n$ clusters in He droplets of about $\langle N_{He} \rangle = 1.2 \times 10^4$ (c), 2.1×10^4 (d and e) 2.5×10^6 (f), and 1.4×10^7 (g and h). Although the bands in the spectra of large clusters broaden considerably it is possible to distinguish four broad maxima, similar to the number of bands in the spectrum of single molecules. In order to quantify the results, the spectra of large clusters of $\langle n \rangle \geq 5$ have been fitted by the sum of four Lorentz functions, which were empirically found to give a fair fit in the entire range of the studied cluster sizes.

Figures 8a and b show the cluster size dependence of the band maxima and relative areas of the peaks. Solid and dashed horizontal lines show the corresponding results in crystalline and in amorphous phases of ammonia, respectively.[66,67]

In small clusters of $n = 2, 3$ the frequency of the stretching ν_1 and ν_3 bands decreases and that of the bending $2\nu_4$ bands increases with the size. This is a typical behavior of the stretching and bending bands upon formation of the hydrogen bond, which is well documented.[68,69] This trend changes abruptly at the cluster size of $n = 4$, which indicates the onset of the strong coupling of the different vibrational modes. In the

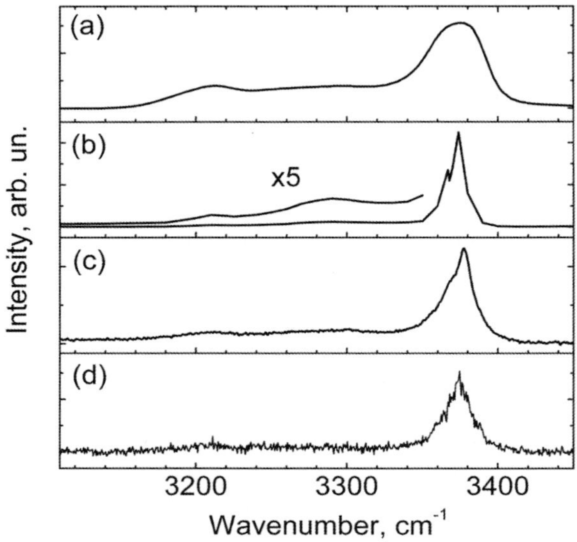

Fig. 9. Absorption spectra of condensed phases of ammonia and of the large ammonia clusters: (a) in amorphous film deposited at $T = 20$ K; (b) in crystalline film deposited at 90 K; (c) in clusters of about 10^4 molecules in an aggregation cell at $T = 37$ K; (d) in depletion spectrum of clusters of about 10^4 molecules in He droplets at $T = 0.37$ K.

range of larger clusters of $n = 5$–10 the relative intensity of the ν_1 band continues to decrease whereas the total intensity of the $2\nu_4$ bands remain approximately constant. The fact that the combined intensity of the $2\nu_4$ bands in clusters consisting of 5–10 molecules is essentially the same as that for the ν_1 and ν_3 bands shows that all four modes are strongly coupled.

The spectra of the larger clusters ($n > 10$) have a well-defined ν_3 band and broad spectrum on its low-frequency side. Figure 8b shows that the relative intensity of the ν_3 band gradually increases from 30% at $\langle n \rangle = 5$ to about 77% at $\langle n \rangle = 10^4$. For comparison, the ν_3 band in the spectrum of ammonia crystal, amorphous solid, and liquid accounts for 87%, 63%, and 45% of the total intensity in the 3 μm spectral range, respectively.[60,70] Figure 9 shows the comparison of the spectrum of the largest ammonia clusters obtained in this work in panel (d) with the spectra of amorphous ammonia solid at $T = 20$ K[67]; crystals at $T = 90$ K[66]; and of clusters of about 10^4 ammonia molecules obtained in a

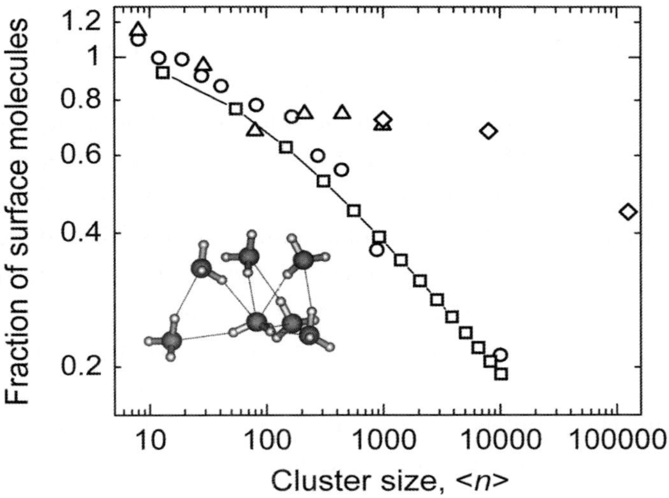

Fig. 10. Fraction of the surface molecules in ammonia cluster versus cluster size: Circles, this work; triangles, in free jet; diamonds, in cooling cell. The results of calculations for a compact cluster are shown by squares.

cooling cell,[71] in panels (a), (b), and (c), respectively. The spectrum of large ammonia clusters obtained in He droplets (trace d) is very similar to the spectrum of free ammonia clusters (trace c) from the cooling cell experiments, but has stronger ν_3 band. The spectrum of the $\langle n \rangle = 10^4$ clusters obtained in this work is also very similar to the spectrum of crystalline ammonia (trace b).

The gradual increase in the relative intensity of the ν_3 mode upon increase of the cluster size in the range of $n > 10$ must reflect some change in HB molecular coordination. The insert in Fig. 10 shows the nearest neighbors of an ammonia molecule in the crystal.[71,72] Each ammonia molecule is a part of the three hydrogen-bonded rings of total seven molecules, which apparently causes an enhancement and diminishing of the of the ν_3 and ν_1 bands, respectively. The relative intensity of the ν_3 band acts as a sensitive probe of the hydrogen-bonding coordination of the NH_3 molecules in clusters.

Clusters differ from the bulk phase in that they contain a larger number of the molecules on the surface.[73] The surface molecules have an incomplete coordination, similar to that in small clusters; therefore, they will contribute to the intensity of the ν_1 and $2\nu_4$ bands. For an estimate we assumed that the relative intensity of the ν_3 band of the surface molecules is the same as in the small-sized clusters of $\langle n \rangle \approx 20$, i.e., about 40% of the total intensity in the studied spectral range. Taking the relative intensity of the ν_3 band in the spectrum of the inner part of the cluster equal to that in ammonia crystal of 87%,[60,70] we calculated the ratio of surface molecules to the total number of molecules in the cluster, F. Figure 10 shows the comparison of the experimental results with calculations for a close-packed cluster. It is seen that the fraction of the surface molecules in ammonia clusters follows that for a close-packed cluster within the error limits. Therefore, we concluded that the clusters formed in He droplets at $T = 0.37$ K have a compact structure characteristic of close packing, and not a fractal which will have a smaller density and slower n dependence. Moreover, the results show that molecules inside clusters have a coordination of hydrogen bonds similar to that in the crystal. This might be surprising in view of the amorphous structure of the ammonia film which was obtained by the deposition of ammonia gas on a surface that was kept at $T = 20$ K.[74] However, the attachment of the molecules onto the cold film of ammonia and onto the clusters in He droplets differ by the kinetic energy of the aggregating molecules. During the film growth, the incoming molecules have a temperature of about 300 K. On the other hand, upon the capture of a molecule by helium droplets both the translational and rotational energy of the molecules relaxes rapidly to the ambient temperature of about 0.37 K.

Thus, in He droplets the molecules approach the surface of the cluster slowly and interaction energy plays the major role in determining the attachment of a new molecule to the surface of the cluster. The important role of the precooling in He droplets was demonstrated previously in experiments on growth of water[75] and methanol[76] clusters. Formation of unusual chain clusters of HCN molecules was demonstrated in He droplets.[77] The proposed mechanism involves orientation and guiding of the HCN molecules by the long-range dipole–dipole interaction. Our results suggest a similar but three-dimensional self-assembly of large NH_3 clusters in He droplets.

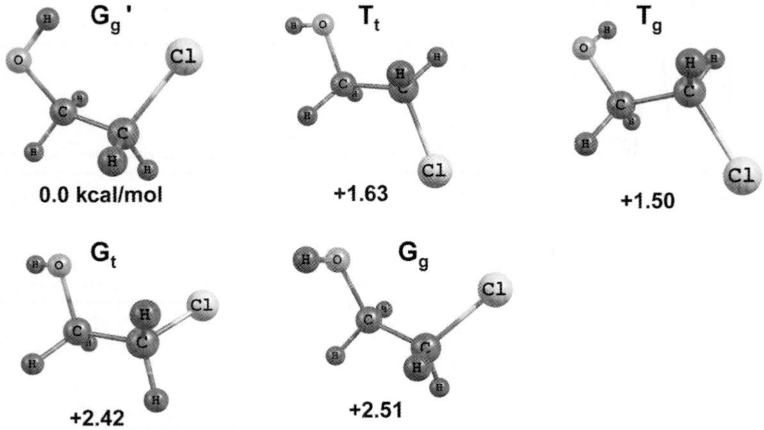

Fig. 11. The structure of the conformers of 2-chloroethanol molecule calculated at the CCSD/cc-pVDZ level of theory. Numbers are relative energies in units of kcal mol^{-1}, which take into account different zero-point energy of the conformers.

5. Interconversion Enthalpy of Molecular Conformers

Experiments on trapping of higher energy conformers of organic molecules have previously been carried out in solid matrix (see Ref. 78, and references therein). It was concluded that the conformational distribution in the vapor remains unchanged in the matrix if conformers are separated by barriers of ~2 kcal mol^{-1} or higher. The abundance of the conformers in matrix can in principle be used to acquire the interconversion enthalpy, ΔH, of the conformers. However, due to the effects of conformer relaxation, there are few examples of using matrix spectra for measurements of ΔH as was done for glycine.[79–83]

Our previous study of the electronic spectra of aminoacids captured in He droplets revealed a number of peaks assigned to different conformers.[84] Different conformers of nucleotides have also been observed in He droplets in the group of R. Miller,[85,86] where the measured angles between the permanent and vibrational transition dipole moments have been used for assignment of conformers. Here we use He droplets for measurements of interconversion enthalpy of conformers of 2-CLE molecules. For this purpose 2-CLE molecules having different

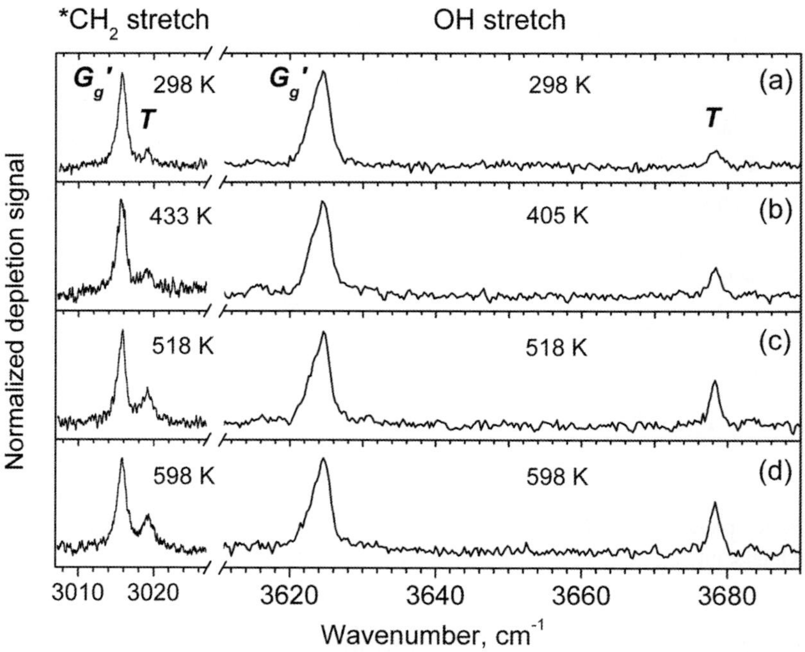

Fig. 12. $*CH_2$ stretch and OH stretch bands measured at different temperatures of the captured 2-CLE molecules as indicated in each panel. Laser linewidth is 0.08 cm^{-1} and 1 cm^{-1} in the case of $*CH_2$ and OH bands, respectively.

temperatures are captured by the droplets and their infrared spectra are obtained.[87]

Figure 11 shows the structure and the relative energy of the conformers calculated using the PC GAMESS *ab initio* package.[88] Five conformers of 2-CLE have been found in agreement with previous studies.[89,90] The lowest energy *gauche* conformer, G_g', has the OH group pointing toward the Cl atom. The two higher energy *trans*-conformers T_g and T_t differ by orientation of the OH group and have very similar energies. The frequency of the OH stretch in the G_g' conformer was calculated to be about 55 cm^{-1} lower than that in the T_t and T_g conformers, which is consistent with the formation of the internal hydrogen bond in the G_g' conformer. On the other hand, the calculated conformational shifts in all four of the CH_2 stretching bands are about or less than 10 cm^{-1}, which is comparable with the accuracy of the calculations.

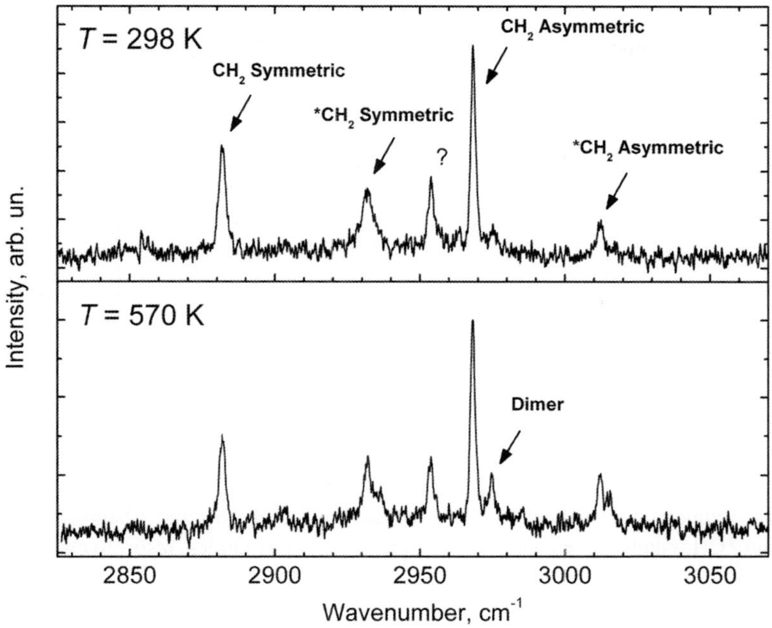

Fig. 13. CH_2 stretching bands of 2-CLE molecules in helium droplets captured at $T = 298$ K and $T = 570$ K as indicated.

Figure 12 shows spectra of the 2-CLE molecules in the range of asymmetric stretch of the CH_2Cl group (*CH_2) and OH stretch, which have been obtained at temperatures in the pickup cell, increasing from 298 to 598 K in panels (a) and (d), respectively. The spectrum in the OH range measured at room temperature (a) has a strong band at 3624.5 cm^{-1} and about a factor of 10 weaker band at 3678.3 cm^{-1}. Based upon the results of the calculations and previous works,[89,91,92] we assigned the bands to the G_g' and T_t/T_g conformers, respectively. Apparently, the T_t and T_g conformers are not resolved, and will be referred as T conformers. The spectra in Fig. 12 show that the relative intensity of the peak T increases with temperature, in agreement with the expected larger abundance of the T conformers at higher temperature.

Figure 13 shows the spectrum of the 2-CLE molecules in the CH_2 stretching range of 2800–3050 cm^{-1}. The bands at 2885.4, 2935.5, 2971.8, and 3015.6 cm^{-1} are assigned to CH_2 symmetric and

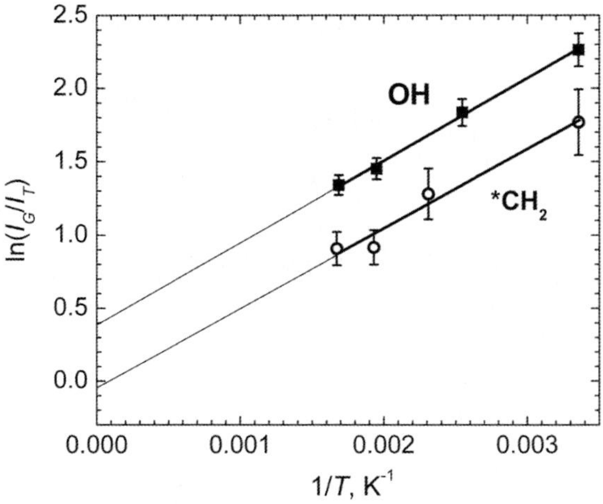

Fig. 14. Temperature dependence of the intensity ratio of the G_g' and T bands. Results for OH and $*CH_2$ bands are shown by solid squares and open circles, respectively. Lines are the results of the least squares fits.

antisymmetric stretches of CH_2OH and CH_2Cl groups of the 2-CLE molecules. An additional band at 2957.6 cm^{-1} remains unassigned. Similar spectra have previously been observed in the gas phase, in cryogenic matrices, and in cryosolutions.[89,92] With the exception of the $*CH_2$ band, the CH bands do not reveal any splitting, indicating that the bands of the G_g' and T conformers have very similar frequencies. Spectra of the $*CH_2$ band at different pickup temperatures are shown in Fig. 12, where the stronger and weaker bands at 3015.6 cm^{-1} and 3019.3 cm^{-1} are ascribed to the G_g' and T conformers, respectively.

Figure 14 shows semilogarithmic plots of the ratio of the integrated intensities of the G_g' and T bands versus inverse temperature as obtained for the OH and $*CH_2$ bands. The results for the two bands show linear dependences having the same slopes but different intercepts.

The enthalpy of interconversion of the conformers, $\Delta H = H_T - H_G$, can be obtained from the results in Fig. 14 using van 't Hoff's equation, as described previously.[93] From the slope of the measured $\ln(I_G/I_T)$ versus $1/T$ plot in Fig. 14 the values of $\Delta H = (1.12 \pm 0.04)$ kcal mol^{-1} and $\Delta H = (1.1 \pm 0.1)$ kcal mol^{-1} are obtained from the OH and $*CH_2$ bands, respectively (i.e., the same within the error bars, as expected). ΔH can be used as an estimate for the enthalpy of the internal hydrogen bond formation in 2-CLE.[91] The magnitude of ΔH obtained in this work is close to the previous results obtained in the gas phase[94] (1.65 ± 0.45) kcal mol^{-1} and (1.20 ± 0.1) kcal mol^{-1} from the analysis of the intensity of the OH and CCl bands, respectively, and the average value for the same bands in liquid xenon (1.19 ± 0.1) kcal mol^{-1}.[89] Electron diffraction measurements in the gas phase gave larger value of $\Delta H = (2.4 \pm 0.2)$ kcal mol^{-1}.[95] Because 2-CLE is a small molecule, *ab initio* calculations are expected to give relatively accurate results. Previously, ΔH for the T_g and T_t conformers have been obtained to be 1.55 and 1.72 kcal mol^{-1} (Ref. 89) and 1.3 and 1.1 kcal mol^{-1} (Ref. 90) via MP2 calculations with the 6-31G** and 6-31++G(d,p) basis sets, respectively. The results from Ref. 89 are close to the present calculations (see Fig. 11).

The values of intercepts in Fig. 14 depend on the difference in entropy and the ratio of the absorption cross sections of the conformers, for which contributions cannot be disentangled without further assumptions. If we assume that the infrared intensities of the $*CH_2$ bands in the G and T conformers are the same, it follows that the entropy change upon interconversion is $\Delta S = 0$ and that the intensity of the OH band of the G conformers is a factor of 1.6 larger as compared with the T conformers.

Observation of a linear van 't Hoff dependence in Fig. 14 is consistent with the freezing of the gas phase conformer's population upon trapping in He droplets.[96] It has been shown that the outcome of the cooling depends on the rate of conformer interconversion above the barrier and on the vibrational relaxation rate.[97,98] The characteristic time for vibrational cooling in a free jet experiment amounts to about 10^{-7} s.[98] Measurements of the linewidths in He droplets suggest that the relaxation time of large amplitude torsional vibrations, which are important for isomerization, can be as short as 10^{-12} s,[4] which facilitates vertical relaxation of the conformers. Present experiments suggest that conformers of 2-CLE remain unrelaxed upon cooling in He droplets in spite of the relatively low barrier for the $T \rightarrow G$ conversion which was calculated to be about 1.5 kcal mol^{-1}.[99] Thus, the barrier height is lower than the vibrational energy content of 2-CLE molecules at 600 K, of about 5 kcal mol^{-1}. It appears that trapping in He droplets is promising for isolation of conformers separated by small barriers. It will be interesting to expand these measurements to larger molecules with even lower barriers, and those having richer conformer's landscape, to explore the generality of this observation.

6. Conclusions

A helium droplet may be viewed as a "personal" nanomatrix for each individual molecule or a cluster, "powered" by evaporative cooling of the droplet material. In the droplets, a high spectroscopic resolution is achieved, while an isothermal low-temperature environment is maintained by evaporative cooling at T = 0.37 K. Most important, the superfluid helium facilitates binary encounters, and absorbs the released binding energy upon recombination. In addition, the pickup process follows relatively narrow Poisson cluster size distribution. These advantages make helium droplet a versatile tool for growth and spectroscopic interrogation of molecular complexes having from few up to about 10^4 particles. Moreover, rapid cooling of the dopants in helium droplets facilitates stabilization of weakly bound metastable species, such as molecular conformers. The examples discussed in this review comprise a small fraction of the large variety of fascinating experiments on matrix isolation in superfluid helium droplets.

Acknowledgments

This material is based upon work supported by the NSF under grants CHE 0513163 and CHE 0809093.

References

1. J. P. Toennies and A. F. Vilesov, *Annu. Rev. Phys. Chem.* **49**, 1 (1998).
2. C. Callegari, K. K. Lehmann, R. Schmied and G. Scoles, *J. Chem. Phys.* **115**, 10090 (2001).
3. F. Stienkemeier and A. F. Vilesov, *J. Chem. Phys.* **115**, 10119 (2001).
4. J. P. Toennies and A. F. Vilesov, *Angew. Chem. Int. Ed.* **43**, 2622 (2004).
5. F. Stienkemeier and K. K. Lehmann, *J. Phys. B: At. Mol. Opt. Phys.* **39**, R127 (2006).
6. M. Y. Choi, G. E. Douberly, T. M. Falconer, W. K. Lewis, C. M. Lindsay, J. M. Merritt, P. L. Stiles and R. E. Miller, *Int. Rev. Phys. Chem.* **25**, 15 (2006).

7. M. Hartmann, R. E. Miller, J. P. Toennies and A. Vilesov, *Phys. Rev. Lett.* **75**, 1566 (1995).
8. S. Grebenev, M. Hartmann, M. Havenith, B. Sartakov, J. P. Toennies and A. F. Vilesov, *J. Chem. Phys.* **112**, 4485 (2000).
9. D. R. Tilley and J. Tilley, *Superfluidity and Superconductivity,* Institute of Physics Publ. (1990).
10. H. Hoshina, D. Skvortsov, B. G. Sartakov and A. F. Vilesov, *J. Chem. Phys.* **132**, 074302 (2010).
11. R. Sliter and A. F. Vilesov, unpublished results.
12. M. Hartmann, N. Portner, B. Sartakov, J. P. Toennies and A. F. Vilesov, *J. Chem. Phys.* **110**, 5109 (1999).
13. S. Grebenev, J. P. Toennies and A. F. Vilesov, *Science* **279**, 2083 (1998).
14. E. R. Dobbs, *Helium Three*. Oxford University Press, Oxford (2000).
15. J. Harms, M. Hartmann, J. P. Toennies, A. F. Vilesov and B. Sartakov, *J. Mol. Spectrosc.* **185**, 204 (1997).
16. K. Nauta and R. E. Miller, *J. Chem. Phys.* **113**, 9466 (2000).
17. M. N. Slipchenko and A. F. Vilesov, *Chem. Phys. Lett.* **412**, 176 (2005).
18. K. E. Kuyanov, M. N. Slipchenko and A. F. Vilesov, *Chem. Phys. Lett.* **427**, 5 (2006).
19. K. Nauta and R. E. Miller, *Chem. Phys. Lett.* **350**, 225 (2001).
20. K. Nauta and R. E. Miller, *J. Chem. Phys.* **115**, 10254 (2001).
21. K. Nauta and R. E. Miller, *J. Chem. Phys.* **115**, 8384 (2001).
22. H. Hoshina, J. Lucrezi, M. N. Slipchenko, K. E. Kuyanov and A. F. Vilesov, *Phys. Rev. Lett.* **94**, 195301 (2005).
23. J. Tang, Y. J. Xu, A. R. W. McKellar and W. Jager, *Science* **297**, 2030 (2002).
24. J. Tang, A. R. W. McKellar, E. Mezzacapo and S. Moroni, *Phys. Rev. Lett.* **92**, 145503 (2004).
25. Y. J. Xu, W. Jager, J. Tang and A. R. W. McKellar, *Phys. Rev. Lett.* **91**, 163401 (2003).
26. Y. Xu, N. Blinov, W. Jager and P.-N. Roy, *J. Chem. Phys.* **124**, 081101 (2006).
27. A. R. W. McKellar, Y. J. Xu and W. Jager, *Phys. Rev. Lett.* **97**, 183401 (2006).
28. Y. Kwon, P. Huang, M. V. Patel, D. Blume and K. B. Whaley, *J. Chem. Phys.* **113**, 6469 (2000).
29. F. Paesani and K. B. Whaley, *J. Chem. Phys.* **121**, 5293 (2004).
30. F. Paesani and K. B. Whaley, *J. Chem. Phys.* **121**, 4180 (2004).
31. S. Paolini, S. Fantoni, S. Moroni and S. Baroni, *J. Chem. Phys.* **123**, 114306 (2005).
32. K. Kuyanov-Prozument, M. Y. Choi and A. F. Vilesov, *J. Chem. Phys.*, **132**, 014304 (2010).
33. M. N. Slipchenko, K. E. Kuyanov, B. G. Sartakov and A. F. Vilesov, *J. Chem. Phys.* **124**, 241101 (2006).

34. M. N. Slipchenko, B. G. Sartakov, A. F. Vilesov and S. S. Xantheas, *J. Phys. Chem.* **A111**, 7460 (2007).
35. M. F. Vernon, D. J. Krajnovich, H. S. Kwok, J. M. Lisy, Y. R. Shen and Y. T. Lee, *J. Chem. Phys.* **77**, 47 (1982).
36. K. Liu, J. D. Cruzan and R. J. Saykally, *Science* **271**, 929 (1996).
37. F. Huisken, M. Kaloudis and A. Kulcke, *J. Chem. Phys.* **104**, 17 (1996).
38. R. Frochtenicht, M. Kaloudis, M. Koch and F. Huisken, *J. Chem. Phys.* **105**, 6128 (1996).
39. J. B. Paul, C. P. Collier, R. J. Saykally, J. J. Scherer and A. OKeefe, *J. Phys. Chem.* **A101**, 5211 (1997).
40. F. N. Keutsch and R. J. Saykally, *Proc. Natl. Acad. Sci. USA* **98**, 10533 (2001).
41. C. J. Burnham, S. S. Xantheas, M. A. Miller, B. E. Applegate and R. E. Miller, *J. Chem. Phys.* **117**, 1109 (2002).
42. K. Ohno, M. Okimura, N. Akai and Y. Katsumoto, *Phys. Chem. Chem. Phys.* **7**, 3005 (2005).
43. D. D. Nelson, G. T. Fraser and W. Klemperer, *Science* **238**, 1670 (1987).
44. S. Sutzer and L. Andrews, *J. Chem. Phys.* **87**, 5131 (1987).
45. J. G. Loeser, C. A. Schmuttenmaer, R. C. Cohen, M. J. Elrod, D. W. Steyert, R. J. Saykally, R. E. Bumgarner and G. A. Blake, *J. Chem. Phys.* **97**, 4727 (1992).
46. T. A. Beu and U. Buck, *J. Chem. Phys.* **114**, 7848 (2001).
47. K. S. Kim, B. J. Mhin, U.-S. Choi and K. Lee, *J. Chem. Phys.* **97**, 6649 (1993).
48. S. S. Xantheas and T. H. Dunning Jr., *J. Chem. Phys.* **99**, 8774 (1993).
49. G. M. Chaban, J. O. Jung and R. B. Gerber, *J. Phys. Chem.* **104**, 2772 (2000).
50. E. M. Mas, R. Bukowski and K. Szalewicz, *J. Chem. Phys.* **118**, 4386 (2003).
51. H. G. Kjaergaard, A. L. Garden, G. M. Chaban, R. B. Gerber, D. A. Matthews, and J. F. Stanton, *J. Phys. Chem.* A **112**, 4324 (2008).
52. J. Bournay and Y. Marechal, *J. Chem. Phys.* **59**, 5077 (1973).
53. C. Laush and J. M. Lisy, *J. Chem. Phys.* **101**, 7480 (1994).
54. M. van Thiel, E. D. Becker and G. C. Pimentel, *J. Chem. Phys.* **27**, 486 (1957).
55. G. C. Pimentel, M. O. Bulanin and M. van Thiel, *J. Chem. Phys.* **36**, 500 (1962).
56. M. Hartmann, R. E. Miller, J. P. Toennies and A. F. Vilesov, *Science* **272**, 1631 (1996).
57. M. Lewerenz, B. Schilling and J. P. Toennies, *J. Chem. Phys.* **102**, 8191 (1995).
58. I. Kleiner, L. R. Brown, G. Tarrago, Q.-L. Kuo, N. Picque, G. Guelachvili, V. Dana and J.-Y. Mandin, *J. Mol. Spectrosc.* **193**, 46 (1999).
59. J. M. Flaud and C. Camy-Peyret, *J. Mol. Spectrosc.* **55**, 278 (1975).
60. J. R. Ferraro, G. Sill and U. Fink, *Appl. Opt.* **34**, 525 (1980).
61. S. Kuma, M. N. Slipchenko, K. E. Kuyanov, T. Momose and A. F. Vilesov, *J. Phys. Chem.* **A33**, 10046 (2006).
62. S. Kuma, M. N. Slipchenko, T. Momose and A. F. Vilesov, *Chem. Phys. Lett.* **439**, 265 (2007).

63. H. G. Kjaergaard, T. W. Robinson, D. L. Howard, J. S. Daniel, J. E. Headrick and V. Vaida, *J. Phys. Chem.* **A107**, 10680 (2003).
64. V. Mozhayskiy, M. Slipchenko, V. K. Adamchuk and A. F. Vilesov, *J. Chem. Phys.* **127**, 094701 (2007).
65. M. N. Slipchenko, B. G. Sartakov, and A. F. Vilesov, *J. Chem. Phys.*, **128**, 134509 (2008).
66. G. Sill, U. Fink and J. R. Ferraro, *J. Opt. Soc. Am.* **70**, 724 (1980).
67. H. Wolff, H.-G. Rollar and E. Wolff, *J. Chem. Phys.* **55**, 1373 (1971).
68. A. G. Jefferey, *An Introduction to Hydrogen Bonding*. Oxford University Press, New York (1997).
69. S. Scheiner, *Hydrogen Bonding. A Theoretical Perspective*, Oxford University Press, New York (1997).
70. C. W. Robertson and D. Williams, *J. Opt. Soc. Am.* **63**, 188 (1973).
71. M. Jetzki, A. Bonnamy and R. Singorell, *J. Chem. Phys.* **120**, 11775 (2004).
72. A. W. Hewat and C. Riekel, *Act. Cryst.* **A35**, 569 (1979).
73. B. M. Smirnov, *Chem. Phys. Lett.* **232**, 395 (1995).
74. J. S. Holt, D. Sadoskas and C. J. Pursell, *J. Chem. Phys.* **120**, 7153 (2004).
75. K. Nauta and R. E. Miller, *Science* **287**, 293 (2000).
76. M. Behrens, R. Frochtenicht, M. Hartmann, J. G. Siebers, U. Buck and F. C. Hagemeister, *J. Chem. Phys.* **111**, 2436 (1999).
77. K. Nauta and R. E. Miller, *Science* **283**, 1895 (1999).
78. A. J. L. Jesus, M. T. S. Rosado, I. Reva, R. Fausto, A. E. S. Eusebio and J. S. Redinha, *J. Phys. Chem.* **A112**, 4669 (2008).
79. I. D. Reva, A. M. Plokhotnichenko, S. G. Stepanian, A.Y. Ivanov, E. D. Radchenko, G. G. Sheina and Y. P. Blagoi, *Chem. Phys. Lett.* **232**, 141 (1995).
80. A. Y. Ivanov, A. M. Plokhotnichenko, V. Izvekov, G. G. Sheina and Y. P. Blagoi, *J. Mol. Struct.* **408**, 459 (1997).
81. W. F. Hoffman, A. Aspiala and J. S. Shirk, *J. Phys. Chem.* **90**, 5706 (1986).
82. H. Takeuchi and M. Tasumi, *Chem. Phys.* **70**, 275 (1982).
83. M. Räsänen, H. Kunttu and J. Murto, *Laser Chemistry* **9**, 123 (1988).
84. A. Lindinger, J. P. Toennies and A. F. Vilesov, *J. Chem. Phys.* **110**, 1429 (1999).
85. M. Y. Choi, G. E. Douberly, T. M. Falconer, W. K. Lewis, C. M. Lindsay, J. M. Merritt, P. L. Stiles and R. E. Miller, *Int. Rev. Phys. Chem.* **25**, 15 (2006).
86. M. Y. Choi and R. E. Miller, *J. Am. Chem. Soc.* **128**, 7320 (2006).
87. D. Skvortsov and A. F. Vilesov, *J. Chem. Phys.* **130**, 151101 (2009).
88. A. A. Granovsky, PC GAMESS; Moscow State University, Moscow, Russia (accessed Jun 2003).
89. J. R. Durig, L. Zhou, T. K. Gounev, P. Klaeboe, G. A. Guirgis and L. F. Wang, *J. Mol. Struct.* **385**, 7 (1996).
90. S. X. Tian, N. Kishimoto and K. Ohno, *J. Phys. Chem.* **A107**, 53 (2003).

91. A. Kovacs and Z. Varga, *Coordination Chem. Rev.* **250**, 710 (2006).
92. M. Perttila, J. Murto and L. Halonen, *Spectrochim. Acta* **A34**, 469 (1978).
93. A. R. Potts and T. Baer, *J. Chem. Phys.* **105**, 7605 (1996).
94. P. Buckley, P. A. Giguere and M. Schneider, *Can. J. Chem.* **47**, 901 (1969).
95. A. Almenningen, L. Fernholt and K. Kveseth, *Acta Chem. Scandinavica* **A31**, 297 (1977).
96. A. R. Potts and T. Baer, *J. Phys. Chem.* **A101**, 8970 (1997).
97. T. S. Zwier, *J. Phys. Chem.* **A110**, 4133 (2006).
98. D. A. Evans, D. J. Wales, B. C. Dian and T. S. Zwier, *J. Chem. Phys.* **120**, 148 (2004).
99. J. Murto, M. Räsänen, A. Aspiala and T. Lotta, *J. Mol. Struct.*: *Theochem* **17**, 99 (1984).

CHAPTER 8

CRYOGENIC SOLUTIONS AS A TOOL TO CHARACTERIZE RED-AND BLUE-SHIFTING C−H···X HYDROGEN BONDING

Wouter A. Herrebout and Benjamin J. van der Veken

Department of Chemistry, University of Antwerp, Groenenborgerlaan 171, B-2020 Antwerp, Belgium

E-mails: wouter.herrebout@ua.ac.be; benjamin.vanderveken@ua.ac.be

During recent years, solutions in liquefied inert gases have proven to be an ideal medium to study molecular complexes involving C−H···X (with X = O, N, F, π, etc.) red- or blue-shifting hydrogen bonds. In this chapter, the experimental setups for the infrared and Raman studies of cryosolutions are described, and the general methodologies used to study weakly bound molecular complexes are discussed. The methods are illustrated using data obtained for a variety of C−H···X hydrogen-bonded complexes involving the mixed haloforms $CHCl_xF_{3-x}$ and halothane $CF_3CBrClH$.

1. Introduction

During the last decades, matrix isolation spectroscopy has been one of the most active fields in low-temperature vibrational spectroscopy.[1] Although this technique remains unsurpassed for a variety of problems including, e.g., the study of free radicals and other unstable species, it has several disadvantages. Because it is difficult to produce sufficiently thick samples and because of strong scattering of the infrared (IR) beam by thicker matrixes, this type of spectroscopy is rarely used to explore weak bands. Moreover, the spectra of matrix isolated species are often complicated by various effects: bands may show a multiplet structure which can arise from rotation of the solute molecule in its trapping site, from the presence of different trapping sites in the matrix, from aggregation of solute molecules, etc. This matrix splitting is often difficult to distinguish from other spectral effects that are of interest for our studies and that arise from the formation of molecular complexes or from the occurrence of multiple conformations. Also, since species are studied in a rigid, solid medium, no direct information on the thermodynamic properties of the species studied is accessible.

A wealth of information, complementary to that derived using solid matrixes, can be obtained by studying solutions of the molecules dissolved in liquefied inert gases, such as rare gases, N_2, or O_2. Apart from the relatively low temperatures used during these experiments and the weak solute–solvent interactions, a major advantage of these solutions is their transparency in a broad spectral interval, ranging from the far-IR to the ultraviolet. Therefore, in contrast to solid matrices, the solutions in cryogenic solvents can be used for the study of weak phenomena. As the cryosolutions in general are in thermodynamical equilibrium, they are ideally suited to determine thermodynamical and other physical properties.

The main obstacle to using liquefied inert gases as solvents is their limited solvatibility: only a small number of molecules dissolve well in liquid argon (LAr). In contrast, substantially more compounds dissolve rather well in liquid krypton (LKr) and even better in liquid xenon (LXe). Fortunately, because of their transparency, longer optical pathlengths can be used so that very often also for the less soluble species experimental information can be obtained.

Over the last few decades, many experiments have been performed in which liquid rare gases are used as solvents and the results have comprehensively been summarized in earlier reviews.[2–10] The items covered not only studies of isolated molecules, but also described studies on solute–solute and solute–solvent interactions, and on chemical and photochemical reactions. In this chapter, we will concentrate on experimental results that have been obtained for weakly bound molecular complexes held together by blue- or red-shifting C−H···X hydrogen bonds. The experimental results, obtained by studying the IR and Raman spectra of a variety of cryosolutions, are compared with theoretical data derived from *ab initio* calculations and Monte Carlo simulations.

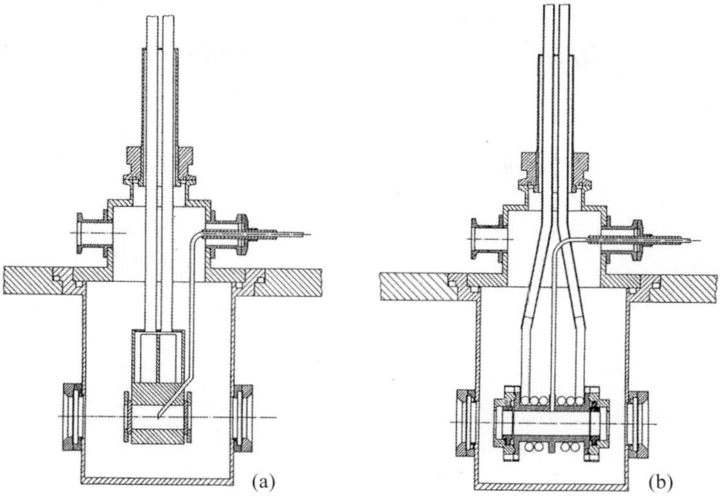

Fig. 1. Typical cells used for liquefied noble-gas cryostats: (a) low-pressure cryostat with clamped-on windows, and (b) cryostat with high-pressure window holders.

2. Experimental Setups

The general constructions of cells for study of cryogenic solutions have been discussed in detail in several reviews.[2,6,8,10–15] Here, we will, therefore, focus on some of the experimental setups presently used in our laboratory.

2.1. *Infrared spectroscopy of cryosolutions*

The setup used for the IR study of cryosolutions consists of three main parts: (i) a pressure manifold needed for filling and evacuation of the cell and for monitoring the amount of solute gas used in a particular experiment, (ii) the actual liquid cell, and (iii) additional equipment to control the temperature of the cell.

Two different cell designs that are presently used are shown in Fig. 1. In both cases, the actual cell is suspended, via a stainless steel liquid nitrogen feedthrough, in a vacuum shroud, to avoid condensation of H_2O and CO_2 onto the outside of the cell. The vacuum shroud is equipped

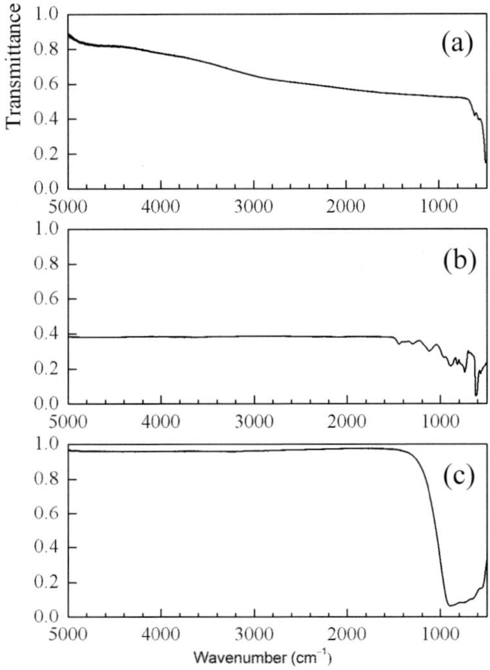

Fig. 2. Transmittance of classical window materials in the mid-IR region: (a) ZnSe, (b) Si, and (c) CaF_2. The thickness of all windows is 5 mm. To avoid interference, a Si window with 0.5° wedge was used.

alternatively with quartz or CaF_2 windows for the near-IR region, with KBr windows for the mid-IR region, and with polyethylene windows for the far-IR region, and is hermetically sealed to the vacuum spectrometer.

The first type of cryostats, based on a low-pressure design, is shown schematically in Fig. 1a. For these cells, the IR window is fixed at the outside of the cell body using an indium gasket. Because of the different expansion coefficients of the window material and the material used to construct the cell body, and because the pressurized solvent inside the cell tends to break the indium seal, the preparation of the latter is very critical. The cells are developed to withstand an internal pressure of 15 bars at 80 K and can be used in different temperature intervals, ranging from 93 to 125 K for LAr, from 120 to 155 K for LKr, and from 152 to 223 K for LXe. The minimal pathlength that has been achieved is 5 mm.

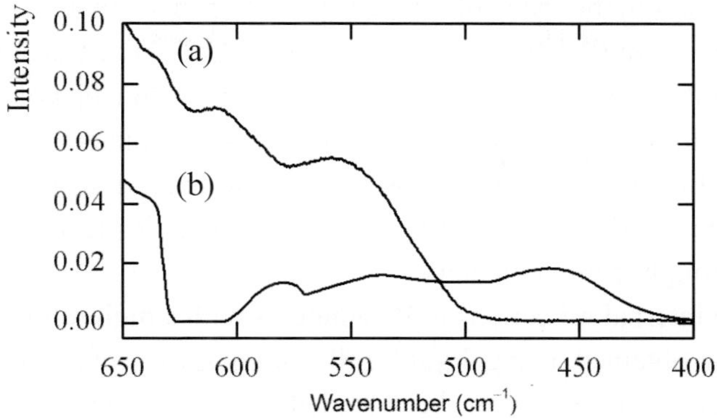

Fig. 3. Single-beam spectra of a liquid cell equipped with ZnSe (a) and Si (b) window materials. Due to the complete absorption by Si, ZnSe is preferred for measurements around 600 cm^{-1}. However, due to complete absorption by ZnSe, Si windows are required for measurements below 500 cm^{-1}. The single-beam spectra were recorded using KBr windows in the vacuum shroud.

The second type of cryostats, based on a high-pressure design similar to those described in Ref. 13 is shown schematically in Fig. 1b. In these cells, the IR windows are fitted into the cell body, using two window holders. The resulting window holders are then sealed to a central cell body. These cells have the advantage that the pressurized solvent inside the cell tends to improve the seal between the IR window and the window holder. Because these cells involve a central cell body and two different window holders, the minimal pathlength that can be obtained is sufficiently larger than that described above.

In principle, any window material of sufficient strength can be used in liquid cells, given that simple rules for selecting the thickness are followed.[16] A popular window material often used in the construction of liquid cells is Si, because this material is quite strong and transparent through most of the IR region, including the far-IR. Unfortunately, Si has a fairly high index of refraction, so that substantial reflection occurs and the actual transmittance of the cell is rather low.

The spectral characteristics of the commonly used window materials Si, ZnSe, and CaF_2 are compared in Figs. 2 and 3. CaF_2 provides excellent transmittance for frequencies above 1000 cm^{-1}, and it is commonly used for experiments in the mid- and near-IR regions. Due to the differences in the index of refraction, the transparency for Si and ZnSe in general is significantly smaller than that of CaF_2. The transmittance of ZnSe is further limited to approximately 500 cm^{-1} while that of Si shows a significant loss below 1000 cm^{-1} and an almost complete absorption near 600 cm^{-1}. In contrast to ZnSe, Si shows a significant transmittance in the far-IR region. CaF_2, ZnSe, and Si windows have their specific advantages and are strongly complementary.

Due to internal reflections in IR windows with a high refractive index, IR spectra obtained using liquid cells often show severe interference fringes. These fringes can be avoided by using windows that are slightly wedged, typically by 0.5° or 1°. Alternatively, the fringes can be avoided by using windows coated with antireflection layers.[17]

The different cryocells described above are cooled with bursts of liquid nitrogen (LN_2) from a slightly pressurized Dewar. In all cases, the LN_2 flow is regulated by a solenoid valve, which is controlled via a Pt-100 thermoresistor embedded in the cell body. The temperature of the solution is measured using a second Pt-100 thermoresistor located close to the solution.

Cells similar to those described above can also be constructed by combining a cell body and a double-stage cryostat involving closed-cycle helium cooling, or by combining a cell body with a LN_2 Dewar, heating cartridge, and proportional–integral–derivative (PID) controller. A typical cell of the first type, recently developed for experiments in liquid Ne,[18,19] is shown in Fig. 4. The setup consists of a cell similar to that shown in Fig. 1a, which is connected to the cold head of a Leybold ROK 10–300 double-stage closed-cycle helium refrigerator. The temperature of the cell is controlled using a set of two Si diodes, located at the top and at the bottom of the cell body. A typical cell of the second type is shown in Fig. 5. The cell body, characterized by pathlength of 10 mm and equipped with ZnSe windows, is mounted below a LN_2 Dewar and is equipped with a Sunrod electric minicartridge heater. The temperature of the cell body is measured using a Pt-100 thermoresistor, and the heating cartridge is controlled using a Eurotherm 3504 PID controller.

Fig. 4. Cryostat used for the study of solutions in LNe, LAr, LKr, and LXe. The cell is equipped with wedged Si windows and has a pathlength of 40 mm. See also Colour Insert.

Fig. 5. Cryostat used for the study of solutions in LAr, LKr, and LXe. The cell has a pathlength of 10 mm and is equipped with ZnSe windows. See also Colour Insert.

The main advantages of designs shown in Figs. 4 and 5 are (i) the improvement in temperature stability and (ii) the possibility to conduct experiments at temperatures below 77 K. The combination with the closed-cycle helium refrigerator, however, also has several disadvantages. Because double-stage cryostats show severe mechanical vibrations, the cryostat cannot be sealed hermetically to a vacuum spectrometer, so that weak absorptions due to water vapor and CO_2 cannot be completely avoided. The use of this type of cryostat also results in a much more complicated setup which is less flexible than those with LN_2 cooling. Finally, the cooling power of closed-cycle cryostats is significantly smaller than that of LN_2, resulting in much longer times to cool down the cell.

For the low-pressure cells, filling and evacuating of the cell is performed using a 1/8″ stainless steel tube brazed into the cell body and connected to the pressure manifold. For the cryostats based on the high-pressure design, two different 1/8″ tubes are brazed into the cell body. One tube is used to connect the cell with the pressure manifold, while the second tube is connected to a safety relief valve. The pressure manifolds are equipped with bellow valves and allow the handling of pressures up to 145 bars. The manifolds are connected to a forepump–diffusion pump combination and equipped with a pirani gauge, a capacitance manometer and a Bourdon-type high-pressure meter. The pirani gauge is used to check the vacuum in the manifold and in the actual cell, while the capacitance manometer is used to monitor the amount of gas used in a particular experiment. The Bourdon manometer is used to control the pressure of the solvent gas. All pressure meters can be isolated from the manifold using bellow valves, which is essential in the case of the pirani and capacity gauges when the manifold is put under high pressure. The manifolds also have several inlets through which compounds and solvent gases can be submitted to the system.

In a typical experiment, the evacuated cell is positioned in the spectrometer and vacuum shroud, and the cell and the pressure manifold are evacuated. Subsequently, the solutions are prepared in the following way. After cooling the cell to the required temperature, some amount of the solute gas, monitored by the capacitance manometer, is admitted into an isolated part (with accurately known volume) of the pressure manifold. Upon opening of the valve connecting the actual cell and the pressure manifold, the vapors condense in the cell or on the walls of the filling tube. If necessary, this procedure is repeated for other species. The manifold and the cell are then pressurized with the solvent gas, which immediately starts to condense. This condensation takes part in the cold part of the filling tube and in the actual cell, allowing the deposited compounds to dissolve. When the filling and evacuation of a liquid cell is followed visually, the condensation of the rare gas into the liquid cell is quite vigorous: the turbulence created while filling the cell suffices to homogenize the solution so that no additional stirring is required. After

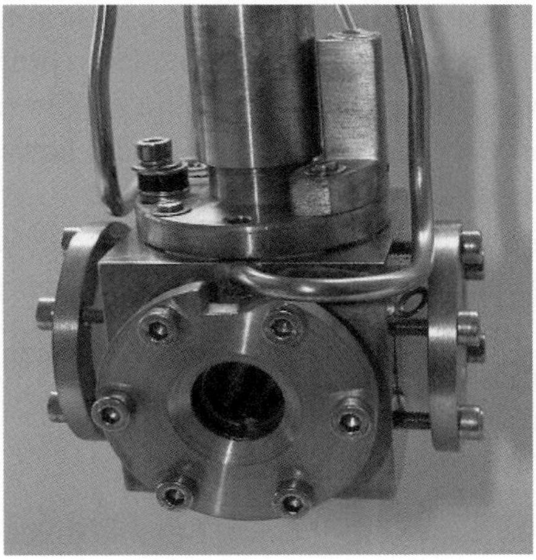

Fig. 6. Liquid cell used to study the Raman spectra of cryosolutions. See also Colour Insert.

stabilizing the temperature, the solution is ready for spectroscopic investigation.

An important characteristic of a solution is the concentration of the solutes. Unfortunately, because it cannot be verified if the compound deposited into the cell has completely dissolved and because the level of the solvent in the filling tube is not precisely known, an accurate determination of the actual concentrations is very difficult.

Because cryosolutions often are studied in cryostats with long pathlengths, special attention should be paid to possible impurities in the solvent gases used. Experiments with LAr or LN_2 as a solvent can be performed by using gases with a stated purity of 99.9999%. The highest purity available for Kr or Xe is substantially less, resulting in several weak absorption bands due to, amongst others, CO_2 and SiF_4.

2.2. *Raman spectroscopy of cryosolutions*

To record Raman spectra of cryosolutions, a high-pressure liquid cell has been recently constructed.[20] The cell body shown in Fig. 6 is machined from a brass block, is equipped with four quartz windows at right angles, and is attached to the cold head of a CRYO Industries of America RC102 continuous flow cryostat. As before, the actual cell is suspended in a vacuum shroud while filling and evacuating of the cell is performed using two separate 1/8″ stainless steel tubes brazed into the cell body.

To maximize the number of scattered photons, the cell is aligned within a multipass arrangement that allows the incident laser beam to pass through the same liquid several times. Because of the double set of windows through which the laser beam has to pass, experience shows that the increase of the signal is limited to about a factor of 5.

The usual stronger lasing lines of an argon–ion laser can be used for Raman excitation, with the power of the incident laser beam typically set between 0.5 and 1.5 W. The frequencies are calibrated using Ne and Hg emission lines, and depending on the setup used, are expected to be accurate to better than 0.2–0.5 cm^{-1}.

3. General Methodology Used for Studying Weakly Bound Molecular Complexes

Since the early reports on blue-shifting C−H···X hydrogen bonds, the spectroscopic and structural properties of these types of interactions have been the subject of various theoretical[21–40] and experimental[41–46] studies. In contrast to classical hydrogen bonds, blue-shifting hydrogen bonds are characterized by a strengthening of the C−H bond, and a concomitant blue shift of its stretching frequency. In addition, in a number of cases, an intensity decrease of the corresponding C−H stretching mode is observed. The C−H bond length changes and the resulting frequency shifts are explained as the result of opposite (repulsive and attractive) forces which occur when the H atom of the C−H bond penetrates the electron rich region of the Lewis base. The changes of the IR intensity are correlated with the dipole moment functions of the proton donor under study. In the following paragraphs, typical results obtained by dissolving mixtures of C−H proton donors such as CHF_3, $CHCl_3$, the mixed haloforms CHF_xCl_{3-x}, and halothane, $CF_3CBrClH$, and a variety of oxygen, nitrogen, and fluorine containing Lewis bases in LAr, LKr, and/or LXe and by studying these solutions using IR and/or Raman spectroscopy are summarized.[47–63] It is interesting to note that different methodologies applied in the study of complexes of C−H proton donors such as CHF_3, C_2H_2, CH_3CCH, and CF_3CCH in cryogenic solutions have been reviewed previously by Tokhadze and co-workers.[64-69] The ideas reported in these papers will not be repeated here. These authors also

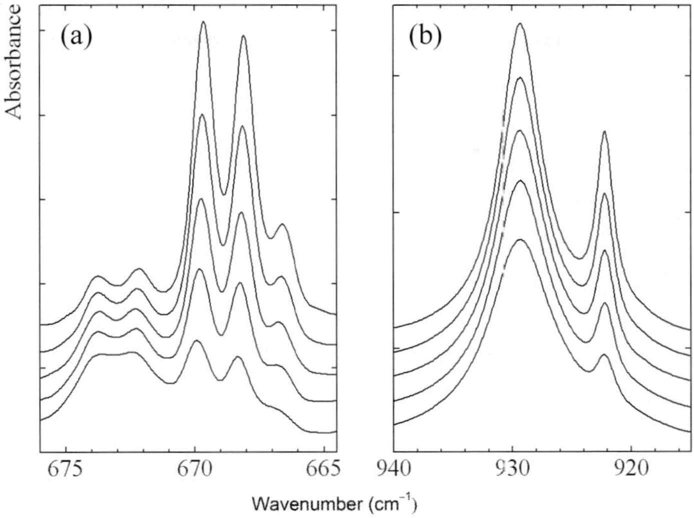

Fig. 7. Symmetric CCl_3 stretching (a) and symmetric C–O–C stretching (b) regions for a solution in LKr containing dimethylether and chloroform. From top to bottom, the temperature of the solution increases, from 123 to 150 K. Reprinted with permission from Ref. 52. Copyright (2004) American Chemical Society.

studied the properties of the proton donors in LN_2, liquid CO, and liquid CO_2, while in the following paragraphs only results obtained using LAr, LKr, and LXe as solvents will be described.

3.1. *General methodology and subtraction procedures*

The study of complexes formed in cryosolutions is divided into several phases. In the first step, spectra of single-monomer solutions of the compounds under study are recorded to investigate their solubility, and to see if auto-complexes such as dimers, trimers, etc. are formed. Next, spectra of mixed solutions are recorded and new bands, not present in the single-monomer spectra, are taken to signal the formation of complexes. These studies typically involve the recording of spectra of solutions with different monomer concentrations, and the recording of spectra at different temperatures. The analysis often requires also the combination of cells with different pathlengths and/or window materials.

As an example, spectra obtained for a temperature study of a solution in LKr containing dimethyl ether (DME) and $CHCl_3$ are shown in Fig. 7. The results show that new bands due to the formation of a complex are observed for both the symmetric CCl_3 stretching region of $CHCl_3$ (Fig. 7a) and the symmetric C–O–C stretching region of DME (Fig. 7b). Due to the weakness of the complex, the new bands are only slightly shifted from the corresponding monomer bands, resulting in considerable overlap. To separate the contributions from the monomers and complex, subtraction techniques are used, in which spectra of single monomer solutions, recorded at similar concentrations and at identical temperatures, are rescaled and subtracted from the mixed-monomer spectra. The results of such analyses obtained for the spectra in Fig. 7 are given in Fig. 8. The data show that for monomer $CHCl_3$ and for the complex, three overlapping bands due to CCl_3 stretching modes in different $^{35}Cl/^{37}Cl$ isotopologues are present. The results for the C–O–C stretching region are relatively straightforward, with monomer and complex bands appearing near 930 and 922 cm^{-1}.

It is of interest to note that, by using subtracting procedures, the band areas of the monomer and complex bands can be accurately determined using a simple numerical integration and that the inaccuracies inherently connected to resolving strongly overlapping bands with least square band fitting procedures can be avoided. This result is important as accurate band areas of monomer of complex bands are required during further analyses including, e.g., the confirmation of the stoichiometry of the complex, the determination of the standard complexation enthalpy, and the determination of the IR-intensity ratio $\varepsilon_{complex}/\varepsilon_{monomer}$.

3.2. *Stoichiometry*

The stoichiometry of the complexes is established from isothermal spectra in which the concentrations of the solutes are systematically varied in dilute solutions. Using the Lambert–Beer law combined with the equilibrium constant for the formation of a complex $A_m{\cdot}B_n$

$$mA + nB \leftrightarrow A_m{\cdot}B_n \tag{1}$$

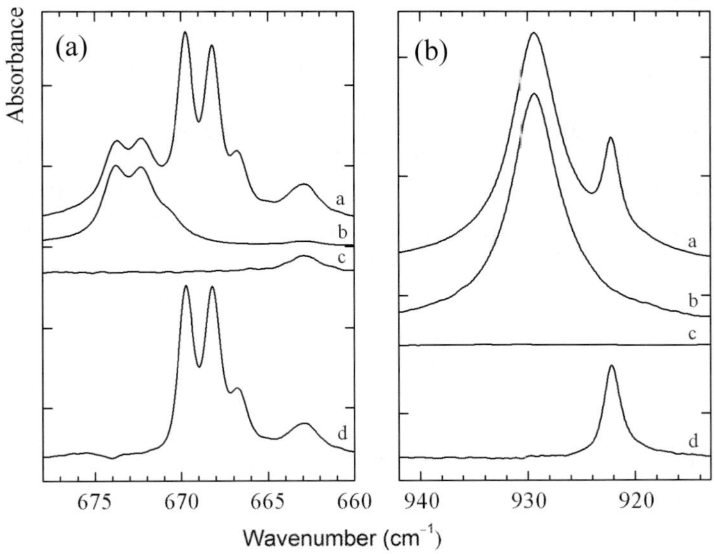

Fig. 8. Decomposition of an experimental spectrum of a solution in LKr containing dimethyl ether and chloroform, at 135 K: (a) symmetric CCl_3 stretching region; (b) symmetric C–O–C stretching region. Trace a shows the original spectra recorded for a solution of dimethyl ether and chloroform. Traces b and c are the rescaled monomer spectra showing the contributions of monomer dimethyl ether (trace b) and monomer chloroform (trace c). Trace d shows the isolated spectra of the complex obtained by subtracting traces b and c from trace a. The weak feature at 663 cm^{-1} observed in traces a, c, and d, is due to a small amount of CO_2. Reprinted with permission from Ref. 52. Copyright (2004) American Chemical Society.

it can be shown that the band area of a complex band, $I_{A_m \cdot B_n}$, is linearly related to the product of the *m*th power of the monomer band area I_A and the *n*th power of the monomer band area I_B:

$$I_{A_m \cdot B_n} = C(I_A)^m (I_B)^n, \tag{2}$$

where C is a constant related to the equilibrium constant. Equation (2) shows that for an isothermal concentration study, the plot of $I_{A_m \cdot B_n}$ versus $(I_A)^x(I_B)^y$ yields a linear dependence only when $x = m$ and $y = n$. Hence, by plotting the complex band area $I_{A_m \cdot B_n}$ versus the product of the monomer band areas $(I_A)^x(I_B)^y$ for various integer values of x and y, the stoichiometry of the complex can be established.

3.3. *Relative stability*

The standard complexation enthalpy ΔH^o of the complexes is derived from temperature studies in which the spectra are recorded at a variety of temperatures. For traditional solvents, the complexation enthalpy for a complex can be obtained by plotting the logarithm of the band area ratio of $I_{A_m \cdot B_n} / (I_A)^m (I_B)^n$ versus the inverse temperature, the relation between both quantities, derived from the van 't Hoff isochore, being given as

$$\ln\left[\frac{I_{A_m \cdot B_n}}{(I_A)^m (I_B)^n}\right] = -\frac{\Delta H^o}{RT} + c^{st}. \tag{3}$$

Due to the pronounced temperature variation of the density of liquid rare gases, this analysis is not directly applicable for cryosolutions. Corrections to account for the changes in liquid density have been described by Bertsev *et al.*[70] and Van der Veken.[71] These studies show that for an ideal solution, the equilibrium constant of a complex $A_m \cdot B_n$ can be written as

$$K_{eq} = \frac{I_{A_m \cdot B_n}}{(I_A)^m (I_B)^n} \frac{\varepsilon_A^m \varepsilon_B^n}{\varepsilon_{A_m \cdot B_n}} \frac{\rho^{n+m-1} d^{n+m-1}}{M_a^{n+m-1}}, \tag{4}$$

where $I_{A_m \cdot B_n}$, I_A, and I_B are the areas of bands assigned to the complex and the monomer species, $\varepsilon_{A_m \cdot B_n}$, ε_A, and ε_B are the infrared intensities, ρ, and M_a are the average density and the average molar mass of the solution, and d is the pathlength of cell. For dilute solutions, combination of expression [Eq. (4)] with $-RT \ln K_{eq} = \Delta G^o$ and $\Delta G^o = \Delta H^o - T\Delta S^o$ leads to the expression

$$-\frac{\Delta H^o}{RT} = \ln\left[\frac{I_{A_m \cdot B_n}}{(I_A)^m (I_A)^n}\right] + (n+m-1)\ln\rho + c^{st} \tag{5}$$

with

$$c^{st} = \ln\left[\frac{\varepsilon_A^m \varepsilon_B^n}{\varepsilon_{A_m \cdot B_n}}\left(\frac{d}{M_a}\right)^{n+m-1}\right] - \frac{\Delta S^o}{R}. \tag{6}$$

These equations can be used to determine the complexation enthalpy ΔH^o.

From Eq. (5) it can be deduced that the influence of the solvent density on the slope of a van ‘t Hoff plot can be corrected for by using two different procedures. In the first, direct method for each temperature studied, the values for $\ln[I_{A_m \cdot B_n} / (I_A)^m (I_B)^n]$ and $(n + m - 1)\ln \rho$ are calculated separately and the resulting sum is plotted against the inverse temperature. The value for the complexation enthalpy is then obtained by calculating the slope of the corresponding linear regression line.

The second, indirect method is based on the fact that in the temperature interval of interest the logarithm of the density of the cryosolvents is very nearly inversely related to the temperature:

$$\ln \rho = a + \frac{b}{T}, \tag{7}$$

where a and b are temperature-independent constants. With this, Eq. (5) can be written as

$$\ln\left[\frac{I_{A_m \cdot B_n}}{(I_A)^m (I_B)^n}\right] = -\left[\frac{\Delta H^{\circ} + (n + m - 1)Rb}{R}\right]\frac{1}{T} - \left(c^{st} + (n + m - 1)a\right). \tag{8}$$

This relation shows that the slope of the van ‘t Hoff plot obtained by plotting $\ln[I_{A_m \cdot B_n}/(I_A)^m(I_B)^n]$ versus $1/T$ equals $-[\Delta H^{o} + (n + m - 1)Rb]/R$, and not $-\Delta H^{o}/R$ as in Eq. (3). In the resulting method, a van ‘t Hoff plot is constructed by plotting the logarithm of the band area ratio $\ln[I_{A_m \cdot B_n}/(I_A)^m(I_B)^n]$ versus $1/T$, and the slope of the regression line is calculated. Subsequently, the complexation enthalpy is obtained by correcting the slope of the regression line by the term $(n + m - 1)Rb$. The values for Rb to be used during these calculations are 0.46 kJ mol^{-1} for LAr, 0.62 kJ mol^{-1} for LKr, and 0.79 kJ mol^{-1} for LXe.[71]

3.4. *Intensity ratios* $\varepsilon_{\mathrm{complex}}/\varepsilon_{\mathrm{monomer}}$

The formation of a C−H···X hydrogen bond not only influences the frequency of the C−H stretching fundamental, but also changes its infrared intensity. The physical origin of this change will be discussed in detail below. It is of interest to note here, however, that experimental information on the intensity ratios $\varepsilon_{\mathrm{complex}}/\varepsilon_{\mathrm{monomer}}$ can also be derived from the temperature studies if the solubility of the compounds under study is not affected in the temperature interval studied. In such case, the total concentration of a compound as a monomer and in the complex remains constant so that an increase in the amount of complex automatically leads to a decrease of the monomer concentration, i.e.,

$$c_{\text{complex}} + c_{\text{monomer}} = c^{\text{st}} \tag{9}$$

$$\Delta c_{\text{complex}} = -\Delta c_{\text{monomer}} \tag{10}$$

Combining Eq. (10) with Beer's law results in:

$$\frac{\Delta I_{\text{complex}}}{\varepsilon_{\text{complex}}\, d} = -\frac{\Delta I_{\text{monomer}}}{\varepsilon_{\text{monomer}}\, d}, \tag{11}$$

where $I = \Delta I\,(T) - I\,(T_{\text{R}})$ and T_{R} is a reference temperature. This changes Eq. (11) in

$$I_{\text{complex}}(T) = -\frac{\varepsilon_{\text{complex}}}{\varepsilon_{\text{monomer}}} I_{\text{monomer}}(T) + c^{st}. \tag{12}$$

Because of the pronounced temperature dependence of the solvent densities, the above assumption again is not valid for cryosolutions. The effect of the changing solvent densities, however, can be accounted for by substituting the band areas I_{complex} and I_{monomer} by the density-corrected values defined as

$$I_i^{\text{corr}} = I_i \frac{\rho(T_{\text{R}})}{\rho(T)}, \tag{13}$$

where $\rho(T_{\text{R}})$ and $\rho(T)$ refer to the density of the solvent at a reference temperature T_{R} and at the actual temperature T, respectively.

To determine the intensity ratio $\varepsilon_{\text{complex}}/\varepsilon_{\text{monomer}}$ for a specific doublet, in the first step of the actual analysis, for each temperature studied, the band areas of the monomer and complex bands are determined, and corrections for density changes are applied. The corrected values are then plotted against each other and the intensity ratio is derived from the slope of the linear regression line.

3.5. *Ab intio calculations and statistical thermodynamics*

The rationalization of experimental data is often supported by a variety of quantum chemical calculations, in which the structural, thermodynamic, and spectroscopic properties of the species under study are predicted. The description of the methodologies and basis sets to be used in these studies are beyond the scope of this chapter. It suffices to note that in order to assess the calculated values, the predicted complexation energies should be converted into complexation enthalpies and that for weak molecular complexes, the statistical thermodynamical corrections for zero-point vibrational and thermal influences used for these conversions yield a vapor phase complexation enthalpy that, in absolute values, typically is 2–3 kJ mol^{-1} smaller than the initial complexation energy. It is also worth noting that the vibrational spectra are traditionally predicted using the double harmonic approximation, and that effects due to mechanical and/or optical anharmonicities often are neglected. This can lead to severe errors, as in most C−H proton donors studied significant anharmonic couplings exist[55,65,66,72] between the C−H stretching and the C−H bending modes.

3.6. *Monte Carlo-free energy perturbation simulations*

Although solutions in liquefied rare gases have been described as a pseudogas phase,[73–75] it is now generally accepted that in the cryosolutions significant solute–solvent interactions can occur, and that these interactions can influence the relative stability of the complexes studied. The *ab initio* complexation energies and the vapor phase complexation enthalpies therefore cannot directly be compared with the experimental values derived from cryosolutions. To estimate the solvent effects, and to predict the complexation enthalpies in the cryosolutions, for each species involved, the Gibbs energy and the enthalpy of solvation are derived using Monte Carlo free energy perturbation (MC-FEP) calculations.[76–78] In these simulations, the solute–solvent interactions are described using a Lennard–Jones function between each solvent atom and each atom of the solute molecule, and an additional term that

Table 1. Complexation shifts (cm^{-1}), and infrared intensity ratios for the complexes of dimethyl ether, acetone, and oxirane with CHF_3, $CHClF_2$, $CHCl_2F$, and $CHCl_3$.[48,49]

Complex	Experiment (LKr, 135 K)		*ab initio* (MP2/6-311++G**)	
	$\Delta\nu_{CH}$	$\varepsilon_{complex}/\varepsilon_{monom}$	$\Delta\nu_{CH}$	$\varepsilon_{complex}/\varepsilon_{monom}$
DME·CHF_3	+18.7	0.09 (2)	+17.7	0.18
DME·$CHClF_2$[a]	+14.0	0.89 (6)	+13.1/+16.6	1.51/1.45
DME·$CHCl_2F$	+4.8	26 (1)	+10.7	16.03
DME·$CHCl_3$	−8.3	56 (3)	−4.5	1179.04
AC·CHF_3	+27.3	0.08 (1)	+30.2	0.19
AC·$CHClF_2$[a]	+24.8	0.61 (4)	+28.0/+27.8	1.22/1.27
AC·$CHCl_2F$	+16.1	4.2 (2)	+17.8	18.68
AC·$CHCl_3$	+1.4	58 (4)	+8.3	828.02
OX·CHF_3	+23.9	0.13 (1)	+38.2	0.22
OX·$CHClF_2$[a]	+20.7	0.48 (2)	+19.2/+30.9	0.54/0.22
OX·$CHCl_2F_2$[a]	+13.9	10 (1)	+22.3/+17.0	4.57/8.14
OX·$CHCl_3$	+1.3	69 (7)	+5.6	573.96

[a]For DME·$CHClF_2$, AC·$CHClF_2$, OX·$CHClF_2$, and OX·$CHCl_2F$, the *ab initio* calculations lead to two different conformations with a barrier separating the different minima in the order 0.5 kJ mol^{-1}. For completeness, the calculated properties of both conformations are given.

accounts for the polarization of the solvent atoms by the solute. The latter is calculated in a noniterative first-order approximation as[79,80]

$$E_{pol} = -\frac{1}{2}\vec{\mu}_{ind}\cdot\vec{E} = -\frac{1}{2}\alpha_{solvent}\left(\vec{E}\cdot\vec{E}\right), \qquad (14)$$

where $\alpha_{solvent}$ is the polarizability of the rare gas used as solvent, and $\vec{E}$ is the electric field generated by the solute. The partial charges on the different nuclei used to mimic the electric field of the solute are optimized so as to reproduce the dipole moment and the electric potential around the molecule.[81]

In the first step of the calculation, the Gibbs energies of solvation for monomers and complexes are obtained at different temperatures, varying between 93 and 143 K for LAr, 119 and 179 K for LKr, and 171 and 224 K for LXe. In the second step, the resulting values are combined and the enthalpies of solvation $\Delta_{sol}H$ are extracted using the expressions $\Delta_{sol}S = -\partial(\Delta_{sol}G)/\partial T$ and $\Delta_{sol}H = \Delta_{sol}G + T\Delta_{sol}S$. Inspection of the calculated values shows that for all complexes studied, the enthalpies of solvation of the complexes are significantly smaller than those of the monomers and that in the solutions studied the complexes are destabilized with respect to the monomers. The destabilization typically varies between 2.5 and 7.5 kJ mol^{-1}.

4. Complexes of Dimethyl Ether, Acetone, Oxirane with CHF_3, $CHClF_2$, $CHCl_2F$, and $CHCl_3$

4.1. *Vibrational spectra*

The complexation shifts and intensity ratios for the haloform C−H stretches obtained for the complexes formed between the haloforms CHF_xCl_{3-x} (with $x = 0, 1, 2$, and 3) and the oxygen-containing Lewis bases dimethylether–d_6 (DME–d_6), acetone–d_6 (AC–d_6) and oxirane–d_4 (OX–d_4)[47–50] are summarized in Table 1. The spectral data leading to the results for AC–d_6 are shown in Fig. 9. The fully deuterated bases were used to avoid overlap of the ν_{CH} fundamentals appearing in the proton donors and acceptors.

For the complexes with DME, the haloform C−H stretch changes in a systematic way from blue shifting for the CHF_3 complex to red shifting for the $CHCl_3$ complex, while its intensity is changing from a decrease to an increase in comparison with that of the monomer haloform. The combined data for AC and OX show that for all haloforms a blue shift is observed, which increases, for either Lewis base, with the number of fluorine atoms in the haloform. As for DME, the IR intensities of the ν_{CH} mode were observed to increase for $CHCl_3$ and $CHCl_2F$, while for $CHClF_2$ and CHF_3, a systematic decrease was found. Comparison of the experimental and *ab initio* data shows that for all complexes, the predicted values reproduce the experimental trends with increasing fluorination, albeit that the quantitative agreement leaves to be desired.

4.2. *Relative stability*

The experimental complexation enthalpies[52] for the complexes observed in LAr and/or LKr, and the predicted values for the gas phase, LAr, and LKr

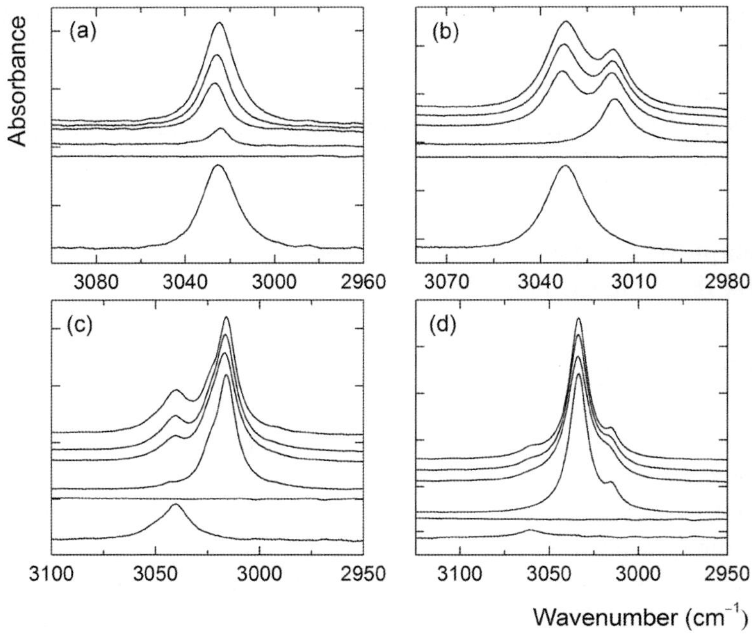

Fig. 9. IR spectra of cryosolutions in the haloform C–H stretching region. (a) The upper three spectra are of a solution in LKr-containing AC–d_6 and $CHCl_3$ at 134, 151, and 169 K (from top to bottom); the next two spectra are of a solution containing $CHCl_3$ or AC–d_6 at 134 K. The bottom trace is the spectrum of the 1:1 complex, at 134 K, isolated by subtraction. (b) The top three spectra are of a solution LKr containing AC–d_6 and $CHCl_2F$ at 138, 150, and 163 K (from top to bottom); the next two traces are the spectra of a solution containing only $CHCl_2F$ or AC–d_6, at 138 K; the bottom trace is the spectrum of the 1:1 complex, at 138 K, isolated by subtraction. (c) The top three spectra are of a solution in LKr containing AC–d_6 and $CHClF_2$ at 134, 156, and 172 K (from top to bottom); the next two traces are the spectra of a solution containing only $CHClF_2$ or AC– d_6, at 134 K; the bottom trace is the spectrum of the 1:1 complex, at 134 K, isolated by subtraction. (d) The top three spectra are of a solution in LKr containing AC–d_6 and CHF_3 at 134, 148, and 163 K (from top to bottom); the next two traces are the spectra of a solution containing only CHF_3 or AC–d_6, at 134 K; the bottom trace is the spectrum of the 1:1 complex, at 134 K, isolated by subtraction.

are summarized in Table 2 For each of the Lewis bases, the complexation energy increases by approximately 20%, when passing from CHF_3 to $CHCl_3$. This trend is not observed for ΔH°(LAr), ΔH°(LKr) , ΔH^{o}_{exp}(LAr), or ΔH^{o}_{exp}(LKr).

Table 2. Complexation energies and enthalpies in gas (g) and solutions (kJ mol^{-1}), for the complexes of dimethyl ether, acetone, and oxirane with CHF_3, $CHClF_2$, $CHCl_2F$, and $CHCl_3$.[47,52]

			T = 110 K			**T = 135 K**		
	Ab initio		Calc.		Exp.	Calc.		Exp.
	ΔE^a	ΔE^b	$\Delta H°(g)$	$\Delta H°(LAr)$	$\Delta H°(LAr)$	$\Delta H°(g)$	$\Delta H°(LKr)$	$\Delta H°(LKr)$
DME/CHF_3	−17.3	−17.9	−15.7	−11.2(4)	−12.5(2)			
DME/$CHClF_2$[c]	−17.5	−18.9	−16.7	−11.9(6)	−12.6(1)	−16.5	−11.1(3)	−12.5(1)
	−17.5	−18.5				−16.1	−10.7(3)	
DME/$CHCl_2F$	−18.1	−20.0	−16.3	−11.5(6)		−17.4	−9.2(4)	−12.2(2)
DME/$CHCl_3$	−19.0	−21.2	−17.7	−10.4(5)	−12.9(6)	−18.5	−8.5(4)	−11.3(1)
AC/CHF_3	−17.2	−18.6				−16.0	−10.8(2)	−10.9(3)
AC/$CHClF_2$[c]	−17.1	−19.6				−17.0	−10.5(2)	−12.5(1)
	−17.3	−19.4				−17.8	−11.5(2)	
AC/$CHCl_2F$	−17.9	−20.4				−17.7	−9.6(2)	−12.9(1)
AC/$CHCl_3$	−18.6	−22.6				−19.8	−10.8(3)	−13.7(1)
OX/CHF_3	−16.9	−18.2				−16.0	−10.0(2)	−11.6(2)
OX/$CHClF_2$[c]	−17.1	−19.6				−17.3	−11.0(1)	−12.0(4)
	−17.3	−19.5				−17.7	−10.8(1)	
OX/$CHCl_2F$[c]	−17.6	−20.4				−18.0	−9.7(1)	−11.8(3)
	−17.9	−21.0				−18.9	−10.9(2)	
OX/$CHCl_3$	−18.5	−21.4				−18.8	−9.4(2)	−10.8(3)

[a]CP-corrected complexation energy obtained at the MP2/6-311++G(d,p) level.
[b]CP-corrected complexation energy obtained at the MP2/aug-cc-PVTZ level.
[c]For DME.$CHClF_2$, AC·$CHClF_2$, OX·$CHClF_2$ and OX·$CHCl_2F$, the results obtained for both conformations are given.

4.3. *Analysis of the complexation shifts*

Insight into the complexation behavior of the C–H stretching modes and the different trends observed for DME, AC, and OX, was obtained[51] by using a perturbative approach of vibrational frequencies similar to that originally proposed by Buckingham.[82–84] In the original approach, the intermolecular potential energy U is expanded as a power series in the dimensionless normal coordinates Q_1:

$$U(Q_1) = U(0) + \left(\frac{\partial U}{\partial Q_1}\right)_0 Q_1 + \frac{1}{2}\left(\frac{\partial^2 U}{\partial Q_1^2}\right)_0 Q_1^2. \tag{15}$$

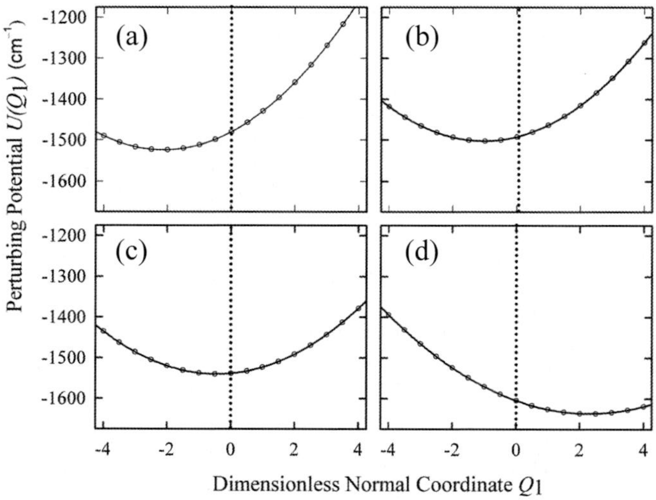

Fig. 10. Perturbing potentials $U(Q_1)$ obtained for DME·CHF_3 (a), DME·$CHClF_2$ (b), DME·$CHCl_2F$ (c), and DME·$CHCl_3$ (d).

The resulting complexation shift, obtained by assuming that the anharmonicities of the stretching mode are dominated by the third-order correction $\alpha_{111}\, Q_1^3$ is given by

$$\Delta\nu_{CH} = \frac{1}{2}\left(\frac{\partial^2 U}{\partial Q_1^2}\right)_0 - \frac{3\alpha_{111}}{\nu_1}\left(\frac{\partial U}{\partial Q_1}\right)_0. \qquad (16)$$

To avoid limitations related to the use of perturbation theory and to be able to account for higher order terms including, e.g., the quartic anharmonicities of the stretching modes, the original approach was further elaborated.[51] In the resulting model, the C−H stretching mode in a specific complex is described by a total Hamiltonian H_t of the form

$$H_t = H_a + U(Q_1), \qquad (17)$$

where H_a is the Hamiltonian that describes the anharmonic ν_{CH} mode in the monomer proton donor, and $U(Q_1)$ is the perturbing potential. The anharmonic Hamiltonian is approximated as

$$H_a = H_h + k_{3a}Q_1^3 + k_{4a}Q_1^4, \qquad (18)$$

Table 3. Complexation shifts (cm^{-1}), calculated for the complexes of dimethyl ether, acetone, and oxirane with CHF_3, $CHClF_2$, $CHCl_2F$, and $CHCl_3$.[51]

Dimethylether	CHF_3	$CHClF_2$	[a]	$CHCl_2F$		$CHCl_3$
$\Delta\nu_1$	13.4	6.4	10.9	2.4		−9.4
$\Delta\nu_2$	12.5	12.3	12.4	10.7		8.0
$\Delta\nu_3$	1.0	1.0	0.2	0.0		−0.1
$\Delta\nu$	26.9	19.7	23.5	13.1		−1.5
Acetone	CHF_3	$CHClF_2$	[a]	$CHCl_2F$		$CHCl_3$
$\Delta\nu_1$	21.4	15.9	17.6	8.4		−1.5
$\Delta\nu_2$	20.0	22.1	20.3	17.4		15.3
$\Delta\nu_3$	1.5	1.9	1.6	1.3		1.1
$\Delta\nu$	43.0	39.9	39.4	27.0		14.9
Oxirane	CHF_3	$CHClF_2$	[a]	$CHCl_2F$	[a]	$CHCl_3$
$\Delta\nu_1$	31.1	11.5	21.9	12.2	9.8	−1.5
$\Delta\nu_2$	16.4	15.0	16.8	16.9	14.3	12.1
$\Delta\nu_3$	0.04	0.8	0.5	0.9	0.7	0.7
$\Delta\nu$	47.5	27.4	39.3	29.9	24.9	12.1

[a]For DME·$CHClF_2$, AC·$CHClF_2$, OX·$CHClF_2$, and OX·$CHCl_2F$, the results obtained for both conformations are given.

where H_h is the harmonic Hamiltonian, and k_{3a} and k_{4a} are the cubic and quartic force constants for the CH oscillator. The perturbing potential is given by

$$U(Q_1) = U(0) + \sum_{n=1}^{3} k_{nu} Q_1^n, \tag{19}$$

In the first step of the calculation, the anharmonic stretching frequencies of monomers and complexes are obtained from a perturbation calculation involving H_a and H_t, respectively, and the wavenumber shifts are calculated. In the second step, the shifts are related to the expansion coefficients of the perturbing potential, using the approximate equation

$$\Delta\nu = \nu_{\text{complex}} - \nu_{\text{monomer}}$$

$$\approx \sum_{n=1}^{3} k_{nu} \langle 1|Q_1^n|1\rangle - \langle 0|Q_1^n|0\rangle = \sum_{n=1}^{3} \Delta\nu_n, \tag{20}$$

where $|0\rangle$ and $1|\rangle$ are the ground and first excited state wave functions, respectively, of the Hamiltonian H_a. The expressions in square brackets

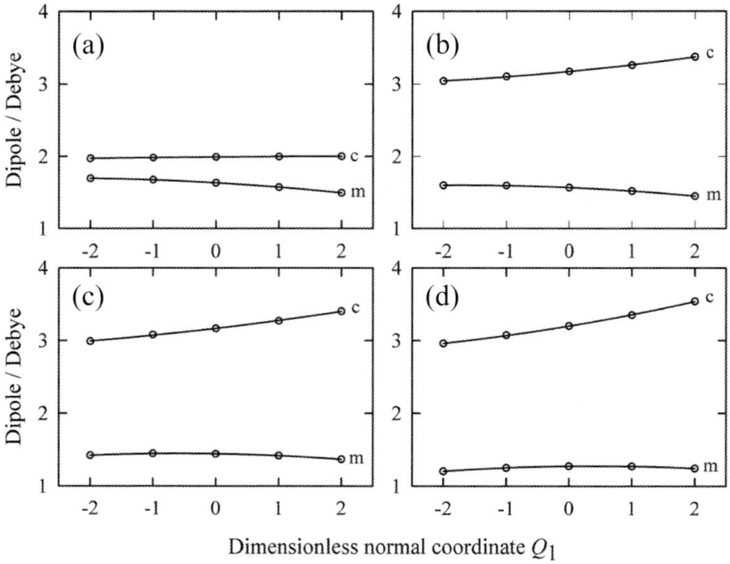

Fig. 11. Dipole moment functions for (a) CHF_3 and OX·CHF_3, (b) $CHClF_2$ and OX·$CHClF_2$, (c) $CHCl_2F$ and DME·$CHCl_2F$, and (d) $CHCl_3$ and OX·$CHCl_3$. The dipole gradients for the monomers are −0.050, −0.038, −0.014, and +0.010 Debye, while those for the complexes are +0.007, +0.082, +0.102, and +0.144 Debye. The predicted intensity ratios $\varepsilon_{complex}/\varepsilon_{monomer}$ are 0.020, 0.207, 53.1, and 207.4.

in Eq. (20) always have positive values, so that the shifts $\Delta\nu_n$ have the sign of the corresponding k_{nu}.

The perturbing potentials obtained for the complexes involving DME are shown in Fig. 10. The shifts obtained for the complexes with DME, AC, and OX are collected in Table 3. It can be seen that for the complexes with CHF_3, $CHClF_2$, and $CHCl_2F$, positive values are observed for k_{1u} ($\Delta\nu_1$) and k_{2u} ($\Delta\nu_2$), while for the complexes with $CHCl_3$, a negative value is observed for k_{1u} ($\Delta\nu_1$) and a positive value is obtained for k_{2u} ($\Delta\nu_2$). Comparison of the calculated values and the experimental data in Table 1 shows that the model correctly predicts the direction of the complexation shifts for all the complexes studied, but that the quantitative agreement leaves to be desired. The most likely reason for the poor quantitative agreement is the absence of averaging of the shifts over the van der Waals modes of the complexes.[51]

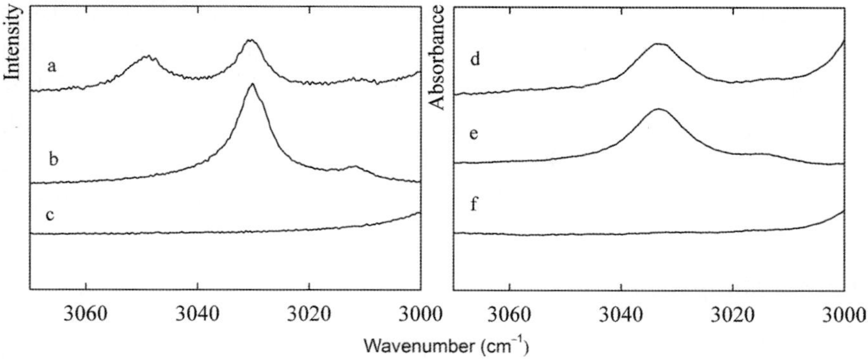

Fig. 12. Vibrational spectra in the C–H stretching region, recorded from solutions in LKr at 131 K containing mole fractions of 8×10^{-4} of each of the solutes. (a) Raman spectrum of a solution containing DME and CHF_3; (b) Raman spectrum of a solution containing CHF_3; (c) Raman spectrum of a solution containing DME; (d) IR spectrum of a solution containing DME and CHF_3; (e) IR spectrum of a solution containing CHF_3; (f) IR spectrum of a solution containing DME.[51]

4.4. *Analysis of the intensity ratios*

In a simplified model, the intensities of the monomer proton donors and those of the complexes are related to the squares of the dipole gradients $(\partial\vec{\mu}_{\text{proton donor}}/\partial Q_1)_0$ and $(\partial\vec{\mu}_{\text{complex}}/\partial Q_1)_0$. The two are related as

$$(\partial\vec{\mu}_{\text{complex}}/\partial Q_1)_0 = (\partial\vec{\mu}_{\text{proton donor}}/\partial Q_1)_0 + (\partial\vec{\mu}_{\text{ind, protondonor}}/\partial Q_1)_0 \\ + (\partial\vec{\mu}_{\text{ind,proton acceptor}}/\partial Q_1)_0 \qquad (21)$$

in which $\vec{\mu}_{\text{proton donor}}$ is the permanent dipole moment of the monomer involved, and $\vec{\mu}_{\text{ind, proton donor}}$ and $\vec{\mu}_{\text{ind, proton acceptor}}$ are the induced dipole moments generated by the electric field of the partner molecule. The induced dipole moment of the proton acceptor traditionally gives rise to a small negative derivative with respect to the C–H bond elongation.[25] The induced dipole moment of the haloform yields a positive and dominating contribution.[25]

The dipole moment functions for the haloforms studied and those for the complexes with OX are shown in Fig. 11. It can be seen that for CHF_3

the negative dipole gradient for the monomer almost completely vanishes upon complexation, while for $CHClF_2$ the negative dipole gradient for the monomer is converted into a positive one, the absolute value of which is quite similar to that observed for the monomer. For the complexes with $CHCl_3$ and $CHCl_2F$, a negligible small dipole gradient is observed for the monomers, while for the complexes a relatively large value is predicted.

4.5. *Raman spectroscopy*

The strong decrease of the IR intensity observed for a variety of C−H···O hydrogen-bonded complex suggests that in some cases blue-shifting hydrogen bonding may escape detection in the IR region. Because the Raman intensities of the C−H stretching modes engaged in blue-shifting hydrogen bonding do not differ significantly from that of the monomers, experimental information on these complexes can be achieved by using Raman spectroscopy. To illustrate the advantages of Raman spectroscopy, the IR and Raman spectra obtained from identical solutions in LKr containing DME and CHF_3 and those of the solutions containing only monomers are compared in Fig. 12. The conditions used in the experiments are such that for the mixed solutions complex bands proving the occurrence of complexes are hardly visible in the IR spectrum.[50] In contrast, in the Raman spectra of the mixed solution, a new band due to the C−H stretching mode in the complex is clearly observed, at 3051.8 cm^{-1}.

5. Complexes of Ammonia and Trimethyl Amine with CHF_3, $CHClF_2$, $CHCl_2F$, and $CHCl_2$

Because of the limited solubility of NH_3 in LAr and LKr, and the formation of NH_3 clusters in LXe,[54] information on C−H···N hydrogen bond complexes is preferentially obtained using trimethyl amine (TMA) as Lewis base.[58,70] The IR spectra of mixed solutions in LXe containing NH_3 and CHF_3 and pyridine (PYR) and CHF_3, however, are of great interest as they give rise to typical examples of so-called pseudo blue-shifting hydrogen bonds. In this type of complexes, the observed blue shift is not caused by a change in the strength of the C−H bond, but is due to changes in the $v_1/2v_4$ Fermi resonance interactions appearing in the proton donor.

Table 4. Influence of the $\nu_1/2\nu_4$ Fermi resonance on the complexation shifts (cm^{-1}) for $NH_3 \cdot CHF_3$, $PYR \cdot CHF_3$, $N(CD_3)_3 \cdot CHF_3$, and $(CD_3)_2O \cdot CHF_3$.[55]

		LAr (90 K)		**LXe (188 K)**		
	CHF_3	$DME \cdot CHF_3$	$TMA \cdot CHF_3$	CHF_3	$NH_3 \cdot CHF_3$	$PYR \cdot CHF_3$
Experiment						
ν_4	1374.5(2)	1390.2(5)	1405.9(5)	1372(1)	1391.4(5)	1388.8(5)
ν_1	3036.7(2)	3054.8(2)	3013.9(3)	3030.5(3)	3038.1(5)	3033.3(5)
$\Delta\nu_1$		+18.1(2)	−22.8(3)		+7.6	+3
$2\nu_4$	2704.7(5)	2737(1)	2749(1)	2700.2(5)	2741(1)	2739(2)
Corrected for $\nu_1/2\nu_4$ Fermi resonance						
ν_1°	2990.2(3)	3005.3(3)	2946.7(5)	2983.6(3)	2983.3(5)	2977.6(5)
$\Delta\nu_1^\circ$		+15.1(4)	−43.5(6)		−0.3(6)	−6.0(6)
Intensity ratio (R)						
R_{exp}	28(5)	1.5(2)	1.5(2)	36(5)	2.3(3)	2.6(3)
R_{calc}	14	1.9	1.9	12.6	2.2	2.2

The characteristic frequencies and complexation shifts derived from the solutions and the results obtained by correcting the observed frequencies for changes in the Fermi resonance are summarized in Table 4. The data not only include the results for $NH_3 \cdot CHF_3$ and $PYR \cdot CHF_3$, but also include the data for the complexes of CHF_3 with $(CD_3)_2O$ and $(CD_3)_3N$. The results, obtained by assuming an effective cubic force constant α_{144}^{eff} of $-161(6)$ cm^{-1}, show that the Fermi-corrected value for ν_1 in the PYR complex is red shifted from the monomer value, by $-6.0(6)$ cm^{-1}, while for the NH_3 complex the corrections reduce the experimental blue shift to a value well below the threshold that allows the type of interaction to be identified. The data further shows that for the DME complex the Fermi correction has only little influence on the ν_1 frequency and does not change the sign of the complexation shift. For the TMA complex, for which a significant red shift of -23 cm^{-1} is observed,[70] the correction for Fermi resonance substantially increases that shift, and, evidently, also does not change its sign. For all Lewis bases studied, evidence for the changes in the Fermi resonance between ν_1 and $2\nu_4$ was also inferred from the changes in the relative intensities of the ν_1 and $2\nu_4$ vibrations.[55] The experimental intensity ratios were rationalized using a first-order perturbation description of the intensities.

It is of interest to note that recent room temperature gas-phase infrared studies of the complex of NH_3 with $CHCl_3$ showed that also for

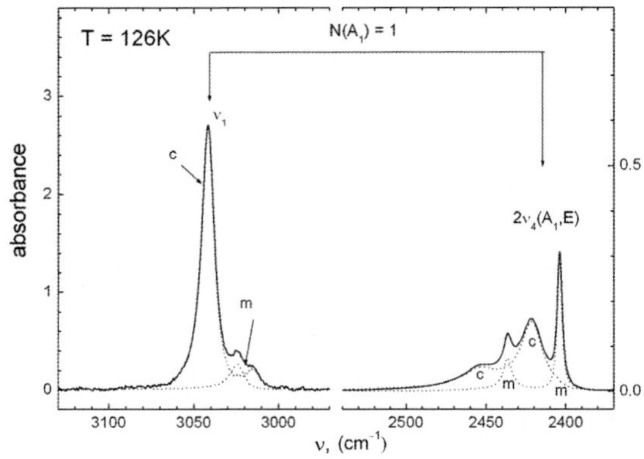

Fig. 13. IR spectra of mixed solutions in LKr, at 126 K, containing CD_3F and $CHCl_3$. The solid line refers to the spectrum of the mixed solution; the dashed lines represent the spectra of monomer $CHCl_3$ (m) and of the complex (c). The $\nu_1/2\nu_4$ Fermi resonance is referred to as $N(A_1) = 1$. Reprinted with permission from Ref. 52. Copyright (2005) Elsevier.

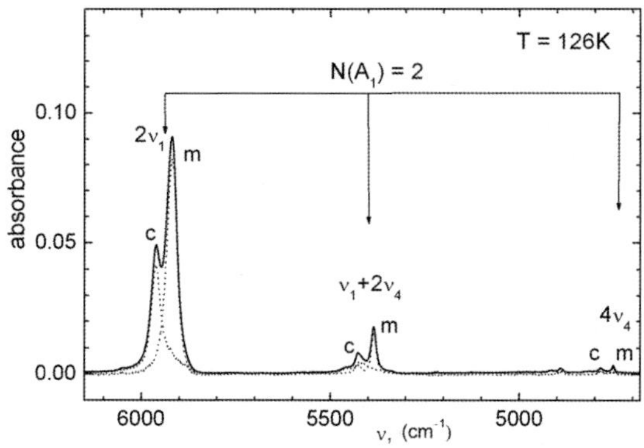

Fig. 14. IR spectra of mixed solutions in LKr, at 126 K, containing CD_3F and Cl_3. The solid line refers to the spectrum of the mixed solution; the dashed lines represent the spectra of monomer $CHCl_3$ (m) and of the complex (c). The $2\nu_1/\nu_1{+}2\nu_4/4\nu_4$ Fermi resonance pattern is referred to as $N(A_1) = 2$. Reprinted with permission from Ref. 52. Copyright (2005) Elsevier.

this species significant corrections for the $\nu_1/2\nu_4$ Fermi resonance had to be accounted for.[85] Therefore, it can be generally stated that, in order to correctly interpret the complexation shifts for the C–H proton donors, the observed frequencies should be corrected for possible Fermi resonances.

The corrections typically are less important for blue-shifting hydrogen bonds, in which the C−H stretching mode is significant blue-shifted, but are definitely required for complexes in which the stretching fundamental is significantly red-shifted.

6. Complexes of Methyl Fluoride-d_3 with CHF_3, $CHClF_2$, $CHCl_2F$, and $CHCl_3$

During recent years, interactions in which a C−H donor binds to a C−F acceptor have been the subject of experimental[7,57,86] and theoretical studies.[87–92] The IR spectra of mixed solutions in LKr containing the haloforms CHF_3, $CHClF_2$, and $CHCl_2F$ and $CHCl_3$ and/or CD_3F have been described.[56,57] For all haloforms, significant blue shifts were observed for the C−H stretching fundamental ν_1, for the first and second overtones $2\nu_1$ and $3\nu_1$, and for a variety of combination bands involving ν_1 or $2\nu_1$. For CHF_3 and $CHClF_2$, the intensity of the fundamental ν_1 was observed to decrease by a factor of 5, while for $CHCl_2F$ and $CHCl_3$ the intensity was observed to strongly increase. In contrast, the intensities of $2\nu_1$ and $3\nu_1$ reveal only modest changes.

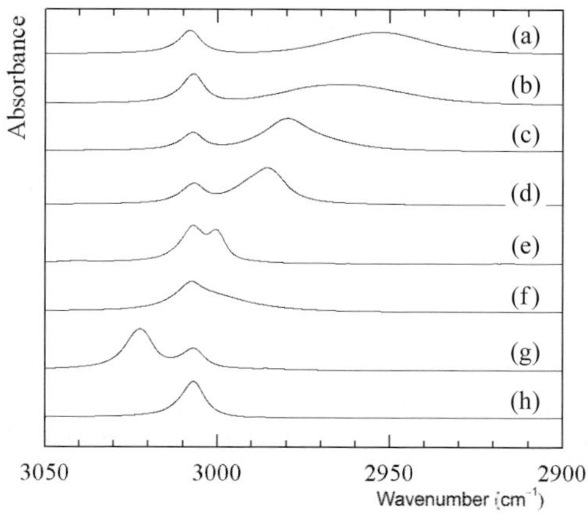

Fig. 15. IR spectra of solutions in LKr, at 123 K, containing mixtures of halothane and various Lewis bases: (a) mixture with ammonia; (b) mixture with dimethyl sulphide–d_6; (c) mixture with tetrahydrofurane–d_8; (d) mixture with dimethyl ether–d_6; (e) mixture with ethene–d_4; (f) mixture with acetone–d_6; (g) mixture with methyl fluoride–d_3; (h) solution containing only halothane.

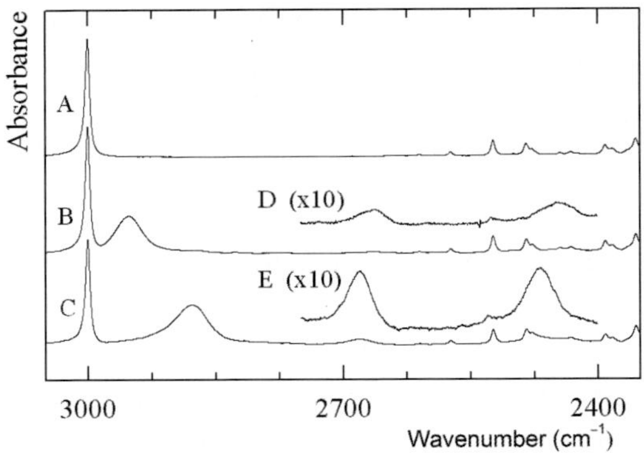

Fig. 16. IR spectra of solutions in LXe, at 188 K, containing mixtures of halothane with NH_3 and $N(CD_3)_3$. Trace A refers to a solution containing only halothane. Traces B and C correspond to a mixed solution containing NH_3 and $N(CD_3)_3$, respectively. Traces D and E show weaker details in the spectra of the complexes obtained by subtracting traces A, B, and C. The complex bands assigned to the v_1 stretching fundamental are observed near 2952 and 2880 cm^{-1}. The transition near 2664, 2446 cm^{-1}, 2680 and 2468 cm^{-1} observed in traces D and E are explained as the result of Fermi resonances involving the C–H stretching fundamental and the overtones of the C–H bending vibrations $2v_2$ and $2v_3$.

As illustrated in Figs. 13 and 14, the $v_1/2v_4$ resonances in CHF_3 and $CHCl_3$ typically result in a doublet for the v_1 fundamental region (v_1, $2v_4$) and a triplet for the $2v_1$ overtone region ($2v_1$, v_1+2v_4, and $4v_4$). The Fermi-corrected complexation shifts for v_1 are +25.7 cm^{-1} for $CD_3F \cdot CHF_3$ and +18.4 cm^{-1} for $CD_3F \cdot CHCl_3$. These values compare favorably with the values of +24 and +18 cm^{-1} deduced from the vibrational spectra, and support the above conclusion that for strongly blue-shifted hydrogen bonds, the effect of the $v_1/2v_4$ Fermi resonance on the complexation shift is negligible.

7. Molecular Complexes Involving the C–H Bond in Halothane

Although anesthetics are used daily all over the world, a lot is still to be learned about the mechanisms of their interactions.[93,94] To gain more insight in this matter the formation of C–H···X bonded complexes between halothane, $CF_3CBrClH$, and a variety of proton acceptors has been studied.[60–63] Typical spectra obtained for solutions containing halothane and a variety of proton acceptors are summarized in Figs. 15 and 16.

Table 5. Complexation shifts (cm^{-1}), and intensity rations obtained[60–63] for the C–H stretching mode in halothane. The experimental data were obtained for solutions in LKr, at 123 K, or LXe, at 176 K. The calculated values were obtained using *ab initio* at the MP2/6-311++G(d,p) level.

Complexes	**Experimental**			**Calculated**	
	Solvent	$\Delta\nu$	$\varepsilon_{com}/\varepsilon_{mon}$	$\Delta\nu$	$\varepsilon_{com}/\varepsilon_{mon}$
Methylfluoride–d_3	LKr	+15.4(1)	1.39(7)	+20.1	*7.1*
Acetone–d_6	LKr	−3.7(15)	8.8(14)		
Ethene–d_4	LKr	−7.2(4)	2.1(2)	−5.4	8.6
Dimethylether–d_6	LKr	−19(2)	6.5(1)	−11.5	30.4
Tetrahydrofuran–d_8	LKr	−26(2)	11(1)	−27.1	38.3
Dimethylsulfide–d_6	LKr	−43(2)	4.5(3)	−35.7	22.2
Ammonia	LXe	−64(3)	8.7(8)	−64.2	51.1
Trimethylamine–d_9	LXe	−126(6)	17.7(21)	−150.3	106.5

Table 6. Experimental and predicted complexation obtained for a variety of complexes involving halothane studied in LKr and/or LXe.[60–63]

Complexes	**Solvent**	ΔH_{exp} **(LKr, LXe)**	ΔH_{calc} **(LKr, LXe)**
Methylfluoride–d_3	LKr	−8.2(3)	–6.3(16)
Acetone–d_6	LKr	−15.5(8)	–
Ethene–d_4	LKr	−5.0(2)	–5.4(2)
Dimethylether–d_6	LKr	−12.3(4)	–10.2(17)
Tetrahydrofuran–d_8	LKr	−15.7(3)	–
Dimethylsulfide–d_6	LKr	−10.7(1)	–
Ammonia	LXe	−17.0(5)	–
Trimethylamine–d_9	LXe	−16.6(4)	–18.8(25)

The complexation shifts and intensity ratios derived from the experimental data and those obtained from MP2 calculations are collected in Table 5. The complexation enthalpies derived from the cryosolutions and the corresponding values derived by combining the results from *ab initio* calculations, statistical thermodynamics and Monte Carlo simulations are summarized in Table 6. The data for halothane covers the complete range of C–H···X hydrogen bonds, starting with a strong blue shift of approximately 15.4 cm^{-1} for CD_3F, and ending with strong red shifts of −64 and −126 cm^{-1} for NH_3 and $N(CD_3)_3$. The complexation enthalpies varies from −5.0(2) kJ mol^{-1} for the weak C−H···π-bonded complex with C_2D_4 and −17.0(5) for the complex with NH_3.

Apart from the intense complex bands at 2952 and 2880 cm^{-1}, assigned to the strongly red-shifted ν_{CH} stretching vibration, in the spectra of the mixed solutions containing NH_3 and $N(CD_3)_3$ two weaker features due to the 1:1 complexes with NH_3 and $N(CD_3)_3$ can be observed to appear, near 2664 and 2446 cm^{-1}, and near 2680 and 2468 cm^{-1}. These features are explained as the result of Fermi resonance interactions involving the C−H stretching fundamental ν_1 and the first overtones of the C−H bending vibrations ν_2 and ν_3. The resonance patterns observed are not yet fully understood, and are the subject of ongoing research activities.

8. Conclusions

In this chapter, typical results for blue- and red-shifting C−H···X hydrogen-bonded complexes obtained by dissolving selected proton donors and acceptors in liquefied rare gases have been summarized. The results show that by recording the infrared and Raman spectra of cryogenic solutions, a broad range of experimental information is obtained, that cannot be obtained by using other spectroscopic techniques, and that can be used to evaluate and further develop methodologies used to describe and to predict the structural, thermodynamic, and vibrational properties of weakly bound molecular complexes.

Acknowledgment

It is a pleasure to acknowledge the contributions of our co-workers, whose names are evident from the references cited herein. Without their excellent work, this review could not have been written. The research presented was carried out with financial support from the Fund for Scientific Research (FWO-Vlaanderen), and from the Flemish Community, through the Special Research Fund.

References

1. E. B. Willson and L. Andrews., In: J. M. Chalmers and P. R. Griffiths (editors), *Handbook of Vibrational Spectroscopy*; Vol. 2, Wiley, Chichester (2002).
2. M. O. Bulanin, *J. Mol. Struct.* **19**, 59 (1973).
3. J. J. Turner, M. Poliakoff, S. M. Howdle, S. A. Jackson and J. G. McLaughlin, *Faraday Discuss. Chem.* Soc. **86**, 271 (1988).
4. Y. M. Kimel'fel'd. In: J. R. Durig (editor), *Vibrational Spectra and Structure,* Vol. 19, Elsevier, Amsterdam (1992).
5. K. G. Tokhadze and N. A. Tkhorzhevskaya, *J. Mol. Struct.* **270**, 351 (1992).
6. M. O. Bulanin, *J. Mol. Struct.* **347**, 73 (1995).
7. R. J. H. Clark and R. E. Hester (editors), *Molecular Cryospectroscopy,* Wiley, Chichester (1995).
8. M. O. Bulanin, S. Velasco and A. C. Hernandez, *J. Mol. Liquids* **70**, 107 (1996).
9. B. J. van der Veken. In: R. Fausto (editor) *Low Temperature Molecular Spectroscopy,* Kluwer Academic Publishers, Dordrecht (1996).
10. M. O. Bulanin, In: J. M. Chalmers and P. R. Griffiths (editors), *Handbook of Vibrational Spectroscopy,* Vol. 2, p. 1329, Wiley, Chichester (2002).
11. R. R. Andréa, H. Luyten, H. A. Vuurman, D. J. Stufkens and A. Oskam, *Appl. Spectrosc.* **40**, 1184 (1986).
12. A. J. Rest, R. G. Scurlock and M. F. Wu, *J. Phys. E: Instrum.* **21**, 1102 (1988).
13. V. V. Bertsev, In: *Molecular Cryospectroscopy,* R. J. H. Clark and R. E. Hester, (Eds.), Wiley, Chichester, 1995; Vol. 23, p. 1.
14. M. Poliakoff and J. J. Turner, In: R. J. H. Clark and R. E. Hester (editors), *Molecular Cryospectroscopy,* Vol. 23, p. 275, Wiley, Chichester (1995).
15. S. S. Nabiev and P. G. Sennikov, *Infrared Phys. Tech.* **41**, 63 (2000).
16. D. C. Harris, *Infrared Window and Dome Materials*, SPIE -The International Society for Optical Engineering (1992).
17. I. S. Gainutdinov, E. A. Nesmelov and I. B. Haibullin, *Interference Covers for Optical Devices* (in Russian), FEN (2002).
18. W. A. Herrebout, B. J. van der Veken, A. P. Kouzov and M. O. Bulanin, *Phys. Rev. Lett.* **92**, 023002 (2004).
19. W. A. Herrebout, B. J. van der Veken and A. P. Kouzov, *Phys. Rev. Lett.* **101**, 093001 (2008).
20. W. A. Herrebout, N. Nagels and B. J. van der Veken, *Chem. Phys. Chem.* **10**, 3054 (2009).
21. R. K. Castellano, *Curr. Org. Chem.* **8**, 845 (2004).
22. A. J. Barnes, *J. Mol. Struct.* **704**, 3 (2004).
23. G. R. Desiraju, *Chem. Comm.* **24**, 2995 (2005).
24. P. Hobza, R. Zahradnik and K. Muller-Dethlefs, *Coll. Czech. Chem. Comm.* **71**, 443 (2006).

25. A. D. Buckingham, J. E. Del Bene and S. A. C. McDowell, *Chem. Phys. Lett.* **463**, 1 (2008).
26. J. Joseph and E. D. Jemmis, *J. Am. Chem. Soc.* **129**, 4620 (2007).
27. J. S. Murray, M. C. Concha, P. Lane, P. Hobza and P. Politzer, *J. Mol. Mod.* **14**, 699 (2008).
28. P. Hobza, *Chemicke Listy* **102**, 884 (2008).
29. I. Hyla-Kryspin, G. Haufe and S. Grimme, *Chem. Phys.* **346**, 224 (2008).
30. M. T. N. Hue, N. D. Pham and T. Zeegers-Huyskens, *Theochem.* **897**, 48 (2009).
31. A. Karpfen and E. S. Kryachko, *J. Phys. Chem.* **A113**, 5217 (2009).
32. Q. Z. Li, Z. B. Liu, J. B. Cheng, W. Z. Li, B. A. Gong and *J. Z. Sun, Theochem.* **896**, 112 (2009).
33. T. T. Nguyen, T. H. Tran and M. T. Nguyen, *J. Phys. Chem.* **A113**, 3245 (2009).
34. S. Parveen, A. K. Chandra and T. Zeegers-Huyskens, *J. Phys. Chem.* **A113**, 6182 (2009).
35. P. Rodziewicz, K. S. Rutkowski and S. M. Melikova, *Pol. J. Chem.* **83**, 1075 (2009).
36. K. S. Rutkowski, A. Karpfen, S. M. Melikova and P. Rodziewicz, *Pol. J. Chem.* **83**, 965 (2009).
37. S. Scheiner, *J. Phys. Chem.* **B113**, 10421 (2009).
38. M. V. Vener, A. N. Egorova, D. P. Fomin and V.G. Tsirelson, *J. Phys. Org. Chem.* **22**, 177(2009).
39. Y. Yang and W. J. Zhang, *Acta Chimica Sinica* **67**, 599 (2009).
40. X. Zarate, M. C. Daza and J. L. Villaveces, *Theochem.* 893, 77 (2009).
41. C. D. Keefe, E. A. L. Gillis and L. MacDonald, *J. Phys. Chem.* **A113**, 2544 (2009).
42. A. Kolocouris, *J. Org. Chem.* **74**, 1842 (2009).
43. K. Mizuno,Y. Masuda, T. Yamamura, J. Kitamura, H. Ogata, I. Bako, Y. Tamai and T. Yagasaki, *J. Phys. Chem.* **B113**, 906 (2009).
44. A. Mukhopadhyay, M. Mukherjee, P. Pandey, A. K. Samanta, B. Bandyopadhyay and T. Chakraborty, *J. Phys. Chem.* **A113**, 3078 (2009).
45. L. Q. Zhang, Y. Wang, Z. Xu and H. R. Li, *J. Phys. Chem.* **B113**, 5978 (2009).
46. B. Tolnai, J. T. Kiss, K. Felfoldi and I. Palinko, *J. Mol. Struct.* **924**, 27 (2009).
47. B. J. van der Veken, W. A. Herrebout, R. Szostak, D. N. Shchepkin, Z. Havlas and P. Hobza, *J. Am. Chem. Soc.* **123**, 12290 (2001).
48. S. N. Delanoye, W. A. Herrebout and B. J. van der Veken, *J. Am. Chem. Soc.* **124**, 7490 (2002).
49. S. N. Delanoye, W. A. Herrebout and B. J. van der Veken, *J. Am. Chem. Soc.* **124**, 11854 (2002).
50. T. Van den Kerkhof, A. Bouwen, E. Goovaerts, W. A. Herrebout and B. J. van der Veken, *Phys. Chem. Chem. Phys.* **6**, 358 (2004).
51. W. A. Herrebout, S. N. Delanoye and B. J. van der Veken, *J. Phys. Chem.* **A108**, 6059 (2004).

52. S. N. Delanoye, W. A. Herrebout and B. J. van der Veken, *J. Phys. Chem.* **A109**, 9836 (2005).
53. W. A. Herrebout, S. N. Delanoye, B. U. W. Maes and B. J. van der Veken, *J. Phys. Chem.* **A110**, 13759 (2006).
54. K.S. Rutkowski, W.A. Herrebout, S. M. Melikova, P. Rodziewicz, B. J. van der Veken and A. Koll, *Spectrochimica Acta* **A61,** 1595 (2005).
55. W A. Herrebout, S. M. Melikova, S. N. Delanoye, K. S. Rutkowski, .D. N. Shchepkin and B. J. van der Veken, *J. Phys. Chem.* **A109**, 3038 (2005).
56. K. S. Rutkowski, P. Rodziewicz, S. M. Melikova, W. A. Herrebout, B. J. van der Veken and A. Koll, *Chem. Phys*. **313**, 225 (2005).
57. K. S. Rutkowski, W. A. Herrebout, S. M. Melikova, B. J. van der Veken and A. Koll, *Chem. Phys.* **354**, 71 (2008).
58. K. S. Rutkowski, A. Karpfen, S. M. Melikova, W. A. Herrebout, A. Koll, P. Wolschann and B. J. van der Veken, *Phys. Chem. Chem. Phys.* **11**, 1551 (2009).
59. K. S. Rutkowski, S. M. Melikova, P. Rodziewicz, W. A. Herrebout, . B. J. van der Veken and A. Koll, *J. Mol. Struct.* **880**, 64 (2008).
60. B. Michielsen, W. A. Herrebout, and B. J. van der Veken, *Chem. Phys. Chem* **8**, 1188 (2007).
61. B. Michielsen, W. A. Herrebout and B. J. van der Veken, *Chem. Phys. Chem* **9**, 1693 (2008).
62. J. J. J. Dom, B. Michielsen, B. U. W. Maes, W. A. Herrebout and B. J. van der Veken, *Chem. Phys. Lett.* **469**, 85 (2009).
63. B. Michielsen, J. J. J. Dom, W. A. Herrebout and B. J. van der Veken (unpublished).
64. K. G. Tokhadze, In: R. J. H. Clark and R. E. Hester, (editors), *.Molecular Cryospectroscopy,* Vol. 23, p. 133, John Wiley & Sons, Chichester (1995).
65. S. M. Melikova, K. S. Rutkowski, P. Rodziewicz and A. Koll, *Pol. J. Chem.* **76**, 1271 (2002).
66. S. M. Melikova, K. S. Rutkowski, P. Rodziewicz and A. Koll, *Chem. Phys. Lett.* **352**, 301 (2002).
67. K. S. Rutkowski, S. M. Melikova, D. A. Smimov, P. Rodziewicz and A. Koll, *J. Mol. Struct.* **614**, 305 (2002).
68. S. M. Melikova, K. S.Rutkowski, P. Rodziewicz and A. Koll, *J. Mol. Struct.* **645**, 295 (2003).
69. S. M. Melikova, K. S. Rutkowski, P. Rodziewicz and A. Koll, *J. Mol. Struct* **705**, 49 (2004).
70. V. V. Bertsev, N. S. Golubev and D. N. Shchepkin, *Opt. Spectrosk.* **40**, 951 (1976).
71. B. J. van der Veken, *J. Phys. Chem.* **100**, 17436 (1996).
72. S. M. Melikova, A. J. Inzebejkin, D. N. Shchepkin and A. Koll, *J. Mol. Struct*. **552**, 273 (2000).
73. A. Brock, N. Mina-Camilde and C. Manzanares I, *J. Phys. Chem.* **98**, 4800 (1994).
74. P. G. Sennikov, *J. Phys. Chem.* **98**, 4973 (1994).

75. M. Tacke, P. Sparrer, R. Teuber, H. J. Stadter and F. Schuster, *J. Mol. Struct.* **349**, 251 (1995).
76. P. Kollman, *Chem. Rev.* **93**, 2395 (1993).
77. R. M. Levy and E. Gallicchio, *Ann. Rev. Phys. Chem.* **49**, 531 (1998).
78. W. L. Jorgensen, In: P. von Ragué Schleyer (editor), *Encyclopedia of Computational Chemistry,* Vol. 2, p. 1754, Wiley, West Sussex (1998).
79. W. L. Jorgensen, N. A. McDonald, M. Selmi and P. R. Rablen, *J. Am. Chem. Soc.* **117**, 11809 (1995).
80. S. Madurga, J. C. Paniagua and E. Vilaseca, *Chem. Phys.* **255**, 123 (2000).
81. C. M. Breneman and K. B. Wiberg, *J. Comput. Chem.* **11**, 361 (1990).
82. A. D. Buckingham, *Trans. Farad. Soc.* **56**, 753 (1960).
83. S. A. C. McDowell and A. D. Buckingham, *J. Am. Chem. Soc.* **127**, 15515 (2005).
84. S. A. C. McDowell and A. D. Buckingham, *Mol. Phys.* **103**, 257 (2005).
85. M. Hippler, *J. Chem. Phys.* **127**, 084306 (2007).
86. S. Blanco, S. Melandri, P. Ottaviani and W. Caminati, *J. Am. Chem. Soc.* **129**, 2700 (2007).
87. E. S. Kryachko and S. Scheiner, *J. Phys. Chem.* **A108**, 2527 (2004).
88. P. Rodziewicz, K. S. Rutkowski, S. M. Melikova and A. Koll, *Chem. Phys. Chem.* **6**, 1282 (2005).
89. P. Rodziewicz, S. M. Melikova, K. S. Rutkowski and F. Buda, *Chem. Phys. Chem.* **6**, 1719 (2005).
90. P. Rodziewicz, K. S. Rutkowski, S. M. Melikova, A. Koll and F. Buda, *Chem. Phys. Chem.* **7**, 1221 (2006).
91. E. S. Kryachko and A. Karpfen, *J. Chem. Phys.* **329**, 313 (2006).
92. P. C. Nam, M. T. Nguyen and T. Zeegers-Huyskens, *Theochem.* **821**, 71 (2007).
93. C. Sandorfy, *Coll. Czech. Chem. Comm.* **70**, 539 (2005).
94. K. Pluhackova and P. Hobza, *Chem. Phys. Chem.* **8**, 1352 (2007).

CHAPTER 9

LOW-TEMPERATURE INFRARED SPECTROSCOPY OF SURFACE SPECIES

Alexey Tsyganenko

V. A. Fock Institute of Physics, St. Petersburg State University
198504 St. Petersburg, Russia
E-mail: tsyg@photonics.phys.spbu.ru

Application of low-temperature infrared spectroscopy to the studies of molecular adsorption, hydrogen bonding, properties of surface sites and molecules in the field of surface charges, lateral interactions between the adsorbed molecules, chemical reactions in heterogeneous systems, and linkage isomerism of surface species is considered.

1. Introduction

Studies of surface complexes at low temperatures, besides unique information about the structure of surface compounds and mechanisms of catalytic and photocatalytic reactions, provide valuable data on intermolecular interactions where the solid surface plays the role of an interacting partner. The spectra of surface species are not complicated by the effect of environment and can be studied in a broad range of temperatures, not limited by points of condensation or crystallization as it is in solutions and matrices. Changes in the spectra of adsorbed molecules caused by interaction with the surface shed light on the nature of surface sites and the effect of the surface field on the molecule.

In this chapter, the technique of low-temperature spectral studies of adsorbed molecules is described and its application is illustrated by several examples.

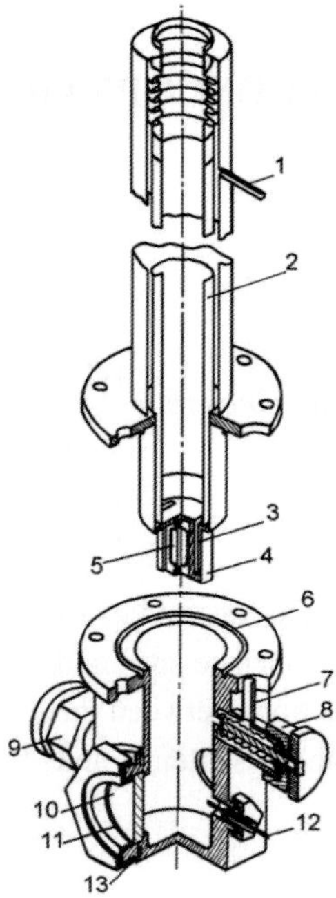

Fig. 1. Cell for spectral studies of surface species at 80-673 K: 1 – evacuation outlet; 2 – volume for coolant; 3 – passage for heating elements; 4 – sample holder; 5 – sample; 6 – teflon O-ring; 7 – outlet to a vacuum pump; 8 and 9 – valves; 10 – window; 11 – indium gaskets; 12 – thermocouple; 13 – passage for thermostating.

2. Experimental Technique

A cell for infrared (IR) studies of adsorbed species has to meet a number of requirements. It should be leak-tight, allow heating the sample *in vacuo* up to at least 400 °C during pretreatment, and cool the sample to desirable temperatures when recording the spectra. In addition, it is often important to monitor the gas pressure and temperature of the sample in the course of measurements. In the cell shown in Fig. 1, the samples can be heated up to ~400 °C or cooled in the measurement position and the spectra can be obtained *in situ* at the conditions of treatment. Such cells are widely used in adsorption experiments; however, the minimum temperature here is limited due to a poor thermal contact between the sample and the coolant.

In the other type of cells,[1,2] the sample can be moved from the cooled part to the place where it can be heated (Fig. 2). The cell body is made of stainless steel, the central part of it can be cooled by liquid nitrogen, and the volume with the sample is separated from the thermoinsulating

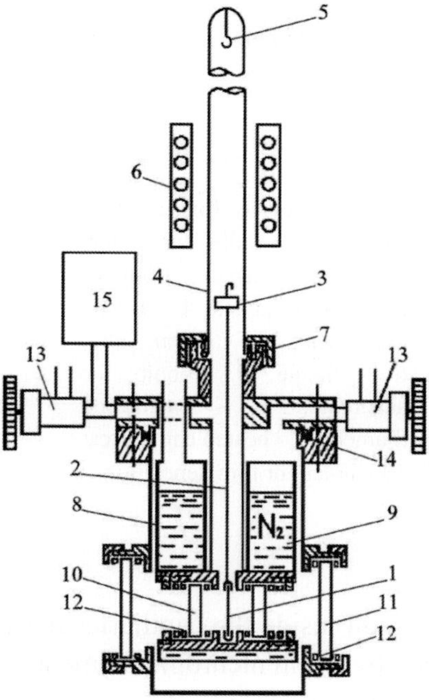

Fig. 2. Stainless-steel cell for studying IR spectra of adsorbed molecules at variable temperatures. 1 – sample; 2 – sample holder; 3 – anchor of magnetic stainless steel; 4 – quartz tube; 5 – hook to hang the sample; 6 – heater; 7 – Viton O-ring; 8 – cell body; 9 – coolable part of the cell; 10 – inner windows; 11 – outer windows; 12 – indium gaskets; 13 – valve; 14 – Viton gasket; 15 – pressure gauge.

volume by inner windows of BaF_2 or ZnSe. In this cell, the spectra of surface species can be obtained at 55–400 K in the presence of gas or even submerged in liquid oxygen or other cryogenic solvents at 77 K. Helium gas (~0.5 Torr), admitted to the inner volume provides a better thermal contact of the sample with the coolant in the absence of surrounding gas or solvent. The elevation of temperature is limited by the melting point of indium gaskets. To cool the sample below 77 K, down to ~55 K, the evaporating nitrogen is pumped from the coolant volume. To achieve lower temperatures, a cell for liquid helium temperatures was developed.[3] To combine spectroscopic and adsorption measurements, the cells are equipped with pressure gauges.

The measurements are usually done with self-supporting pellets of pressed disperse powders. To study the films of volatile oxides such as WO_3 or MoO_3, a device for film deposition can be attached instead of the quartz tube.[4] Porous ice films can be prepared by condensation of water vapor directly onto the inner windows by means of the device shown in Fig. 3. Water vapor is sprayed from the capillaries placed between the

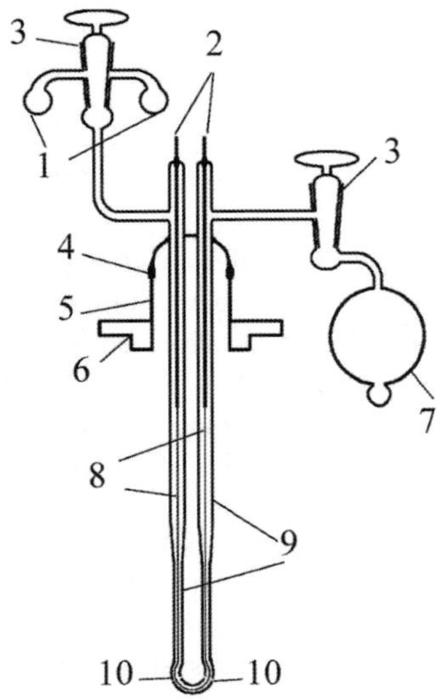

Fig. 3. Device for ice film studies. 1 – bulbs for water; 2 – molybdenum leads-in; 3 – glass stopcock; 4 – kovar-glass seal; 5 – kovar tube; 6 – flange; 7 – bulb for adsorbate; 8 – nichrome filament; 9 – heated capillaries; 10 – holes for film deposition.

cold windows. To avoid water condensation inside the capillaries during film deposition, the capillaries are heated by a thin nichrome filament.

For adsorption, gases are added into the cell at 77 K (CO, CH_4) or at slightly higher temperature (ethene, CO_2). Less volatile compounds are sprayed from a special glass bulb through the capillaries onto the sample at 77 K and they diffuse through the pores inside on raising the temperature. Otherwise, compounds soluble in liquid oxygen (such as O_3) can be brought in contact with the sample from the solution. In this way, ozonolysis of adsorbed ethene on ice surface was studied at 60–77 K when the pressure of reagents in the gas phase was negligible. Even less volatile species (for instance, NH_3) can be codeposited with water through the capillaries from a pre-prepared mixture.

3. Frequency Changes upon Adsorption

IR spectroscopy provides direct information about the active sites of solid surfaces. Some of them have vibration frequencies in the transparency region of the adsorbent and hence can be observed in the spectra of pure materials if the surface dangling bonds are saturated by extrinsic atoms or groups formed during sample preparation or due to contact with atmosphere. These surface bonds include O–H on oxide surface, Si–H on silicon surface, and M=O on surface of transition-metal oxides. Other sites, such as coordinately unsaturated metal atoms or oxygen ions of oxide surfaces, are not seen in the spectra of pure adsorbents but can be detected by adsorption of test molecules. The test molecules should have at least one observable absorption band sensitive to adsorption.

Nitrogen-containing bases, such as ammonia and pyridine, have been the most popular test molecules to study electron-accepting centers of oxides, which form a coordinate bond with molecules possessing lone pairs of electrons and are considered in chemistry as Lewis acid sites. For ammonia, donation of the lone pair of electrons to coordinately unsaturated metal cations (σ-coordinate bond) results in a great increase of the frequency of the symmetric bending mode δ_s. Interaction with the strongest Lewis sites of alumina surface shifts this band to 1320 cm^{-1} or even higher (by >350 cm^{-1}).[5] This blue shift can be explained by the confinement effect when the motion of atoms in the molecule is restricted by the surface. In general, the bands of adsorbed molecules can be higher or lower in frequency compared to the gas-phase values. This fact shows that the adsorption-induced modification of the molecular force field has a complex effect on their frequencies.

The band shift of surface hydroxyls caused by adsorption ($\Delta\nu_{OH}$) depends on the proton-accepting (basic) properties of the adsorbed molecule and on the proton-donating ability of the OH group. Using the frequency shifts of different molecules absorbed on the same silanol groups, the basicity of various adsorbates can be compared. The analysis can be extended to a variety of molecules that form weak bonds with hydroxyl groups at low temperatures. Adsorption of CO and N_2 shifts the band of the Si–OH groups of aerosil by about −90 and −30 cm^{-1}, respectively.[6] This method can also be used to estimate the basic properties of such reactive molecules as ozone, which is not generally easy because it reacts with organic indicators. It has been shown that when ozone is adsorbed on silica, it shifts the band of OH groups down in energy by 80–100 cm^{-1} showing the basicity close to that of CO.[7,8]

Studies of surface OH groups on oxides using the same adsorbate show that the proton-donating ability of hydroxyls varies greatly and changes even for different types of hydroxyl groups of the same adsorbent.[9] Considering the most acidic OH groups of each oxide, the proton-donating ability of surface OH groups decreases as follows: TiO_2 (anatase) > BeO, HfO_2 > TiO_2 (rutile), Ta_2O_5 > SiO_2 > In_2O_3 > Al_2O_3 > ZnO, Ga_2O_3, ZrO_2, NiO >> MgO, CaO. This order remains in effect for various adsorbates although the $\Delta\nu_{OH}$ values are quite different, which was also shown for benzene that does not form strong complexes with Lewis acid sites.[9] The same relative acidities were found for perturbation by CO and N_2 or even for submerging the sample in liquid N_2 or O_2.[6]

The frequencies of adsorbed molecules can also be used for acidity characterization. CO adsorbed on OH groups at 77 K shows a frequency increase from 2143 cm^{-1} for the gas phase up to ~2175 cm^{-1} for the most acidic OH groups of zeolites, which protonate ammonia or pyridine and hence indicate the Brønsted acidity. The frequency shifts of test molecules are probably a better measure of acidity than $\Delta\nu_{OH}$ because they do not depend on the individual properties of hydroxyl groups. CO adsorption at 77 K is an informative test for acidic sites. The CO frequency is high enough not to interfere with the bulk absorption and it is quite sensitive to different interactions.[10,11] Adsorption on surface cations (Lewis acid sites) increases ν_{CO} up to 2220 cm^{-1} or higher.

Homonuclear diatomic molecules in the gas phase are not active in IR absorption, but interaction with surface lowers the symmetry of such molecules and adsorption-induced bands of H_2 or N_2 are observed in the spectra of various samples at 77 K. Hydrogen adsorption on oxides or zeolites decreases the vibrational frequencies, and the H_2 bands are often accompanied by satellites at higher frequency, which can be assigned to combinations of the stretching mode with low-frequency vibrations of H_2 with respect to the surface.[12] The H_2 absorption intensity induced by an electrostatic field can be calculated, and it was used to estimate the strength of surface electrostatic fields. Molecular adsorption of nitrogen is similar to that of CO. On metal surfaces, both molecules form carbonyl-like species with a lower frequency than in the gas phase, whereas the interaction with cationic sites shifts the band to higher frequencies.[13]

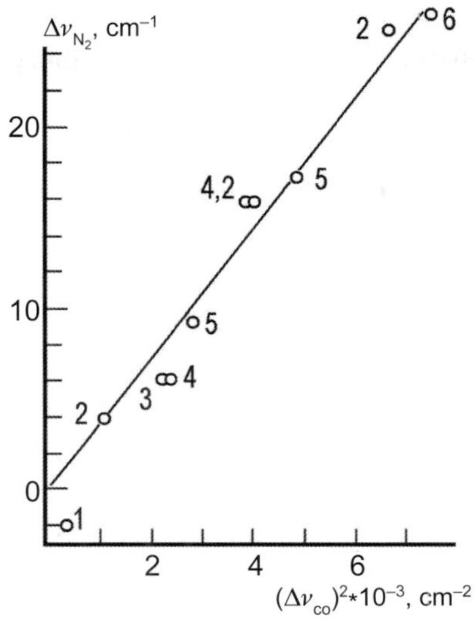

Fig. 4. Relation between $\Delta\nu_{N_2}$ and $(\Delta\nu_{CO})^2$ observed for CO and N_2 adsorbed on (1) SiO_2, (2) H–mordenite, (3) ZnO, (4) BeO, (5) TiO_2, and (6) silica–alumina.

Simple diatomic molecules are very convenient to analyze factors influencing the adsorption-induced shift. CO adsorption on cationic sites can be considered as σ-coordination by the weakly antibonding lone pair of the C atom slightly increasing the frequency. If, however, the cation of transition metal has *d* electrons, the back donation to the antibonding π-orbitals of the molecule results in a frequency decrease typical of metal carbonyls. CO adsorption on ionic solids can be described by an electrostatic model. The frequency shift is due to the vibrational Stark effect when the vibrational frequency is changed in the electric field of the cation. The frequency increases with respect to the gas phase when CO interacts with cations via the C atom, whereas the opposite field direction for adsorption via the O atom results in a frequency decrease.

For homoatomic molecules, the frequency shift should not depend on the field direction. In other words, it should be an even function of the field strength and in the expansion $\Delta\nu = K_1E + K_2E^2 + \ldots$ the linear term is zero. Thus, the shift is proportional to E for CO adsorbed on cationic sites whereas its proportionality to E^2 is expected for N_2, which suggests a linear dependence of $\Delta\nu_{N_2}$ on $(\Delta\nu_{CO})^2$. Figure 4 illustrates this dependence for CO and N_2 adsorbed on the same sites of oxides.[13]

The electrostatic field draws the electron density from the optimal position for the chemical bond. For homoatomic molecules, it follows that the coefficient K_2 is negative and the frequency decreases for both the linear and side-on complexes with ions of any sign. This conclusion agrees with the red shifts found for molecular hydrogen on different adsorbents. The same red shifts should occur if CO forms side-on complexes oriented perpendicular to the site-molecule axis. As seen below, this takes place upon interaction with the anionic sites.

Nitrogen seems to be an exception of this rule, showing an increase of the frequency up to 30 cm^{-1} upon adsorption on cationic sites. This fact can be explained by the effect of confinement mentioned above. The interaction with the surface increases the frequency, despite a slight decrease in the force constant of the molecule due to the quadratic vibrational Stark effect in the electric field of the cation.

4. Band Intensities of Adsorbed Molecules

In addition to the frequencies, adsorption changes the band intensities of adsorbed species. Vibrations, which are forbidden in IR absorption, become active because of the symmetry lowering and the relative intensity of IR-active vibrations can also be changed. For molecules adsorbed on metals, this is probably due to the "surface selection rule" first suggested by Pearce and Sheppard.[14] They have shown that only vibrations which give dipole changes perpendicular to the metal surface are IR active because the dipole image induced under the surface has the same direction. In contrast, the vibrations parallel to the surface have low absorbance because of the opposite directions of the dipole and its image.

For oxides, the integrated extinction coefficient ε allows in principle to obtain the concentration of surface sites from the spectra of adsorbed test molecules. However, the data about ε and its dependence on the molecular vibrational frequency are highly controversial even for the best-studied low-temperature CO adsorption. According to the early work by Seanor and Amberg,[15] CO interaction with Lewis acid sites increases ε compared to the gas-phase value, this increase being the greater, the higher the frequency. Other authors report that ε dramatically decreases for weakly bonded CO molecules on some zeolites and metal oxides while increases with frequency. A small increase of ε with the CO frequency was found for titania.[16] For CO adsorbed on metals or NiO, the frequency decreases due to the back donation of *d* electrons of metal to the antibonding orbitals of CO. In this case, a strong increase of ε was found either in direct measurements or indirectly from the band shifts caused by lateral interactions in the adsorbed layer.[17]

The low values of ε observed previously for weakly adsorbed molecules are caused by inhomogeneous distribution of molecules over the surface due to the heating of the central part of the sample by the IR beam.[18] To avoid this effect, small samples with minor temperature gradients were used. The amount of adsorbed CO was determined by accurate measurement of the gas pressure in a special dosing volume before adsorption or from the pressure jump during sample transfer from the cooled part of the cell to the tube kept at elevated temperatures when all adsorbed CO turns into gas filling the known volume.

In the spectra of some samples (TiO_2, SiO_2, etc.), two sites of adsorbed CO with distinct bands coexist. In these cases, the amount of adsorbed molecules was measured for two values of coverage when the concentration ratios for the two adsorption forms were different as much as possible. Supposing that each form of adsorption has a certain absorption coefficient (independent of coverage), two measurements with two different values of coverage lead to a system of two equations with two unknowns. In this way, the absorption coefficient of CO adsorbed on the OH groups and cationic sites of oxides was measured.[19] The obtained values of ε are always smaller than for CO in the gas phase (2.6 cm μmol^{-1}). This behavior of ε in the presence of an electric field is predicted by the electrostatic model discussed in more detail below.

IR absorption of homoatomic molecules (H_2 or N_2) is forbidden in the gas phase, but becomes active in the adsorbed state. For ionic adsorbents, this absorption is induced by the electric field E and can be expressed by the formula[20]

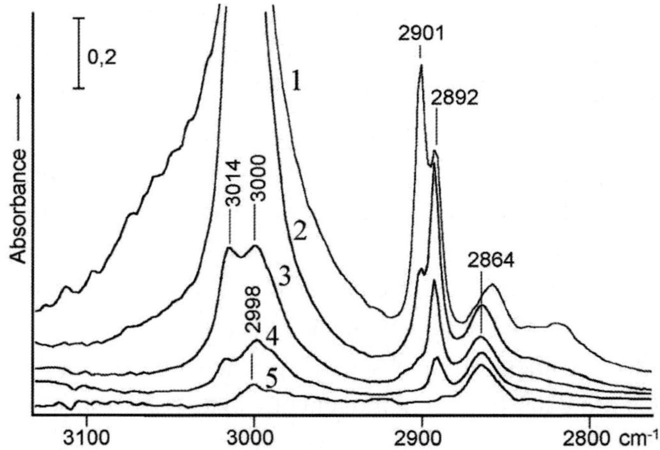

Fig. 5. Spectra of methane adsorbed on ZnY zeolite as a function of coverage. The coverage decreases from 1 to 5.

$$\varepsilon = \frac{\pi}{3cm}\left(\frac{\partial \alpha}{\partial q}\right)_0^2 E^2, \tag{1}$$

where m is the reduced mass, c is the velocity of light, and $\partial\alpha/\partial q$ is the derivative of the polarizability with respect to the normal coordinate q. In this way, Sheppard and Yates estimated the surface field of porous glass from the band intensity of adsorbed hydrogen.[20] For metal oxides or zeolites, similar measurements are complicated by the presence of different forms of adsorption because the band intensity corresponds to the total amount of adsorbed gas including weakly adsorbed molecules with low IR absorption.

The activation of otherwise IR-inactive transitions depends on the adsorption site. For adsorbed methane, the symmetric stretching vibration ν_1 is activated in the adsorbed state (in addition to the asymmetric C−H stretching mode ν_3). Figure 5 shows the C−H stretching region of CH_4 adsorbed on ZnY zeolite. The ν_1 band at 2864 cm^{-1} due to molecules adsorbed on the strongest sites at very low surface coverage (curve 5) is even more intense than the ν_3 band. The appearance of other ν_1 bands at 2892 and 2901 cm^{-1} at higher amounts of methane adsorbed on weaker sites correlates with the growth of much stronger ν_3 bands at

3014 and 3000 cm^{-1}. The integrated absorption coefficients of the v_1 bands at 2901 and 2864 cm^{-1} differ by almost two orders of magnitude.

The absorption intensity induced by an electrostatic field depends dramatically on the orientation of the molecule with respect to the field, and this can be used to obtain information about the geometry of adsorbed complexes. This analysis was done by Smirnov *et al.* for the bands of simultaneous transitions,[21] i.e., the transitions when one absorbed quantum is used for vibrational excitation of both the molecule and the silanol group on which it is adsorbed. Such bands were detected for CO, N_2, and O_2 adsorbed on OH and OD groups of thick silica samples, and the experimentally observed intensities were compared with the calculations for the linear or perpendicular geometries of adsorbed complexes. For CO and N_2 adsorbed on the OH and OD groups, the values calculated for the linear complexes are close to the experimental data. For O_2, the observed intensities are at least one order of magnitude weaker. It was concluded that CO and nitrogen form linear complexes with the Si–OH groups, whereas oxygen prefers the perpendicular geometry of weak H-bonded complexes. This analysis, however, cannot distinguish between two linear configurations, −OH···CO versus −OH···OC. The evidence for coexistence of both the structures is provided by intensity measurements at variable temperatures (see Section 9).

5. Bandwidth of Adsorbed Species

The width and shape of absorption bands of surface species as well as their temperature dependence shed light on the dynamics of vibrational excitation. The natural linewidth of a free molecule in the gas phase is determined by the lifetime of the vibrationally excited state through the Heisenberg relation $\Delta\varepsilon\Delta t = h\Delta v\Delta t \sim \hbar$, and it is rather small (~10^{-5} cm^{-1}) in normal conditions for vibrational bands of medium intensity. Collisions cause additional homogeneous broadening of rotational lines, while translational movement of molecules results in Doppler broadening. The bands in the vibrational spectrum of molecules on surfaces usually have no rotational structure. Only in few cases of physisorbed diatomics, indications of rotation are found. Most of the

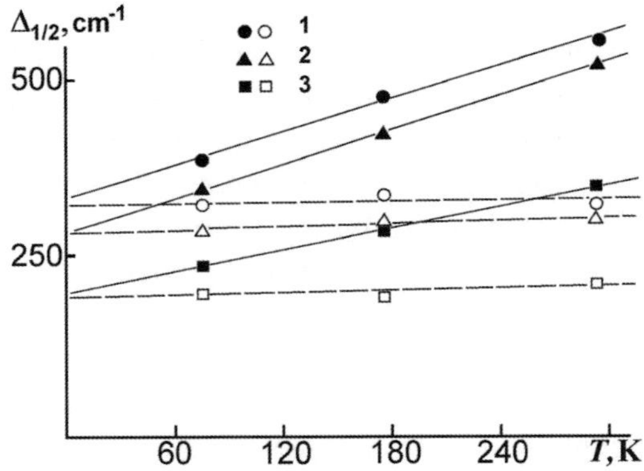

Fig. 6. Temperature dependence of the half-width of the Si–OH groups perturbed by adsorbed ammonia. Dotted lines show the corresponding dependences after subtraction of the anharmonic coupling contribution. Lines 1–3 correspond to pretreatment at 773, 923, and 1073 K.

bands exhibit inhomogeneous broadening caused by different environment of individual molecules. The line positions in the rotational structure of vibrational bands for gas-phase molecules are temperature independent. In contrast, it was demonstrated by Peri that the band of surface silanol groups shifts to lower frequencies and broadens on heating the sample.[22] Furthermore, the Si–OH band perturbed by adsorbed ammonia shifts from 3050 to 2850 cm^{-1} upon cooling the sample from 300 to 4 K and the half-width decreases from ca. 600 to 370 cm^{-1}.[23]

To understand the mechanism of this phenomenon, the OH and OD bands of aerosil perturbed by ammonia were studied at variable temperatures.[24] It appears that the bandwidth is a nearly linear function of temperature with almost the same slope for different pretreatment temperatures (Fig. 6). The observed temperature dependence is due to the weakening of hydrogen bonding upon thermal excitation of the low-frequency vibrations of the adsorption complex, which can be described in the approximation of anharmonic vibrational coupling.[25] According to this model, the temperature dependences of the band position and bandwidth of the perturbed OH groups are connected. The contribution of anharmonic coupling to the bandwidth can be extracted from the experimental band positions at different temperatures. If we subtract this contribution from the experimental bandwidth, the remaining part is almost independent of temperature, suggesting that the temperature dependence of the bandwidth of surface complexes is entirely due to anharmonic coupling. This explanation probably concerns not only the H-bonded complexes, but other surface species as well. The red shift and broadening of the bands of silanol groups reported by Peri[22] can be explained by thermal excitation of two bending OH vibrations and the stretching Si–O mode.

The remaining part of the bandwidth, which does not depend on the sample temperature, decreases with the increasing pretreatment temperature, from 215 cm^{-1} for the 773 K pretreatment to 140 cm^{-1} for 1073 K; thus, at least a part of the width is from inhomogeneous broadening due to different environments of the surface complexes. The other part of the bandwidth should include also homogeneous broadening caused by a vibrational lifetime limited by the processes of phase and energy relaxation.[26] Analysis of the experimental bands of the free OD stretching vibrations of silica surface was done by the method of autocorrelation functions, yielding relaxation times of 5.4 and 3.9 ps for spectra recorded at 80 and 293 K, respectively.[26] However, the energy relaxation time measured directly by time-resolved spectroscopy for the OH and OD groups of disperse silica was much longer, about 150–200 ps.[27] It follows that the main reason of the band broadening is the phase relaxation via interaction with the low-frequency vibrations.[26]

It should be noted that not all processes cause the band broadening in the spectra of surface species. A band broadened due to the inhomogeneity of the adsorbed layer can be narrowed by the dynamic interaction between the adsorbed molecules to a limit set by the lifetime of the excited state.[28]

6. Thermodynamics of Adsorption

Spectroscopy at variable temperatures combined with pressure measurements allows to study thermodynamics of adsorption. Usually, the heat of adsorption is measured by microcalorimetry. It yields differential heats as a function of coverage, from which the average integral heat can be derived. Alternatively, the spectra can be measured over a temperature and pressure range. Then, the Clausius–Clapeyron equation is applied to the data obtained at the same surface coverage to yield the isosteric heat (or enthalpy) of adsorption. It is believed that for low pressures of adsorbate, the heat, energy, and enthalpy of adsorption do not differ markedly, and we can compare these values measured by different methods.

Measurements at variable temperatures provide data sufficient to calculate the isosteric heat of adsorption on distinct surface sites. The occupation of different surface sites can be estimated from the ratio of the intensity of the corresponding band to the maximum intensity of this band at saturation $\theta_r = I/I_{max}$. For the cell shown in Fig. 2, it is convenient to remove the coolant and record the spectra during warming up when θ_r changes with T and P. Measurements with different amounts of gas in the central volume finally give the values P as a function of T for given θ_r. The isosteric heat Q of adsorption is then obtained from the slope of a plot of $\ln(P)$ versus $1/T$ based on the equation

$$\ln(P) = -\frac{Q}{RT} - \frac{\Delta S^\circ}{R}, \qquad (2)$$

derived from the Clausius–Clapeyron equation for adsorption–desorption equilibrium.[29]

This method was applied to estimate the heat of CO chemisorption on thermally activated CaO.[1] For the species reversibly formed at 77 K, the isosteric heat of adsorption was 30 kJ mol^{-1}. The reversibility of adsorption means here that the adsorbed species exist at this temperature in equilibrium with gas and can be removed from the surface by pumping. The spectroscopically measured enthalpy of nitrogen adsorption on the acidic OH groups of H–ZSM–5 zeolite $\Delta H^\circ = -19.7(\pm 0.5)$ kJ mol^{-1} is in fair agreement with the adsorption heat of -19 kJ mol^{-1} determined microcalorimetrically.[30] For the IR-active dioxygen species on NiO with the O-O vibration band at 1490 cm^{-1} assigned to a complex of singlet oxygen (usual triplet O_2 has the frequency of 1555 cm^{-1}),[31] the isosteric heat of adsorption decreases from 13 to 6.5 kJ mol^{-1}

with the increasing coverage. Even a stronger dependence on coverage was found for CO adsorbed on ZnO where the decrease of adsorption heat is caused by strong repulsive lateral interaction.[32]

In addition to isosteric heat, the plot of ln(*P*) versus 1/*T* provides the entropy of adsorption $\Delta S°$ for certain adsorbed species. This value reflects the loss of mobility of molecules upon adsorption, and the data obtained for CO on silica indicated free rotation of the molecules adsorbed on siloxane oxygen atoms.[33]

7. Isotopic Methods in Spectral Studies of Adsorption

It is not easy to recognize the structure of adsorbed species. Some bands are usually hidden by the bulk absorption of adsorbent. Comparison with the known spectra of hypothetical compounds in the solid, liquid, and gas phases is often insufficient to assign bands that appear together and probably belong to the same surface complex. Additional information about the structure and composition of surface species can be obtained from isotopic substitution.

Adsorption of CO_2 on hydroxylated oxides leads among others to a band at ca. 1220 cm^{-1}. Various CO-containing species can absorb in this region, but this band does not appear for adsorption on a deuterated surface. Such sensitivity to H/D exchange assigns it to the bending vibration δ_{OH} of surface bicarbonate HCO_3^- structures, resulting from the reaction of CO_2 with the surface OH groups.[34]

The isotopic shift for diatomics can be calculated from the reduced masses. For the $C^{16}O/C^{18}O$ substitution, the calculated shift of 52 cm^{-1} is in perfect agreement with the experimental value for weakly adsorbed CO.[1] An additional band of CO adsorbed on CaO is observed at 2038 cm^{-1} with a shift of only 26 cm^{-1}. This band cannot be due to CO molecules, and it is assigned to the ketene group in dioxoketene $O = C = CO_2^{2-}$ ion.[1]

The number of spectral components for partial isotopic substitution enables us to count the atoms of this element in the compound in question. Three equidistant bands with the strongest central component are observed after adsorption of O_2 enriched with ^{18}O isotope on NiO, where usual $^{16}O_2$ gives rise to one band at 1490 cm^{-1}. It follows that there

$$CO + O^{2-} \xrightarrow{CO} CO_2^{2-} \xrightarrow{CO} C_2O_3^{2-} \xrightarrow{CO,\, O^{2-}} CO_3^{2-} + (CO)_n^{2-}$$

Scheme 1.

are two equivalent oxygen atoms in the species identified as singlet dioxygen.[31]

For ozone adsorption on silica, a band appears at 2107 cm^{-1}, which is split into six components if O_3 is prepared from a $^{16}O_2/^{18}O_2$ mixture.[7,35] The spectrum consists of two triplets with more intense central bands, revealing two equivalent O atoms. The interval between the triplets is caused by isotope substitution of the third (central) O atom in O_3, which has the geometry of an isosceles triangle as in the gas phase. This is not the case of ozone coordinately bound to CeO_2 surface when one of ozone bands is split into four bands of almost the same intensity.[36] This splitting can arise only in the structure of three unequivalent atoms. Thus, coordinately bonded ozone is linked to the surface cation M by a terminal oxygen atom: M ← O–O–O.

In the spectrum of MgO, a band at 1472 cm^{-1} appears after the CO adsorption at 77 K.[1] Addition of isotopic mixtures containing ^{13}C splits this band into two components while ^{18}O substitution splits it into three components, the central one being the most intense. The compound, thus, contains one C atom and two equivalent oxygens and was identified as "carbonite" CO_2^{2-} ion.[1] The same ions formed upon CO interaction with basic oxygens of La_2O_3 show slightly different shifts on $C^{18}O$ adsorption and after addition of $C^{16}O$ to an ^{18}O-enriched sample.[37] In these ions, the surface and CO oxygen atoms are not equivalent and do not scramble after adsorption. Thus, isotopic substitution in the low-temperature spectral studies reveals the CO transformations on basic oxides. The first step is the formation of carbonite ions, which then react with the next CO molecules producing dioxoketene, and it leads to more complex products of disproportionation (Scheme 1).[1]

If isotopes are randomly distributed in the adsorbed species, the structure can be analyzed using the band intensities upon adsorption of isotopically mixed compounds. Ammonia adsorption can lead to NH, NH_2 groups, molecular NH_3, and NH_4^+ ions, all having the NH stretching bands in the 3500–3000 cm^{-1} region. To determine the number of H atoms in the NH_x species, the band intensities versus the deuteration degree *a* are inspected. For NH and OH groups, the intensity decreases linearly with increasing *a*, the NH_2 band intensity decreases as $(1 - a)^2$, the bands of molecular ammonia as $(1 - a)^3$, etc. This idea was used to find the δ_s band of ammonia adsorbed on silica.[23] The NH_3 band overlaps with the strong absorption of the absorbent, but deuteration leads to a band at 950 cm^{-1} in a narrow window of transparency. The dependence of this band intensity on *a* assigns it to NHD_2, yielding the δ_s value of 1125 cm^{-1} for NH_3, in agreement with the data on the combination bands.

8. Lateral Interaction between Adsorbed Molecules

Lateral interactions between adsorbed molecules were first observed for CO on Pt surface by Eischens *et al.*[38] who noticed a gradual shift of the absorption band with coverage. Interaction between molecules adsorbed on oxide surfaces, which is reflected in the dependence of band positions on surfaces coverage and in the complex structure of these bands, has been observed for CO adsorbed on ZnO[39,40] and some other oxides.[32]

In the spectrum of CO adsorbed on ZnO at 77 K, a band at 2192 cm^{-1} appears when the first doses of adsorbate reach the surface. As the coverage increases, this band decreases and additional CO bands rise at lower frequencies. The frequency of coordinately bonded CO molecules is evidently affected by the occupation of the adjacent sites, and the bands of the previously adsorbed and new molecules are shifted down in energy. For the highest coverage, a narrow band at 2168 cm^{-1} remains in the spectrum. At the lowest and highest coverage, a single band of coordinately bonded CO dominates, i.e., adsorption geometry is well defined in these extreme cases. The observed changes cannot be explained by sequential occupation of different sites and reveal the interaction between adsorbed molecules. The isosteric heat of adsorption decreases from 10 to 1.5 kcal mol^{-1} with the increasing coverage due to strong repulsive interaction between the adsorbed molecules.[32]

For metals, the coverage affects the CO frequency mainly due to dynamic dipole coupling.[38] The coupling occurs for species with close absorption frequencies so that it can be distinguished from the static effect using isotope dilution. If ^{13}CO contains small admixture of usual

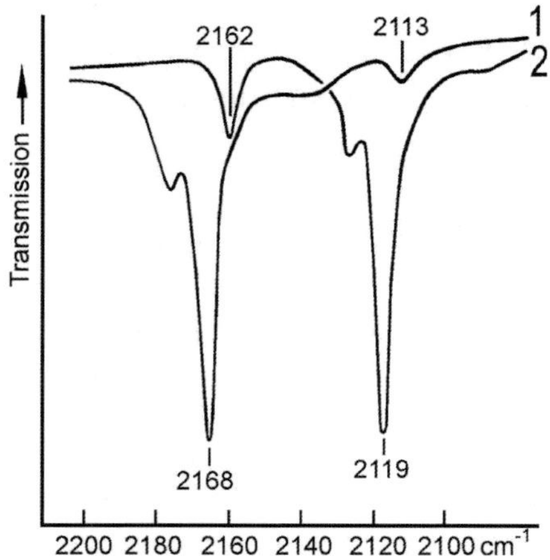

Fig. 7. IR spectrum of CO with (1) 1% and (2) 85% of ^{13}C isotope adsorbed on ZnO at coverage of 2.0 molecules nm^{-2} and $T = 77$ K.

^{12}CO, the admixed molecules do not participate in the coupling interaction; hence, the frequency difference between the band of pure ^{12}CO at saturation coverage (v_s) and the band of ^{12}CO admixed in ^{13}CO (v_a) is a measure of dynamic dipole coupling (Δv_{dyn}).[41]

Figure 7 shows the results of the experiment with ZnO. The v_a band of the admixed ^{12}CO molecules at 2162 cm^{-1} is lower than v_s (2168 cm^{-1}) by $\Delta v_{dyn} = 6$ cm^{-1}. The same dynamic shift of 6 cm^{-1} is observed for ^{13}CO. The results obtained for a number of other metal oxides show that the dipole coupling always exists and shifts the frequencies upward, but this effect only partly compensates the large red shift due to the static interaction. The static shift (Δv_{st}) is the difference between the frequency v_0 of CO obtained at low coverage and the frequency v_a of CO diluted by the molecules with the other carbon isotope at maximum coverage. For ZnO, $\Delta v_{st} = v_0 - v_a = 2192 - 2162 = 30$ cm^{-1}. The total shift, observed with the coverage increase up to θ_{max}, is a sum of the two factors: $v_0 - v_s = \Delta v_{st} - \Delta v_{dyn}$.

It is impossible to experimentally distinguish the effect of direct interaction between adsorbed molecules and the effect mediated by the surface. However, the frequency shifts caused by direct interaction between the adsorbed molecules can be estimated theoretically. The dynamic shift $\Delta\nu_{\text{dyn}}$ can be determined using the approach suggested by Hammaker *et al.*[42] Taking into account that $\Delta\nu_{\text{dyn}} \ll \nu_{\text{CO}}$, their expression is transformed to the formula for the shift[43]:

$$\Delta\nu_{\text{dyn}} = \nu_s - \nu_a = \frac{1}{2}\alpha_v \nu_a T_{\text{ds}}. \tag{3}$$

Here α_v is the vibrational polarizability of the interacting molecules, and T_{ds} is the dipole sum $\Sigma_j r_{ij}^{-3}$. Taking α_v = 0.057 Å^3 of the CO gas and T_{ds} = 0.026 Å^{-3} for saturation (2.0 molecules nm^{-2}), we obtain $\Delta\nu_{\text{dyn}}$ = 1.6 cm^{-1}, in contrast to the experimental shift of 6 cm^{-1}.[43]

The static shift due to the direct intermolecular interaction can be calculated from pure electrostatics as a result of the vibrational Stark effect in the field of surrounding molecules:

$$\Delta\nu_{\text{st}} = K_{\text{st}} E = K_{\text{st}} \mu T_{\text{ds}}, \tag{4}$$

where μ is the dipole moment of the molecule and K_{st} is the Stark coefficient. The calculated static shift for CO on ZnO at saturation coverage does not exceed −3 cm^{-1} taking into account the increase of μ on adsorption, which is very different from observed $\Delta\nu_{\text{st}}$ = −30 cm^{-1}.

Thus, both dynamic and static shifts are not the result of the direct interaction only, but they are enhanced by the adsorbent surface. Apparently, this enhancement is due to the polarizability of the solid in the field of a dipole, which decreases with the distance similarly for the dynamic and static interactions. In this case, a correlation between static and dynamic shifts is probable. For the direct interaction of identical molecules, the dipole sum T_{ds} is the same for static and dynamic shifts and their ratio is independent of the geometry of the adsorbed layer:

$$\frac{|\Delta\nu_{\text{st}}|}{\Delta\nu_{\text{dyn}}} = \frac{2K_{\text{st}}\mu}{\alpha_v \nu_a}. \tag{5}$$

For free CO, this ratio is 0.23, while experiment with ZnO gives 30:6 = 5. Such a great difference cannot be explained by the increase of CO dipole in the field of the cation, and it shows that the ability of the oxide to enhance the interaction between molecules is greater for static than for dynamic interaction. The same difference was observed for a

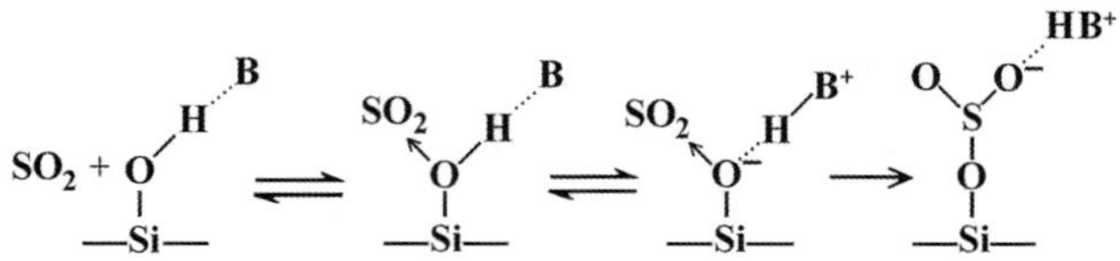

Scheme 2.

number of other oxides and alkali halides[44] and explained by the frequency dependence of polarizability of ionic solids.[43] Polarizability is known to be higher for static fields than in the optical region due to the contribution of the ionic constituent.[43] It has been concluded that it is the surface relaxation induced by adsorption, i.e., the displacement of surface ions on occupation of the adjacent sites, which strongly enhances repulsive static interaction between the adsorbed molecules.[43]

The flexibility of oxide surfaces is also responsible for the enhancement of attractive interactions between coadsorbed basic and acidic molecules. For CO adsorption on silanol groups of silica surface, the band of H-bonded species appears at 2156–2160 cm^{-1}, corresponding to moderate acidity, which is not high enough to protonate strong bases, such as 2,6-dimethylpyridine (DMP). However, when DMP is coadsorbed with gaseous acids (SO_2 or NO_2), the bands at 1650–1630 cm^{-1} appear typical of protonated base $DMPH^+$. This effect of the “induced Brønsted acidity”[45] can be explained by simultaneous interaction of the OH groups with two different molecules. While the electron-donating molecule forms a hydrogen bond with the proton of the OH group, the electron-accepting molecule interacts with the electron lone pair of the oxygen atom of the hydroxyl group (Scheme 2). The formation of a coordinate bond with the oxygen atom increases acidity of the OH group to the extent that proton transfer occurs, indicating Brønsted acidity.

The same scheme illustrates the increase of basic properties of the hydroxyl oxygen atom induced by the base. SO_2 is a weak acid, which reversibly interacts with SiO_2 surface at 300 K, and it becomes chemisorbed in the presence of preadsorbed base. Thus, the base interaction with the proton of the OH group increases the negative charge or basicity on the oxygen atom, which strengthens the coordinate bond between it and the Lewis acid or the hydrogen bond with the adsorbed proton-donating molecules. This effect of "induced basicity" has been recently confirmed by the spectra of adsorbed fluoroform,[46] revealing very strong basic sites after preliminary adsorption of ammonia on silica.

Addition of CO_2 or ozone to a sample with ammonia or DMP preadsorbed on silica does not lead to proton transfer; however, the resulting ν_{OH} shifts become greater than those observed at any coverage by only one of the coadsorbed compounds. This demonstrates the strengthening of the hydrogen bond with the base. The additional shift can be used to estimate the Lewis acidity of such molecules as ozone that is stronger than that of CO_2 and close or slightly lower than the acidity of SO_2.[46] If the acidity of surface hydroxyls increases as a result of the action of acidic molecules on oxygen atoms of the OH groups, the increase of electron-accepting properties of surface cations after adsorption of acidic molecules on neighboring oxygen sites is expected. Such induced Lewis acidity has been detected in recent studies of the influence of CO_2 (or SO_2) on IR spectra of CO adsorbed on CaO pretreated at 972 K.[47] The band of coordinately bound CO, observed at the lowest coverage, shifts from ca. 2165 cm^{-1} for CO on pure CaO up to 2179 (or 2184 cm^{-1}) when the sample is exposed to CO_2 (or SO_2). This shift reflects the increase of Lewis acidity due to chemisorption of acidic molecules. The same mechanism explains superacidity of sulfated zirconia[48] or other halogenated or sulfated oxides.

9. Linkage Isomerism on Adsorption

In the spectra of CO adsorbed on zeolites, a weak band shifted to low frequencies with respect to the band of a free molecule (LF band) is often found along with the usual high-frequency band of molecules bound to the cations (HF band). Various hypotheses have been advanced concerning the origin of the LF band. For Na zeolites, the LF band is very close to the ^{13}CO band. However, the LF band increases with the temperature when the HF band decreases (see Fig. 8a), which is inconsistent with assignment to the isotope band.[49]

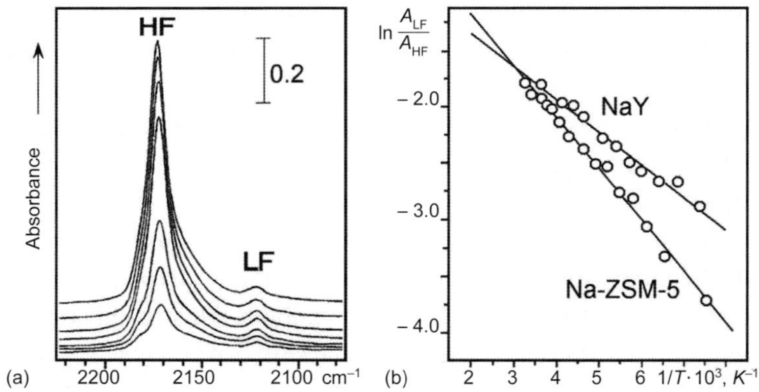

Fig. 8. (a) Evolution of IR spectrum of CO adsorbed on NaY zeolite with temperature in the 167–233 K interval and (b) van 't Hoff plots for CO on NaY and NaZSM-5.

As shown in Ref. 49, the logarithm of the ratio of the LF and HF band intensities changes linearly with $1/T$ (see Fig. 8b). This dependence can be expected when the LF band corresponds to an excited state with excitation energy of ~kT. The origin of excitation was explained by the existence of two isomeric CO complexes, where CO is bound to the cation via the C or O atom.[49] These two forms are in thermodynamic equilibrium, and the concentration ratio depends on the temperature and isomerization enthalpy $\Delta H°$ following the van 't Hoff equation written as

$$\ln\left(\frac{A_{LF}}{A_{HF}}\right) = -\frac{\Delta H^\circ}{RT} + \frac{\Delta S^\circ}{R} + \ln\left(\frac{\varepsilon_{LF}}{\varepsilon_{HF}}\right). \tag{6}$$

Here the concentration ratio of the substances in equilibrium is expressed in terms of the LF and HF band intensities (A_{LF} and A_{HF}) and the extinction coefficients of these surface species ε_{LF} and ε_{HF}. From the slope of the van 't Hoff plot (Fig. 8b), the CO/OC isomerization enthalpy $\Delta H°$ can be obtained. The slopes are different for the two lines and yield 3.8 and 2.4 kJ mol^{-1} for NaZSM-5 and NaY zeolites, respectively. Hence, CO interacts differently with the same Na^+ cation presumably depending on its local environment influenced by the zeolite structure.

Linkage isomerism of adsorbed CO was found for Y and ZSM-5 zeolites with Li, K, Rb, Cs, Ca, and Sr cations and for the H-forms of zeolites, silica, and silicalite where the hydroxyl groups form hydrogen bonds with CO molecules.[50–53] However, for X zeolites or metal oxides, where CO forms coordinate bonds with surface cations, linkage isomerism was not detected. Using a simple electrostatic model, it is possible to answer why a dipole molecule forms linear complexes with a cation by both ends, why it does not do this when adsorbed on metal oxides or X zeolites with the same cations, and how the formation of each isomeric state affects the frequency and absorbance of CO.[53] With this approximation, the adsorption energy ε_{ads} is $(1/2)\alpha_{||}E^2 \pm \mu_0 E$, and $(1/2)\alpha_{||}E^2$ for the linear M–CO and M–OC and side-on complexes with a surface cation (M), respectively. The contribution of the induced dipole to the total adsorption energy is always positive while that of the permanent dipole of molecule μ_0 depends on the orientation. For CO with $\mu_0 = 0.11$ D, both C- and O-bonded structures appear energetically favorable for $E > 2\mu_0/\alpha_{||}$. Because the parallel component of the polarizability is larger than the perpendicular component (for CO, $\alpha_{||} = 2.33$ Å^3 and $\alpha_{\perp} = 1.8$ Å^3), the two states are separated by an energetic barrier for the field strength $E > 2\mu_0/(\alpha_{||} - \alpha_{\perp})$ and the O-bonded state becomes metastable. The $\Delta\varepsilon$ value can be calculated as the reorientation energy of the CO dipole in the field of the cation E, which can be found from the frequency shift caused by the Stark effect upon adsorption:

$$\Delta\varepsilon = 2\mu_0 E = \frac{2\mu_0 \Delta v}{K_{st}}, \tag{7}$$

where $K_{st} = 4.29 \times 10^{-9}$ cm^{-1} V^{-1} m for CO.

A good agreement was found between experimental ΔH°_{exp} determined from the van 't Hoff plots and the electrostatic model.[53] However, if the molecule is considered as a polarizable dipole, the agreement between experimental and computational absorption energies is rather poor. The agreement is much better if the quadrupole momentum of the molecule is taken into account. In addition to the small dipole moment, CO has a relatively large linear quadrupole where both ends of the molecules are negatively charged and the positive charge is in the middle. Interaction of the quadrupole and cations increases the

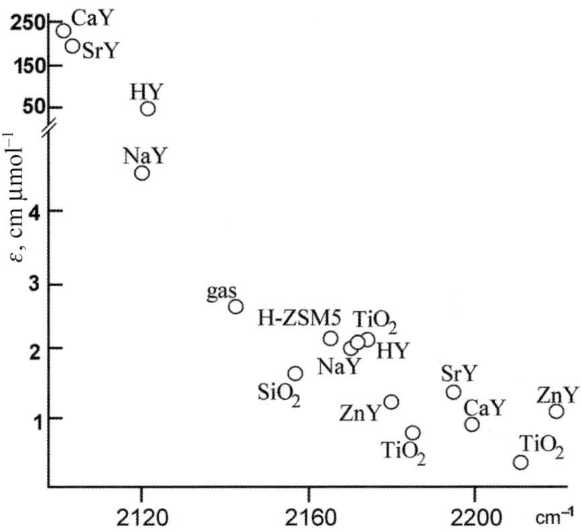

Fig. 9. Integrated absorption coefficient of CO adsorbed on oxide adsorbents as a function of the stretching frequency.[19]

adsorption energy substantially for both C- and O-bonded configurations and the barrier between the two states grows.

The quadrupole interaction also explains spectral features of CO adsorption on basic sites. For the side-on adsorption on anions, the contribution of the quadrupole–anion interaction to ε_{ads} is positive and comparable with the energy of adsorption on the cations. If the effective charge on oxygen is −1.5, CO interaction with this anion is stronger than in the linear complexes with Cs, Rb, or K. For zeolites X or for oxides with a similar charge on oxygen atoms, the barrier between C- and O-bonded states can disappear if the transition between the two states occurs via the side-on complex with a neighboring oxygen ion. This might be a reason why CO/OC isomerism has not been observed for metal oxides. The side-on interaction with oxygen accounts for the band with the frequency as low as 2125 cm^{-1} for CO adsorbed on CsX zeolite.[33] Similar CO interaction with even more basic oxygen atoms of CaO or other basic oxides leads finally to the formation of surface "carbonite" CO_2^{2-} ions with a greater decrease of ν_{CO}.[1]

For the M–CO complexes, the electrostatic model predicts an increase of the CO frequency and a decrease of the absorption

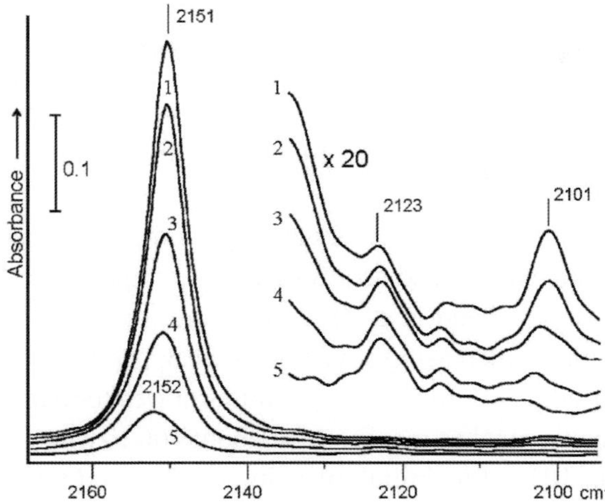

Fig. 10. IR spectra of CO adsorbed on KBr sample at different temperatures: (1) 77 K, (2) 90 K, (3) 95 K, (4) 100 K, and (5) 110 K.[54]

coefficient. For the O-bonded complexes, the situation is opposite, i.e., the frequency decreases and the absorbance is enhanced. It is not possible to measure directly the concentration of the LF isomer. However, the $\varepsilon_{LF}/\varepsilon_{HF}$ ratio can be found from the intersection points of the van 't Hoff plot extrapolated to $1/T = 0$ if we can estimate the changes of isomerization entropy.[33] For this purpose, the experimental correlation between the entropy and enthalpy of CO adsorption was considered and $\varepsilon_{LF}/\varepsilon_{HF}$ was found to be always above 1 and grow with $\Delta H°$. By measuring ε_{HF} as in Section 4, the values of ε_{LF} were obtained. The experimental values of ε_{HF} and ε_{LF} are shown in Fig. 9 for different oxide adsorbents. The observed decrease of the absorbance with the increasing CO frequency agrees with the theory.

Thus, linkage isomerism is expected for the ionic adsorbents where the electric field of anions is weak compared to that of cations. This condition can be achieved if the charge of anions is lower than that of oxygen, as it is for metal halides or compounds with large ionic radii of anions. It means that the isomerism is more probable for bromides than for fluorides or chlorides. The field of anions can be even weaker for polyatomic anions (e.g., SO_4^{2-}). Here the double negative charge is distributed over four oxygen ions. Zeolites with high Si/Al ratio are the ultimate system in this row because the negative charge (equal to the charge of cations) is almost homogeneously distributed between oxygen atoms of the framework, and the field of cations is close to that of bare ions in free space. Thus, it is not by chance that the linkage isomerism was first found for zeolites.

To test predictions of the electrostatic model, the spectrum of CO adsorbed on KBr was studied as a function of temperature (Fig. 10).[54] At 77 K, an intense ν_{CO} band is at 2152 cm^{-1} and it slightly shifts to lower frequencies with the increasing coverage. A weak band due to ^{13}CO isotope is seen at 2101 cm^{-1}. For higher temperatures, a weaker band at 2123 cm^{-1} grows in intensity, reaches maximum at about 100 K when the 2152-cm^{-1} band has about 10% of its initial intensity at 77 K, and finally both bands disappear at ~150 K. This is exactly the behavior expected for the band of the M–OC species.

Recently, linkage isomerism has been also found for CN^- ions which form on zeolites as a result of dissociative adsorption of HCN.[55] The spectrum of CsX zeolite was measured from ca. 80 up to 500 K by means of the cell shown in Fig. 1. In addition to the band of CN^- ions at 2150 cm^{-1} (usual C-bonded ions), a band at 2138 cm^{-1} grows with the temperature. The lower frequency band is assigned to the ions bound to Cs^+ cations via the N atom.

10. Surface Properties of Disperse Water Ice

Properties of water ice surface are of a great interest for the chemistry of the atmospheres of Earth and other planets and for understanding the mechanism of photoprocesses in interstellar clouds. In the previous studies of amorphous ice surface, the dangling OH bands have been detected, which shift to lower frequencies on adsorption of such gases as CO,[56] halogenated hydrocarbons,[57] etc., indicating the formation of weak hydrogen bonds with adsorbed molecules.

The spectra of ice films in the presence of gaseous adsorbate have been studied, and some results are shown in Fig. 11. The broad intense bands at 2442 and 3303 cm^{-1} are due to OD and OH stretching vibrations

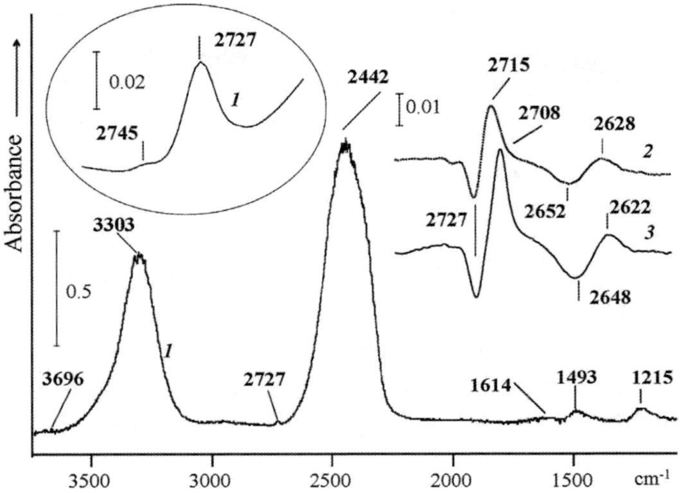

Fig. 11. IR spectrum of deuterated water ice at 77 K (1) and difference spectra in the region of dangling OD groups observed after addition of nitrogen (2), and methane (3).

of bulk water and the bending vibrations of H_2O, HDO, and D_2O are seen at 1614, 1493, and 1215 cm^{-1}. The weak bands at 3696 and 2727 cm^{-1} belong to the dangling OH and OD groups. Perturbation of these groups by adsorbed N_2 and methane is seen in the difference spectra (curves 2 and 3). Besides the shifts of the dangling OD band to 2715 (N_2) and 2708 cm^{-1} (methane), the decrease of absorption near 2650 cm^{-1} and the increase at 2628–2622 cm^{-1} occur in the presence of gaseous adsorbates. These changes at the high-frequency edge of the bulk absorption can be interpreted as an evidence for H-bond strengthening in the surface layer of ice particles induced by reversible adsorption.[58] This effect is close to the induced acidity of silica described above and means that surface properties of ice are modified in the presence of other molecules.

Acidic properties of pure water ice and mixed ices of codeposited NH_3–water and HCN–water mixtures has been studied with adsorbed CO as a test molecule.[59] Acidity of dangling hydroxyls was found to be weaker than that of silanol groups of silica surface (CO band at 2153 cm^{-1} for ice and 2156–2160 cm^{-1} for silica). Another band of adsorbed CO at 2137 cm^{-1} was attributed to molecules bound to unsaturated surface oxygen atoms. Admixed ammonia or HCN have no noticeable effect on the position of adsorbed CO bands but change their relative intensities, hence showing that the proportion between the numbers of protonic and oxygen sites is influenced by acidic or basic dopants in the bulk of ice.

11. Conclusions

Surface of disperse solids is an excellent medium for low-temperature spectral studies of intermolecular interactions, molecules in strong electric fields, etc. Spectral studies of adsorption in a wide temperature range provide valuable information on the studied systems and supply fundamental knowledge of various phenomena at interfaces.

Acknowledgments

This work was partially supported by the Federal Agency of Science and Innovations, Contract No. 02.740.11.0214.

References

1. M. A. Babaeva, D. S. Bystrov, A. Yu. Kovalgin and A. A. Tsyganenko, *J. Catal.* **123**, 396 (1990).
2. C. O. Areán, O. V. Manoilova, A. A. Tsyganenko, G. T. Palomino, M. Peñarroya Mentruit, F. Geobaldo and E. Garrone, *Eur. J. Inorg. Chem.* **7**, 1739 (2001).
3. A. A. Tsyganenko, *Instrum. Exp. Tech.* **23**, 265 (1980).
4. N. A. Sekushin and A. A. Tsyganenko, *Russian J. Phys. Chem.* **61**, 82 (1987).
5. A. A. Tsyganenko, D. V. Pozdnyakov and V. N. Filimonov, *J. Mol. Struct.* **29**, 299 (1975).
6. A. A. Tsyganenko, A. V. Khomenya and V. N. Filimonov. *Adsorbtsiya i adsorbenty*, p. 86, No. 4, Kiev, Naukova Dumka, 1976, (in Russian).
7. K. M. Bulanin, A. V. Alexeev, D. S. Bystrov, J.-C. Lavalley and A. A. Tsyganenko, *J. Phys. Chem.* **98**, 5100 (1994).
8. L. Mariey, J. Lamotte, P. Hoggan, J.-C. Lavalley, K. M. Bulanin and A. A. Tsyganenko, *Chem. Lett.*, 835 (1997).
9. N. E. Tret'yakov and V. N. Filimonov, *Kinet. Catal.* **13**, 735 (1972).
10. T. A. Rodionova, A. A. Tsyganenko and V. N. Filimonov. *Adsorbtsiya i adsorbenty,* p. 33, No. 10, Kiev, Naukova Dumka (1982) (in Russian).

11. M. I. Zaki and H. Knozinger, *Mater. Chem. Phys.* **17**, 201 (1987).
12. S. Yu. Maslov, L. A. Denisenko, A. A. Tsyganenko and V. N. Filimonov, *React. Kinet. Catal. Lett.* **20**, 273 (1982).
13. S. M. Zverev, K. S. Smirnov and A. A. Tsyganenko, *Kinet. Catal.* **29**, 1251 (1988).
14. H. A. Pearce and N. Sheppard, *Surface Sci.* **59**, 205 (1976).
15. D. A. Seanor and C. H. Amberg, *J. Chem. Phys.* **42**, 8 (1965).
16. C. Morterra, E. Garrone, V. Bolis and B. Fubini, *Spectrochimica Acta* **A43,** 1577 (1987).
17. E. E. Platero, D. Scarano, G. Spoto and A. Zecchina, *Faraday Disc. Chem.* Soc. **80**, 183 (1985).
18. K. S. Smirnov and A. A. Tsyganenko, *Opt. Spectrosc.* (*USSR*) **60,** 407 (1986).
19. E. V. Kondratieva, O. V. Manoilova and A. A. Tsyganenko, *Kinet. Catal.* **49**, 451 (2008).
20. N. Sheppard and D. J. C. Yates, *Proc. Roy. Soc.* **A238**, 69 (1956).
21. K. S. Smirnov, M. A. Nikolskaya and A. A. Tsyganenko, *Opt. Spectrosc.* (*USSR*) **62**, 743 (1987).
22. J. B. Peri, *J. Phys. Chem.* **70,** 2937 (1966).
23. A. A. Tsyganenko and M. A. Babaeva, *Opt. Spectrosc.* (*USSR*) **54**, 665 (1983).
24. A. Yu. Pavlov and A. A. Tsyganenko, *Opt. Spectrosc.* (*USSR*) **77**, 21 (1994).
25. D. N. Schepkin. *Anharmonic Effects in the Spectra of H-Bonded Complexes*. Dep. VINITI, p. 86, No. 7511-1387 Leningrad (1987) (in Russian).
26. A. A. Tsyganenko, K. S. Smirnov, E. P. Smirnov and A. Yu. Pavlov, *In: Physics, Chemistry and Technology of Surface. Issue*. I. p. 65, Kiev, Naukova Dumka (1993).
27. M. P. Casassa, E. J. Heilweil, J. C. Stephenson and R. R. Cavanagh, *J. Electron. Spectrosc. Relat. Phenom.* **38**, 257 (1986).
28. Yu.A. Tsyganenko, V. A. Ermoshin, M. R. Keyser, K. S. Smirnov and A. A. Tsyganenko, *Vibr. Spectrosc.* **13**, 11 (1996).
29. J. H. De Boer, *The Dynamical Character of Adsorption*, Oxford, Clarendon Press (1953).
30. C. O. Areán, O. V. Manoilova, G. T. Palomino, M. R. Delgado, A. A. Tsyganenko, B. Bonelli and E. Garrone, *Phys. Chem. Chem. Phys.* **4**, 5713 (2002).
31. A. A. Tsyganenko and V. N. Filimonov, *Spectrosc. Lett.* **13**, 583 (1980).
32. A. A. Tsyganenko, L. A. Denisenko, S. M. Zverev and V. N. Filimonov, *J. Catal.* **94**, 10 (1985).
33. A. A. Tsyganenko, E. V. Kondratieva, V. S. Yanko and P. Yu. Storozhev, *J. Mater. Chem.* **16**, 2358 (2006).
34. Y. M. Grigorev, V. N. Filimonov and D. V. Pozdnyakov, *Russ. J. Phys. Chem.* **46**, 186 (1972).
35. K. M. Bulanin, J.-C. Lavalley and A. A. Tsyganenko. *Colloid. Surface Physicochem. Eng. Aspect.* **A101**, 153 (1995).

36. K. M. Bulanin, J. C. Lavalley, J. Lamotte, L. Mariey, N. M. Tsyganenko and A. A. Tsyganenko, *J. Phys. Chem.* **B102**, 6809 (1998).
37. A. A. Tsyganenko, J. Lamotte, J. P. Gallas, J.-C. Lavalley, *J. Phys. Chem.* **93,** 4179 (1989).
38. R. P. Eischens, S. A. Francis and W. A. Pliskin, *J. Phys. Chem.* **60**, 194 (1956).
39. G. L. Griffin and J. T. Yates, *J. Chem. Phys.* **77**, 3751 (1982).
40. L. A. Denisenko, A. A. Tsyganenko and V. N. Filimonov, *React. Kinet. Catal. Lett.* **25**, 23 (1984).
41. F. M. Hoffman, *Surf. Sci. Rep.* **3**, 107 (1983).
42. R. M. Hammaker, S. A. Francis and R. P. Eischens, *Spectrochim. Acta* **21**, 1295 (1965).
43. A. A. Tsyganenko and S. M. Zverev, *React. Kinet. Catal. Lett.* **36**, 269 (1988).
44. E. E. Platero, D. Scarano, G. Spoto and A. Zecchina, *Disc. Faraday Soc.* **80**, 193 (1985).
45. A. A. Tsyganenko, E. N. Storozheva and O. V. Manoilova, *Catal. Today* **70**, 59 (2001).
46. E. N. Storozheva, V. N. Sekushin and A. A. Tsyganenko, *Catal. Lett.* **107**, 185 (2006).
47. A. A. Tsyganenko and K. V. Voronina (unpublished).
48. A. V. Ivanov and L. M. Kustov, *Rus. Chem. Bul.* **49**, 39 (2000).
49. A. A. Tsyganenko, E. E. Platero, C. O. Areán, E. Garrone and A. Zecchina, *Catal. Lett.* **61**, 187 (1999).
50. C. O. Areán, O. V. Manoilova, A. A. Tsyganenko, G. T. Palomino, M. Penarroya Mentruit, F. Geobaldo and E. Garrone, *Eur. J. Inorg. Chem.* 1739 (2001).
51. A. A. Tsyganenko, P. Yu. Storozhev and C. O. Areán, *Kinet. Catal.* **45**, 530 (2004).
52. C. O. Areán, G. T. Palomino, A. A. Tsyganenko and E. Garrone, *Intern. J. Mol. Sci.* **3**, 764 (2002).
53. P. Y. Storozhev, V. S. Yanko, A. A. Tsyganenko, G. T. Palomino, M. R. Delgado and C. O. Areán, *Appl. Surface Sci.* **238**, 390 (2004).
54. A. A. Tsyganenko and R. A. Belykh (unpublished).
55. A. A. Tsyganenko, *High Energ. Chem.* **42**, 610 (2008).
56. J. P. Devlin and V. Buch, *J. Phys. Chem.* **99**, 16534 (1995).
57. N. S. Holmes and J. R. Sodeau, *J. Phys. Chem.* **A103**, 4673 (1999).
58. A.V. Rudakova, M. S. Poretskiy, I. L. Marinov and A. A. Tsyganenko, *Opt. Spectrosc.* **109**, 708 (2010).
59. A. V. Rudakova, V. N. Sekushin, I. L. Marinov and A. A. Tsyganenko, *Langmuir* **25**, 1482 (2009).

CHAPTER 10

PHOTOLYSIS AND RADIOLYSIS OF WATER ICE

Robert E. Johnson

Department of Materials Science and Engineering and Department of Astronomy, University of Virginia, Charlottesville, VA 22904, USA
E-mail: rej@virginia.edu

The study of the chemical changes produced in ice by electrons, ions, and photons has been stimulated to a large extent by its relevance to problems in astronomy, but it is also of interest as a model system for solid-state chemistry. In this chapter, the physical and chemical processes induced by the incident ions, electrons, or photons are reviewed as is their relevance to the irradiation of icy bodies in the outer solar system. An overview is given of the considerable laboratory database on the radiolysis and photolysis of water ice and recent models on the decomposition of ice into H_2 and O_2 are reviewed and discussed.

1. Introduction

Radiolysis and photolysis of liquid water has been studied for years as a model for radiation biology. Considerable interest in the radiation chemistry occurring in ice followed the experiments of Brown, Lanzerotti, and co-workers on irradiation of icy bodies in the outer solar system and interstellar medium.[1,2] In this chapter, the chemical pathways initiated by the irradiation of ice are outlined. Although the chemical processes occurring in water are a guide, the production of defects, the presence of a substrate and a vacuum interface, and trapping of radicals in ice can result in important differences.

Early experiments on desorption from ice focused on the easy-to-detect ions, and chemical changes were often determined by melting irradiated samples.[3] Radiation products in the sample were also identified spectroscopically[4] and by electron spin resonance experiments.[5] Brown *et al.*[6–8] showed that the chemistry initiated near the vacuum interface of an ice by ionizing radiation could cause, with relatively high-efficiency, molecular ejection. This process is often referred to as electronically induced sputtering or, if the yield is small, electronically induced desorption.

The surface binding energy acts as a filter for the ejecta so that the most volatile species are lost more efficiently and the near-surface composition of the irradiated sample is altered.[9] For instance, studying the radiation-induced decomposition of a low-temperature ice sample Reimann *et al.*[10] found that H_2 is formed and lost preferentially. Therefore, with increasing fluence oxygen-rich species become important constituents in the region penetrated by the radiation.[11–15]

The incident radiation also produces defects, which, under continued irradiation can lead to the formation of voids and percolation pathways to the surface.[16] This can affect both the escape rate and trapping of molecules and radicals.[17–21] It is especially interesting that many of the chemical changes and ejection processes seen in the laboratory are directly relevant to telescopic observations of the ice-rich surfaces of the moons of the giant planets in the outer solar system, and ice grains in the interstellar medium. For instance, reflectance spectra show that the icy surfaces of these moons, which are continuously exposed to particle radiation, not only contain trapped H_2O_2, but also trapped O_2 and O_3.[14,22,23] In addition, these bodies have very tenuous H_2 and O_2 atmospheres produced by the radiation-induced decomposition of ice.[14,24] That H_2O_2 and O_2 are produced in ice was seen in very early experiments[3] and is, of course, expected based on radiolysis in liquid water.[25,26] However, the production of O_2 at low-radiation doses suggests that there can be additional chemical pathways in the solid. These early laboratory and space observations have led to considerable experimental and modeling efforts in both the planetary science and radiation effects communities. Following a brief review of the space observations, those efforts and my understanding of the present state of knowledge are presented.

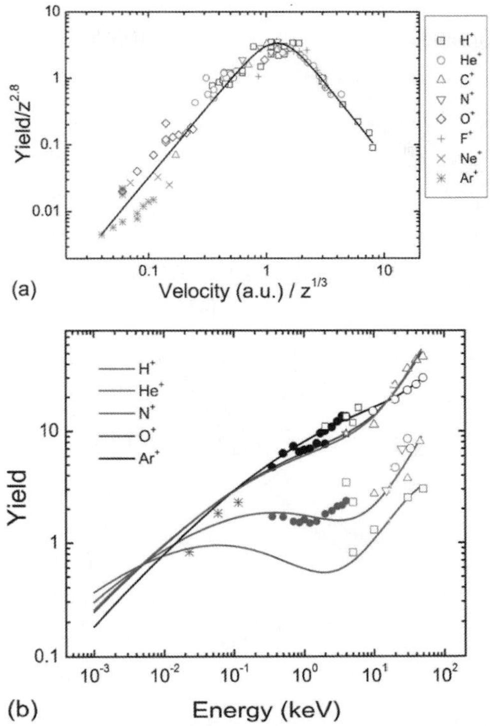

(a)

(b)

Fig. 1. (a) Yield, Y, in equivalent water molecules ejected from ice per incident fast ion versus the ion velocity in atomic mass units (2.18×10^8 cm s^{-1}) $T < 100$ K: ejection is primarily due to electronic excitations and ionizations. Yield and velocity are scaled by nuclear charge, Z, of ion from Johnson *et al.*[14]; line gives fit from http://www.people.virginia.edu/~rej. (b) Y in lower energy regime includes electronic and knock-on sputtering processes: summary of data, see Fama *et al.*[30]; lines are analytic expressions in that paper: electronic contribution indicted by upturn above ~1 keV amu^{-1}. Lower energies dominated by knock-on sputtering. See also Colour Insert.

2. Summary of the Space Observations

During the transit of the Jovian system by the Voyager I spacecraft, it became clear that the ice covered moons of Jupiter were imbedded in a flux of ions and electrons with energies from electronvolts to megael ectronvolts.[14] This flux is especially intense at the dehydrated moon Io which has frozen SO_2 and S_8 on its surface. It is also intense at Europa

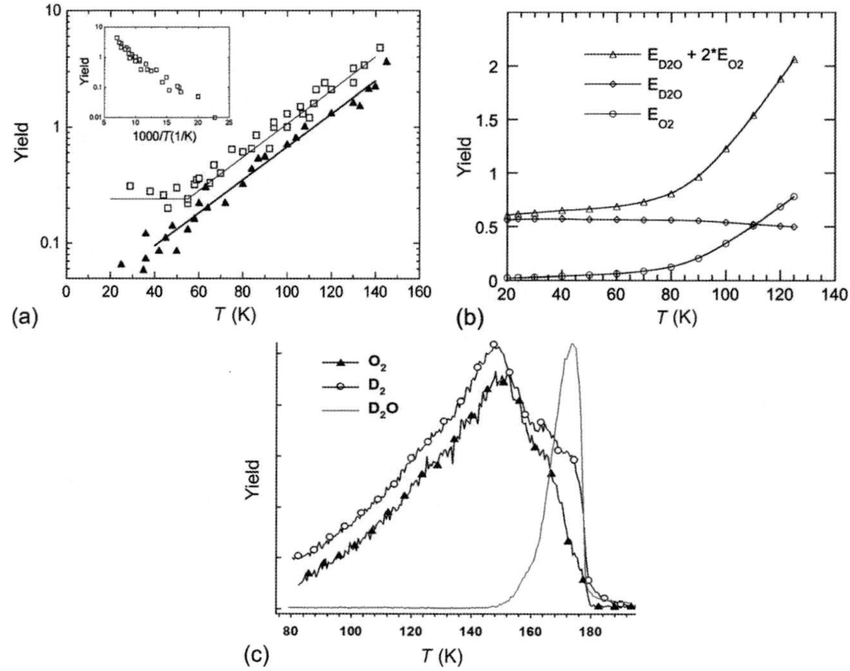

Fig. 2. Gas-phase ejecta due to irradiation: (a) Saturation yields for D_2 (squares) and O_2 (diamonds) due the 1.5-MeV He^+ on ice versus sample T adapted from Brown *et al.*[7] Inset: D_2 yield minus the constant yield at low T versus $1/kT$, giving an average activation energy of 0.03 eV. (b) D_2O and O_2 yields versus T for 87-eV electrons on ice at 100 K from Petrik and Kimmel[29]: as also shown in Ref. 17 the total yield in equivalent water molecules is roughly the sum of the D_2O and twice the O_2 yield: see also Ref. 7. (c) Relative yield versus temperature for incident electrons in steady state, showing the correlation between the production of O_2 and D_2 in steady state.[15] Also shown: sublimation of D_2O.

and Ganymede, moons that are primarily covered by water ice with temperatures ~80–140 K. This realization led Brown, Lanzerotti, and co-workers to initiate a series of experiments on energetic plasma ion irradiation of low-temperature ice.[1,2,7] Three results from these experiments have defined the subsequent extensive experimental effort.

First, they found that sputtering, the ejection of molecules from the surface due to incident radiation, was produced not only by knock-on collisions, the usual sputtering process, but also by the electronic excitations and ionizations produced in ice, as seen by a summary of data in Fig. 1. This electronic sputtering process is related to electronically stimulated desorption from surfaces and to the radiolysis of ice.[27,28] Second, they also found that the yield could be very large (~1000 H_2O/ion for ~100-keV O^+) and is even significant for incident MeV protons. Finally, they showed that the ejecta was predominantly whole molecules, not radicals, with H_2O dominating the ejecta at low temperature (below ~110 K) and H_2 and O_2 increasing in importance with increasing temperature as seen in Fig. 2a for energetic He^+ on D_2O ice. In fact, above ~120–130 K the H_2 and O_2 produced by the decomposition of ice dominate the mass loss. Similar results were obtained eventually for incident electrons[29] and lower energy ions[30] as seen in Figs. 2b and c.

Such experiments have led to a number of predictions on the radiation effects produced on the icy moons of Jupiter[11,31] many of which were confirmed by space observations. For instance, as predicted, O_2 was eventually detected in the very thin atmospheres of Europa and Ganymede, two icy moons of Jupiter[32] and the chemical composition of the surface was seen to be altered.[14] Surprisingly, O_2 was also seen trapped in inclusion in these icy surfaces,[33] an observation that *preceded* the laboratory detection of O_2 *in* irradiated ice. That is, in addition to the observation of gas-phase oxygen, two well-known transitions for interacting pairs of O_2 molecules, ${}^1\Delta_g + {}^1\Delta_g \leftarrow {}^3\Sigma_g^- + {}^3\Sigma_g^-$ were observed in the reflectance spectra of Ganymede[34,35] and Europa.[36,37] These absorption bands were not due to atmospheric O_2, which would require very high pressures or very long line-of-sight columns of O_2. The positions of the two absorption features for the O_2 dimer at 577.2 and 627.5 nm[38,39] indicate that the O_2 is locally dense[34] with band shapes and relative band depths slightly distorted from solid or liquid O_2. These bands are due to inclusions of O_2 trapped within the satellite's icy surface[16,34] with possible distortions in band shape due to presence of other species[40] or cage effects.[41]

In a related set of observations, the Hartley band of O_3 was identified in the ultraviolet (UV) reflectance spectra of Ganymede[42] and certain icy satellites of Saturn.[43] Although the band shape suggests additional absorbing species are present,[23,40,44] trapped O_3 is consistent with the presence of radiolysis in trapped oxygen inclusions.[16,45] In fact, the more readily observable O_3 feature is a marker for O_2 inclusions.[22] The O_3-like feature in the UV is superimposed on a very broad UV absorption feature extending from about 0.4 μm to shorter wavelengths, a feature seen on most of the icy satellites. This broad UV absorption has been attributed to another radiolytically produced oxidant, H_2O_2,[22,46] but might also include absorption by trapped HO_2 and OH.[4,22] Identification of specific radiolytic products in the UV is difficult because the bands overlap and are broad. However, the presence of trapped H_2O_2 was confirmed at Europa by the presence of a band in the infrared observed by the Galileo spacecraft.[23,46] The amount seen in the infrared is roughly consistent with the amount suggested by the slope in UV reflectance spectrum.

Although each of the identifications of the products of radiolysis or photolysis (OH, H_2O_2, O_2, O_3) has considerable uncertainty, the reflectance spectra of the icy satellite surfaces taken as a whole are self-consistent. That is, the irradiated surfaces of these satellites contain trapped molecules that are expected due to radiolysis or photolysis of ice.[13,14,22] Consistent with this picture, oxygen-rich molecules are formed by irradiation of an ice containing trace amounts of sulfur (SO_2 and a sulfate) and carbon (CO_2 and a carbonate).[13,14,23,47,48] The sulfate is likely a hydrated sulfuric acid.[49,50] These observations indicate that both volatile and refractory oxygen-rich molecules are readily formed by radiolysis and photolysis in these icy surfaces.[13,23]

It is now generally agreed that the radiolysis by energetic ions and electrons chemically and physically alters the surface ice and causes decomposition producing H_2 and O_2. What is particularly exciting is the more recent realization that these bodies very likely have underground oceans.[51] This is the case as ice is a good thermal insulator and the moons have internal radiogenic and tidal heat sources. The presence of oceans has led to speculation on possible geologic transport of radiation-produced oxidants from the surface to the putative subsurface oceans as a means of sustaining aerobic processes.[52–55] Not surprisingly, the trapping and transport of the products of radiolysis and photolysis of ice is now of considerable astrophysical interest[13] and missions by both NASA and ESA are planed to visit Jupiter and its satellites with an emphasis on understanding the implications of the radiolytic processing of the surface of Europa.

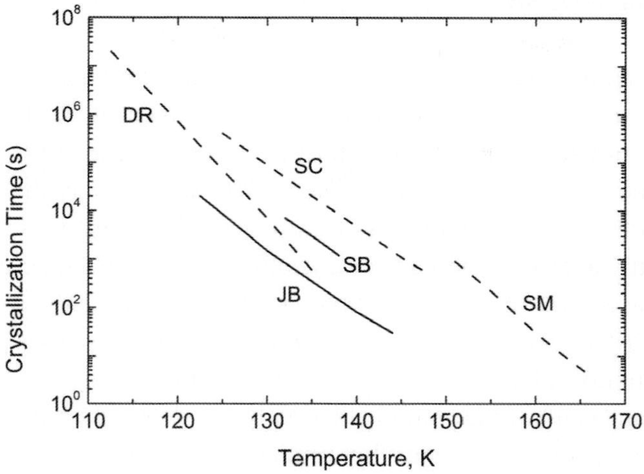

Fig. 3. Summary of data for crystallization times for ASW ice from Baragiola[62]: lines are fits: JB[66] start of transition; SB[67] 63% crystallized; DR,[68] SC,[69] SM[70] end of transformation.

Radiation-induced alteration of ice is also of interest for ices in the interstellar medium[56–60] and in the debris discs of young stellar objects and protostars.[61] Studies of the irradiation of ice have also been suggested as a model system for radiation-induced decomposition of water in underground containers of radioactive materials. Below I describe the radiation chemistry and chemical pathways based on recent reviews.[13,15,23,62–64] The principal results of the considerable laboratory database are outlined followed by a description of models for the chemical kinetics induced by ionizing radiation. The proposed pathways at low doses and their relationship to those thought to be dominant in liquid water are then examined.

3. Summary of Laboratory Results

The laboratory data on the radiolysis and photolysis of ice is now extensive, but the chemical pathways occurring in the solid at low doses are not yet agreed upon. Unlike in the liquid or gas, structure plays a critical role in the photolysis and radiolysis of ice. Results for incident ions, electrons, and UV photons are first summarized.

3.1. *Sample structure*

Low-temperature ice can have a variety of structures depending on how it is formed.[18,19,64,65] The emphasis here, as in much of the literature, is on amorphous solid water (ASW) formed by vapor deposition with some experiments on thin, relatively defect free samples. With increasing temperature, an ASW sample passes through a glass transition at ~120 K. It eventually anneals forming polycrystalline ice with average grain sizes that increase with the annealing time and temperature. However, the as-deposited ASW and annealed ASW are typically microporous[66] although very thin uniform films can be grown.[65] Since defects are produced by the radiation and can be made mobile thermally, the defect density can change by increasing temperature or by irradiation.[20,60] Structural changes in vapor-deposited samples occur in the temperature regime (50–150 K) with the time for the changes depending strongly on the sample temperature (e.g., see Fig. 3) and/or radiation dose.

Icy surfaces formed in the very distant regions of the solar system or the interstellar medium are often assumed to be ASW, but the icy surfaces of the moons of Jupiter and Saturn, although highly porous have been annealed over very long periods of time resulting in grains that are much larger than those in typical laboratory samples. On exposure to a plasma flux, radiation-induced defects and thin amorphous surface layers are formed.[71–73] In addition, the icy bodies in the outer solar system are subject to meteoroid impacts, irradiations, sublimation and redeposition, and, possibly at Europa, the flow of liquid water onto the surface. Because of the differences between laboratory samples, and their difference from surfaces in space, there are considerable uncertainties in the application of laboratory results to observations. For instance, since laboratory ice samples can have porous pathways to the surface, it has been difficult to produce the observed oxygen inclusions (bubbles) in ice which led to incorrect conclusions on O_2 formation.[74,75] However, Johnson and Jesser[16] have pointed out that gas inclusions are readily formed in solids exposed to penetrating radiation, which is clearly the case for ices in space.

ASW can have stable defects, inclusions and porous pathways to the vacuum interface and the incident radiation alters the defect structure of

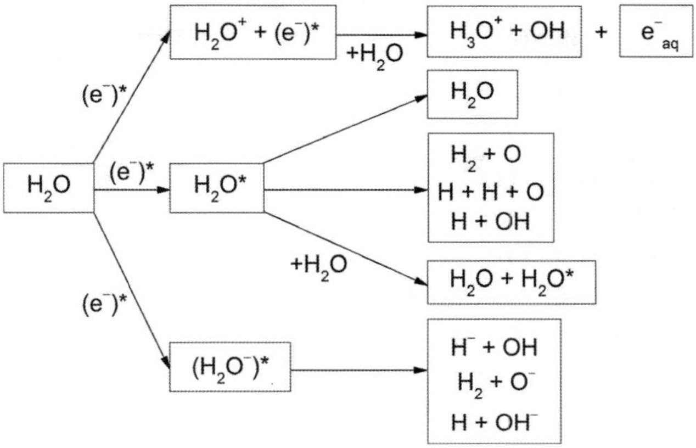

Fig. 4. Suggested radiation products in the liquid phase due to photon and electron impact and electron attachment. Data from Garret *et al.*[63]

the solid and the porosity. Since surfaces play an important role in the chemical evolution of irradiated ice,[29,76] comparisons with liquid water is only very rough. For instance, molecular reorientation is restricted in the solid so that a critical species in the liquid, the trapped electron, appears to play a minor role in the solid. However, a so-called precursor electron, cooled but not fully trapped and mobile may be important.[77] By contrast, radical radiation products are trapped and quite stable affecting the dose dependence of the chemical yield. In spite of these important differences, the extensive literature on radiolysis and photolysis of water in the liquid and gas phase can be a guide to understanding the experiments on ice.

3.2. *Radiation products and desorption*

Incident photons, fast ions, and electrons ionize and/or excite the molecules in ice with heavy ions producing multiple ionizations.[26] In addition, slow ions deposit a significant fraction of their energy in momentum transfer collisions. These events produce bond breaking and hot recoils leading to desorption, as seen in molecular dynamics studies,[78] or resulting in solid-state chemistry and the formation of new molecules.[63] As described above, radiation products have been identified by electron-spin resonance (ESR), UV-visible luminescence and absorption, and IR

Table 1. Enthalpy changes ($-\Delta H$) for exothermic reaction.

Reaction	$-\Delta H$ **(eV)**
$H + OH \rightarrow H_2O$	5.12
$O + O \rightarrow O_2$	5.12
$O + H_2 \rightarrow H_2O$	5.04
$H + H \rightarrow H_2$	4.48
$O + H \rightarrow OH$	4.40
$O + H_2O_2 \rightarrow O_2 + H_2O$	3.68
$OH + HO_2 \rightarrow O_2 + H_2O$	3.12
$O + OH \rightarrow HO_2$	2.73
$\rightarrow H + O_2$	0.72
$O + HO_2 \rightarrow O_2 + OH$	2.40
$OH + OH \rightarrow H_2O_2$	2.17
$H + O_2 \rightarrow HO_2$	2.00
$H + HO_2 \rightarrow OH + OH(0.92)$	1.67
$\rightarrow H_2 + O_2(0.08)$	2.47
$\rightarrow O + H_2O(0.02)$	1.75
$OH + H_2O_2 \rightarrow HO_2 + H_2O$	1.28

absorption spectroscopy and, when molecules are ejected, by mass spectrometry.

Incident UV radiation can cause direct dissociation, primarily leading to H + OH. In the gas phase there is a ~10% branching ratio to O + H_2. Incident electrons with energies less than the band gap can produce H_2,[79] so the initial products are OH, H, H_2, and O. The H and OH are both seen in ice by ESR and, in the case of OH, by its UV absorption band at ~0.28 μm (blue-shifted from the gas phase)[4] and, possibly, the UV-excited luminescence bands at 0.42 and 0.34 μm (red-shifted from the gas phase).[80] The presence of O atoms in the ice lattice was suggested by luminescence bands associated with the formation of excited O_2.[81] More recently, it was identified as an irradiation of product due to X-rays[82] and from proton irradiation using isotopically labeled H_2O.[45] H and O atoms ejected from the surface have also been detected.[79,83,84]

When the incident radiation ionizes water molecules, H_2O^+ and electrons are the primary products (e.g., see Fig. 4). H^+, H_2^+, and $H^+(H_2O)_n$ are seen to be desorbed from ice by electron impact with a threshold ~22 eV consistent with multiple ionizations.[85] As in the gas or liquid, proton transfer from H_2O^+ should rapidly lead to H_3O^+ + OH.

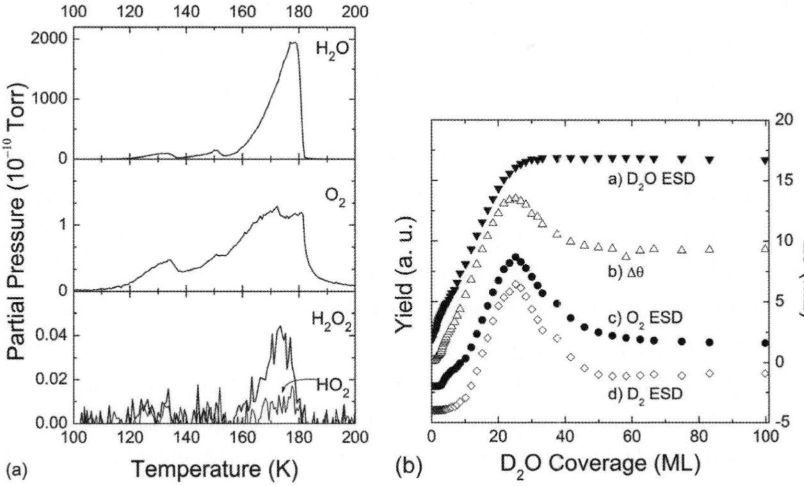

Fig. 5. Ejected radiation products: (a) Programmed thermal (~1.5° min^{-1}) desorption yield of a 7-μm-thick 100 K ice film irradiated by 200-keV H^+ (1.5×10^{15} ions cm^{-2}) indicating irradiation products trapped in ice: from Baragiola.[62] (b) Electron simulated desorption (ESD) yield versus thickness in monolayers (ML) for incident 87 eV, $T = 100$ K, 9.5×10^{15} e cm^{-2}: $\Delta\theta$ is the change in thickness in ML; O_2 and D_2 integrated yields grow with thickness until the film is thicker than the mean electron penetration depth, this indicates that the vacuum and substrate interface contribute to the production of O_2 and D_2: from Petrik and Kimmel.[29]

Other products, such as O, are produced following multiple ionizations (e.g., $H_2O^{+2} + 2H_2O \rightarrow 2H_3O^+ + O$).[86] Although it is essential for describing the radiation-induced chemistry, the fate of ionized species in ice at early times is not well understood. Holes and electrons are seen via mobility experiments,[87] but, since thin films of ASW do not appear to charge significantly, electron or hole transport to the substrate must be efficient.[29,76] Neutralization of H_2O^+ or H_3O^+ by dissociative recombination also leads to the formation of primary products in Fig. 4 (OH, H, H_2, O) with the nature of the initial excitation event determining the relative amounts of each. Therefore, only these neutral products are considered in the following, independent of their formation process which is an important caveat.

The primary radiation products can react to reform water. This process is more efficient in the solid due to the so-called cage effect,[41,88] which results in lower product yields than in the gas or liquid. Transiently energized species that are just produced or made mobile by warming, can diffuse and react forming H_2, H_2O_2, HO_2, and O_2 (Table 1). Such products are seen after irradiation by temperature-programmed desorption (e.g., see Fig. 5a). Although the principal dissociation products are H and OH, the dominant gas-phase species *ejected* from an irradiated ice are those that have the lowest binding energies to the ice matrix: H_2, O_2, and H_2O (e.g., Figs. 2 and 5b). Since surface binding acts as a filter, the yield of H_2 from ice is significant, although the gas-phase production cross section of H_2 is only ~10% of the principal dissociation cross section.[13] Since this volatile product preferentially escapes into the vacuum, the stoichiometry of the irradiated layer changes slightly[9] affecting the subsequent chemistry.[13]

Trapped OH, H_2O_2, and HO_2 produced in UV photolysis were suggested to be observed in the IR,[89] although those results have been questioned. Incident fast protons produced H_2O_2 in ice[90] as well as HO_2 and HO_3 in ice with O_2[40]. Also, O, OH, O_2, H_2O_2, and HO_2 have been seen trapped in ice irradiated by soft X-rays at 20 K[82] and HO_2 and H_2, H_2O_2, and O_2 were seen to evolve from irradiated samples on warming.[3,91] H_2O_2 is suggested by a broad UV absorption feature (<0.25 μm), which leads to its dissociation,[92] with absorption by HO_2 contributing via a broadband at 0.23 μm.[4] The radical HO_2 is thought to be the dominant ESR signal when ice is warmed above 77 K[93,94] with the equivalent species at much lower temperatures suggested to be O_2^-.[5] The trapped electron absorbs in the visible (~650 nm) and is often referred to as e-vis in irradiated ice. Electrons have also been suggested to trap as OH^-[81,95] and can initiate reactions by dissociative attachment: i.e., $H_2O^- \rightarrow H^- + OH$.[91,96]

The formation of O_2 *in* ice was suggested by luminescence studies following photolysis,[81] by ion scattering experiments,[21] and by soft X-ray experiments.[82] Since O_2 can be trapped in ice,[40] O_3 must certainly form,[45] although it has not yet been directly detected in irradiation of a *pure* laboratory ice despite its strong absorption bands in the UV at ~0.25 μm and in the IR at ~10 μm. This is likely due to insufficient buildup in the O_2 present as inclusions, which can be enhanced in the surface layers by codeposition of H_2O.[21]

Energetic ion and electron bombardment causes ejection of H_2O as well as the formation and ejection of H_2 and O_2 (see Fig. 2). Excited H and O have also been seen in the ejecta but positive ions are, typically, a very small fraction depending on the incident radiation, with ejected H^-, O^-, and OH^- seen in low-energy electron irradiations.[91,97,98]

In the radiolysis of liquids and solids the energetic electrons ejected from an atom or molecule, called secondary electrons, can produce regions in which there is a high density of excitations often called spurs. In addition to the secondary electrons, heavy energetic ions produce a "track" of high excitation density along their path through the solid forming a roughly cylindrical region of ionized and fragmented species.[58,86,99,100] Multiple excitations in a track or spur can, in principal, directly produce secondary species such as H_2O_2, HO_2, and O_2. Since the resulting damage is considerable and the transient effective track temperatures are high, the local porosity can be altered and excited species, radicals, and volatile products can diffuse along the track as discussed earlier.[10,101,102]

3.3. *Trapping and diffusion*

The density of defects, voids, and grain boundaries in laboratory samples varies with temperature and irradiation time (fluence, dose). In addition, defect mobility and molecular reordering change the population and morphology of the traps, so that the structure, diffusion length, and trapping probability are interrelated. Major changes to the composition of the ice appear to occur around 80 K, where OH becomes mobile.[40,103–105] Near 100 K, considerable structural reorientation occurs on laboratory time scales. Therefore, the decay of trapped OH is greatly enhanced at a little over 100 K and, in the 100–115 K range, OH is effectively detrapped. At low temperatures the presence of OH limits the population of the trapped electron (e-vis), presumably due to OH^- formation. But the concentration of e-vis also drops as the sample is warmed above ~120 K. In contrast to these species, the production and loss of O_2 from an ice sample dramatically increases with temperature in this temperature region.

At low temperatures, H atoms appear to be mobile and disappear in ESR spectra above ~50 K. The trapped proton is also thought to be seen in irradiated ice.[106] The rate of detrapping increases as ice is warmed above 100 K and falls off again above ~125 K. The nature of H^+ in the ice lattice is not clear and is often assumed to be H_3O^+.

There are a variety of trapping sites for each species.[107] These vary in depth (activation energy), so that at temperatures greater than 100 K some H remains trapped. Energy spectra suggest that the H_2 and O_2 ejected from ion tracks are primarily thermally accommodated,[108] consistent with diffusion out of the irradiated sample, although desorption experiments produce vibrationally hot H_2,[109] presumably directly from the surface layer. In addition, depending on the temperature, a fraction of the O_2 and H_2 can remain trapped,[21,102,110] if they do not have pathways to the vacuum interface.[16] Weak trapping is associated with orientation defects, whereas deep traps are associated with vacancies or substitution defects.

Making defects mobile by annealing or irradiation produces vacancy clusters or voids. Voids form at grain boundaries and interfaces and can nucleate in a large grain.[16] This reduces the average porosity by reducing the surface to volume of the pores.[60,103] Voids can act as traps in which species such as H_2 and O_2, can have, roughly, gas-phase characteristics or, when full, exhibit solid-state characteristics. What might be more important, these voids, like the grain and the vacuum interfaces, provide *interior* surfaces on which reactions can proceed. That is, products, such as O, H, or OH, or excited species can migrate to an exterior or interior surface where they can react, possibly forming H_2 and O_2.[29] At temperatures above ~40 K, O_2 readily diffuses through micropores in ASW with irradiation producing pathways to the surface.[10,101] In the absence of a connection to the vacuum interface, O_2 inclusions can form with H_2 accumulating at lower T.[102] For incident heavy ions with low-penetration depths, O_2 has been seen as accumulating in the highly damaged near-surface region.[21] An uncontrolled variable in many experiments is the defect concentration and the porosity of the sample, which determine the amount of interior void and pore space as well as the ability to retain volatiles.

3.4. *Reactions*

Reaction schemes that mimic those in the liquid are typically used even though solid-state reactions are known to be site specific and often occur preferentially at defects and interfaces. Other than recombination to H_2O, reactions can be unimolecular, as in the direct production of H_2 at a surface. They also can be bimolecular as in the interaction of excited or dissociated neighbors in an ion track, or by mobile species reacting with other mobile species or with previously trapped species. Such reactions are often approximated by rate equations in which it is assumed that the initially excited species have decayed to radicals and the radicals have low densities and are, roughly, thermally equilibrated with the lattice. For high densities of radicals, the exothermic energy release can result in hot-particle reactions, a process typically ignored when modeling photolysis and radiolysis data.

Reaction rates are often assumed to have the Arrhenius form: $k = A\exp(-D_E/kT)$, where D_E is the activation barrier (e.g., inset in Fig. 2a) and A is a proportionality constant determined by the interaction volume and the lattice vibration frequency. The decay of species with time can be exponential because many reactions are bimolecular and follow second-order kinetics. A complication is that reaction rates can depend on the rate of de-trapping, which is difficult to unravel due to the spatial distribution of traps, voids, and grain boundaries.

Two principal ranges of activation energies play a role in reaction schemes (Table 2): $D_E \sim 0.3$–0.5 eV, connected with lattice reordering, defect mobility, and OH mobility, and $D_E < 0.1$ eV connected with interstitial H or H^+ mobility. The decay times of interest for disappearance of a trapped species in the laboratory are of the order of 10^{-9} s (for fluorescence) to times of the order of seconds or minutes (for chemical processes or phosphorescence). The much lower irradiation fluxes in space imply that on icy satellite surfaces much longer times could be of interest. Therefore, the temperature at which primary trapped products "disappear" in laboratory experiments will be higher than on the icy surfaces in space. For example, species which de-trap in minutes in the laboratory at 100 K would take about 10^3 times longer at 80 K if the

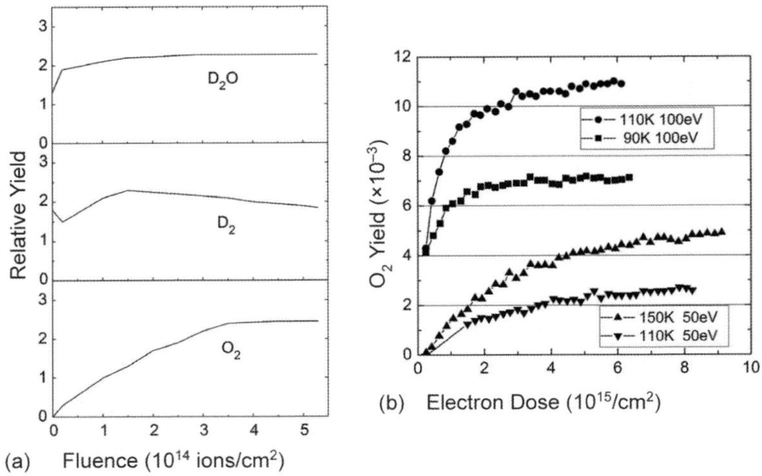

Fig. 6. Relative yield versus fluence (dose). (a) 1.5-MeV Ne^+ in D_2O ice at 20 K: ejected D_2O, D_2, and O_2 from D_2O ice (from Reimann *et al.*[10]); and (b) O_2 yield for incident electrons: energies and temperatures indicated (from Sieger *et al.*[112]).

trap depth were 0.3 eV. Since surfaces in space can be stable over very long times, the effective disappearance of primary products in the surface of an icy satellite can occur at much lower temperatures than in the laboratories.

3.5. *Radiation-induced decomposition*

The lack of a good description of decomposition is a key issue in the radiation-induced chemistry of ice. It was shown early on that the production and loss of H_2 from ice was prompt even at very low temperatures, but the production and loss of O_2 was found to depend on fluence.[10] In addition, the yield of O_2 produced from vapor-deposited ice samples at low-radiation doses depends on the temperature and the radiation dose (not the dose rate) for temperatures below ~140 K.[10,110–112] At very low doses, the yield is seen to increase roughly linearly with dose until saturation occurs (Fig. 6). These yields also depend on the sample formation temperature and, possibly, crystallinity, consistent with the discussion of sample structure and defect density.

The observed fluence dependence suggests that a precursor species is required,[112] as is the case for liquid water.[100] H_2O_2, HO_2, and trapped O have been recently examined as possible precursors in ice, as discussed below. Remarkably, in experiments at both high and low excitation densities, the O_2 yield exhibits a nearly linear dependence at low fluences. This led to a simple two-step model involving a trapped precursor.[112] Recent studies using 87-eV electrons on high quality, thin samples have shown that the vacuum and substrate interfaces and the substrate play an important role in decomposition[29,113] and the linear dependence of the O_2 yield appears to breaks down at low temperatures (below ~80 K) at small fluences.[110,111]

A threshold of ~6.5 eV and a cross section ~10^{-20} cm^2 was found for O_2 production using low-energy electrons with a similar threshold for ejection from the surface of $O(^3P, ^1D)$ and H_2.[79,109,114] A higher threshold, ~8.5 eV, was seen by Petrik and Kimmel.[29] In photolysis an O_2 production threshold was reported at ~5.3 eV,[81] but may be related to the dissociative attachment threshold at ~5 eV,[91] which produces a precursor to O_2 formation.

4. Models for Decomposition of Ice

Here it is assumed that O_2 and H_2 are formed by related processes. Although it is seen in Fig. 6a that D_2 is promptly ejected from D_2O ice and the O_2 signal rises slowly, after an incubation dose the O_2 and H_2 yields are correlated.[10] These yields exhibit effective activation energies over temperatures ~40–120 K that are relatively small (~0.03 eV) (see Fig. 2a). The integrated O_2 electron stimulated desorption (ESD) yields in Fig. 6b were described by a double exponential with "activation energies" of ~0.05–0.06 eV and ~0.002 eV. Such energies are likely related to the mobility of protons or hydrogen in ice or to a low energy threshold structural re-arrangement.

The yield variation with thickness in Fig. 5b indicates the vacuum-interface and, for thin films, the internal surface at the substrate play a role in decomposition,[76,109,113] as well as percolation from depth for energetic ion irradiation.[101] Since the yields also depend on the excitation density, the comparison of laboratory results for different radiations types, sample thicknesses, and formation temperatures has been difficult.

Hydrogen peroxide and related species formed by incident radiation are not readily ejected into the gas phase, but can be detected via absorption bands as trapped species and are seen on vaporizing the irradiated ice samples. Hydrogen peroxide forms from the principal dissociation product, OH, and the production yield appears to also correlate with H_2 formation. H_2O_2 and/or HO_2 have been suggested as potential precursor species for O_2,[110–112,115] consistent with what occurs in liquid water.[63] In matrix isolation studies, trapped O can be converted to hydrogen peroxide and hydrogen peroxide converted into trapped O by the incident radiation: i.e., H_2O–O + $h\nu \rightarrow H_2O_2$; $H_2O_2 + h\nu \rightarrow H_2O$–O.[116] Whether this process occurs in an ice matrix is not yet clear since mass spectrometry of warmed samples cannot distinguish between H_2O–O and H_2O_2. Models for decomposition are described in Appendix A, which can be simplified and related to the analytic model below.

4.1. *Decomposition: Analytic model*

The yield of O_2 from ice at low temperature appears to be independent of beam flux and increases with fluence at the low fluencies for *both* low-energy electrons, that make a single excitation for each impact, and fast ions, that make a density of excitations along their path through the solid. At low temperatures, if the beam is turned off and later turned back on with T constant, the measured yield returns to the value it had before the beam was interrupted indicating that a stable alteration occurred. That is, either the ice is chemically altered forming stable precursors for O_2 and/or the observed O_2 was produced and trapped in the solid and then caused to escape by the subsequent radiation.[10] Whereas the former is most likely the case when the incident radiation produces a single excitation, both processes can occur when the excitation density is high.

In describing their low-energy electron experiments, Sieger *et al.*[112] separated the temperature-dependent precursor formation step from the temperature-independent O_2 production step. If N_p is the precursor column density for O_2 formation and loss, and O_2 is formed/lost by electronic excitation of the precursor or by a subsequent reaction, a pair of simple rate equations results. We write these in terms of the fluence, $\Phi = \phi t$, the number of photons, electrons, or ions, incident on the surface per unit where ϕ is the flux and t the irradiation time:

$$\frac{\mathrm{d}N_{\mathrm{p}}}{\mathrm{d}\Phi} = \sigma_{\mathrm{p}}N - (\sigma_{\mathrm{O_2}} + \sigma_{\mathrm{D}})N_{\mathrm{p}}, \tag{1a}$$

$$\frac{\mathrm{d}N^{\mathrm{es}}_{\mathrm{O_2}}}{\mathrm{d}\Phi} = \sigma_{\mathrm{O_2}} N_{\mathrm{p}}. \tag{1b}$$

Here σ_{p} is the precursor production cross section, σ_{D} the precursor destruction cross section, and $\sigma_{\mathrm{O_2}}$ is the O_2 production cross section. N is that column density of H_2O penetrated by the radiation that can affect production and escape of O_2. Therefore, $\mathrm{d}N^{\mathrm{es}}_{\mathrm{O_2}}/\mathrm{d}t$ gives O_2 escape flux, which for the shallow penetration depths of the low-energy electrons ($\sim 3 \times 10^{15}$ H_2O cm^{-2}) was assumed to equal the net O_2 production rate. If thermal- or radiation-enhanced diffusion from depth occurs, as for penetrating radiation,[21,101] the trapping and subsequent destruction of O_2 can be included using an escape depth and a destruction rate. Alternatively the effective depth, N, from which excitations can contribute to O_2 formation and escape can have a temperature dependence or excitation density dependence. Since the precursor density is small the destruction of H_2O in N can be ignored in Eqs. (1). For instance, in X-ray experiments at low temperatures the number of dissociated H_2O in steady state is <10% of the irradiated column.[82]

From Eqs. (1a) and (1b), the fraction of precursors c_{p} in the penetrated column after time t is $c_{\mathrm{p}} = N_{\mathrm{p}}/N = (\sigma_{\mathrm{p}}/\sigma)[1 - \exp(-\sigma\Phi)]$. Here $\sigma = (\sigma_{\mathrm{O_2}} + \sigma_{\mathrm{D}})$ is the net precursor loss cross section. The O_2 yield Y can now be written as

$$Y = \left(\frac{\mathrm{d}N^{\mathrm{es}}_{\mathrm{O_2}}}{\mathrm{d}\Phi}\right) = \sigma_{\mathrm{O_2}}\, c_{\mathrm{p}}N = Y_\infty\,[1 - \exp(-\sigma\Phi)], \tag{2a}$$

$$Y_\infty = (\sigma_{\mathrm{p}}N)\left(\frac{\sigma_{\mathrm{O_2}}}{\sigma}\right). \tag{2b}$$

The steady-state yield, Y_∞, is seen to be proportional to the precursor production efficiency ($\sigma_{\mathrm{p}}N$) times the ratio of O_2 production and loss to precursor destruction. At low dose the destruction cross section drops out and Y is linear in Φ: $Y \rightarrow (\sigma_p N)(\sigma_{O_2} \Phi)$. The yield is determined by precursor production followed by O_2 production and escape.

The form in Eq. (2a) gives good fits to the low-energy electron data in Fig. 6b. Since the steady-state yield varies with sample temperature, at least one of the parameters in Y_∞ depends on T, suggesting there is an activation barrier to a critical reaction or that diffusive transport is important. Experiments also indicate that the sample formation temperature and the radiation history are important. For instance, a sample irradiated at an elevated temperature and then cooled, and then again irradiated, initially exhibits an elevated O_2 yield which gradually returns to the yield expected at the lower temperature.[29] This is consistent with the precursor formation/trapping step being temperature dependent. Although Sieger *et al.* set $\sigma_D = 0$ (i.e., $\sigma = \sigma_{O_2}$),[112] precursor destruction processes other than O_2 formation are certain to be important.

4.2. *Decomposition pathways*

In Eq. (1a) the precursor is assumed to form in a single excitation event at low doses. This led to the suggestion of trapped O as a precursor.[13,15] As suggested for liquid water,[25] other proposed precursors (e.g., H_2O_2 and HO_2) require two or more events, e.g., direct dissociation: $2H_2O \rightarrow 2H + 2OH \rightarrow H_2 + H_2O_2$ *and* $OH + H_2O_2$ (or $H_2O_2^*$) $\rightarrow H_2O\ (H) + HO_2$;[82,110] or electron attachment: $2(H_2O)^- \rightarrow 2H^- + 2OH \rightarrow 2H^- + H_2O_2$ and $H_2O_2^- \rightarrow H^- + HO_2$.[91] Rate equations including such precursors are considered in Appendix A and indicate that the form for the yield in Eq. (2a) is roughly recovered *except* at low fluences and low T. Near steady state, the difference from linearity is exhibited as an offset in fluence:

$$Y \sim Y_\infty \left[1 - \exp(-\sigma\Phi) - \frac{\sigma}{\sigma_r}\right], \ \Phi >> \frac{1}{\sigma_r}. \tag{2c}$$

As above, σ is the precursor destruction cross, Y_∞ is given in Appendix A and $1/\sigma_r$ is the fluence offset. If peroxide formed by reaction of two OH is the precursor, then σ_r is the sum of the cross sections for the loss of OH by the formation of peroxide and by recombination to water.

Laffon *et al.* used a multistep model to describe the evolution of trapped molecules on warming a sample irradiated to saturation at 20 K, but did not study the fluence dependence.[82] However, an offset in the fluence dependence was seen in recent experiments using 87-eV electrons on ASW films for T below ~50 K.[76] On average, ~3 ionization/excitation events are produced over a mean depth of 10–15 monolayers in these experiments. The observed offset at 20 K in Ref. 76 Fig. 1, ~5 s (or $\Phi \sim 10^{14}$ electrons cm^{-2}), implies that $\sigma_r \sim 10^{-14}$ cm^2 in Eq. (2c). This corresponds to an excitation density in the near-surface region of $\sim 10^{21}$ cm^{-3}, or an average distance between excited molecules ~10 Å, prior to the nearly linear increase in O_2 loss. This offset was interpreted as being due to the production of a precursor that requires multiple events (Appendix A). Since it disappears at higher temperatures, it is temperature dependent; possibly consistent with their observation that diffusive transport to the surface is involved in the production of O_2.

Petrik *et al.* also showed that adding layers of H_2O with the O isotopically labeled on a 100 K sample suppressed the contribution from O_2 formed from the underlying water molecules.[110] The e-folding thickness of ~1.7 ML indicates that for electron irradiation, the near-surface water molecules primarily contributed to the formation and escape of O_2 even though excitations at larger depth contribute to the decomposition. The e-folding distance for D_2 desorption was somewhat larger.[117] On the other hand, adding layers of H_2O_2 increased the O_2 yield, consistent with the suggestion that a change in the stoichiometry of the ice correlates with O_2 production.[22] They also showed that the observed fluence dependence was not consistent with H_2O_2 as the final precursor. Similar to other models for ice[82] and water,[100] they proposed that HO_2, formed from the added H_2O_2, was the final precursor. Petrik *et al.*[111] used a set of rate equations for the decomposition of ice to fit the significant ESD database using 87-eV electrons. They assumed that an H reaction with or excitation of H_2O_2 produced the final precursor HO_2. However, a potential precursor, O_t (H_2O–O) discussed below, can in principal also be produced by excitation of H_2O_2.[116] Their rate equations are given in the appendix and the reaction rates extracted by fits to their experimental data are given in Table A1.

4.3. *Trap density*

The role of trapping sites was not explicitly included in the models discussed above even though it was shown that *transport* to the vacuum interface and the substrate interface affect the decomposition rate. In addition, a key observation in all experiments on radiation-induced decomposition of ice is the following: although the ejection of water molecules is independent of temperature, the yield of O_2 from an irradiated ice sample *increases* with increasing T (below ~140 K) (Figs. 2a and b), while the density of defects, voids, and grain boundaries produced on formation or by the irradiation typically *decreases* with increasing T.

This result could imply that escape of newly formed O_2 is determined by *its* trapping efficiency. That is, if with increasing T a newly produced O_2 escapes from depth more readily, it is less the likely to be destroyed by the subsequent radiation. The competition between escape and destruction was suggested to play a role for incident ions producing high excitation densities[10,101] and appears to occur for low-energy, heavy ion irradiation.[21] However, it is less likely to apply to the low-energy electron experiments.

The observed temperature dependence can also result if a diffusion length critical to O_2 formation increases with increasing temperature. Since diffusion to an interface is important,[29,110] such a temperature dependence would not be surprising. Diffusion to a trapping site was included in a model in which trapped oxygen, O_t, was assumed to be the principal precursor at low fluence.[13,15] Although this model has not been confirmed experimentally, it is described here as a model system that accounts for both trapped and mobile species.

In this overly simplified model, O_2 is formed by the reaction of a new product, such as O or OH, with O_t, or by electronic excitation of O_t. The kinetics of the production of O_2 by O-atom reactive scattering is similar to the production of CO_2 in a CO/H_2O ice mixture[118] and the formation of O_3 in an O_2/H_2O mixture.[45] In this model, direct or indirect dissociation to H_2 + O (following excitation, ionization, or attachment) is given by the effective cross section σ_{d2} with H_2 assumed to be primarily lost from the ice samples at the temperatures above ~40 K. The O produced can be in the ground state or in an excited state which quenches in the ice matrix releasing energy. It eventually traps according to the reaction rate $k_{O,t}$ as an oxygen–water molecule complex, $O–(H_2O)_n$, referred to here as O_t. O_2 is then produced on excitation of O_t (σ_{O_t}), by the production of OH (σ_{d1}) which diffuses to and reacts with O_t (k_{OH,O_t}), or by the production

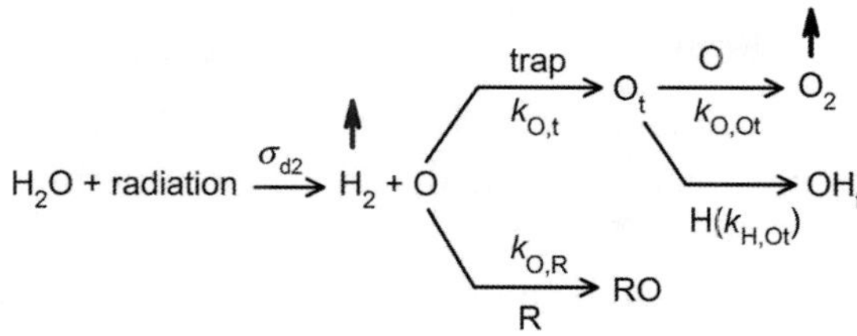

Fig. 7. Chemical pathways for a simplified model in which decomposition is initiated by the production of transiently mobile O which can trap at a defect or interact with previously trapped species as in Johnson *et al.*[13,15] Competing reactions involving mobile H and trapped trace reactive species or other radiation products R are included [see Eqs. (A.1)–(A.4)]. Results are given in Eqs. (2c). Reactions leading to H_2O_2 or HO_2 proceed in parallel.

of a subsequent O which reacts with O_t (k_{O,O_t}). These processes have similar rate equations, which is also the case if a newly produced O reacts with trapped OH to produce O_2. However, trapped OH is not a viable final precursor since its presence decreases at higher temperatures leading to the formation of H_2O_2 and HO_2.

The effect of trapping sites in this model is examined by assuming O_t is the final precursor, and a subsequent absorption event produces a *mobile* O ($\sigma_{d2}\phi$) that can find and react with O_t: $O + O_t \rightarrow O_2$ described by k_{O,O_t}.[13,15] In this model the principal dissociation channel, $H_2O \rightarrow H + OH$ (σ_{d1}), produces mobile H which can react with and destroy O_t (k_{H,O_t}). Mobile H and O can also be removed by other reactants R: e.g., $R + H \rightarrow RH$ ($k_{H,R}n_R$); $R + O \rightarrow RO$ ($k_{O,R}n_R$), where R is a contaminant or a radiation product such as OH. O_t might also be produced and lost by excitation of H_2O_2,[116] but such a process, possibly important at saturation, is not included here. The simplified reaction pathways considered are given in Fig. 7. Assuming that the excitations in a depth, N, produce H_2 and O_2 that can escape, a set of rate equations is obtained (see Appendix A). These act in parallel with other equations for H, OH and peroxide. If the interaction of O_t with a newly produced OH or direct excitation of the precursor can lead to O_2, then k_{O,O_t} would be replaced by either k_{OH,O_t} or $[\sigma_{O_2}\phi N_{O_t}]$.

4.4. *Trapped versus mobile species*

Since the mobile species produced by the radiation trap in times short compared to the time scale associated with the incident flux, a background of mobile O does *not* accumulate. That is, unless the trap density is extremely small and the temperature high, only species trapped at defects, pores, or grain surfaces are present before a subsequent radiation-induced event. This is consistent with the lack of a dependence of the O_2 yield on the incident beam flux at fixed T. Therefore, rate equations for trapped O and mobile O and H can be solved and give a yield like that in Eq. (2a).[13,15] For O_t as the precursor and a significant trap density ($n_t >> n_R$)

$$Y_\infty \rightarrow \frac{(\sigma_{d2}N)}{q} \sim (\sigma_{d2}N)\left(\frac{\sigma_{d2}}{\sigma_{d1}}\right)\left[\frac{k_{H,R}n_R}{k_{O,t}n_t}\right]\left[\frac{k_{O,O_t}}{k_{H,O_t}}\right], \tag{3a}$$

with q given in Appendix A. Comparing with $Y_\infty \sim (\sigma_p N)(\sigma_{O_2}/\sigma)$ in Eq. (2a), and assuming O traps efficiently, $(\sigma_p N)$ is replaced by $(\sigma_{d2}N)$. Precursor *destruction* is seen to be determined by the production of H (σ_{d1}) which can quench O_t. This occurs in competition with the loss of mobile H to other reactive species, R, which can include OH. Therefore, in Eqs. (2)

$$\sigma \rightarrow \frac{k_{H,O_t}\, n\sigma_{d1}}{k_{H,R}n_R}, \tag{3b}$$

assuming destruction is much more efficient than O_2 production from the precursor. Similarly, the *effective* O_2 formation cross section, σ_{O2}, requires that a newly formed mobile species, here O (σ_{d2}), reacts with O_t (k_{O,O_t}) *prior* to its being trapped ($k_{O,t}$):

$$\sigma_{O_2} \rightarrow \left[\frac{n\sigma_{d2}k_{O,O_t}}{k_{O,t}n_t}\right]. \tag{3c}$$

In this model system the steady-state yield, Y_∞ in Eq. (3a), depends *inversely* on the density of traps, n_t, which is equivalent to depending directly on the transport distance. That is, if there are competing destruction processes for trapped O, then the freshly produced, mobile O must find a trapped O before it becomes trapped *or* before mobile H destroys the precursor, O_t. Therefore, in the model system the increase in the O_2 yield with increasing T is related to the change in the trap density

Table 2. Parameters for sample data: σ_{d2} is of the order of 10% of σ_{d1}.

Radiation	*E* (eV)	*T* (K)	σ (10^{-16}) cm^2	σ_{d2} (10^{-16})cm^2	E_a (eV)
e[a]	30–100	90–120	0.8–3	~1	0.02–0.03
e[b]	87	20–120	(10)[b]	(~0.1)[b]	0.05–0.06
hν[c, d]	~10	~10	>0.1[c]	0.09[b]	0.03[c]
He^{+} [e]	10^6	40–140	~10		0.03
Ne^{+} [f]	1.5×10^6	10	62	<~35	0.02–0.07

[a]Sieger *et al.*,[112] precursor formation $(0.5–2.0) \times 10^{-18}$ cm^{-2} is proportional to $(E - E_0)$, $E_0 \sim 10$ eV, $N \sim 10^{15}$ cm^2.

[b]Petrik *et al.*,[110] HO_2 as a precursor: σ destruction of HO_2 and $\sigma_{d2} \to \sigma_{d1}$ appendix.

[c]Watanabe *et al.*,[119] saturation of yield of $D_2O \sim 5 \times 10^{-19}$ cm^2.

[d]Westley *et al.*,[56] E_{av} based on D_2O yields, σ based on lowest fluence.

[e]Brown *et al.*,[7] based on lowest fluence.

[f]Reimann *et al.*,[10] σ_{d2} based on $(dE/dx)/W = 92 \times 10^{-15}$ eV cm^2/26 eV.

with temperature, and the small activation energies discussed earlier are related to the reorientation of the lattice leading to the reduction in trap density and/or increase in the transport distance. Dissociation, directly or indirectly, to $H_2 + O$ is consistent with the production of comparable amount of O and H_2 from the surface[109] and was confirmed by Cooper *et al.*[45] by seeding ice with O_2 and measuring the production of O_3 ($O + O_2 \rightarrow O_3$). However, other mobile and trapped species would result in imilar sets of equations. For example, if a newly produced OH can react with O_t to produce O_2, then the term in brackets becomes $[n\sigma_{d1} k_{OH,O_t}/(k_{OH,t} n_t)]$ and the yield remains inversely proportional to n_t. Finally, $k_{O,O_t}/(k_{O,t} n_t)$ can be replaced by mobile O transport lengths and reaction efficiencies giving an equivalent result. Zheng *et al.*[120] suggest that O_2 formation decreases with increasing *T*, but they primarily measure the O_2 evolving on warming the irradiated ice.

4.5. *Comparison with data*

Estimates of the cross sections in the above models based on laboratory data are given in Table 2 adapted from Johnson *et al.*[15] They assumed that $\sigma_{d2} \sim 0.1\sigma_{d1}$ based on the gas-phase cross sections and on the result of Watanabe *et al.*[119] for production of D_2 at low fluences: ~10%–20% of the total absorption cross section with significant uncertainties. A *lower bound* on the O production in ice using experiments in Cooper *et al.*[45] is ~0.004 O per ionization produced by a 0.8-MeV proton, consistent with the above (see also Laffon *et al.*).[82] Since O produced can be in an excited state, the role of the excess energy in an ice matrix also needs to be evaluated. For comparison, a result based on the rate constants obtained by Petrik *et al.* is also given in Table 2.[110] They assume HO_2 is the precursor. Their rate equations are given in Appendix A and their reaction rates are given in Table A.1. For the model in which a single excitation produces a precursor, the sizes of the cross sections for mobile O production, σ_{d2}, are not unreasonable; nor are the values for the net precursor destruction cross section, σ. For the model in which HO_2 is the final precursor, the reaction rates again suggest a reasonable precursor destruction cross section but a rather small value for the OH production cross section. Clearly more detailed analysis is needed.

4.6. *Incident ions: Escape from depth*

Based on the above it is surprising that the O_2 yield versus fluence is also found to vary linearly with fluence at low fluences for fast, heavy ion irradiation. Even though multiple excitations are produced in the high excitation density of an ion track and due to its shower of secondary electrons, the yield can be roughly fit by the expressions in Eqs. (2). For MeV Ne^+ the linear regime extends over a larger range of fluences,[10] but the fit using the simple model is not as good as it is for the low-energy electrons consistent with a more complex interaction. For instance, Y_∞ versus T has two "plateaus": one at very low doses and one at higher doses, both with activation energies in the range ~0.02–0.07 eV. The different plateaus suggest that the track structure is important.[10] That is, for a low dose, loss of O_2 that is formed in an individual track dominates, while at higher doses the fluence is such that that the track cores overlap. Therefore, the quantity N in Eq. (1a) can be complex accounting for the depth from which excitations can cause the production, ejection, and destruction of O_2. This can involve excitation and/or radical transport as well as radiation-induced percolation of O_2 to the surface. Ignoring the differences and fitting the low-dose data for the incident 1.5-MeV Ne^+ at 10 K, parameters are also given in Table 2. The effective destruction cross section, $\sigma \sim 60 \times 10^{-16}$ cm^2, is roughly comparable to the size of the molecular destruction cross section, $\sim 30 \times 10^{-16}$ cm^2 by these ions.

At saturation (steady state) a single new event leading to an O can in principal produce an O_2 from the steady-state density of trapped precursors in the model system described above. Therefore, allowing for changes in sample structure, the yield at saturation should be nearly linear in the excitation cross section.[6,10] This is the case for the low-energy electrons, and it was initially assumed to be the case for the fast ions. However, for a temperature at which decomposition dominates the sputtering of ice, the steady-state yield, Y_∞ appears to be proportional to the square of the electronic energy deposition per unit pathlength in ice.[62] If *deep excitations* dominate[10] the initiation of precursor formation (e.g., $H_2O^{+2} \rightarrow O$ as discussed in Ref. 86), then the production and destruction cross sections in Y_∞ in Eq. (2) scale nonlinearly with $(dE/dx)_e$. If that is not the case, then the ratio $(\sigma_p \sigma_{O_2}/\sigma)$ also varies as $(dE/dx)_e$. Therefore, the quantity N in Y_∞ must also vary with $(dE/dx)_e$ for incident fast heavy ions.[13,15] That is, the depth N from which O_2 can be mobilized to escape would depend on the excitation density $(dE/dx)_e$ giving the quadratic dependence observed for these ions.[13,15] This is consistent with the observation that, depending on temperature, the O_2 yield depends on thickness for samples thinner than the radiation penetration depth.[101]

Based on the proposed model, a description of the radiation-induced percolation depth for escape, N, is required to compare the yields for penetrating radiation with those for nonpenetrating radiation. Since multiple excitations affect the source term and the defects/damage produced by the ions can affect both the density of traps and the percolation of O_2, the low-energy electron and UV photon experiments provide better tests of the chemistry for molecular oxygen production.

5. Summary

The radiation-induced chemistry in ice is of interest as a model system for understanding the differences between radiation chemistry in the liquid and in the solid. Possibly more important, it has been studied because of its relevance to radiation processing of ices in space. Although the first results from the experiments of Brown, Lanzerotti, and co-workers, inspired by Voyager's journey to Jupiter, were presented over 30 years ago it is remarkable that there is still considerable disagreement as to the chemical pathways occurring in ice. In particular, there is little agreement on radiation-induced decomposition of ice, a process which is critical in understanding both the radiation chemistry and irradiation effects seen on the icy bodies in outer solar system.

The brief description of the model systems given here is intended to outline the important issues and to provide a framework for quantitatively comparing results obtained for different incident radiation types and sample formation temperatures. The experiments of Petrik *et al.*[110] and Laffon *et al.*[82] suggest that HO_2 plays a role in the production of O_2 at saturation, although it seems to this author that decomposition at low fluences is still unresolved. Although the nature of the *critical* transport process is still uncertain, excitation or mobile radical transport appears to be occurring[29] possibly due to the so-called precursor electrons.[77] In addition, the temperature dependence of the yield of O_2 has been suggested to be related to a transport length, which was modeled above as inversely proportional to the density of trapping sites.[15] Since the density of defects is in turn affected by the sample formation process[103] and by the radiation history,[60] unraveling the available laboratory data has been challenging.

Although the radiations used in the experiments discussed are primarily ionizing, the role of the secondary electron and its transport has not been considered in any detail. In the low excitation density experiments the principal precursor has not been clearly identified, but transport to the surface region is clearly occurring.[29] In addition, an important caveat in the modeling described is that the initial ionization-induced reactions have been ignored except that they eventually contribute the production of a set of transiently mobile radicals. This omission is likely to be especially important for the high excitation density radiations for which multiple pathways for the formation of O_2 are likely operative in the highly excited track.[10] In addition, when the excitation density is high the depth, N, from which excitations produced by the incident radiation affects the production and escape of O_2. This depth increases with increasing excitation density complicating the comparisons with the incident electron and UV experiments.

In spite of the caveats described above, it is exciting that the recent history of the study of radiolysis and photolysis in ice has been driven to a large extent by the observations of icy bodies in space that are exposed to photon and particle radiation. And although the chemical processes induced by the incident radiation are not fully understood, the significant laboratory database has contributed enormously to our understanding of the evolution of icy bodies in the outer solar system and ice grains in the interstellar medium. This understanding has, to a significant extent, allowed analysis of the Cassini and Galileo data on the icy moons of Jupiter and Saturn[14,23,24,129] and the icy ring particles at Saturn,[127,128] and has helped motivate the new NASA and ESA missions proposed to go Europa and Ganymede.

Acknowledgments

The author acknowledges comments on the manuscript by P. Cooper, G. Kimmel, N. Petrik, J. P. Jay-Gerin, M. Loeffler, R. Carlson, thanks P. Cooper for results in advance of publication, M. Liu for help in preparing the manuscript, and support of NSF Astronomy Division and NASA's Planetary Geology and Geophysics Program.

Appendix A

A.1. *Introduction*

Proposed rate equations on the radiolysis and photolysis of ice are discussed in this appendix with emphasis on decomposition of ice, as it is a key to unraveling the dominant chemical pathways. Integrating such equations over the excitation depth that contributes to production and escape of O_2 gives the results discussed in the text in which number densities are replaced by column densities and rate constants are replaced by effective cross sections. A precursor formed in a single event (e.g., O_t) is first discussed and equations are considered in which the trapped and transiently mobile species are treated separately. This is followed by a discussion of precursors produced by multiple excitation/ionization events (e.g., H_2O_2 or HO_2). These differences bear on the near linearity of the O_2 yield versus fluence.

A.2. *Trapped and mobile species: O as a model precursor*

In the following, all of the radiolytic species in Fig. 4 are not included, so that important products, such as the charged species (e.g., H_3O^+, multiply ionized species, trapped electron), are not explicitly described. Starting with the pathways given in Fig. 6, a simplified set of rate equations for mobile O and H and trapped O was used[13,15]:

$$\frac{dn_O}{dt} = \sigma_{d2}\phi n - k_{O,t}n_O n_t - k_{O,R}n_O n_R - k_{O,O_t}n_O n_{O_t}, \quad \text{(A.1a)}$$

$$\frac{dn_{O_t}}{dt} = k_{O,t}n_t n_O - k_{O,O_t}n_O n_{O_t} - k_{H,O_t}n_H n_{O_t}, \quad \text{(A.1b)}$$

$$\frac{dn_H}{dt} = \sigma_{d1}\phi_n - k_{H,O_t}n_H n_{O_t} - k_{H,R}n_H n_R, \quad \text{(A.1c)}$$

$$\frac{dn_{O_2}}{dt} = k_{O,O_t}n_O n_{O_t}. \quad \text{(A.1d)}$$

Here ϕ is the radiation flux, n is the molecular number density of ice, n_i is the density of species i in the irradiated volume, and n_t is the density of trapping sites. The rate constants are defined in the text and within some critical depth, Δx, excitations can contribute to the formation and escape of O_2 and H_2. This depth is defined as a column, $N = \Delta x n$.

Eqs. (A.1a)–(A.1d), of course, act in parallel with the equations for OH and H_2O_2 discussed below, and a newly formed OH might also react with trapped O to form O_2 ($k_{OH,O_t} n_{OH} n_{O_t}$). Since we assume all transient species have decayed, the pseudo reaction rates above can be related to an effective interaction volume for each produced species divided by the time between impacts. As in Eqs. (1), O_2 destruction processes are ignored for simplicity even though trapped O_2 that does not escape can be dissociated by a subsequent excitation event.[21,101] In addition, if direct excitation of the precursor can produce O_2, then the term ($k_{O,O_t}n_O n_{O_t}$) is replaced by a term like ($\sigma_{O_2} \phi n_{O_t}$). Integrating the above equations over the depth of escape of O_2 from an irradiated sample, the analytic precursor model in Eq. (2) can be recovered.[13,15] That is, c_p, σ_p, σ_{O_2}, and σ are determined using the rate constants and primary dissociation cross sections in Eqs. (A.1). As the density of trapped O, n_{O_t}, is likely small compared with the density of sites at which H can react, then Eq. (A.1d) gives $n_H \sim [\sigma_{d1}\phi n/k_{H,R}n_R]$, where for a typical ϕ the density of mobile H, n_H, is small. The O_2 yield, Y_{O_2}, is equal to $[\Delta x\ (dn_{O_2}/dt)/\phi]$ with Δx the depth from which an O_2 can escape. Assuming $k_{O,t} \sim k_{O,R}$, the yield can be written in the form in Eq. (2a):

$$\frac{Y_\infty}{\sigma_{d2}N} \approx 1 + \frac{\{(q-1) - [(q-1)^2 + 8q\delta]^{0.5}\}}{4\delta},$$

$$q = [k_{O,t}(n_t + n_R)]\frac{k_{H,O_t}n_H}{\sigma_{d2}n\phi k_{O,O_t}}$$

$$\approx \left(\frac{\sigma_{d1}}{\sigma_{d2}}\right)[k_{O,t}(n_t + n_R)]\frac{k_{H,O_t}}{k_{O,O_t}k_{H,R}n_R},$$

$$\delta = 0.5\left(1 + \frac{n_t}{n_t + n_R}\right). \tag{A.2}$$

For large n_t and small ϕ, q can be large and, since it is likely that $n_t >> n_R$, Y_∞ can reduce to the expression given in Eq. (3a). Equation (A.2) can in principal be rewritten in a form more directly related to Eqs. (1). Integrating Eqs. (A.1) over the depth from which excitations cause the production and escape of O_2, Δx, the number densities are replaced by column densities and rate constants by effective cross sections, $[k_i n_j] \rightarrow [\sigma_i \phi]$. Note that N is determined by the distribution and transport of excitation events and the O_2 escape (percolation) depth.

A.3. *Peroxide as a precursor*

As discussed in Johnson *et al.*[13] a simple set of rate equations can be used to describe the formation of a H_2O_2 precursor, which are then related to the rate equations in Eqs. (1). The set of reaction used is: H_2O + radiation → H + OH (σ_{d1}); H + H → H_2 ($k_{H,H}$); H + OH → H_2O($k_{H,OH}$); and OH + OH → H_2O_2 ($k_{OH,OH}$). In the first reaction, production of OH can be by direct excitation or by recombination ($H_2O^+ + e \rightarrow H + OH$). Assuming that the H_2 produced escapes, the rate equations in the irradiated volume are

$$\frac{dn_{OH}}{dt} = \sigma_{d1}\phi n - 2k_{OH,OH}n_{OH}n_{OH} - k_{H,OH}n_H n_{OH}, \tag{A.3a}$$

$$\frac{dn_H}{dt} = \sigma_{d1}\phi n - 2k_{H,H}n_H n_H - k_{H,OH}n_H n_{OH}, \tag{A.3b}$$

$$\frac{dn_{H_2O_2}}{dt} = k_{OH,OH}n_{OH}n_{OH} - \text{destruction processes} \tag{A.3c}$$

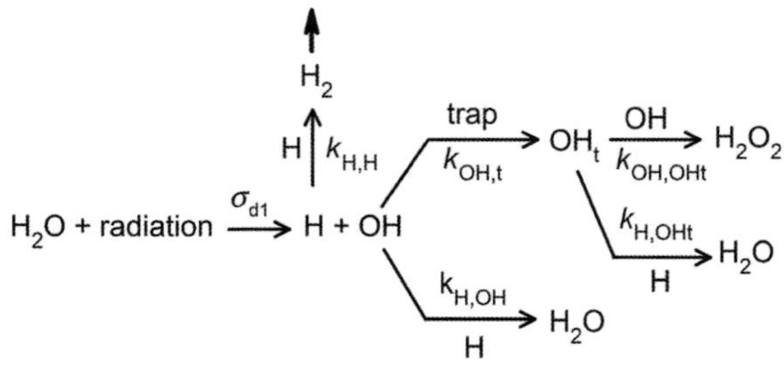

Fig. A.1. Chemical pathways associated with Eqs. (A.4).

with n_{OH} and n_H the density of the dissociation products. Trapped and mobile species are not distinguished in Eqs. (A.3), although for low density of excitations, any OH produced by one particle must be trapped prior to interacting with an OH produced by a subsequent particle. Therefore, the disappearance of trapped OH at elevated T is not explicitly described.

Equations (A.3) can be related to those in Eqs. (1) by integrating over Δx, the depth from which excitations produced can cause the production and loss of O_2. That is, as above, $\Delta x n = N$ in Eq. (1a) and the precursor formation rate, $[\sigma_p \phi N]$ is replaced by $[k_{OH,OH} n_{OH} n_{OH}]\Delta x$. Prompt loss of H_2 in this model implies $k_{H,H} >> k_{OH,OH}$, but in steady state the production of H_2O_2 becomes equal to H_2. Simplifying one obtains

$$[k_{OH,OH}\, n_{OH}\, n_{OH}]\Delta x \rightarrow [\phi_p \sigma_{d1}]\phi N \tag{A.3d}$$

with $f_p = 0.5/[1 + 0.5k_{H,OH}/(k_{H,H}\, k_{OH,OH})^{0.5}]$. Therefore, σ_p in Eq. (1a) is a fraction, f_p, of the total dissociation cross section, σ_{d1}. Since the H_2O_2 fraction in an irradiated ice is small,[13] then $k_{H,OH} >> (k_{OH,OH}\, k_{H,H})^{0.5}$, giving $f_p \rightarrow [(k_{OH,OH}\, k_{H,H})^{0.5}/k_{H,OH}]$.

A.4. *Trapped versus mobile OH: peroxide*

Trapped OH and freshly produced, transiently mobile OH should be treated as separate species.[81] Therefore, assuming H remains mobile until it reacts with trapped species, the OH equation above is replaced by one for trapped OH, n_{OH_t}, and one for mobile OH, n_{OH}. At the relevant fluxes and temperatures, the reaction of two mobile OH is unlikely, so that peroxide is formed from the interaction of a trapped OH with a subsequently produced OH: OH + trap → OH_t (k_{OH_t}); OH + OH_t → H_2O_2 (k_{OH,OH_t}); and H + OH_t → H_2O (k_{H,OH_t}). The reaction pathways are shown in Fig. A.1. Here n_t is the trap density, which at low temperatures is large. Assuming $dn_t/dt \sim 0$ gives:

$$\frac{dn_{OH}}{dt} = \sigma_{d1}\phi n - k_{H,OH}n_H n_{OH} - k_{OH,t}n_t n_{OH} - k_{OH,OH_t}n_{OH_t}n_{OH}, \quad (A.4a)$$

$$\frac{dn_{OH_t}}{dt} = k_{OH,t}\, n_{OH}n_t - k_{OH,OH_t}n_{OH}n_{OH_t} - k_{H,OH_t}n_H n_{OH_t}, \quad (A.4b)$$

$$\frac{dn_H}{dt} = \sigma_{d1}\phi n - 2k_{H,H}\, n_H n_H - k_{H,OH}\, n_{OH}n_H - k_{H,OH_t}\, n_{OH_t}n_H, \quad (A4.c)$$

$$\frac{dn_{H_2O_2}}{dt} = k_{OH,OH_t}n_{OH}n_{OH_t} - \text{destruction processes.} \quad (A.4d)$$

For typical radiation fluxes, ϕ, the diffusion of mobile radicals to traps is relatively fast. Therefore, in a low-temperature solid at low-dose rates a significant steady-state background of diffusing OH does not accumulate. The freshly produced OH rapidly react, so that averaging over the bombardment rate $dn_{OH}/dt \sim 0$ in Eq. (A.4a). Assuming that H also diffuses rapidly and either reacts or forms H_2, then on average $dn_H/dt \sim 0$ also. Therefore, between impacts only species trapped at defects, pores, or grain surfaces are present as suggested by the lack of a dependence on the incident beam flux. Combining Eqs. (A.4a) and (A.4c) with Eq. (A.4b), the rate of change of trapped OH in this limit is

$$\frac{dn_{OH_t}}{dt} = 2k_{H,H}n_H n_H - 2k_{OH,OH_t}n_{OH}n_{OH_t}. \quad (A.4e)$$

Therefore, the production of peroxide, the second term, depends on the loss of H by formation and loss of H_2, but in steady state the production rates are equal.

Due to the cage effect, the recombination of mobile species modifies the size of ($\sigma_{d1}\phi n$). Integrating Eqs. (5b) and (5d) over Δx, the densities in Eqs. (A.4) are replaced by column densities and the rate constants, k_i, involving mobile species, can be written as effective cross sections, $[k_i n_j] \rightarrow [\sigma_i\, \phi]$. The values σ_i give the interaction length for each mobile radical. Combining Eqs. (A.3a) and (A.4b), to eliminate $k_{OH,t}n_{OH}n_t$, the trapped OH column density, N_{OH_t}, and peroxide production rate are estimated from:

$$\frac{dN_{OH_t}}{d\Phi} \approx \sigma_{d1}N - 2\sigma_{OH,OH_t} N_{OH_t} - \sigma_{H,OH_t} N_{OH_t}. \tag{A.5a}$$

$$\frac{dN_{H_2O_2}}{d\Phi} \approx \sigma_{OH,OH_t} N_{OH_t} - \sigma N_{H_2O_2}. \tag{A.5b}$$

Here the σ_i are the particular reaction cross sections, and the precursor destruction processes, including the formation of O_2 (σ_{O_2}), are contained in σ. The second and third terms on the right in Eq. (A.5a) are the loss of OH due to the production of H_2O_2 and H_2O. The interaction length of mobile H is contained in σ_{H,OH_t}. If H_2O_2 is indeed the precursor, then in Eq. (1b) in the $[\sigma_p N]$ becomes $[\sigma_{OH,OH_t} N_{OH_t}]$ from Eq. (A.5b). The precursor destruction process, which gives back two mobile OH, and the production of H_2 can both be accounted for via these cross sections.

The O_2 yield is obtained by replacing N_p in Eq. (1b) by the column of H_2O_2, $N_{H_2O_2}$. Solving Eqs. (A.5a) and (A.5b), the column density of OH_t and the concentration of precursors, c_p, versus fluence is obtained:

$$c_p = f_p\left(\frac{\sigma_{d1}}{\sigma}\right)\left\{[1 - \exp(-\sigma\Phi)] - \left(\frac{\sigma}{\sigma_r}\right)[1 - \exp(-\sigma_r\Phi)]\right\} \tag{A.6a}$$

with

$$f_p = \frac{\sigma_{OH,OH}}{(\sigma_r - \sigma)} \tag{A.6b}$$

with $\sigma_r = (2\sigma_{OH,OH_t} + \sigma_{H,OH_t})$ and $\sigma/\sigma_r < 1$. It is seen that the precursor oncentration with time, c_p, depends on those reaction rates that determine the interactions with trapped OH, σ_r, and the destruction of H_2O_2, σ. Similar to Eqs. (2) the yield can be written as $Y = \sigma_{O_2} c_p N$. If σ_r and σ are comparable, then Y is quadratic in Φ at small Φ. However, for $\sigma_r >> \sigma$ the linear dependence on Φ is obtained with a shift in Φ, as in Eq. (2c). Further, the steady-state yield is now:

$$Y_\infty(T) = \sigma_{O_2}[f_p\,\sigma_{d1}]\frac{N}{\sigma}. \tag{A.6c}$$

That is, $\sigma_p \rightarrow [f_p\sigma_{d1}]$ and the temperature dependence for the incident electron data would be contained in $[\sigma_{O_2} f_p]$ or the ratio $[\sigma_{O_2}\sigma_{2OH}/\sigma_{H,OH}]$.

Table A.1. Rate constants for reactions in Eqs. (A.7) (from Petrik *et al.*[110]).

T (K)	**k_1/ϕ**	**k_2/ϕ $(10^{-13}$ cm$^2)^2$**	**k_3/ϕ 10^{-15} cm^2**	**k_4/ϕ $(10^{-13}$ cm$^2)^2$**	**k_5/ϕ $(10^{-15}$ cm$^2)$**
20	0.25	0.75	1.0	0.03	10
40	0.25	0.75	1.0	0.035	10
60	0.25	1.0	1.0	0.06	10
80	0.35	1.0	2.0	0.1	10
90	0.40	1.0	2.0	0.2	10
100	0.45	1.0	2.5	0.5	10
110	0.55	4.0	3.5	1.0	17
120	0.75	5.0	3.5	2.5	25
130	0.77	10	3.5	3.0	37

Based on the above, the lowest fluence at which O_2 is first detected experimentally also sets a limit on the time scale for mobile–mobile reactions.

A.5. *HO_2 as a precursor*

Laffon *et al.*[82] and Petrik *et al.*[110] considered a set of reactions in which HO_2 is the precursor. The incident radiation dissociates water molecules producing OH (k_1), dissociates H_2O_2 producing two OH (k_3), and dissociates HO_2 producing O_2 (k_5). In addition, 2OH combined to form H_2O_2 (k_2) and OH reacted with H_2O_2 to produce their final precursor, HO_2 (k_4). No distinction is made between mobile and trapped species, so that the rate equations for the column densities in the irradiated volume are[110]:

$$\frac{dN_{OH}}{dt} = k_1 + 2k_3N_{H_2O_2} - k_2N_{OH}^2 - k_4N_{OH}N_{H_2O_2}, \tag{A.7a}$$

$$\frac{dN_{H_2O_2}}{dt} = k_2N_{OH}^2 - k_3N_{H_2O_2} - k_4N_{OH}N_{H_2O_2}, \tag{A.7b}$$

$$\frac{dN_{HO_2}}{dt} = k_4N_{OH}N_{H_2O_2} - k_5N_{HO_2}, \tag{A.7c}$$

$$Y\phi = \frac{dN_{O_2}}{dt} = \alpha k_5 N_{HO_2}. \tag{A.7d}$$

The ESD yield of O_2 versus fluence for a number of temperatures is fit giving the k_i with the production/detection efficiency, α, absorbed into k_5.[110] In Table A.1, the k_i are divided by the typical flux, $\phi = 2 \times 10^{13}$ e cm^{-2} s^{-1}. Unfortunately, the 2 in Eq. (A.7a) was left out, but the authors report privately that it did not significantly affect the fits at low T. Based on this model, $(k_1/\phi) \approx \sigma_{d1}N$, so that it is seen that in Table A.1 that $(\sigma_{d1}N)$ varies from 0.25 to 0.77 with increasing T. As seen in Fig. 5b the irradiation effect exhibited a characteristic thickness ~ 20 ML related to the radiation penetration depth.[110] Therefore, $N \sim 2 \times 10^{16}$ cm^{-2} consistent with $\sigma_{d1} \approx 0.12 \times 10^{-16}$ cm^2 at low T. Surprisingly, this is not at all close to the expected value for the electron impact cross section leading to OH via direct dissociation and/or ionization and recombination. Although the scaling of these quantities is problematic, it suggests that the production of OH might *not* be the initiating process. On the other hand, $(k_3/\phi) \rightarrow \sigma$ in Eq. (A.5b) gives a reasonable H_2O_2 destruction cross section, $\sigma \sim 10^{-15}$ cm^2, by 87-eV electrons. The very large value for conversion of HO_2 into O_2 at low T, $(k_5/\phi) \rightarrow \sigma_{O2} \sim 10^{-14}$ cm^2, might indicate that O_2 can be formed and escape from considerable depths. This is contrary to their results for capping the samples with isotopically labeled H_2O. Alternatively, the excitations produced at large depths produce and eject O_2 in the near surface column of ~2 ML consistent with the results in their Fig. 3. Comparing (k_2/ϕ) and (k_4/ϕ) and assuming the range over which OH can migrate and react is the same in each case, the fit for these parameters are consistent with the formation of H_2O_2 being more efficient than the formation of HO_2.

A.6. *G values: Oxygen and peroxide production*

The dependence of the yield on the thickness of an ice sample for highly penetrating ions suggests that the O_2 yield is roughly related to the amount of energy deposited in the sample. Therefore, the amount of O_2 produced and lost from ice in steady state is often given as a G value, the number of molecules produced per 100 eV deposited (Table A.2). In applying these G values care must be taken, since the yields are temperature, fluence, and excitation density dependent. They are also modified by the presence of proton scavengers, R in the above discussion.

The *G* values for peroxide in Table A.2 are those for production at low fluence, but the *G* values for O_2 are for the production and escape of gas-phase oxygen *at saturation*. Therefore, they depend on the formation and destruction cross sections in the model above. Such *G* values can be used to describe a steady-state source, but are not helpful in determining the ability to produce or trap O_2 at depth. Over the time scales of the experiments, some O_2 is seen to remain trapped as expected.[16] Typically, there are a number of trapping sites, and recent estimates of the activation

Table A.2. *G* values for O_2 and H_2O_2 production.

Radiation^a	**Prod.**	T **(K)**	E_a **(eV)**	**Percent at saturation**	G^b **(prod./100 eV)**
$He^{+(1)}$	O_2				0.15
$\gamma^{(2,3)}$	O_2	77 (300)			0.003 (0)
$Ne^{+}{}_{30\ MeV}{}^{(3)}$	O_2	300			0.031
$He^{2+}{}_{6\ MeV}{}^{(3)}$	O_2	300			0.008
$He^{2+}{}_{29\ MeV}{}^{(3)}$	O_2	300			0.0016
$H^{+}{}_{80\ keV}{}^{(4)}$	O_2	70			0.0002
$H^{+}{}_{200\ keV}{}^{(5)}$	O_2	120			0.0007
$He^{2+}{}_{1.5\ MeV}{}^{(6)}$	O_2	50 (100)	0.03		$(0.7–4) \times 10^{-4}$
$Ne^{+}{}_{1.5\ MeV}$					
(7)	O_2	7		10–15^c	0.03
(8)		70–140	0.05–0.07		0.1
$Ar^{+}{}_{0.1\ MeV}{}^{(9)}$	O_2	30–140	<0.01–0.04		0.006–0.08
		160			0.01
$e_{30–100\ eV}{}^{(10)}$	O_2	40	0.002		0.009
		120	0.05–0.06		0.03
$h\nu_{10.2\ eV}{}^{(11)}$	O_2	50 (100)	0.03		1×10^{-4} (3×10^{-4})
$H^{+}{}_{1\ MeV}{}^{(12)}$	H_2O_2	16 (77)			0.1(<0.01)
+ 10% CO_2		77			0.1
+ 10% O_2		77			0.4
$H^{+}{}_{100\ keV}{}^{(13)}$	H_2O_2	20(80)		0.75 (0.14)	0.72 (0.36)
$C^{+}{}_{30\ keV}$	H_2O_2	16 (77)		1.7 (2.2)	0.23 (0.34)
$N^{+}{}_{30\ keV}$	H_2O_2	16 (77)		1.6 (1.4)	0.21 (0.11)
$O^{+}{}_{30\ keV}$	H_2O_2	16 (77)		1.4 (4.0)	0.26 (0.15)
$Ar^{+}{}_{30\ keV}$	H_2O_2	16 (77)		2.2 (2.0)	0.11 (0.06)

$^{a(1)}$Lefort[121]; $^{(2)}$Ghormley and Stewart[122]; $^{(3)}$Baverstock and Burns[123]; $^{(4)}$Baragiola *et al.*[124]; $^{(5)}$Baragiola and Bahr[125]; $^{(6)}$Brown *et al.*[7] (ion does not stop in sample); $^{(7)}$Benit and Brown[101] (ion does not stop in sample); $^{(8)}$Reimann *et al.*[10]; $^{(9)}$Baragiola[62]; $^{(10)}$Petrik *et al.*[110]; $^{(11)}$Westley *et al.*[56]; $^{(12)}$Moore and Hudson[126]; $^{(13)}$ Lauffler *et al.*[90]
^{b}*G* values are for escaping O_2. For H_2O_2 and O_2 formation, both of which require multiple events, yields after saturation are used. In a few cases yields are obtained by melting the sample. Values in parentheses are those at another temperature.
^{c}Percent of film eroded as O_2 molecules for long-term irradiation.

energy for an O_2 molecule trapped in a substitutional site suggest that it can be trapped for times comparable to the time required for the conversion of amorphous ice into Ih.[17] This is contrary to an earlier belief.[74] To understand the relationship between the G values for O_2 in Table A.2, we note that fast light ions and electrons lose their energy predominantly by ionizations and excitations. The ionization events produce secondary electrons that, in turn, produce further ionizations and excitations. The number of ionization events, N_i, produced by a fast light ion stopping in a water ice is, $N_i = E_0/W$, where E_0 is the initial energy and W is the average energy expended per ionization event, ~27 eV in water ice. Therefore, radiolysis by fast ions is often described as the net effect of the shower of electrons set in motion by the incident ion. In this manner, the G values for production of O_2 from ice by energetic light ions and electrons can be roughly estimated as the sum of the effects of all the secondary electrons using the yield $(Y_{O_2})_e$ obtained from experiments for ~10–100-eV electrons.[112] That is, ignoring initially the production/escape depth, N, an estimate of the number of O_2 produced and ejected can be obtained using $(Y_{O_2})_e$: $G \sim (N_i Y_{O_2}/E_0) \sim (Y_{O_2}/W) \sim (0.006$ $O_2/100$ eV).

This estimate of G is seen to be an order of magnitude larger than the keV proton value for O_2 in Table A.2, but is much smaller than the G value for energetic Ne^+. These differences from the simple estimate are consistent with the observation that the yield in the *saturation* regime for penetrating ions varies *nonlinearly* with excitation density. The difference from the keV proton G value indicates that N is much less that the penetration depth. Therefore, either the excitations at depths can not migrate to the surface and produce O_2 or O_2 formed at depth remains trapped and can be destroyed by subsequent ions. On the other hand, the larger G value for MeV Ne^+ is clearly due to the high density of excitations. This can result in more efficient production of O_2, increased production of damage sites for forming O_2, and/or increased transport (percolation) along the transiently heated track. That is, the ability of the O_2 to percolate to or preferentially form[21] at the surface during ion bombardment can play an essential role in determining the measured yields as discussed above.

References

1. L. J. Lanzerotti, W. L. Brown and J. M. Poate, *Geophys. Res. Lett.* **5**, 155 (1978).
2. W. L. Brown, L. J., Lanzerotti, J. M. Poate and W. M. Augustyniak, *Phys. Rev. Letts*. **40**, 1027 (1978).

3. E.J. Hart and R. L. Platzman, In: M. Erreva and M. A. Forssberg (editors), *Physical Mechanisms in Radiation Biology 1*, p. 99, Academic Press, New York (1961).
4. I. A. Taub and K. Eiben, *J. Chem. Phys.* **49**, 2499 (1968).
5. J. Bednarek and A. Plonka, *J. Chem. Soc. Faraday Trans.* **83**, 3725 (1987).
6. W. L. Brown, W. M. Augustyniak, L. J. Lanzerotti, R. E. Johnson and R. Evatt, *Phys. Rev. Lett.* **43**, 1632 (1980).
7. W. L. Brown, W. M. Augustyniak, E. Simmons, K. J. Marcantonio, L. J. Lanzerotti, R. E. Johnson, J. W. Boring, C. T. Reimann, G. Foti and V. Pirronello, *Nucl. Instrum. Meth.* **198**, 1 (1982).
8. W. L. Brown, W. M. Augustyniak, K. J. Marcantonio, E. N. Simmons, J. W. Boring, R. E. Johnson and C. T. Reimann, *Nucl. Instrum. Meth.* **B1**, 307 (1984).
9. R. R. Rye, T. E. Madey, J. E. Houston and P. H. Holloway, *J. Chem. Phys.* **69**, 1504 (1978).
10. C. T. Reimann, J. W. Boring, R. E. Johnson, J. W. Garrett and K. R. Farmer, *Surf. Sci.* **147**, 227 (1984).
11. R. E. Johnson, *Charged Particle Interaction with Atmospheres and Surfaces*, Springer-Verlag, Berlin (1990).
12. R. E. Johnson, In: B. Schmitt and C. deBergh (editors), *Solar System Ices,* p. 303, Kluwer, Dordrecht, The Netherlands (1998).
13. R. E. Johnson, T. I. Quickenden, P. D. Cooper, A. J. McKinley and C. Freeman, *Astrobiology* **3**, 823 (2003).
14. R. E. Johnson, R. W. Carlson, J. F. Cooper, C. Paranicas, M. H. Moore and M. Wong, In: *Jupiter: Satellites, Atmosphere, and Magnetosphere,* Chap. 20, F. Bagenal (editor), Cambridge University Press, Cambridge (2004).
15. R. E. Johnson, P. D. Cooper, T. I. Quickenden, G. A. Grieves and T. M. Orlando, *J. Chem. Phys.* **123**, 184715 (2005).
16. R. E. Johnson and W. A. Jesser, *Astrophys. J. Lett.* **480**, L79 (1997).
17. P. Ayotte, R. S. Smith, K. P. Stevenson, Z. Dohnálek, G. A. Kimmel and B. D. Kay, *J. Geophys. Res.* **106**, 33387 (2001).
18. G. A. Kimmel, K. P. Stevenson, Z. Dohnálek, R. S. Smith and B. D. Kay, *J. Chem. Phys.* **114**, 5284 (2001).
19. G. A. Kimmel, K. P. Stevenson, Z. Dohnálek, R. S. Smith and B. D. Kay, *J. Chem. Phys.* **114**, 5295 (2001).
20. G. A. Grieves and T. M. Orlando, *Surf. Sci.* **593**, 180 (2005).
21. B. D. Teolis, R. A. Vidal, J. Shi and R. A. Baragiola, *Phys. Rev.* **B72**, 245422 (2005); B. D. Teolis, J. Shi and R. A. Baragiola, *J. Chem. Phys.* **130**, 134704 (2009).
22. R. E. Johnson and T. I. Quickenden, *J. Geophys. Res.* **102**, 10985 (1997).
23. R. W. Carlson, W. M. Calvin, J. B. Dalton, G. B. Hansen, R. L. Hudson, R. E. Johnson, T. McCord and M. H. Moore, In: R. Pappalardo *et al.* (editors), *Europa,* Chap. 13, pp. 283–327 (2009).
24. R. E. Johnson, M. H. Burger, T. A. Cassidy, F. Leblanc, M. Marconi and W. H. Smyth, In: R. Pappalardo *et al.* (editors), *Europa*, Chap. 20, pp. 507–527 (2009).
25. B. C. Garrett *et al., Chem. Rev.* **105**, 355 (2005).
26. J. Meesungnoen and J-P. Jay-Gerin, *J. Phys. Chem.* **A109**, 6406 (2005).
27. R. E. Johnson and W. L. Brown, *Nucl. Instrum. Meth.* **198**, 103 (1982).
28. T. E. Madey, R. E. Johnson and T. M. Orlando, *Surf. Sci.* **500**, 838 (2002).

29. N. G. Petrik and G. A. Kimmel, *J. Chem. Phys.* **123**, 054702 (2005).
30. M. Famá, J. Shi and R. A. Baragiola, *Surf. Sci.* **602**, 156 (2008).
31. R. E. Johnson, L. J. Lanzerotti and W. L. Brown, *Nucl. Instrum. Meth.* **198**, 147 (1982).
32. D. T. Hall, P. D. Feldman, M. A. McGrath and D. F. Strobel, *Astrophys. J.* **499**, 475 (1998).
33. J. R. Spencer, W. Calvin and J. Person, *J. Geophys. Res.* **100**, 19049 (1995).
34. W. M. Calvin, R. E. Johnson and J. A. Spencer, Geophys. *Res. Lett.* **23**, 673 (1996).
35. W. M. Calvin and J. R. Spencer, *Icarus* **130**, 505 (1997).
36. J. R. Spencer and A. Klesman, *Bull. Am. Astron. Soc.* (DPS #33, #47.04.2001).
37. J. R. Spencer and W. M. Calvin, *Astron. J.* **124**, 3400 (2002).
38. A. Landau, E. J. Allis and H. L. Welsh, *Spectrochim. Acta* **18**, 1 (1962).
39. P. D. Cooper, R. E. Johnson and T. I. Quickenden, *Planet Space Sci.* **51**, 183 (2003).
40. P. D. Cooper *et al. J. Phys. Chem.* **A110**, 7985 (2006); *Icarus* **194**, 379 (2008)
41. A. Engdahl and B. Nelander, Phys. *Chem. Chem. Phys.* **4**, 2140 (2002).
42. K. S. Noll, R. E. Johnson, A. L. Lane, D. L. Dominigue and H. A. Weaver, *Science* **273**, 341 (1996).
43. K. S. Noll, T. L. Rousch, D. P. Cruikshank, R. E. Johnson and Y. J. Pendleton, *Nature* **388**, 45 (1997).
44. R. E. Johnson, In: R. Dessler (editor), *Chemical Dynamics in Extreme Envionments*, p. 390, World Scientific, Singapore (2001).
45. P. D. Cooper, M. H. Moore and R. L. Hudson, *J. Geophys Res* (2010) (in press).
46. R. W. Carlson, M. S. Anderson, R. E. Johnson, W. D. Smythe, A. R. Hendrix, C. A. Barth, L. A. Soderblom, G. B. Hansen, T. B. McCord, J. B. Dalton, R. N. Clark, J. H. Shirley, A. C. Ocampo and D. L. Matson, *Science* **283**, 2062 (1999).
47. W. M. Grundy, L. A. Young, J. R. Spencer, R. E. Johnson, E. F. Young and M. W. Buie, *Icarus* **184**, 543 (2006).
48. A. R. Hendrix and R. E. Johnson, *Astrophys. J.* **687**, 706 (2008).
49. R. W. Carlson, R. E. Johnson and M. S. Anderson, *Science* **286**, 97 (1999).
50. R. W. Carlson, M. S. Anderson, R. E. Johnson, M. B. Schulman and A. H. Yavrouian, *Icarus* **157**, 456 (2002).
51. R. T. Pappalardo, W. B. McKinnon and K. Khurana, Europa, Univeristy of Arizona Press, Tucson (2009) (in press).
52. C. F. Chyba, *Nature* **403**, 381 (2000).
53. J. S. Kargel, J. Z. Kaye, J. W. Head III, G. M. Marion, R. Sassen, J. K. Crowley, O. P. Ballesteros, S. A. Grant and D. L. Hogenboom, *Icarus* **148**, 226 (2000).
54. C. F. Chyba and K. P. Hand, *Science* **292**, 2026 (2001).
55. J. F. Cooper, R. E. Johnson, B. H. Mauk, H. B. Garrett and N. Gehrels, *Icarus* **149**, 133 (2001).
56. M. S. Westley, R. A. Bargiola, R. E. Johnson and G. A. Baratta, *Planet. Space Sci.* **43**, 1311 (1995).
57. T. L. Williams, N. G. Adams, L. M. Babcock, C. R. Herd and M. Geoghegan, *Monthly Notices R. Soc.* **282**, 413 (1996).
58. E. M. Bringa and R. E. Johnson, In: V. Pirronello and J. Krelowaky (editors), *Solid State Astrochemistry,* p. 357, Kluwer, Netherlands (2003).
59. S. Andersson and E. F. van Dishoeck, *Astron. Astrophys.* **491**, 907 (2008).

60. U. Raut, M. Famá, M. J. Loeffler and R. A. Baragiola, *Astrophys. J.* **687**, 1070 (2008).
61. A. Grigorieva, Ph. Thébault, P. Artymowicz and A. Brandeker, *Astron. Astrophys.* **475**, 755 (2007).
62. R. A. Baragiola, Planet. *Space Sci.* **51**, 953 (2003).
63. R. A. Baragiola, M. Famá, M. J. Loeffler, U. Raut and J. Shi, *Nucl. Instr. Meth. Phys. Res.* **B266**, 3057 (2008)
64. W. Kuhs, *Physics and Chemistry of Ice,* The Royal Society of Chemistry, 2007.
65. G. A. Kimmel, N. G. Petrik, Z. Dohnálek and D. Kay, *J. Chem. Phys.* **125**, 044713 (2006).
66. P. Jenniskens and D. F. Blake, *Astrophys. J.* 473, 1104 (1996).
67. N. J. Sack and R. A. Baragiola, *Phys. Rev.* **B48**, 9973 (1993).
68. L. G. Dowell and A. P. Rinfret, *Nature* **158**, 1144 (1960).
69. B. Schmitt, S. Espinasse, R. J. A. Grim, J. M. Greenberg and J. Klinger, *ESA SP*-302, 65 (1989).
70. R. S. Smith, C. Huang, E. K. L. Wong and B. D. Kay, *Surf. Sci.* **367**, L13 (1996).
71. G. Strazzulla, G. A. Baratta, G. Leto and G. Foti, *Europhys. Lett.* **18,** 517 (1992); M. Fama, M. J. Loeffler, U. Raut, R.A. Baragiola, *Icarus* **207**, 314.
72. G. Leto and G. A. Baratta, *Astron. Astrophys.* **397**, 7 (2003).
73. R. M. E. Mastrapa and R. H. Brown, *Icarus* **183**, 207 (2006).
74. R. A. Vidal, D. Bahr, R. A. Baragiola and M. Peters, *Science* **276**, 1839 (1997). 75.
75 R. E. Johnson, *J. Geophys. Res.* **104**, 14179 (1999).
76. N. G. Petrik and G. A. Kimmel, *Phys. Rev. Lett.* **90**, 166102 (2003).
77. Q.-B Lu and L. Sanche, *Phys. Rev.* **B63**, 153403 (2001).
78. S. Andersson, S. A. Al-Halabi, G-J Kroes and E. F. van Dishoeck, *J. Chem. Phys.* **124**, 064715 (2006).
79. G. A. Kimmel, T. M. Orlando, C. Vizina and L. Sanche, *J. Chem. Phys.* **101**, 3282 (1994).
80. T. I. Quickdenen, R. A. J. Litjens, C. G. Freeman and S. M. Trotman, *Chem Phys. Lett.*, **114**, 164 (1985).
81. A. J. Matich, M. G. Bakker, D. Lennon, T. I. Quickenden and C. G. Freeman, *J. Phys. Chem.* **97**, 10539 (1993).
82. B. Laffon, S. Lacombe, F. Boumel and Ph. Parent, *J. Chem. Phys.* **125**, 204714 (2006).
83. T. I. Quickenden, R. A. J. Litjens, M. G. Bakker, S. M. Trotman and D. F. Sangster, *Radiat.8 Res.* **115**, 403 (1988).
84. M. G. Bakker, T. I. Quickenden, C. F. Vernon, C. G. Freeman and D. F. Sangster, *Radiat. Phys. Chem.* **32**, 767 (1988).
85. J. Herring, A. Aleksandrov and T. M. Orlando, *Phys. Rev. Lett.* **92**, 187602 (2004).
86. J. Meesungnoen and J.-P. Jay-Gerin, *Radiation Res.* **171**, 379 (2009).
87. J. M. Warman, M. P. deHass and J. B. Verberme, *J. Phys. Chem.* **84**, 1240 (1980).
88. E. I. Tarasova, A. M. Stepanenko, I. Y. Fugol, M. Chergui, R. Schriever and N. Schwentner, *J. Chem. Phys.* **98**, 7786 (1993).
89. P. A. Gerakines, W. A. Schutte and P. Ehrenfreund, *Astron. Astrophys.* **312**, 289 (1996).
90. M. J. Loeffler, U. Raut, R. A. Vidal, R. A. Baragiola and R. W. Carlson, *Icarus* **180**, 265 (2006).

91. X. Pan, A. D. Bass, J.-P. Jay-Gerin and L. Sanche, *Icarus* **172**, 521 (2004).
92. M. Suto and L. C. Lee, *Chem. Phys. Lett.* **98**, 152 (1983).
93. S. L. Siegel, H. Baum, S. Skolnik and J. M. Flournoy, *J. Chem. Phys.* **32**, 1249 (1960).
94. J. Kroh, B. C. Green and J. W. T. Spinks, *J. Am. Chem. Soc.* **83**, 2201 (1961).
95. T. I. Quickenden, A. J. Matich, M. G. Bakker, C. G. Freeman and D. F. Sangster, *J. Chem. Phys.* **95**, 8843 (1991).
96. D. Klyachko, P. Rowntree, and L. Sanche, *Surf. Sci.* **346**, L49 (1996).
97. P. Rowntree, P. L. Parenteau and L. Sanche, *J. Chem. Phys.*, **94**, 8570 (1991).
98. X. Pan, H. Abdoul-Carime, P. Cloutier, A. D. Bass and L. Sanche, Radiat. *Phys. Chem.* **72**, 193 (2005).
99. K. A. Long and H. G. Paretzke, *J. Chem. Phys.* **95**, 1049 (1991), and references herein.
100. J. Meesungnoen and J.-P. Jay-Gerin, *J. Phys. Chem.* **A109**, 6406 (2005).
101. J. Benit and W. L. Brown, *Nucl. Instrum. Methods Phys. Res.* **B46**, 448 (1990).
102. G. A. Grieves and T. M. Orlando, *Surf. Sci.* **593**, 180 (2005).
103. B. Rowland, M. Fisher and J. P. Devlin, *J. Chem. Phys.* **95**, 1378 (1991).
104. F. E. Livingston, J. A. Smith and S. M. George, *J. Phys. Chem.*, **A106**, 6309 (2002).
105. M. J. Loeffler, B. D. Teolis and R. A. Baragiola, *Astrophys. J.* **639**, L103 (2006).
106. U. Eckner, D. Helmreich and H. Engelhardt, In: S. J. Jones and L. N. Gold (editors), *Physics and Chemistry of Ice*, E. Whalley, R. Soc. of Can., Ottawa (1973).
107. P. J. Woolridge and J. P. Devlin, *J. Chem. Phys.* **88**, 3086 (1988).
108. R. A. Haring, R. Pedrys, D. J. Oostra and A. E. DeVries, *Nucl. Inst. Meth.* **B5**, 483 (1984).
109. G. A. Kimmel and T. M. Orlando, *Phys. Rev. Lett.* **75**, 2606 (1995).
110. N. G. Petrik, A.G. Kavetsky and G. A. Kimmel, *J. Phys. Chem.* **B110**, 2723 (2006).
111. N. G. Petrik, A.G. Kavetsky and G. A. Kimmel, *J. Chem. Phys.* **125**, 124702 (2006).
112. M. T. Sieger, W. C. Simpson and T. M. Orlando, *Nature* **394**, 554 (1998).
113. M. C. Akin, N. G. Petrik and G. A. Kimmel, *J. Chem. Phys.* **130**, 104710 (2009).
114. T. M. Orlando and M. T. Sieger, *Surf. Sci.* **528**, 1 (2003).
115. X. Pan, A. D. Bass, J.-P. Jay-Gerin and L. Sanche, *Icarus* **172**, 521 (2004).
116. L. Khriachtchev, M. Pettersson, S. Jolkkonen, S. Pehkonen and M. Räsänen, *J. Chem. Phys.* **112**, 2187 (2000).
117. N. G. Petrik and G. A. Kimmel, *J. Chem. Phys.*, **121**, 3736 (2004).
118. N. Watanabe and A. Kouchi, *Astrophys. J.* **567**, 651 (2002).
119. N. Watanabe, T. Horii and A. Kouchi, *Astrophys. J.* **541**, 772 (2000).
120. W. Zheng, D. Jewitt and R. I. Kaiser, *Astrophys.* J.**639**, 534(2006); ibid., **753**(2006); *Chem. Phys. Lett.* **435**, 289 (2007)
121. M. Lefort, In: M. Haissinsky (editor), *Actions Chimiques et Biologiues des Radiations*, Vol. 1, p. 203, Masson, Paris (1955).
122. J. A. Ghormley and A. C. Stewart, *J. Am. Chem. Soc.* **78**, 2934 (1956).
123. K. F. Baverstock and W. G. Burns, *Nature* **260**, 316 (1976).
124. R. A. Baragiola, C. L. Atteberry and D. A. Bahr, *J. Geophys. Res.* **104**, 14183 (1999).

125. R. A. Baragiola and D. A. Bahr, *J. Geophys. Res.* **103**, 25865 (1998).
126. M. H. Moore and R. L. Hudson, *Icarus* **145**, 282 (2000).
127. R. E. Johnson., M. Fama, M. Liu, R. A. Baragiola, E. C. Sittler, Jr and H. T. Smith, *Planet. Space Sci.* **56**, 1238 (2008)
128. R. E. Johnson, R. E. Johnson, J. G. Luhmann, R. L. Tokar, M. Bouhram, J. J. Berthelier, E. C. Siler, J. F. Cooper, T. W. Hill, H. T. Smith, M. Michael, M. Liu, F. J. Crary and D.T. Young, *Icarus* **180**, 393 (2006).
129. B. D. Teolis, G. H. Jones, P. F. Miles, R. L. Tokar, B. A. Magee, J. H. Waite, E. Roussos, D. T. Young, F. J. Crary, A. J. Coates, R. E. Johnson, W.-L. Tseng and R. A. Baragiola, *Science*, accepted (2010).

CHAPTER 11

COOL INTERSTELLAR PHYSICS AND CHEMISTRY

Alexander G. G. M. Tielens
Leiden Observatory, P.O. Box 9513, 2300 RA Leiden, The Netherlands
E-mail: tielens@strw.leidenuniv.nl

Louis J. Allamandola
Space Sciences & Astrobiology Division, MS 245-6, NASA Ames Research Center, Moffett Field, CA 94035, USA
E-mail: Louis.J.Allamandola@nasa.gov

The infrared (IR) spectra of molecular cloud material in the interstellar medium of galaxies are dominated by absorption features due to simple molecules such as H_2O, CH_3OH, NH_3, CO_2, and CO in cryogenic molecular ices. These ices are an important reservoir of material and play an important role in the organic inventory of regions of star and planet formation. Observational studies have revealed the presence of several independent ice components, often along the same line of sight. These interstellar ices are formed by accretion and reaction on grain surfaces and further processed by ultraviolet (UV) photons as well as highly energetic cosmic ray ions. Thermal processing by nearby newly formed stars is also important. In this chapter, we will review the characteristics, formation, and evolution of interstellar ices with the emphasis on relevant laboratory studies in the area of IR spectroscopy of low-temperature molecular mixtures, the surface chemistry of simple species relevant to astrophysical environments, and analog studies of the energetic processing of ices near newly formed stars.

1. Introduction

Over the last decade or so, a quiet revolution has taken place in astronomy. While for hundreds of years, humans have pondered whether the Earth and the Solar System were unique, now some 350 planets orbiting other stars are known in our own galactic neighborhood alone.[1] Almost all of these exoplanets are giant planets like Jupiter rather than Earth-size planets, but the recently launched Kepler satellite[2] will survey small portions of our galaxy and is expected to add hundreds of terrestrial planets to this growing list of galactic real estate. Subsequent missions will characterize these planets and search for the chemical signposts of life. These missions are designed to address the all-important question "are we alone?"

Star and planet formation is a common process in the interstellar medium of galaxies. This process starts in the deep and cold interiors of so-called molecular clouds, when gravity overwhelms the supporting thermal pressure forces and collapse ensues.[3] While a central core quickly forms — the nascent star — because of centrifugal forces, much of the material will form a surrounding disk which slowly accretes onto the star. In this disk small (submicron) dust grains can coagulate and grow. When they have grown to sizes of ~1 cm, these dustballs will settle in a thin disk where further growth can take place, leading eventually to planetesimals, cometesimals, and then terrestrial and gaseous planets.[4]

This physical evolution during the star formation process is accompanied by strong chemical evolution. Ion–molecule and neutral–neutral radical reactions are at the basis of a rich chemistry in the gas phase of molecular clouds. Deep within the protected cores of these clouds, temperatures can drop to as low as 6 K. At such temperatures, atoms and molecules will collide with and stick to a grain surface, forming an ice mantle coating.[5] Subsequently, several processes may act to modify the mantle. Atoms can thermally hop from site to site over the ice matrix or undergo quantum mechanical tunneling, eventually reacting with another atom or molecule. Water and methanol are believed to form by this route, adding H atoms to an O atom and CO molecule on the grain surface, respectively.[6,7] Close to a forming protostar, increasing temperatures and a higher flux of UV or X-ray photons lead to high-energy processing of the ice mantle as well as thermal processing and sublimation of the more volatile species. Studies of laboratory analogs have shown that energetic processing of simple ices leads to formation of more complex species[8,9] such as OCN^-, NH_2CHO, and C_2H_5OH and an

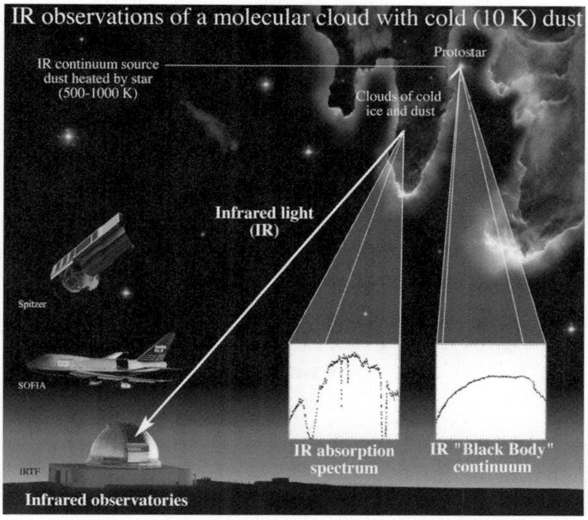

Fig. 1. Measuring the IR absorption spectra of interstellar clouds. A background star (protostar) serves as the IR source, the cloud is the sample, and the telescope gathers the light and sends it to a monochromator or spectrometer. The advent of airborne IR telescopes in the 1970s and orbiting telescopes in the 1980s made it possible to avoid IR absorptions by atmospheric H_2O, CO_2, and so on, opening a new window into the Cosmos. Figure courtesy of Dr. Jason Dworkin. See also Colour Insert.

organic residue.[10] Investigating how the composition and structure of ice mantles vary as a function of physical environment is vital to understanding chemical evolution in the envelopes of young stars, protoplanetary disks, and comets and hence the understanding of the organic inventory available to newly forming planets.

2. Spectroscopy of Interstellar Ice

Infrared (IR) spectroscopy is the premier tool to study the composition of interstellar ice. Some 30 years ago, matrix isolation techniques were adopted to study the spectroscopy and chemistry of mixed molecular ices rather than species isolated in inert matrices to interpret the spectra of interstellar clouds. IR spectra of protostars embedded in, or of background stars located behind, molecular clouds reveal broad absorption bands due to simple molecular ices (Figs. 1 and 2).

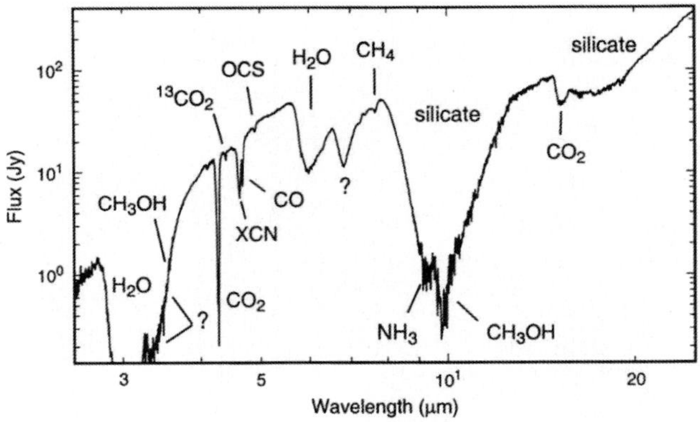

Fig. 2. The mid-IR spectrum of the protostar W33A observed with the short wavelength spectrometer on-board the Infrared Space Observatory shows a variety of absorption features. Except for the 10 and 18-μm silicate bands, these features are due to simple molecules in an ice mantle. Reproduced with permission from Ref. 11.

Comparisons of these spectra with the spectra of laboratory analogs provide direct identification of the solid compounds present in space and allow measurement of accurate abundances and thus the composition of the ices. Subtle profile details may reflect the interaction of these molecules with the matrix in which they are embedded.

Systematic studies of large samples of sources provide then the inventory of ices in different regions and allow inferences on the origin and evolution of ices in space either because abundances vary in a systematic way, reflecting the underlying chemistry, or because variations in the profiles indicate the importance of variations in the molecular ice environment. Table 1 provides the chemical make up of interstellar ices along the lines of sight through several molecular clouds.

2.1. *Composition of interstellar ice*

The most striking aspect of the IR spectrum in Fig. 2 is the apparent simplicity of the composition of interstellar ices. Many species are highly hydrogenated (H_2O, NH_3, CH_4), but the spectra also show mixed oxidation/reduction patterns (CO, CO_2, H_2CO, CH_3OH, CH_4). Furthermore, the abundance of these species is very different in the ice than in the gas phase of dark clouds. The H_2O/CO ratio is quite telling: This ratio is less than 10^{-4} in the gas in dark clouds[20] compared to ~4 in the ice (see Table 2). This preference for saturated species in the ice and this abundance difference between gas phase and ice betrays a different chemical origin and emphasizes the importance of chemistry on the surfaces of interstellar grains. Clearly, H_2O ice is a major reservoir of oxygen formed by hydrogenation of O and/or via O_2 on grain surfaces.[6] The observed, low abundances of solid NH_3 and CH_4 (relative to H_2O) suggests that, because of gas-phase ion-molecule reactions, most of the nitrogen and carbon in the accreting gas is locked up in N_2 and CO rather than in N and C atoms. Molecular nitrogen is inert at the low temperatures ($T < 100$ K) appropriate for interstellar grain-surface chemistry. Hydrogenation of CO leads to the formation of the carbon species H_2CO, CH_3OH, and — after an intermediate oxidation step — to HCOOH while direct oxidation results in CO_2. Finally, except for the 6.85-μm band[a] (1460 cm^{-1}), which consists of a blended set of features whose identifications are unclear (see Section 2.5), no major unidentified bands are present. Weaker features superimposed on some of the strong bands reveal that other species are present as well but at lower abundances. Clearly, grain (surface) chemistry — and more generally, ice chemistry — heavily favors the formation of relatively simple species.

However, this is not the full story. UV photolysis (see Section 3.2) or thermal processing (see Section 3.4) is expected to lead to the formation of more complex species at concentrations too low to be readily observable among the strong IR bands of simple ices. When these ices evaporate (after heating by a newly formed star), these traces of complex chemistry may become observable using the much more sensitive technique of emission spectroscopy in the microwave regime (e.g., rotational transitions). Before reviewing this chemistry, a few examples of the information that can be extracted from interstellar IR spectra will be documented.

---

[a] Time-honored tradition has it that astronomical bands are "named" by wavelength rather than by frequency. For convenience, we will generally provide the wavenumber conversion.

Table 1. The composition of interstellar ice

Species	Quiescent Cloud[a]	Low-Mass Protostar[b]	High-Mass Protostar[c]	Comets[d]
H_2O	100	100	100	100
CO (total)	25	5	8	23
CO (H_2O ice)	3	1	6	–
CO (pure CO)	22	4	2	–
CO_2 (total)	21	19	13	6
CO_2 (H_2O ice)	18	?	9	–
CO_2 (CO mix)	3	?	2	–
CO_2–CH_3OH complex	–	–	1.2	–
CO_2 (pure crystalline)	–	–	1.0	–
CH_4	<3	<1.4	1.5	0.6
CH_3OH	<2	<5	18	2.4
H_2CO	–	–	6	1.1
HCOOH	<1	<1	7	0.09
OCS	<0.2	<0.1	0.2	0.4
NH_3	<8	<11	15	0.7
OCN^-	<0.5	<0.2	3.5	0.1[e]

[a]Observations toward the star, Elias 16, behind the Taurus cloud; taken from Refs. 12 and 13.
[b]Observations toward the low-mass protostar, Elias 29, in the ρ Oph cloud taken from Refs. 14–16.
[c]Observations toward the luminous protostar, W33A, taken from Refs. 16–18.
[d]Ice abundances inferred from measurements of the gas-phase composition measured for the coma of comet Hale-Bopp at 1 AU from the Sun taken from Ref. 19.
[e]Taken as HNCO.

2.2. *CO in interstellar ice*

Studies of the profile of the CO ice band illustrate the complexity inherent to interstellar ice bands and the power of IR spectroscopy in revealing the intricacies of interstellar ice compositions. Indeed, astronomical observations have revealed clear variations in the detailed profile of the CO ice band near 2136 cm^{-1} (4.67 μm; Fig. 3). All profiles can be decomposed into mixtures of two intrinsically very different profiles: a narrow profile centered at 2139.9 cm^{-1} and a much broader and, generally, much weaker feature at 2136.5 cm^{-1} (Fig. 3). There is also evidence for a weaker band at 2143.7 cm^{-1}(4.66 μm) in these profiles. In addition, there are two bands — that often blend into one feature — at higher frequencies (~2170 cm^{-1}).[18,21,22]

Extensive laboratory studies have demonstrated the sensitivity of the characteristics of this band (peak position, width) to the environment of CO molecules in the matrix.[23–25] The two main bands are "solidly" attributed to CO in ices. The narrow strong band at 2136 cm^{-1} is due to pure solid CO or CO mixed in with other nonpolar ice molecules. The much broader component at 2139.9 cm^{-1} is due to CO mixed in with very polar species, in particular H_2O. The very weak band at 2143.7 cm^{-1} (4.66 μm) was originally attributed to the longitudinal optical (LO) component of the vibrational transition in pure crystalline CO ice that appears when the background source is linearly polarized.[18] However, more likely[16] it is due to a mixture of CO:CO_2 at a ratio of ~ 1:1. For the two weak bands at shorter wavelengths, only that at 4.60 μm (2175 cm^{-1}) may be related to solid CO, specifically (traces of) CO chemisorbed to the bare surfaces of (silicate) grains.[26] The other band is unrelated to CO and will be discussed in Section 2.3.

These observations reveal a very diverse set of CO adsorption sites: CO-rich, H_2O-rich, and CO_2-rich ices and chemisorbed surfaces. That much of the solid CO is not mixed in with polar molecules (i.e., H_2O) is a very robust conclusion.[18,21,22] This surprising conclusion implies that along most lines of sight there are, at least, two independent ice components: a water-rich ice dominating the bulk of the interstellar ice (e.g., responsible for the strongest interstellar ice band, at 3 μm, due to H_2O ice; Fig. 2) with a small amount of CO admixed in and a second ice component containing only nonpolar molecules such as (most of) the solid CO. The other absorption bands are clearly present but because they are weak and blended with other bands, their properties and interpretation are less certain. The presence of the two major and independent interstellar ice components (CO-rich and H_2O-rich ices) along most sight-lines may reflect thermal processing (volatile CO sublimates at much lower temperatures than H_2O) or differences in the chemical history between these two species (H_2O ice is formed on grain surfaces in the presence of atomic H which also reduces CO to H_2CO and CH_3OH). The importance of thermal processing of interstellar ices is

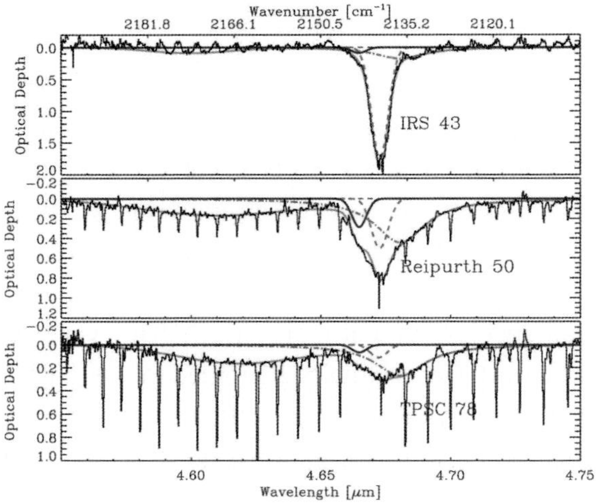

Fig. 3. The 4.5–4.75 μm spectra of three low-mass protostars (IRS 43, Reipurth 50, TPSC 78) show the relatively broad solid CO absorption at about 4.67 μm and a forest of gas-phase CO rotational–vibrational transitions. The profile of the interstellar solid CO band consists of several independent components: A narrow band at 2139.9 cm^{-1} (4.67 μm) dominates the top spectrum. The profile in the bottom panel is dominated by a much broader band peaking at 2136.5 cm^{-1} (4.68 μm). The middle panel is a combination of these two extreme profiles. In addition, the solid CO band is accompanied by a weak band at somewhat shorter wavelengths (2143 cm^{-1}; 4.66 μm). Finally, the much broader and weaker band at ~ 4.61 μm actually consists of two components at 4.62 μm (2165 cm^{-1}) and 4.60 μm (2175 cm^{-1}). The figure is adapted from Ref. 18. See also Colour Insert.

attested to by the observed profiles of the solid CO_2 ice bands (see Section 2.4). Thermal processing is discussed further in Section 3.4.

2.3. *The OCN^- ion in interstellar ice*

The solid CO spectra (see Fig. 3) reveal the presence of an absorption component at 4.62 μm (2165 cm^{-1}). A band at this position was noted early on in UV photolysis experiments of ice mixtures containing CO and NH_3.[27] Subsequent studies suggested that this band was due to the anion, OCN^-.[8,28,29] The use of electron donor and acceptor dopants unambiguously established the involvement of an anion, OCN^-, in this laboratory band. By inference, the interstellar band has been assigned to

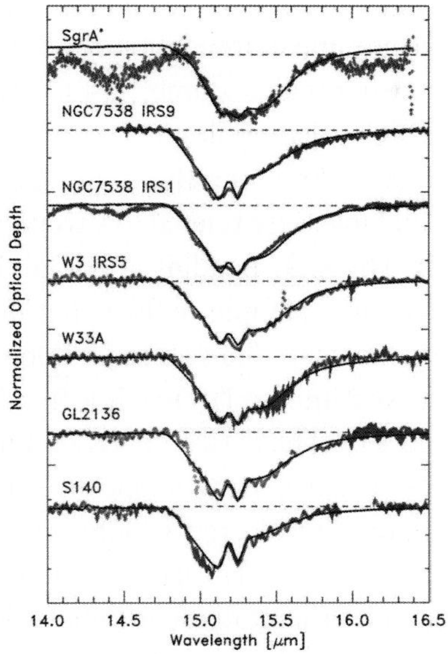

Fig. 4. The 14–16.5 μm (600–700 cm^{-1}) spectra of the CO_2 bending mode near 15 μm (650 cm^{-1}), illustrating the thermal evolution of interstellar ices. The sources are displayed in order of increasing *gas* temperature (from top to bottom). Solid lines are fits based upon laboratory data for an ice mixture $H_2O:CH_3OH:CO_2$ = 1:1:1 at various ice temperatures. Laboratory *ice* temperatures are from top to bottom: 115, 119, 145, 136, 112, 117, and 136 K. Taken from Ref. 30. See text for details.

OCN^- as well, an assignment which is now "universally" accepted. While the early experiments implicated UV photolysis in producing this species (see Section 3.2), subsequent experiments revealed that thermal processing of ices containing HNCO and NH_3 can also produce this anion (see Section 3.4).

2.4. *CO_2 in interstellar ice*

As with CO, extensive laboratory studies have shown that the solid CO_2 band's position and profile reflects the nature of the ice in which it is trapped.[30–32] Observations show that solid CO_2 is an important component of interstellar ices in star-forming regions as evidenced by its strong stretching and bending modes (see Fig. 2).[16,30,33–36] These bands are common in sources embedded in or located behind dense clouds. In addition, the weak overtone and combination bands at 3600 and 3708 cm^{-1} have been detected in one particular source, S140,[37] as well as the stretching mode of the very rare $^{13}CO_2$ isotopologue ($^{12}C/^{13}C \sim 90$) in a number of sources.[35] The main bands show variations in peak position and profile from source to source and these are related to variations in the composition of the ice (see Fig. 4). The CO_2 bending mode at 15 μm is particularly revealing. Analysis of a large sample reveals that this band contains five independent CO_2 components.[16] The two most important components are due to traces of CO_2 mixed into an H_2O-rich ice and to an approximately 1:1 mixture of CO_2 and CO. These two bands dominate most sources. In addition, some sources reveal narrow subpeaks due to pure, crystalline CO_2. A few sources show a weak, red shoulder that has been identified with a 1:1 complex of CO_2 with CH_3OH.[30,31] The last component is very weak, more uncertain, and difficult to identify but may be due to CO_2 very diluted in a CO matrix.

The interstellar H_2O:CO_2:CO system is thus very complex with the simultaneous presence along most sight-lines of a H_2O-rich mixture with some CO and CO_2 mixed in (both at the ~0.2 fractional abundance), a CO-rich mixture with perhaps a small trace of CO_2 (~0.04), and a CO_2:CO mixture with a (very uncertain) mixing ratio in the range of 2:1. The spectral components due to pure CO_2 and to CO_2–CH_3OH complex are also particularly interesting and provide a sensitive diagnostic of the CO_2 ice environment. Extensive laboratory studies on a variety of astrophysically relevant mixtures have shown that these two components represent the effects of thermal processing of ice mixtures containing CO_2, H_2O, and CH_3OH in a ratio of approximately 1:1:1.[31,38] Figure 4 illustrates this with a comparison of laboratory spectra during this thermal processing with a selected set of interstellar spectra. Upon heating to ~100 K, the ice matrix rearranges and molecules start to be mobile within the ice matrix and, at that point, the narrow peaks and the shoulder form in the spectra. The two narrow peaks represent then the formation of "pure" CO_2 clusters in a H_2O/CH_3OH ice environment, while the shoulder is due to a direct complex between CO_2 and CH_3OH. The miscibility of (fluid) water and carbon dioxide with methanol is of interest to supercritical fluid chromatography[39] and with ethanol this has

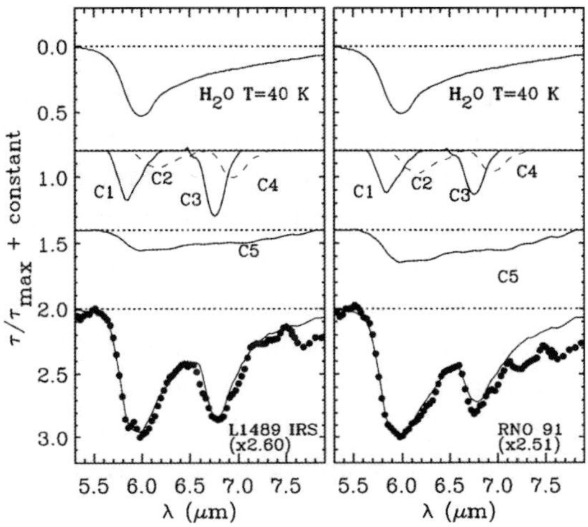

Fig. 5. The 5.3–7.9 μm (1265–1890 cm^{-1}) spectra of the protostars, L1489 and RN091, are dominated by the interstellar 6.0 and 6.85 μm absorption bands (1650 and 1460 cm^{-1}). These bands contain a number of independent absorption components. Besides the water band at 6.0 μm (1650 cm^{-1}), additional five components (here labeled C1 through C5) have been identified. Together, these produce good fits to the observed absorption features as illustrated in the bottom panels (dots: interstellar data; lines: linear combinations of the five components). Taken from Ref. 15.

been long studied near Reims.[40] These studies have revealed that methanol is well miscible with H_2O and CO_2, but H_2O does not mix well with CO_2 except at low concentrations. In ternary mixtures, the presence of methanol enhances the miscibility of CO_2 in H_2O. However, the ice studies have not been placed within this context, yet. A detailed analysis of the stretching mode of $^{13}CO_2$ ice at 4.38 μm corroborates this conclusion.[35] These results provide important evidence for extensive ice segregation and thermal processing in the envelopes around massive protostars.

2.5. *Complex molecules in interstellar ice*

Observations of protostars in the 5–8 μm (1250–2000 cm^{-1}) range reveal two prominent bands, at 6.0 and 6.85 μm (1650 and 1460 cm^{-1}), whose detailed profiles shows variations from source to source (see Fig. 2).[41,42]

With the Spitzer space telescope, a large data set has become available which allows a systematic unravelling of these bands into multiple components.[15] After subtraction of a contribution of H_2O ice to the 1650 cm^{-1} band, an additional five components can be identified empirically (see Fig. 5). A number of carriers have been identified as contributing to these bands. Generally, these estimated contributions are based upon unambiguous assignments of bands elsewhere in the spectra. These include methanol (contributing 5%–20% to the C3 component), formic acid (contributing 50%–100% of the C1 component), formaldehyde (contributing 10%–35% to the C1 component), and ammonia (contributing 10%–50% of the C2 component).[15] These species are all expected from simple grain-surface chemistry schemes (see Section 3.1.).

However, even after accounting for these species, the five components are not fully accounted for (in particular, for components C3, C4, and C5), revealing the presence of other absorbers in interstellar ices. The "excess" absorption in the C3 and C4 components (not explained by known species) may be carried by the ion NH_4^+.[8] This ion may be the counter-ion to the OCN^- ion identified as the carrier of the interstellar 4.62 μm (2165 cm^{-1}) band (Section 2.3). The two separate components would then reflect thermal variations in peak profile of this band. Thermal processing may be a key ingredient in the formation of the OCN^- and NH_4^+ ions (Section 3.4). However, the temperature variations implied (heating up to 200 K and salt formation) are difficult to reconcile with the observations.[15] The C5 component is enigmatic. One possibility would be a blend of absorptions due to various anions (NO_2^-, NO_3^-, and HCO_3^-). These anions are not produced by thermal processing but may instead be a sign of UV processing (Section 3.2). Alternatively, UV processing of interstellar ices produces an organic residue (Section 3.2) whose blended absorption features produces an absorption band not unlike the C5 component.[15] In addition to complex species produced within the ice by energetic processing, polycyclic aromatic hydrocarbons (PAHs) are also likely to be important ice constituents (Section 3.3) and to contribute to these features as well.

3. Chemistry of Interstellar Ices

Over the history from early molecular cloud assemblage to dense core formation to gravitational collapse and the formation of the star and surrounding planetary disk with icy planetesimals, interstellar ices undergo a varied set of physical and chemical processes that have a profound influence on the chemical composition. This starts with the accretion and reaction of gas-phase species on preexisting silicate or 'soot' grains forming simple molecules (Section 3.1). These ices can be processed in various ways. Highly energetic (~100 MeV/nucleon) cosmic ray ions can penetrate these clouds and directly interact with these ices. These cosmic rays can also electronically excite gaseous H_2, which then decay through emission in the Lyman–Werner bands ($^1\Sigma_g \leftarrow {}^1\Sigma_u$ and $^1\Sigma_g \leftarrow {}^1\Pi_g$). Some of these UV photons can be absorbed by the ices initiating a rich photochemistry (Section 3.2). Near the surface of the cloud, UV photons penetrating from nearby stars will also contribute to this photolysis process. The photolysis of large molecules such as PAHs is of particular interest (Section 3.3). Finally, the newly formed star can heat its environment, promoting the diffusion and reaction of reactive species produced by photolysis as well as polymerization reactions (Section 3.4).

3.1. *Grain-surface chemistry*

Interstellar grains provide a surface on which accreted species can meet and react and to which they can donate excess reaction energy.[5] Grain surface chemistry is thus governed by the accretion rate — which sets the overall timescale for the process — and the surface migration rate — which governs the reaction network. We will briefly discuss these "steps" and contrast and compare them with laboratory studies.

Interstellar dust clouds are very tenuous ($n \approx 10^4$ cm^{-3}, mostly molecular hydrogen) and at low temperatures ($T \approx 10$ K) and pressures ($P \approx 10^{-17}$ atm). Because timescales are measured in hundreds of thousands to tens of millions of years, accretion and reaction is (still) of importance. Interstellar grains range in size from 50 to 3000 Å and have some 5000 to 10^7 sites. Thus, a typical interstellar flux of atomic hydrogen of 10^5 species cm^{-2} s^{-1} on a single grain corresponds to one gas atom accreting every couple of months (for a 50-Å grain) to one H atom every day (for a 1000-Å grain). Often, one reactive species is present on an interstellar grain surface waiting for a coreactant to arrive. Thus, surface chemistry on interstellar grains occurs in the diffusion limit where transport of material (from the gas) to the surface controls the chemistry. In contrast, in the laboratory, an ultra high vacuum (UHV) background pressure of $P \sim 10^{-13}$ atm still corresponds to a flux of 10^{11} species cm^{-2} s^{-1}, and surface chemistry is studied at much higher fluxes ($\approx 10^{13}$ species cm^{-2} s^{-1}) than interstellar conditions. Moreover, experiments are generally conducted on bulk samples and hence large surface concentrations of reactants are built up and chemistry occurs in the reaction limit where reaction among accreted species is the limiting factor.

3.1.1. *The composition of the accreting gas*

The chemical routes in interstellar grain-surface chemistry are partly controlled by the composition of the accreting gas. While most of the hydrogen is in the form of H_2, which has limited reactivity at the low temperatures ($T \approx 10$ K) of the interstellar medium, cosmic rays maintain a low abundance of atomic hydrogen which is very reactive. In the gas, carbon is largely in the form of the very stable molecule, CO, but cosmic-ray-produced UV photons will break out a small fraction of the carbon into atomic form. The oxygen not locked up in CO will be mainly in atomic form. Nitrogen is thought to be mainly in the form of N_2. This species is stable against cosmic-ray-produced UV photons, but cosmic rays will also ionize helium atoms and ionized helium can break the N_2 bond. Ammonia, the next most abundant nitrogen-bearing species, is about 3–4 orders of magnitude less abundant than N_2. Table 2 summarizes the composition of the accreting gas.

Table 2. Composition of accreting gas.

Species	Gas–Phase Abundance[a]	Ice Abundance[g]	Ice/Gas
H_2	0.5	–	
He[b]	0.2	–	
H[c]	$2/n$	–	
CO	8 (−5)	2.5 (−5)	0.3
O[d]	2.4 (−4)	–	
C[e]	8 (−7)	–	
N_2[f]	8 (−5)	?	
H_2CO	2 (−8)	–	
CH_3OH	2 (−9)	~2 (−6)	~1 (3)
OH	3 (−7)	–	
H_2O	<7 (−9)	1 (−4)	>1 (4)
NH_3	2 (−8)	8 (−6)	4 (2)
HCN	2 (−8)	?	

[a]Gas-phase abundances relative to all hydrogen nuclei derived from observations of the molecular cloud, TMC-1, unless otherwise noted (data taken from the compilation in Ref. 5).
[b]Cosmic abundance of helium.
[c]Theoretical estimate of atomic hydrogen abundance from balance of cosmic ray dissociation and grain-surface formation.
[d]Cosmic abundance of oxygen minus oxygen locked up in silicates and in gaseous CO.
[e]Theoretical estimate of atomic carbon abundance from balance of photodissociation by cosmic-ray-produced UV photons and CO formation.
[f]Cosmic abundance of nitrogen.
[g]Ice abundances in quiescent clouds. High abundances of CH_3OH, H_2CO, and NH_3 have been detected in ices toward luminous protostars (see Table 1), but that may reflect the effects of energetic processing in these environments rather than grain-surface chemistry.

3.1.2. *Binding energies and surface migration*

Surface migration is set by the binding energy of the species to the surface, and for ice surfaces binding will occur in physisorped sites with typical binding energies of 400–4000 K. At 10 K, species can in essence then be divided into "migrators" and trapped species depending on whether the surface migration timescale is much faster than the accretion timescale of coreactants or not. The former species can quickly scan a

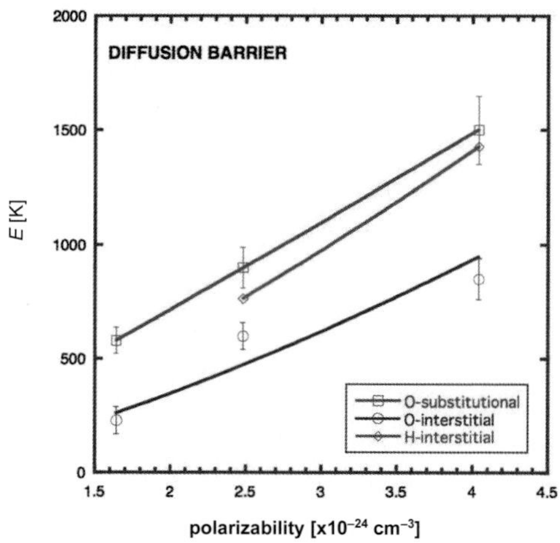

Fig. 6. Barriers against migration for oxygen and hydrogen atoms trapped in noble-gas matrices as a function of the polarizability of the matrix. Migration barriers have been estimated from molecular thermoluminescence, laser-induced fluorescence, or electron paramagnetic resonance. The data are taken from Refs. 50–53. See also Colour Insert.

grain surface for a coreactant, while the latter have to await the arrival of coreactants; in principle, either through surface migration (Langmuir–Hinshelwood mechanism) or through direct reaction with an incoming gaseous species (Eley-Rideal mechanism) but under the conditions of the interstellar medium, only the former is really of importance.

A number of experimental studies have been performed relating to the mobility and recombination of hydrogen atoms to form H_2 on ice surfaces.[43–46] While these studies have not always been interpreted in the same way, the consensus emerging from these studies is that surface mobility is much affected by the structure and surface morphology of the ice. Specifically, H_2O deposited at 10 K forms high-density amorphous solid water[6,47] with a very porous open structure and a distribution of binding sites for H_2 ($E_b \approx 500$–700 K). Upon warm-up to temperatures above 60 K, this ice ("continuously") anneals to low-density amorphous solid water with low porosity[47] and somewhat lower binding energy for H_2 ($E_b \approx 400$–600 K). The measured H_2 binding energies on porous amorphous water are somewhat lower than the calculated binding energy of 900–1100 K.[44]

The binding energy of atomic H is generally not measured directly, but from polarizability arguments one might expect a binding energy that is ≈15% lower than for H_2. Detailed theoretical calculations yield a binding energy of 500–800 K on porous amorphous solid water for H.[49] On both types of ice surfaces, the H_2-recombination experiments do provide direct evidence that atomic H is highly mobile and migrates readily even at 8 K with a barrier against migration on porous amorphous solid water estimated to be only 240 K.[46] Much of the molecular hydrogen formed on porous amorphous solid water is, however, retained in the pores and, in Temperature Programmed Desorption experiments, not released until warm-up in a random-walk process of sublimation and condensation in the pores. Because of the high porosity, H atoms can "penetrate" several monolayers into the ice and the density of surface sites is several times (5–10) larger than the nominal 2×10^{15} sites cm^{-2}.

There are few experiments on the mobility of other important reactants on interstellar ice analogs. The polarizability of atomic oxygen is very similar to that of atomic H. However, the binding energy of atomic oxygen to hexagonal ice has been calculated to be much higher, 1320 K.[54] Molecular thermoluminescence experiments involving oxygen in noble-gas matrices (see Ref. 55) have also been used to provide insight in binding and mobility of atomic oxygen on interstellar ices.[6] This data suggests that, in these matrices, atomic O behaves similar to atomic H (Fig. 6). These data are however notoriously difficult to interpret. Specifically, the data reveal that atomic oxygen recombination occurs as a first-order reaction despite that these O atoms often have to diffuse some 300 Å. It has been suggested that this enhanced mobility reflects the thermal excitation of atomic oxygen from the triplet to the singlet state and $O(^1D)$ is expected to be much more mobile in view of its smaller interaction radii.[53,56] Alternatively, the mobility of oxygen trapped in interstitial or substitutional sites may be triggered by local phonon excitation, releasing the atom, which then may "ride" this phonon wave "ballistically" until a coreactant is encountered. Besides these uncertainties associated with the binding energy and migration barrier, it is also unknown whether atomic oxygen (similar to atomic H) can penetrate deeply into the porous structure of amorphous solid water.

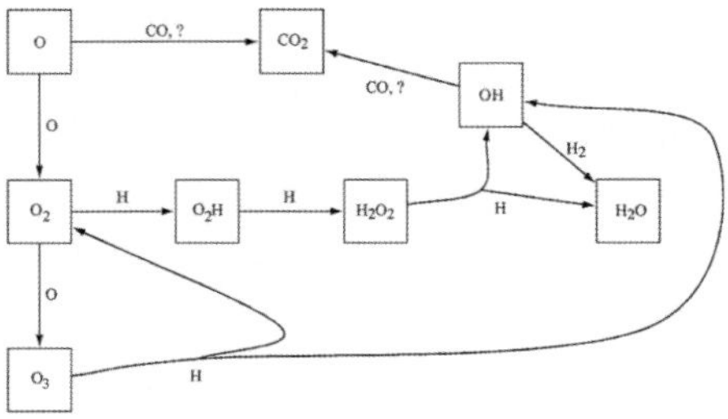

Fig. 7. Grain-surface chemical routes toward H_2O formation involve reactions of H with O_2 and O_3.[6] In principle, an accreted O can remain on the surface until an H arrives and reacts to form OH and then H_2O. However, O will likely react with trapped O_2 to form O_3 which is then hydrogenated. Molecular oxygen can be lost to hydrogenation (through the peroxide channel) as well leading to H_2O.

Finally, in evaluating these data, it should also be kept in mind that because of the high abundance of H_2 in molecular clouds ($H_2/H \geq 10^4$), sites with the highest E_b in the distribution of binding sites on interstellar grains will preferentially bind H_2, and they are then effectively blocked to H.[57,58] Evaluating the balance between accretion and sublimation, sites that bind H_2 more strongly than ≈500 K (corresponding to a surface coverage of ~0.2) will be passivated this way for typical interstellar densities and grain sizes.[6] This aspect is not fully probed in the diffusion and reaction experiments.

3.1.3. *Oxygen/water reaction network*

Models of interstellar grain surface chemistry include reactions of migrating species (generally, atomic H, D, C, N, and O) among themselves and with trapped species (particularly O_2, O_3, CO, H_2CO).[6,59–61] Relevant routes are shown in Figs. 7 and 8. In these studies, reactions in the diffusion limit are considered where a single (mobile) species is scanning the surface for a coreactant.[b] In such an environment, reactions involving substantial activation barriers can still occur. This is particularly true for atomic H (and D) which is expected to efficiently tunnel through barriers

[b] But see Ref. 62 for a different scenario.

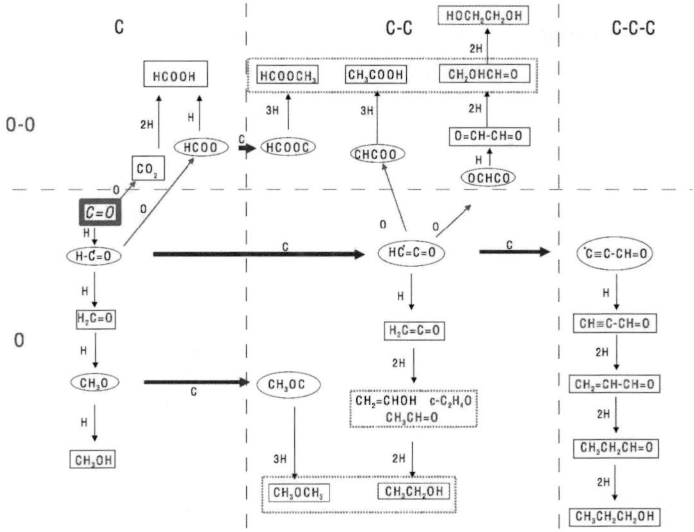

Fig. 8. Hydrogenation of CO enables a rich organic chemistry. Hydrogenation, oxidation, and carbon addition reactions are indicated with black (vertically), red (diagonally), and blue (horizontally) arrows, respectively. See text for details. Taken from Ref. 72 after Ref. 6. See also Colour Insert.

as high as 3400 K.[6] For water formation, reactions of H with O_2 and O_3 are then important (Fig. 7). Essentially, accreted atomic oxygen migrates in these models and will react to form molecular oxygen. Once a small amount of molecular oxygen is present, further accreted O will react to form O_3. Atomic H will then preferentially react with O_3, forming OH which reacts immediately with H_2 present on the surface to form H_2O. This reforms O_2, and the oxidation–hydrogenation cycle can repeat itself. The barrier for hydrogenation of O_2 is higher than that for O_3; nevertheless, once enough O_2 has built up, the hydrogen peroxide channel to water formation channel will also open up. Experimental determination of these activation barriers and diffusion rates is of fundamental importance.

A number of UHV surface chemistry experiments have been performed to study the formation of H_2O on interstellar ice analogs. While the results provide support for the chemical routes involved, direct comparison has to be done with care in view of the difference in surface concentrations of reactants (for instance, diffusion versus reaction limit).

As expected, these studies reveal efficient water formation from direct recombination of atomic H and O on porous water ice and other cryogenic matrices.[63,64] The hydrogenation steps in the peroxide route have been investigated in Refs. 65 and 66 and shown to form water readily. The first step in the ozone channel (the reaction of O with O_2) is known to occur readily in cryogenic matrices.[67–69] The hydrogenation step has not yet been investigated but has only a small barrier in the gas phase (470 K), and H-tunneling reactions should proceed readily.

In water formation, some of the oxygen is lost through reactions of CO forming CO_2 (see Fig. 7). Thermoluminescence studies provide experimental support that ground-state atomic oxygen [$O(^3P)$] can diffuse and react with ground-state CO ($X^1\Sigma^+$) in low-temperature matrices to form CO_2 ($^3B_2$).[55] Supposedly, the matrix enables this spin forbidden reaction to occur. The reaction of an isotopically labeled atomic oxygen beam with a H_2O–CO mixture has been studied under UHV conditions.[70] Inefficient (<2.5% of the atomic O) formation of CO_2 was observed after warm-up and ascribed to migration of the reactants through the porous structure; the remainder was probably lost to O_2 formation or sublimation. Efficient reaction of migrating O atoms with other species (for instance, O_2) than CO probably also limited CO_2 formation in Ar/CO/O_2 matrix experiments.[67] In view of the low efficiency of the reaction of O with CO, other routes to CO_2 on interstellar ice surfaces have been explored, specifically the reaction of OH with CO. The hydroxyl radical will be formed in an excited state on grain surfaces and likely migrate a few sites before its excess energy is dissipated. During this migration process, OH may encounter a nearby-trapped CO and react to form CO_2.[71]

3.1.4. *Formaldehyde/methanol reaction network*

The reaction of H with CO on interstellar grain surfaces truly opens up organic chemistry in interstellar space. In the gas phase of interstellar clouds, most of the carbon is locked up in CO which is very unreactive under interstellar conditions. A small amount (≈0.01) of carbon is broken out through UV photolysis and, in the process of reformation of CO, hydrocarbons are (temporarily) formed in the gas phase. Generally, these are acetylene derivates such as C_nH and $HC_{2n+1}N$. While their large dipole moment and their (relatively) small partition function make the pure rotational transitions of these linear species readily detectable in the gas phase, abundance-wise they are not particularly important organic reservoirs in molecular clouds. As for oxygen chemistry on ice surfaces, the importance of hydrogenation (of CO, this time) was realized early on,[6] based largely upon the expected low-energy barrier for the gas-phase three-body reaction ($E_a \approx 1000$ K) and supported by thermoluminescence studies on cryogenic matrices.[73] The relevant reaction network is shown in Fig. 8. The resulting formyl radical can readily react with subsequent accreted atoms such as H (leading to formaldehyde and methanol formation), O (producing ultimately formic acid), N (the formamide channel), and of course C (leading to ketene and ethanol). Experimentally, only the hydrogenation channel has been studied in detail in UHV set-ups and shown to be very efficient.[64,74–77] While the initial studies only observed efficient H_2CO formation, later studies[77,78] demonstrated that this reflected the adopted H flux and that, under interstellar conditions, full hydrogenation of CO to CH_3OH is expected to be efficient. At low temperatures, these reactions are assisted by tunneling[77] and this may be particularly relevant for interstellar conditions. Finally, it should be mentioned that *ab initio* quantum chemical calculations predict barriers for the H + CO reaction of 300–360 K, depending on the number of clustered H_2O molecules,[79] and this is somewhat less than inferred from the experiments (~500 K).[77] More disturbing is, though, that the barriers calculated for the reaction of H with H_2CO are substantially higher (~850–1000 K) than those for the reaction with CO, and this is contrary to the experiments where hydrogenation of H_2CO seems easier than that of CO. As an interesting aside, HCO, the doorway into this chemistry, was the first radical studied by IR spectroscopy in an inert matrix[80] and, with uncanny insight, the putative interstellar relevance was foreshadowed in this article.

3.1.5. *Deuterium chemical routes*

Reactions of CO with the deuterium isotope may provide additional information on the chemical routes involved as well as — through the large mass difference — probe the importance of tunneling of H atoms

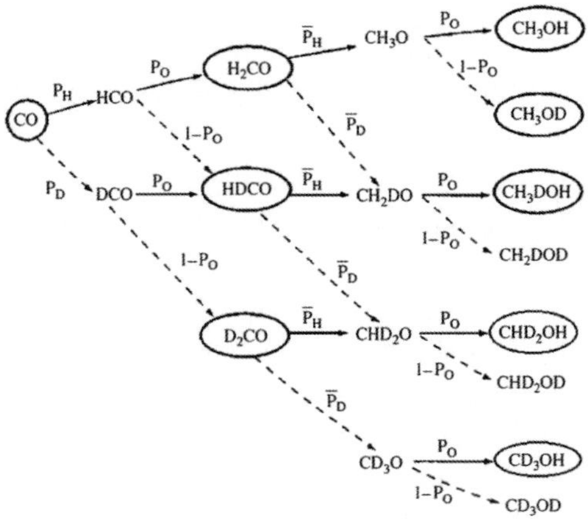

Fig. 9. The reactions involved in the formation of deuterated formaldehyde and methanol on grain surfaces. This is essentially the CO hydrogenation-branch in Figure 8. Species that have been observed in the warm gas around the low-mass protostar, IRAS 16293, are enclosed in an ellipse. Reactions of radicals with H and D have a probability set by the relative accretion rates, which are assumed in this figure to add up to 1. The reactions of H and D with CO and H_2CO possess activation barriers and their reaction probabilities will depend on the accretion rates, surface concentrations of (all) reactants, and the barriers involved. Taken from Ref. 85.

for grain-surface chemistry in molecular clouds. In addition, observationally, interstellar molecules often show extremely large deuterium fractionation effects. Typically, observed abundance ratios of a deuterated species compared to their hydrogenated counterpart are 0.1;[81] much higher than the elemental abundance ratio $D/H = 1.5 \times 10^{-5}$ in the interstellar medium. Indeed, doubly and even triply deuterated species have been detected at appreciable abundances (ND_3/NH_3 and $CD_3OH/CH_3OH \sim 10^{-3}$–$10^{-2}$).[82,83] This high deuterium fractionation reflects the small zero-point energy difference between deuterated species and their parent species (200–500 K) which, at the low temperatures of the interstellar medium (10 K), drives the reaction to high deuteration. Now, such high deuterium fractionations have also been observed in warm (100–300 K) gas near young protostars where the zero-point energy difference should have little influence. In this case, it is generally thought that the high deuterium fractionation reflects an earlier, much colder phase (before the new star formed), and this "heritage" has been preserved in ices. Once the new star forms, it will heat up its environment, sublimating the ices and releasing the stored deuterium fractionations into the gas phase where they will remain "observable" for a kinetic timescale of ~10^4 years.[7,84–86]

Theoretically, the important role of interstellar grain-surface chemistry in the deuteration of species such as CH_3OH, H_2CO, H_2O, and NH_3 has been long recognized.[87] Figure 9 shows the surface chemistry involved in the hydrogenation/deuteration of CO.[85] While there is a dependence on the reaction barriers involved, deuterium fractionation on grain surfaces is largely driven by the relative accretion probability of atomic H and D from the gas. The atomic D/H ratio in the gas phase is very much enhanced above the elemental abundance ratio because almost all the hydrogen nuclei are locked up in molecular hydrogen while, in contrast, the atomic D/HD ratio is high due to gas-phase ion– molecule reactions.

Experimentally, the importance of deuteration in the hydrogenation of CO on grain surfaces has been investigated by Refs. 88 and 89. These studies determined the reaction rate of deuteration of CO relative to the hydrogenation reaction (but in an indirect way) to be 0.08. This result has then implications for the potential energy curve involved.[89] It should be noted that in these studies most of the D (H) atoms are lost to recombination reactions forming D_2 (H_2). On interstellar grains, a single atom will be on the grain looking for an available reaction partner (CO, H_2CO, O_2, O_3, etc.) and loss through reaction with H is unimportant. Likewise, rather than the competition between H and D for CO, the key question is the reaction rate of D with CO relative to all other species it can react with (H_2CO, O_2, O_3, etc.), which depends on the activation barriers involved and the abundances; however, these other systems have not yet been investigated. Hence, this experimental data cannot be directly translated to interstellar conditions and the implications for interstellar chemistry have not yet been assessed.

These experimental studies also revealed another route toward deuterium fractionation of methanol: exchange reactions of D with various isotopomers of methanol. Exposure of methanol to beams of atomic D shows that isotope exchange on the methyl group is rapid.[90]

Substitution on the OH group, on the other hand, is not observed. Nor is the backward reaction with atomic H and deuterated methanol observed to occur under the experimental conditions. In this way, the deuterium fractionation can be driven beyond what the reaction scheme in Fig. 9 would predict. Finally, hydrogen/deuterium exchange has been studied in mixed H_2O/CD_3OD ices.[91–93] In this case, rapid exchange is observed on the hydroxyl group forming HDO and CD_3OH at temperatures above 120 K before crystallization of H_2O ice. Such H/D exchange has been studied for other hydrogen-bonding systems (H_2O/D_2O and NH_3/D_2O) and is related to proton transfer and ion defects.[94–96] Since heating and sublimation of ices near newly formed stars is a key ingredient in models for the warm gas in these environments, these reactions may be very relevant as well. Both these types of reactions can result in very different deuterium fractionation on the groups within one molecule or between molecules in the same ices. Observationally, differences in deuteration between the methyl and hydroxyl groups of methanol and between methanol and water have been observed around some protostars.

3.2. *Photochemistry of interstellar ices*

In this section, we focus on the photochemical processes that take place within the bulk of the ice. Astronomically, this is particularly important in regions of star and planet formation within clouds, and the boundaries of dense molecular clouds with the general diffuse interstellar medium where the UV-rich interstellar radiation field can penetrate. Although the new species produced by photochemistry within the ice represent only a few percent of the ice, energetic processing is important because it produces complex molecules and chemical groups that cannot be made via gas-phase and gas-grain reactions at the low temperatures and pressures of interstellar clouds. The *in situ* production of energetic reaction partners in the solid phase favors chemical complexity and diversity while the kinetics, radiation fields, and low densities of the gas phase and surface favor simplicity.

Interstellar ice composition in protostellar regions is particularly important since this is where new planets are also forming. The material surrounding these star-forming regions dominates the absorption

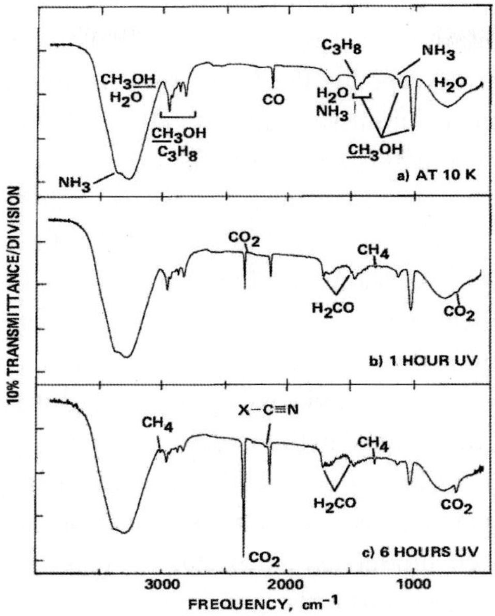

Fig. 10. Infrared spectra showing the photochemical evolution of the interstellar ice analog H_2O:CH_3OH:NH_3:CO:C_3H_8 (100:50:10:10:10) at 10 K. The spectra were taken before (a) and after 1 h (b) and 6 h (c) of UV irradiation. Note the ready formation of CO_2, H_2CO, CH_4, and OCN^- at the expense of CH_3OH. Figure adapted from Ref. 97.

spectrum because the gas and dust is orders of magnitude more concentrated here than it is along the rest of the line of sight. These are the regions in which the OCN^-, CO_2, and CH_3OH bands become important. The presence of methanol in these ices is of pivotal importance since it drives a rich interstellar ice photochemistry,[9,97,98] is important in ice segregation,[99] and plays an important role in gas-phase chemistry after evaporation near a newly formed star.[7,87] The discussion below will focus on the photochemistry of mixed molecular ices comprised of the most abundant interstellar ice species as described in Section 2. There is now a large body of work on ice photochemistry of various ice mixtures, many of astronomical interest,[10,100–102] but the kinetic parameters required to extrapolate the experiments to astrophysical conditions have not been studied in detail. This is an area in which much laboratory work is needed.

Figure 10 shows the spectroscopic evolution of an interstellar ice analog (H_2O:CH_3OH:CO:NH_3:C_3H_6 = 100:50:10:10:10) as a function of UV photolysis. Except for the C_3H_6, this analog reflects the major interstellar ice components associated with protostellar environments (see Table 1). The exposure to UV results in the destruction of several species (particularly methanol) and the creation of others. The simplest and most abundant species include the radical HCO, as well as the molecules H_2CO, CH_4, CO, and CO_2. As shown in Table 1, many of these new species have been identified in interstellar ices. As these species are produced at the expense of methanol,[97] lowering the CH_3OH concentration, all the major interstellar ice ingredients known toward massive protostellar environments are represented in this experiment and at roughly the correct concentrations. However, as described above, the detection of these ice species in molecular clouds does not imply that radiation processing is responsible for their production since most of these simple species can also be formed by gas-phase or gas-grain chemistry. Given that this mixture is an approximation of the interstellar ices and does not include the great overabundance of hydrogen available in molecular clouds, it is remarkable that UV processing does not alter the spectra more. Thus, while observations are consistent with energetic processing, there is no smoking gun. The strongest evidence that energetic processing is important, at least in some locations within dense clouds, is provided by the OCN^- feature, and perhaps by the ubiquity of CO_2. As described in Section 3.1, while CO_2 can be formed by reactions between CO and O or OH on the ice surface, it is also efficiently produced by UV photolysis in realistic mixed molecular ice experiments as illustrated in Fig. 10. The OCN^- feature in protostellar regions is difficult to explain by any of the more abundant species predicted by gas-phase and gas-grain chemical models. Since it is readily made by the radiative processing of laboratory ices containing C, N, and O, it has been considered strong evidence for radiative processing. However, it has recently been shown that this ion can be formed by thermal reactions occurring within the bulk of the ice, provided the necessary precursors are present and at the right concentrations. This is described further in Section 3.4.

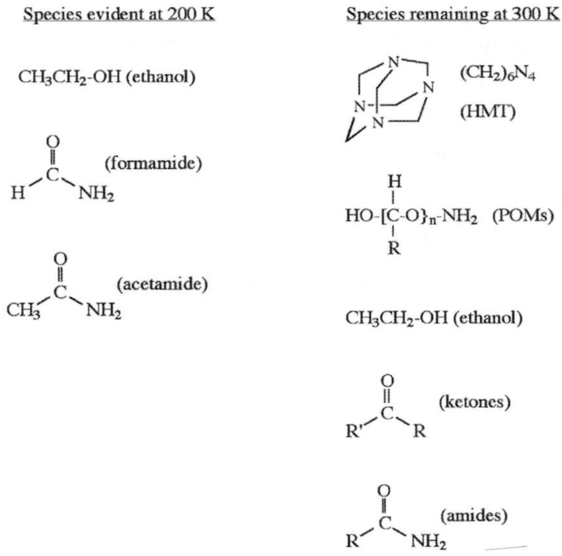

Fig. 11. Compounds produced by the 10-K UV photolysis and subsequent warm-up of the interstellar ice analogs H_2O:CH_3OH:NH_3:CO (100:50:10:10) and (100:50:1:1). HMT: hexamethylenetetramine; POM: polyoxymethylene. Figure courtesy of Dr. Max Bernstein.

Detailed descriptions of the UV-induced chemical evolution of interstellar ice analogs can be found in the literature.[10,97,100–102] The role of methanol in the ice photochemistry discussed here is very critical to the kinds of compounds produced. Figure 10 shows the *in situ* formation of the most abundant, simple photoproducts at 10 K. The full chemistry is, however, far more complex. Upon warm-up to about 200 K, the parent compounds and the most volatile ice components sublime, leaving a mixture of less volatile species on the substrate. Of the overwhelming array of compounds produced from even the simplest starting ice containing H_2O, CH_3OH, NH_3, and CO, only a few have been identified.

The structures of some species that have been identified are presented in Fig. 11. In keeping with their expected low concentration and abundance, clear-cut IR spectroscopic evidence for these types of compounds in interstellar ices is presently lacking although, as mentioned in Section 2.5, some of the weak spectral structure detected in the 2000–1250 cm^{-1} region is consistent with their presence.

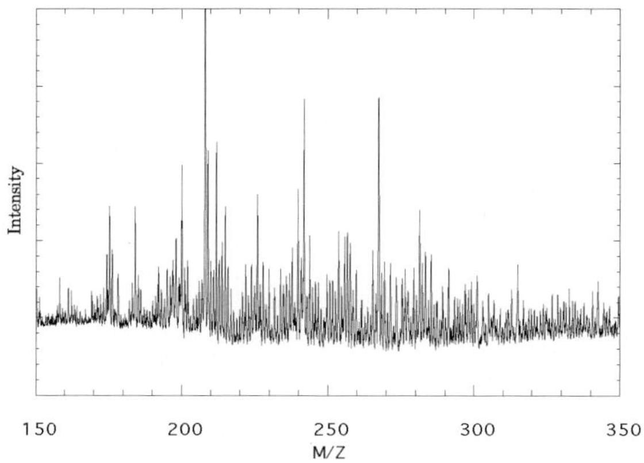

Fig. 12. The laser desorption (L2MS) mass spectrum of an interstellar ice analog residue produced by the photolysis of a H_2O:CH_3OH:NH_3:CO (100:50:1:1) ice. The complexity of the interstellar ice analog residue is clearly evident. Taken from Ref. 104. See also Ref. 105.

Unfortunately, spectral screening by the far more abundant, simpler ice species will likely represent an important, long-term obstacle to making firm identifications. Nonetheless, as shown above, IR spectroscopy can place strong constraints on the types of chemical compounds and subgroups that are present in interstellar ices. Higher quality astronomical spectra than those currently available are necessary to probe the species present at these lower concentrations.

While many of these new species sublime when the irradiated ice is warmed from 200 K to room temperature under vacuum, a thick, sometimes oily, residue of nonvolatile material remains. The residue that remains after the irradiated ice is warmed to room temperature is very important. As an intrinsic part of the inventory of interstellar material, it is plausible that this organic refractory material is associated with that preserved in comets, meteorites, and interplanetary dust particles (IDPs). Since these grow into planets and their satellites and continually bombard these newly formed objects, this type of material likely plays a role in steering their early chemistry.

Remarkably, this residue is rich in the cage molecule hexamethylenetetramine (HMT, $C_6H_{12}N_4$) and polyoxymethylenes (POMs).[9] This contrasts with the organic residues produced by

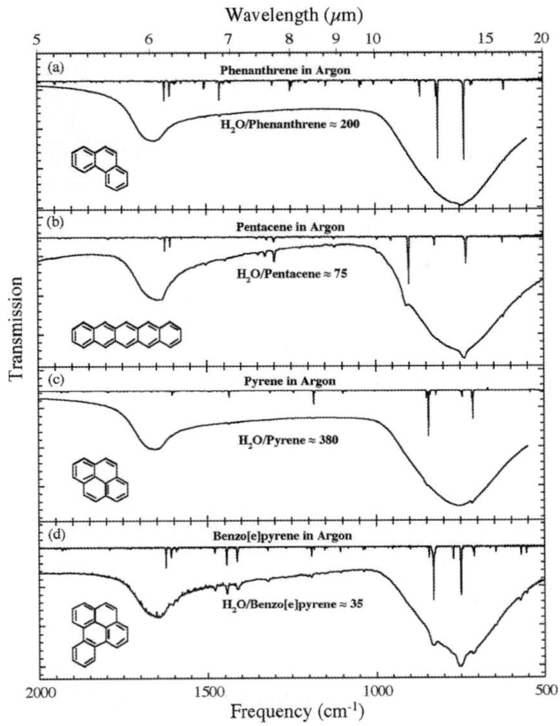

Fig. 13. The 2000–500 cm^{-1} (5.0–20.0 μm) IR spectra of H_2O/PAH ices at 15 K. The concentrations shown here span the range expected for ices in dense clouds. The very broad bands centered roughly at 1650 and 750 cm^{-1} are due to H_2O bending and libration modes. Each H_2O/PAH spectrum is compared to the corresponding spectrum in an Ar matrix. This shows that, at best, astronomical PAHs will contribute weak features to the ice spectra in dense clouds. Taken from Ref. 108.

irradiating mixed molecular ices which do not contain methanol[106] and those produced in thermally promoted polymerization-type reactions in nonirradiated realistic ice mixtures (see Section 3.4).[103,107] The synthesis of HMT by photolysis implies the presence of several other interesting intermediates in these ices as well. For example, a variety of species including nitriles and isonitriles, amino acids, and methylamines are readily formed by HMT hydrolysis, thermalization, or photolysis, all processes which can occur in comets, interstellar ice grains, and asteroids. As illustrated by the mass spectrum of the residue shown in Fig. 12, the

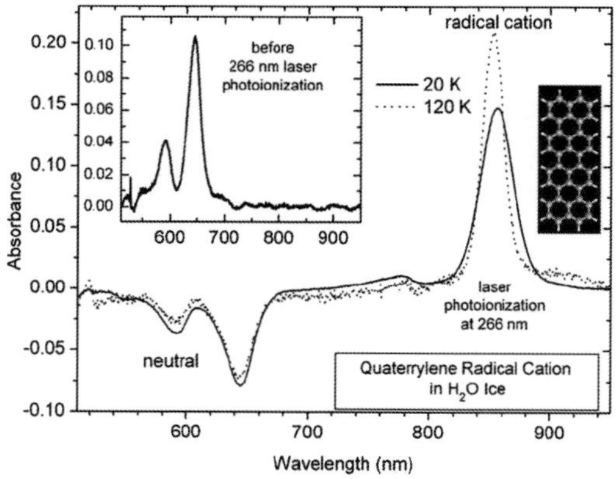

Fig. 14. The 520–920 nm spectrum of quaterrylene (QTR, $C_{40}H_{20}$, upper right insert) in a water ice matrix. The spectrum of the freshly deposited QTR/H_2O (1/500) ice at 20 K is shown in the upper left insert. The full spectrum (solid line) is that of the same ice after photolysis with 266-nm laser light. Negative bands represent depletion of the parent neutral QTR and positive bands represent generation of QTR^+. The dotted line spectrum was recorded at 120 K, after step-wise warm-up over 28 days. The QTR^+ band areas are equal, showing QTR^+ is stable in water ice at this temperature. Taken from Ref. 114.

nonvolatile residue of photolyzed ices containing methanol contains a bewildering array of organic compounds.

3.3. *PAH ice photochemistry*

The mid-IR signature of individual PAHs in the gas phase is common throughout the cosmos.[109–111] In dense clouds, they should condense out of the gas and become part of the mixed molecular ices at concentrations of roughly a tenth of a percent. While these condensed PAHs should contribute weak absorption bands to the spectra of ices in dense molecular clouds, as illustrated in Fig. 13, they will suffer severe blending with the strong absorption bands of the major ice components. Compare the relative strength of the PAH/H_2O ice bands between 2000 and 500 cm^{-1} region in Fig. 13 with the unidentified spectral structure in the interstellar ice spectra shown in Fig. 5.

Mass spectroscopic studies of the complex residues made by the photolysis of interstellar ice analogs containing PAHs showed that

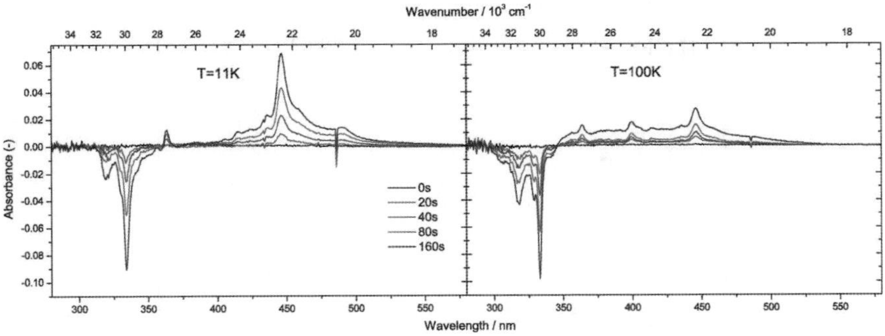

Fig. 15. The VUV-induced spectroscopic changes in pyrene: H_2O ice for two different temperatures as a function of photolysis time. Comparing the spectra from the 11-K ice (left) with those of the 100-K ice (right) shows the critical role that temperature plays in determining photochemical pathways in a PAH-containing ice. The loss of the neutral pyrene (Py, $C_{14}H_{12}$) band complex near 330 nm is similar at both temperatures. However, in the 11-K ice, cation formation (~ 450 nm) is favored over the species responsible for the broad, 350–500 nm absorption and the 400 and 405 nm band carriers formed at higher temperatures. The opposite holds for the 100-K ice. Taken from Ref. 116. See also Colour Insert.

important aromatic species were produced.[112] However, Figure 13 shows that it is not possible to follow PAH/H_2O ice photochemistry via IR spectroscopy as was the case for the mixed molecular ices described in Section 3.2. Optical spectroscopy overcomes these limitations and has revealed that photochemistry of interstellar ice analogs containing PAHs is both important and remarkable.

The photochemistry of naphthalene, pyrene, 4-methylpyrene, and quatterylene in water ice at low temperatures shows that they readily convert to their radical cation (PAH^+) form upon exposure to radiation.[113–116] In contrast to the photolysis behavior common in rare-gas matrices, where at most 15% of the neutral PAH molecules are photoionized (e.g., Ref. 117), over 70% of the neutral PAH molecules are ionized in H_2O ice. Furthermore, these PAH cations remain to remarkably high H_2O matrix temperatures. This is illustrated in Fig. 14 which compares the spectrum of the quaterrylene (QTR, $C_{40}H_{20}$) cation generated in water ice at 20 K to that of the same sample at 120 K after one month, slow, step-wise warm-up. The QTR^+ band diminishes as the ice matrix is warmed to higher temperatures, but the loss is slow up to ~140 K. Although the loss increases at higher temperatures, band

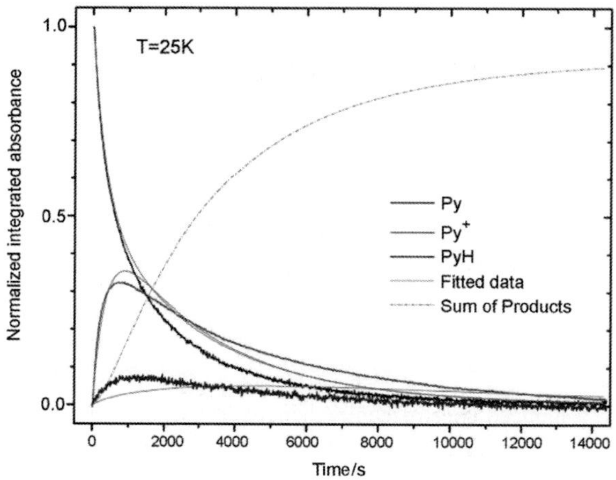

Fig. 16. Behavior of the Py 334 nm, Py^+ 445 nm, and PyH 400 nm bands as a function of VUV irradiation in Py:H_2O ices at 25 K. Taken from Ref. 116. The curves labeled fits are produced by the chemical network described in Ref. 116. See also Colour Insert.

reduction stops when warm-up stops. Remarkably, the band is still evident at 170 K, the temperature at which the ice sublimes.[114] Exploratory studies on QTR also suggested that PAH ionization energy is lowered by at least 2 eV when they are in water ice,[113] a suggestion supported by theoretical studies on several different PAH/H_2O ices.[119]

These initial studies showed that PAHs likely play important but over-looked roles in cosmic ices. A far more detailed understanding of the processes involved is necessary before one can incorporate these ideas into astrochemical models. To provide this critical information, Ref. 116 has started a systematic study of several PAHs in H_2O ice, tracking the *in situ* behavior of each PAH and their photoproducts as a function of ice temperature and time with subsecond responsivity. Figure 15 shows the dramatic effect ice temperature has on the photochemistry of the PAH pyrene (Py, $C_{14}H_{12}$) in water ice, and Figure 16 shows the evolution of each band as a function of photolysis time. The ability to measure spectra simultaneously with photolysis and with subsecond time resolution permits kinetic studies previously inaccessible and provides new insights into the processes occurring within the ice during photolysis. Since these temperatures span the range of dust in dense clouds

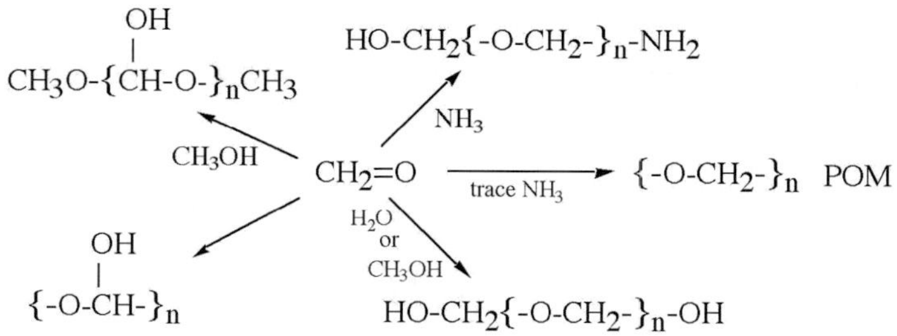

Fig. 17. Various polyoxymethylenes produced by the simple thermally promoted polymerization reactions of unphotolyzed interstellar ice analogs. Any mixed molecular ice containing H_2CO and trace amounts of NH_3 produces residues comprised primarily of these polymers. The scheme is taken from Refs. 103 and 107.

in the transition zones with star- and planet-forming regions and the ranges in the temperatures of icy Solar System bodies beyond the orbit of Jupiter, much experimental work is needed on PAH photochemistry in "reactive matrices" before the fascinating processes at play – both in the laboratory and across the cosmos – can be fully understood.

3.4. *Thermal processing of interstellar ices*

Warming unirradiated condensed mixed molecular ices can also induce physical and chemical change. Studies of the physical effects on ice properties induced by temperature variations form a vast literature that is well beyond the scope of this chapter. However, although the chemical processes promoted by simply warming ices comprised of likely interstellar species have been little studied, these are also likely to be astronomically important. To the best of our knowledge two systems have been studied in some detail, one involving mixed molecular ices containing formaldehyde (H_2CO) and the other focusing on the production of OCN^- from neutral precursors.

Interestingly, some of the complex organic materials produced by irradiation and subsequent warm-up of CH_3OH-containing ices as described above are significantly different from the complex organic materials produced by the 10–150 K warm-ups of unirradiated mixtures containing H_2CO, the most abundant species produced by the photolysis of CH_3OH-containing ices. Thermally promoted reactions were studied in cryogenic ices containing H_2CO, H_2O, CH_3OH, CO, and NH_3.[103,107] Small traces of NH_3 ($NH_3/H_2CO > 0.005$) were found sufficient to convert significant fractions of H_2CO (>40%) into complex organics. However, H_2CO reactions do not proceed without NH_3. Reactions start at temperatures between 40 and 80 K, depending on the ice. While the products are all derivatives of polyoxymethylene (POM), the polymer of H_2CO, POM itself was not made in significant amounts. Thermally promoted reactions in ices containing H_2CO tend to form polymeric structures (Fig. 17) instead of the well-defined refractory compounds which seem to be characteristic of the photochemical pathway shown in Fig. 11. Quantum chemical calculations of the reactions possible in H_2CO- and NH_3-containing ices constrain the possible pathways and suggest interesting new species including tertiary amines.[118]

The second area in which thermally promoted ice reactions of astronomical interest have been investigated focus on OCN^-. This is particularly important since the interstellar 4.62-μm OCN^- band has been taken as evidence for interstellar ice photoprocessing (see Sections 2.3 and 3.2). References 120 and 121 performed thorough and detailed quantum chemical calculations on these systems showing that OCN^- charge-transfer complexes were formed from precursor components HNCO and HOCN with either NH_3 or H_2O. Ammonia and water are common in interstellar ices, while the other two reactants may be formed through grain-surface reactions of the formyl radical with atomic nitrogen followed by isomerization. Thus, it seems plausible that OCN^- can also form in interstellar ices without requiring UV photons. Given the importance of this issue of ice processing, a better understanding of ionization and charge-transfer processes within water-rich ices is another area in which low-temperature ice chemical studies can provide key insight.

4. Astrobiology

Due to the extreme complexity and analytical challenge posed by deeper chemical analysis, and encouraged by recent interest in the application of these results to early earth biochemistry, effort has been redirected from

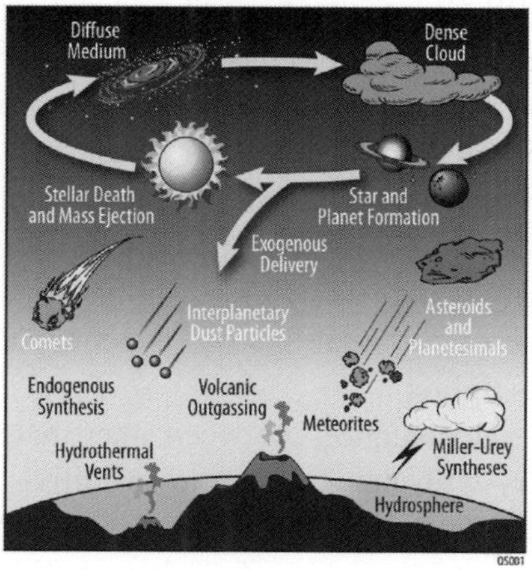

Fig. 18. The cycle of interstellar material leading to the production and evolution of complex organic compounds that can be delivered intact to, and mixed with, those produced on the newly formed planets and moons. There they participate in the primordial chemistry on these young objects. Figure courtesy of Jason Dworkin, NASA Goddard Space Flight Center, taken from Ref. 123. See also Colour Insert.

solely establishing the chemical inventory of species produced in these interstellar–precometary ice analogs to searching for the presence of specific biogenically important species.

Figure 12 shows the laser desorption mass spectrum (L2MS) of the residue from a photolyzed H_2O:CH_3OH:NH_3:CO (100:50:1:1) ice. Mass spectra such as this provide further critical insight into the nature of the residue described in Section 3.2, showing that there are hundreds of compounds produced and the majority are far more complex than the starting materials. The new materials produced are responsible for the envelope spanning the mass range from about 100–350 m/z. Given that none of the simple starting materials of H_2O:CH_3OH:NH_3:CO has a mass greater than 32, nor do they contain a single C–C bond, the complexity and extent of the photoproducts is staggering. Many of these compounds have molecular masses up to ten times larger than that of any of the starting materials. When this mass spectrum is compared with those of two IDPs there is an intriguing similarity between the mass envelope of the interstellar ice analog residue shown in Fig. 12 and the mass envelopes in the IDP spectra.[104] In addition, the high performance liquid chromatography (HPLC) of these residues is similar to that of the carbonaceous material in the Murchison meteorite.[122] This resemblance between the laboratory ice residue and the extraterrestrial organics brought to Earth in IDPs and meteorites raises the question, "Do the compounds in IDPs and meteorites have an interstellar ice-cometary heritage?" Since many IDPs are thought to originate in comets, and comets are often thought to represent "pristine samples" of interstellar material preserved in the frozen outskirts of our Solar System, the connection with interstellar cloud chemistry gets tighter.

Investigating the connection between the materials produced by cosmic ice chemistry with that brought to primitive bodies in the Solar System by IDPs, comets, and meteorites is now an important component of meteoritic and planetary science. As schematically shown in Fig. 18, primordial planets are seeded with these carbonaceous compounds with an interstellar ice heritage. There they take part in the budding chemistry on these young worlds.

Acknowledgments

Over the years, we have been blessed to work with many great students, postdoctorates, and colleagues who were always willing to share their latest insights and discoveries of the universe with us. We particularly acknowledge Adwin Boogert and Jacqueline Keane. LJA very gratefully acknowledges sustained support from NASA's Laboratory Astrophysics and Astrobiology Programs.

References

1. See for up-to-date information on exoplanets: http://exoplanets.org/
2. See http://kepler.nasa.gov/
3. F. H. Shu, F. C. Adams and S. Lizano, *Annu. Rev. Astron. Astrophys.* **25**, 23 (1987).
4. C. P. Dullemond, D. Hollenbach, I. Kamp and P. D'Alessio, In: B. Reipurth, D. Jewitt and K. Keil (editors), *Protostars and Planets V*, The University of Arizona Press is in Tucson, Arizona, The United States. (2007), p. 555.
5. A. G. G. M. Tielens, *The Physics and Chemistry of the Interstellar Medium.* Cambridge University Press, Cambridge (2005).
6. A. G. G. M. Tielens and W. Hagen, *Astron. Astrophys.* **114**, 245 (1982).

7. S. B. Charnley, A. G. G. M. Tielens and T. J. Millar, *Astrophys. J.* **399**, L71 (1992).
8. W. A. Schutte and R. K. Khanna, *Astron. Astrophys.* ***398***, 1049 (2003).
9. M. P. Bernstein, S. A. Sandford, L. J. Allamandola, S. Chang and M. A. Scharberg, *Astrophys. J.* **454**, 327 (1995).
10. J. M. Greenberg, A. Li, C. X. Mendoza-Gomez, W. A. Schutte, P. A. Gerakines, and M. de Groot, *Astrophys. J.* **455**, L177 (1995).
11. E. L. Gibb *et al.*, *Astrophys. J.* **536**, 347 (2000).
12. E. A. Bergin, G. J. Melnick, P. A. Gerakines, D. A. Neufeld and D. C. B. Whittet, *Astrophys. J.* **627**, L33 (2005).
13. C. Knez *et al.*, *Astrophys. J.* **635**, L145 (2005).
14. A. C. A. Boogert, A. G. G. M. Tielens, C. Ceccarelli, A. M. S. Boonman, E. F. van Dishoeck, J. V. Keane, D. C. B. Whittet and T. de Graauw, *Astron. Astrophys*. **360**, 683 (2000).
15. A. C. A. Boogert *et al.*, *Astrophys. J.* **678**, 985 (2008).
16. K. M. Pontoppidan *et al.*, *Astrophys. J.* **678,** 1005 (2008).
17. E. L. Gibb, D. C. B. Whittet, A. C. A. Boogert and A. G. G. M. Tielens, *Astrophys. J. Suppl. Ser.* **151**, 35 (2004).
18. K. M. Pontoppidan *et al.*, *Astron. Astrophys.* **408**, 981 (2003).
19. D. Bockel´ee-Morvan *et al.*, *Astron. Astrophys.* **353**, 1101 (2000).
20. R. L. Snell *et al.*, *Astrophys. J.* **539**, L101 (2000).
21. A. G. G. M. Tielens, A. T. Tokunaga, T. R. Geballe and F. Baas, *Astrophys. J.* **381**, 181 (1991).
22. J. E. Chiar, P. A. Gerakines, D. C. B. Whittet, Y. J. Pendleton, A. G. G. M. Tielens, A. J. Adamson and A. C. A. Boogert, *Astrophys. J.* **498**, 716 (1998).
23. 23. S. A. Sandford, L. J. Allamandola, A. G. G. M. Tielens and G. J. Valero, *Astrophys. J.* **329**, 498 (1988).
24. M. E. Palumbo and G. Strazzulla, *Astron. Astrophys.* **269**, 568 (1993).
25. P. Ehrenfreund, A. C. A. Boogert, P. A. Gerakines, D. J. Jansen, W. A. Schutte, A. G. G. M. Tielens and E. F. van Dishoeck, *Astron. Astrophys*. **315**, L341 (1996).
26. H. J. Fraser, S. E. Bisschop, K. M. Pontoppidan, A. G. G. M. Tielens and E. F. van Dishoeck, *Mon. Not. R. Astron. Soc.* **356**, 1283 (2005).
27. W. Hagen, L. J. Allamandola and J. M. Greenberg, *Astron. Astrophys.* **86**, L3 (1980).
28. R. J. A. Grim and J. M. Greenberg, *Astrophys. J.* **321**, L91 (1987).
29. K. Demyk, E. Dartois, L. D'Hendecourt, M. Jourdain de Muizon, A. M. Heras and M. Breitfellner, *Astron. Astrophys.* **339**, 553 (1998).
30. P. A. Gerakines *et al.*, *Astrophys. J.* **522**, 357 (1999).
31. E. Dartois, K. Demyk, L. d'Hendecourt and P. Ehrenfreund, *Astron. Astrophys*. **351**, 1066 (1999).
32. S. A. Sandford and L. J. Allamandola, *Astrophys. J.* **355,** 357 (1990).
33. T. de Graauw *et al.*, *Astron. Astrophys*. **315**, L345 (1996).
34. L. D'Hendecourt *et al.*, *Astron. Astrophys*. **315**, L365 (1996).

35. A. C. A. Boogert *et al.*, *Astron. Astrophys.* **353**, 349 (2000).
36. G. Zasowski, F. Kemper, D. M. Watson, E. Furlan, C. J. Bohac, C. Hull and J. D. Green, *Astrophys. J.* **694**, 459 (2009).
37. J. V. Keane, A. G. G. M. Tielens, A. C. A. Boogert, W. A. Schutte and D. C. B. Whittet, *Astron. Astrophys.* **376**, 254 (2001).
38. P. Ehrenfreund, E. Dartois, K. Demyk and L. D'Hendecourt, *Astron. Astrophys.* **339**, L17 (1998).
39. S. T. Lee, T. S. Reighart, S. V. Olesik, *Fluid Phase Equilib.* **122**, 223 (1996).
40. J. Robinson, *The History of Winemaking*. Oxford University Press (2006).
41. A. G. G. M. Tielens, L. J. Allamandola, J. Bregman, J. Goebel, F. C. Witteborn and L. B. D'Hendecourt, *Astrophys. J.* **287**, 697 (1984).
42. J. V. Keane, A. C. A. Boogert, A. G. G. M. Tielens, P. Ehrenfreund and W. A. Schutte, *Astron. Astrophys.* **375**, L43 (2001).
43. L. Hornekær, *et al.*, *Science* **302**, 1943 (2003).
44. L. Hornekær *et al.*, *J. Chem Phys.* **122,** 124701 (2005).
45. H. B. Perets *et al.*, *Astrophys. J.* **627**, 850 (2005).
46. E. Matar *et al.*, *Astron. Astrophys.* **492**, L17 (2008).
47. P. Jenniskens and D. F. Blake, *Science* **265**, 753 (1994).
48. F. Dulieu, L. Amiaud, J. H. Fillion, E. Matar, A. Momeni, V. Pirronello and J. L. Lemaire, arXiv:0911.2082v1 (2009).
49. A. Al-Halabi and E. F. van Dishoeck, *Mon. Not. R. Astron. Soc.* **382**, 1648 (2007).
50. K. Vaskonen, J. Eloranta, T. Kiljunen, H. Kunttu, *J. Chem. Phys.* **110**, 2122 (1999).
51. G. B. Gumenchuk. *PhD Thesis*, Technical University Munich (2007).
52. A. Schrimpf *et al.*, *J. Phys. Cond. Matter* **8**, 3677 (1996).
53. A. V. Danilychev and V. A. Apkarian, *J. Chem. Phys.* **99**, 8617 (1993).
54. B. J. Murray. *PhD Thesis,* University of East Anglia (2003).
55. J. Fournier, J. Deson, C. Vermeil and G. C. Pimentel, *J. Chem. Phys.* **70**, 5726 (1979).
56. V. A. Apkarian and N. Schwentner, *Chem. Rev.* **99**, 1481 (1999).
57. T. R. Govers, L. Matera and G. Scoles, *J. Chem. Phys.* **72**, 5446 (1980).
58. A. Schutte *et al.*, *J. Chem. Phys.* **64**, 4135 (1976).
59. T. I. Hasegawa and E. Herbst, *Mon. Not. R. Astron. Soc.* **263**, 589 (1993).
60. T. Stantcheva and E. Herbst, *Astron. Astrophys.* **423**, 241 (2004).
61. O. M. Shalabiea, P. Caselli and E. Herbst, *Astrophys. J.* **502**, 652 (1998).
62. R. T. Garrod and E. Herbst, *Astron. Astrophys.* **457**, 927 (2006).
63. F. Dulieu, L. Amiaud, E. Congiu, J. Fillion, E. Matar, A. Momeni, V. Pirronello and J. L. Lemaire, arXiv:0903.3120 (2009).
64. K. Hiraoka *et al.*, *Astrophys. J.* **498**, 710 (1998).
65. S. Ioppolo, *Astrophys. J.* **686**, 1474 (2008).
66. N. Miyauchi *et al.*, *Chem. Phys. Lett.* **456**, 27 (2008).
67. R. J. A. Grim and L. B. D'Hendecourt, *Astron. Astrophys.* **167**, 161 (1986).

68. T. Kiviniemi, M. Pettersson, L. Khriachtchev, M. Räsänen and N. Runeberg, *J. Chem. Phys.* **121**, 1839 (2004).
69. A. V. Benderskii and C. A. Wight, *J. Chem. Phys.* **104**, 85 (1996).
70. J. E. Roser, G. Vidali, G. Manicò and V. Pirronello, *Astrophys. J.* **555**, L61 (2001).
71. J. Keane. *PhD Thesis*, University of Groningen, 2002.
72. M. A. Requena-Torres, J. Martin-Pintado, Martin, S., M. R. Morris, *Astrophys. J.* **672,** 352 (2008).
73. L. J. van IJzendoorn, L. J. Allamandola, F. Baas and J. M. Greenberg, *J. Chem. Phys.* **78**, 7019 (1983).
74. N. Watanabe *et al.*, *Plan. Space Sci.* **54**, 1107 (2006).
75. N. Watanabe, T. Shiraki and A. Kouchi, *Astrophys. J.* **588**, L121 (2003).
76. N. Watanabe, A. Nagaoka, T. Shiraki, A. Kouchi, *Astrophys. J.* **616**, 638 (2004).
77. G. W. Fuchs *et al.*, arXiv:0906.2292 (2009).
78. H. Hidaka *et al.*, *Astrophys. J.* **614**, 1124 (2004).
79. D. E. Woon, *Astrophys. J.* **569**, 541, (2002).
80. C. N. Huggins and G. C. Pimentel, *J. Chem. Phys.* **60**, 1615 (1956).
81. T. J. Millar, *Space Sci. Rev.* **106**, 73, (2003).
82. F. F. S. van der Tak, P. Schilke, H. S. P. Muller, D. C. Lis, T. G. Phillips, M. Gerin and E. Roueff, *Astron. Astrophys.* **388**, L53 (2002).
83. B. Parise, A. Castets, E. Herbst, E. Caux, C. Ceccarelli, I. Mukhopadhyay and A. G. G. M. Tielens, *Astron. Astrophys.* **416**, 159 (2004).
84. S. B. Charnley, A. G. G. M. Tielens and S. D. Rodgers, *Astrophys. J.* **482**, L203 (1997).
85. P. Caselli, T. Stantcheva, O. Shalabiea, V. I. Shematovich and E. Herbst, *Plan. Space Sci.* **50**, 1257 (2002).
86. P. Caselli, T. I. Hasegawa and E. Herbst, *Astrophys. J.* **408**, 548 (1993).
87. A. G. G. M. Tielens, *Astron. Astrophys***. 119**, 177 (1983).
88. N. Watanabe, H. Hidaka, A. Nagaoka and A. Kouchi, In: J. L. Lemaire and F. Combes (editors), *Molecules in Space and Laboratory.* (2007), p. 52.
89. H. Hidaka, A. Kouchi and N. Watanabe, *J. Chem. Phys.* **126**, 204707 (2007).
90. A. Nagaoka, N. Watanabe and A. Kouchi, *J. Phys. Chem.* **A111**, 3016 (2007).
91. A. Ratajczak, E. Quirico, A. Faure, B. Schnitt and C. Ceccarelli, *Astron. Astrophys.* **496**, L21 (2009).
92. R. Souda, H. Kawanowa, M. Kondo and Y. Gotoh, *J. Chem. Phys.* **119**, 6194 (2003).
93. R. Souda, *Phys. Rev. Lett.* **93**, 235502 (2004).
94. J. E. Bertie and J. P. Devlin, *J. Chem. Phys.* **78**, 6203 (1983).
95. W. B. Collier, G. Ritzhaupt and J. P. Devlin, *J. Phys. Chem.* **88**, 363 (1984).
96. P. J. Wooldridge and J. P. Devlin, *J. Chem. Phys.* **88**, 3086 (1988).
97. L. J. Allamandola, S. A. Sandford and G. J. Valero, *Icarus* **76,** 225 (1988)
98. W. A. Schutte and P. A. Gerakines, *Plan. Space Sci.* **43**, 1253 (1995).

99. D. Blake, L. J. Allamandola, S. Sandford, D. Hudgins and F. Freund, *Science* **254**, 548 (1991).
100. P. A. Gerakines, W. A. Schutte and P. Ehrenfreund, *Astron. Astrophys*. **312**, 289 (1996).
101. P. A. Gerakines, M. H. Moore and R. L. Hudson, *J. Geophys. Res*. **106**, 33381 (2001).
102. W. A. Schutte, L. J. Allamandola and S. A. Sandford, *Adv. Space Res*. **12**, 47 (1992).
103. W. A. Schutte, L. J. Allamandola and S. A. Sandford, *Icarus* **104**, 118 (1993).
104. J. P. Dworkin, J. Seb Gillette, M. P. Bernstein, S. A. Sandford, L. J. Allamandola, J. E. Elsila, D. Ryan McGlothlin and R. N. Zare, *Adv. Space Res*. **33**, 67 (2004).
105. J. M. Greenberg *et al., Astrophys. J.* **531**, L71 (2000).
106. R. Briggs, G. Ertem, J. P. Ferris, J. M. Greenberg, P. J. McCain, C. X. Mendoza-Gomez and W. Schutte, *Orig. Life Evol. Biosph*. **22**, 287 (1992).
107. W. A. Schutte, L. J. Allamandola and S. A. Sandford, *Science* **259**, 1143 (1993).
108. M. P. Bernstein, S. A. Sandford and L. J. Allamandola, *Astrophys. J. Suppl. Ser*. **161**, 53 (2005).
109. J. D. T. Smith *et al*., *Astrophys. J.* **656**, 770 (2007).
110. B. T. Draine *et al*., *Astrophys. J.* **663**, 866 (2007).
111. A. G. G. M. Tielens, *Annu. Rev. Astron. Astrophys*. **46**, 289 (2008).
112. S. F. M. Ashbourn, J. E. Elsila, J. P. Dworkin, M. P. Bernstein, S. A. Sandford and L. J. Allamandola, *Meteor. Plan. Sci.* **42**, 2035 (2007).
113. M. S. Gudipati and L. J Allamandola, *Astrophys. J.* **615**, L177 (2004).
114. M. S. Gudipati and L. J. Allamandola, *Astrophys. J.* **638**, 286 (2006).
115. M. S. Gudipati and L. J. Allamandola, *Astrophys. J.* **596**, L195 (2003).
116. J. Bouwman, D. M. Paardekooper, H. M. Cuppen, H. Linnartz and L. J. Allamandola, *Astrophys. J.* **700**, 56 (2009).
117. D. M. Hudgins, C. M. Bauschlicher, Jr., L. J. Allamandola and J. C. Fetzer, *J. Phys. Chem.* **A104**, 3655 (2000).
118. D. E. Woon, *J. Phys. Chem.* **A105**, 9478 (2001).
119. D. E. Woon and J.-Y. Park, *Astrophys. J.* **607**, 342 (2004).
120. J.-Y. Park and D. E. Woon, *Astrophys. J.* **601**, L63 (2004).
121. D. E. Woon and J.-Y. Park, *J. Phys. Chem.* **A108**, 6589 (2004).
122. J. P. Dworkin, D. W. Deamer, S. A. Sandford and L. J. Allamandola, *Proc. Nat. Acad. Sci.* **98**, 815 (2001).
123. D. Deamer, J. P. Dworkin, S. A. Sandford, M. P. Bernstein and L. J. Allamandola, *Astrobiology* **2**, 371 (2002).

CHAPTER 12

HIGH-RESOLUTION SINGLE-MOLECULE SPECTROSCOPY IN CONDENSED MATTER

Michel Orrit

Leiden Institute of Physics, Huygens Laboratory, Postbus 9504
NL-2300 RA, Leiden, The Netherlands
E-mail: orrit@molphys.leidenuniv.nl

William E. Moerner

Departments of Chemistry and Applied Physics, Stanford University
Mail Code 5080, Stanford, California, 94305 USA
E-mail: wmoerner@stanford.edu

Optical spectroscopy of single impurity molecules in low-temperature solids provides an exquisitely sensitive probe of the structure and dynamics of the specific local environment around the single molecule, unobscured by ensemble averaging. Effects such as spectral diffusion, perturbations by external fields, changes in molecular photophysics, optical switching, dynamics due to amorphous system physics, magnetic resonance of a single molecular spin, and perturbations due to nearby charge transport have all been observed. Because properly chosen single molecules act as two- and three-level quantum systems, a variety of quantum optical effects may be observed, such as photon antibunching, optical Stark effect, and single-photon emission on command. When they carry spin in their ground and excited states, single molecules and other nano-objects open many more experiments in quantum optics and nanophysics, because they are sensitive to magnetic interactions and can serve as readouts for spin systems with long coherence times. As a new class of nanoscale experiments, the ability to optically pump and probe single 2–3-nm-sized molecules in complex systems provides a new window into the local dynamics and heterogeneity of solids.

1. Introduction

Over the past two decades, the power of optical spectroscopy with high-resolution laser sources has been extended into the fascinating domain of individual impurity molecules or defects in solids, where the single molecule (defect) acts as a probe of the immediate local environment with unprecedented sensitivity.[1–5] In these experiments, exactly one molecule hidden deep within a solid sample is probed at a time by selecting the molecule's resonance with tunable laser radiation, which represents detection and spectroscopy of 1.66×10^{-24} moles of material, or 1.66 yoctomoles.[a] Such single-molecule measurements completely remove the normal ensemble averaging that occurs when a large number of molecules are probed at the same time. Thus, the usual assumption that all molecules contributing to the ensemble average are identical can now be directly examined, and strong tests of truly microscopic theories can be completed.

Single-molecule spectroscopy (SMS) in solids exemplifies the push throughout various fields of science to study and understand "all things nano," yet is distinct from the well-established field of spectroscopy of single electrons, atoms, or ions confined in electromagnetic traps.[6–8] The vacuum environment and confining fields of an electromagnetic trap are quite different from the environment of a single molecule in a solid where the lattice constrains the molecule, preventing rotation in most cases, and where the molecule is bathed in the phonon vibrations of the solid available at a given temperature. In comparison to near-field probing of single atoms on surfaces with scanning tunneling microscopy[9,10] or atomic force microscopy,[11] which usually requires a strong bond between the molecule of interest and the underlying surface as well as tolerance of the perturbing forces from the tunneling electrons or the tip, SMS usually operates noninvasively in the optical far-field with a corresponding loss in spatial resolution, but with no loss of spectral resolution. To increase spatial resolution, single-molecule

[a] This fact has prompted the following science pun: Since a single molecule is the smallest unit of a molecular material, a more appropriate unit in this case would be the "guacamole," which is the quantity of moles exactly equal to the inverse of avocado's number (with apologies to Amadeo Avogadro).

imaging has been achieved with near-field optical techniques with 100-nm resolution at room temperature,[12–14] but the spectral resolution which is available under such conditions is three to five orders of magnitude poorer than the high-resolution SMS described here. Superresolution imaging of single molecules beyond the optical diffraction limit was first achieved using spectral selection and point-spread function fitting[15] before the recent developments based on photoswitching at room temperature.[16] Of course, room-temperature single-molecule studies have blossomed for a variety of applications in materials science and biology,[17–21] but these areas are not the focus of this chapter.

This chapter focuses on the optical probing of single molecules, atoms, defect centers, or ions in complex condensed matter environments, but for convenience we will use the term "molecule" to refer to all of these, unless a specific system is under discussion. SMS probes a regime which presents interesting detection challenges as well as exciting opportunities for discovery. The challenges generally derive from the limitations posed by diffraction effects and the difficulty of detecting one single entity in the interfering background arising from the host crystal, polymer, or protein, and methods to overcome these obstacles are described below. The opportunities result from the wealth of new information obtainable regarding the interactions of the single entity with its native environment, in a fashion which is unobscured by the ensemble averaging that plagues conventional experiments.

Why is it useful to study unique, individual molecules in a complex condensed matter host, each of which may be marching to a different drummer? Clearly, standard ensemble measurements which yield the average value of a parameter for a large number of (presumably identical) copies of the system of interest still have great value. But when information is extracted from individual molecules, one by one, a frequency histogram can be constructed of the actual distribution of values for an experimental parameter, rather than just the average (mean) value of the distribution. The shape of the full distribution can be examined to see if there are multiple peaks, or a strongly skewed shape. These details of the underlying distribution become crucially important when the system under study is inhomogeneous as in many condensed matter hosts, where heterogeneity arises from the exact constellation of functional groups, atoms, ions, electrostatic charges, and/or other sources of local fields in the immediate vicinity of each molecule. Similarly, for biomolecules, heterogeneity easily arises if the various individual copies of a protein or oligonucleotide are in different folded states, for example.

Another rationale for single-molecule measurements is that the need for synchronization of many single molecules undergoing a time-dependent process disappears. For example, a large ensemble of molecules undergoing intersystem crossing events must be synchronized in order to measure the triplet lifetime. In the single-molecule regime, intersystem crossing becomes a digital, yes/no process, the observation of which has been termed "quantum jumping" as the system undergoes transitions from triplet state to singlet state and vice versa.[21]

A final reason for the use of single-molecule techniques is the possibility of observing new effects in unexplored regimes. For example, many single-molecule systems studied to date have unexpectedly shown some form of fluctuating, flickering, or stochastic behavior.[22] The absorption frequency of the single molecule can change due to a change in its photophysical parameters or a change in local environment; this behavior has been termed "spectral diffusion."[23] If a fixed-frequency laser is used to probe a spectrally diffusing molecule, the shifting of the absorption in and out of resonance generates amplitude fluctuations in the detected emission signal, which can be analyzed to gain knowledge about the underlying physical mechanism. Such fluctuations are important diagnostics of the single-molecule regime, and they provide unprecedented insight into behavior which is normally hidden by the usual ensemble averaging.

This chapter presents an overview of selected recent advances produced by the spectroscopy of single impurity centers in solids. After presentation of the basic principles in Section 2, typical molecular and impurity-based systems will be summarized. A selection of key experimental approaches enables studies ranging from quantum optics to amorphous system dynamics, reviewed in Section 4. The reader may consult one of several reviews for more information. The significance of these SMS studies are: (i) new physical effects have been observed in the single-molecule regime, and (ii) it is now possible in a single nanoenvironment to directly probe the connection between specific microscopic theories[24,25] of local structure, dynamics, and host–guest interactions and the statistical mechanical averages that are measured in conventional experiments. Several comprehensive reviews of this area may be consulted for details beyond the scope of this article.[4,26–30] In all cases, the primary motivation for observing the behavior of single molecules is to explore this fascinating new regime, because observing the individuals one at a time allows explicit removal of heterogeneity of populations, direct observation of fluctuations and stochastic time dependence, and analysis of the effects of the immediate local environments.

2. Basic Principles

The daunting task of detecting a single absorbing molecule amidst other impurity molecules and trillions of transparent solvent or matrix molecules with laser light (Fig. 1a) is greatly facilitated by ultrasensitive fluorescence techniques which themselves depend upon the photophysical properties of the molecule itself, illustrated in Fig. 1b for a molecule with singlet S and triplet T states. Laser excitation of the allowed single-singlet electronic transition (green) can be followed by fluorescence emission (cyan–yellow–red) or nonradiative processes including intersystem crossing to the triplet. Fluorescence usually represents only a tiny fraction of the incident energy, but gives rise to emitted photons that can be red-shifted with respect to the excitation if the emission terminates on vibrational levels of the ground state (yellowred), termed Stokes-shifted photons. These photons can be efficiently shifted out of an overwhelming number of photons in the exciting beam, so that the photons emitted from a single molecule can be detected with a high signal-to-background ratio if care is taken. Polycyclic aromatic hydrocarbons, e.g., are rigid molecules for which the Stokes shift is often small (less than 1000 cm^{-1}). In a frozen environment at low temperature, their geometry change can be so small that a very narrow line appears in the absorption and fluorescence spectrum. This rearrangement-free optical line is known as the zero-phonon line (ZPL) and is used in many optical experiments at cryogenic temperatures.[31–33] The homogeneous

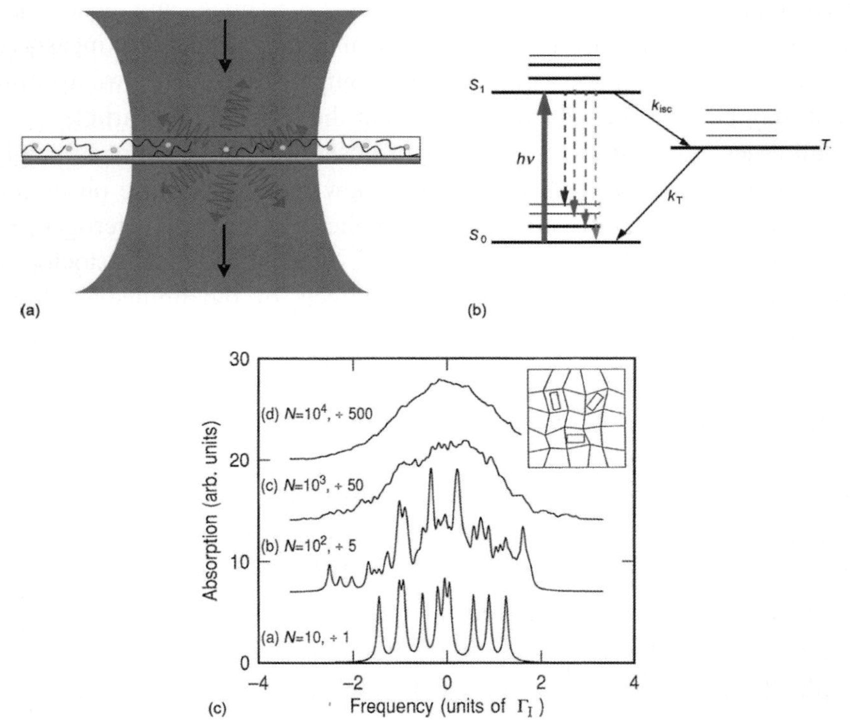

Fig. 1. (a) Schematic of the selection of a single molecule out of many in the same focal volume, (b) typical energy level scheme, and (c) illustration of inhomogeneous broadening. See also Colour Insert.

width $\Delta\nu_H$ of the ZPL in solids can approach the lifetime-limited value of ~10 MHz below 10 K.

There is an additional crucial phenomenon important for SMS which results from the extreme narrowness of zero-phonon transitions: inhomogeneous broadening (Fig. 1c).[34,35] This occurs when $\Delta\nu_H$ becomes narrower than the (mostly static) distribution of resonance frequencies in the solid caused by dislocations, point defects, or random internal strain and electric fields and field gradients (inset). Inhomogeneous broadening is a universal feature of high-resolution spectroscopy of defects in solids and of other spectroscopies where ZPLs are probed such as Mößbauer and magnetic resonance.[31,32] (A dynamic inhomogeneous broadening also occurs for Doppler-broadened lines of gases.) Inhomogeneous broadening facilitates selection of a single impurity, because the different guest molecules in the probe volume take on slightly different resonance frequencies according to the local fields at the location of the impurity. One then simply uses the tunability of a narrowband laser to select different single impurity centers. Naturally, this must be done in a region of the optical frequency spectrum where the number of molecules per homogeneous width is on the order of or less than one. This may be accomplished (i) by tuning the laser wavelength out into the wings of the inhomogeneous line (Fig. 1c, cases with higher doping level), (ii) by using a sample with a very low doping level (Fig. 1c, bottom case), or (iii) by producing a large inhomogeneous linewidth, and all three of these methods have been used. It is not true that single molecules can only be studied in the wings of inhomogeneous lines.

In discussing single-molecule fluorescence, two important quantities must be considered, namely the rate of fluorescence and the total energy or number of photons that a molecule can emit on average. In the case of organic molecules, the rate is often limited by trapping in a dark state, such as the triplet states, where absorption and fluorescence are switched off. Efficient fluorophores-like laser dyes having low trapping rates and/or short lifetimes of the dark states will be easiest to detect. The total amount of fluorescence per molecule is limited in ambient conditions by photobleaching, i.e., by photochemical reactions in the excited state which irreversibly alter the chemical structure of the fluorophore. At cryogenic temperatures, photodecomposition channels are often suppressed, and small reactive molecules such as oxygen or water are frozen and cannot diffuse. Then, the amount of fluorescence is limited only by sudden conformational changes of the host–guest system, which shift the absorption line of the molecule away from resonance with the laser. For some crystal systems at low temperatures, the available amount of fluorescence per molecule can be virtually infinite.[29]

Single-molecule fluorescence signals are rather weak, and can easily be drowned in background and noise. For ease in detection of single molecules, experimentalists must choose the right compound. In order to optimize the emission rate, the probe molecules should possess a strongly allowed absorption transition and a high fluorescence yield. An important parameter characterizing the interaction of a molecule with light is the absorption cross section, which may be regarded as the effective "area" of the molecule for "capture" of an incoming photon. The general expression for the cross section for a randomly oriented molecule is $\sigma = 2\pi (\lambda /2\pi)^2(\gamma_r / \Gamma_{tot})$, where λ is the light wavelength, γ_r the spontaneous (i.e., radiative) fluorescence rate, and Γ_{tot} the total frequency width of the absorption.[29,36,37] For a dye molecule in solution at room temperature, $\gamma_r \approx \Delta\nu_H$ is about 10^{-3} cm^{-1}, which corresponds to a radiative lifetime of a few ns, and Γ_{tot} is about 1000 cm^{-1}, so that the absorption cross section of a single dye molecule is $\sigma \sim 6$ Å^2, a value comparable to the molecular size. For the ZPL of an aromatic molecule at low temperature, the linewidth is dramatically smaller, about $\Gamma_{tot} = 10^{-3}$ cm^{-1}, and the radiative rate of the ZPL is reduced from the total value at room temperature by the Debye–Waller factor (about 0.1). Therefore the absorption cross section is many orders of magnitude larger than at room temperature, up to about $\sigma = 10^6$ Å^2, corresponding to the total physical area of some 100 000 molecules! Another important parameter is the maximum emission rate of a saturated molecule at very high excitation power. For a two-level molecule, this rate is given by $\tau_f^{-1} / 2$, and for a typical fluorescence lifetime τ_f of 5 ns, it reaches 10^8 s^{-1}, which, even with a modest detection yield, gives a very intense signal. Unfortunately, the rate is often limited by the bottleneck effect of dark states such as the triplet (see level diagram of Fig. 1b), which reduces the maximum rate to $\tau_f^{-1}/(2 + k_{ISC}\tau_T)$, where k_{ISC} is the population rate of the dark state from the excited singlet and τ_T the lifetime of the dark state ($k_T = \tau_T^{-1}$ being its decay rate, see Fig. 1b). For this reason, molecules with weak bottlenecks are generally preferred. When the intersystem crossing yield of a guest molecule is extremely small, it can easily be enhanced spectacularly through interaction with the host. Such is the case for terrylene when it is included in an anthracene crystal,[38] and presumably also for perylene when included in a naphthalene crystal.[39] One therefore should choose host molecules with triplet states lying at higher energies than the guest's singlet.[40,41]

Despite the fact that single-molecule signals are relatively strong, struggling against all background sources is the foremost worry of experimentalists who work with single molecules. A first source of background may lie with experimental limitations such as the dark

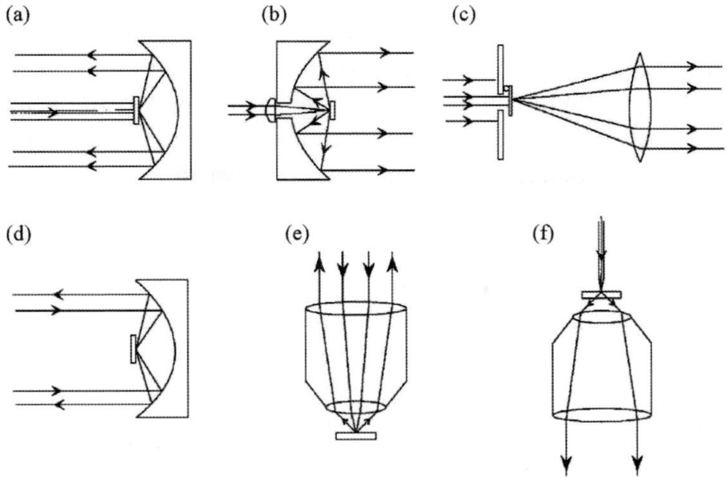

Fig. 2. Optical schematics for low-temperature SMS. In (d) and (e), the same optics focus the excitation light and collect the fluorescence. In (a), (b), (c), and (f), different elements are used for excitation and collection, which requires careful adjustments of the two foci.

counts of the detector, the residual fluorescence from optical parts, particularly from colored glass filters, or residual emission by the excitation source in the spectral range where fluorescence is detected.

But background photons emitted by the sample itself are more difficult to suppress. The two main sources are residual fluorescence from other impurities, and Raman scattering by the solvent or host matrix. Both are proportional to the volume of the illuminated sample. For example, a significant background arises from phonon-assisted absorption by the other probe molecules in cryogenic experiments. Even if their fluorescence yield is very weak, unwanted impurities or molecular components of the solvent, particularly in biological samples, can emit strong residual fluorescence. Residual fluorescence increases very rapidly with the energy of the exciting photon, and thus working in the blue or near-UV part of the spectrum will often give rise to more serious background problems.

Given that the background cross sections are proportional to the excited volume, an optimal signal-to-background ratio will require a minimum excited volume. Various optical designs have been proposed to limit the excited volume (Fig. 2). By means of a lens or a single-mode optical fiber (Fig. 2a), a loosely focused excitation beam is obtained, which leads to volumes of the order of 100 μm^3. The tight focus of a microscope objective (Fig. 2b) illuminates a volume of 1 μm^3 or less, whereas subwavelength illumination through the small aperture of a near-field optical probe (Fig. 2c) gives volumes of about 10^{-2} μm^3 or less.[12] Another useful detection configuration uses total internal reflection to reduce the illuminated volume.[42,43] A molecule radiates fluorescence photons in all directions of space, with the emission pattern of a dipole. To obtain a strong fluorescence signal, it is important to collect fluorescence over as wide a solid angle as possible. Several collection strategies have been used, e.g., parabolic mirrors or microscope objective lenses (see Fig. 2). The same optical component can be used for excitation and collection. This is often the case in confocal designs, where the focal spot is imaged onto a small-area detector, usually a highly efficient avalanche photodiode.

By displaying the signals from single molecules with their background as functions of various control parameters, an image or scan is obtained in which the single molecules appear as discrete maxima or spots. To distinguish individual molecules in the image, their spots must not overlap. Finally, when either the excitation or the analysis device are scanned in wavelength at a fixed spot in the sample, the signals from single molecules also appear as "images" in the spectrum (see, e.g., in Fig. 3), whenever the homogeneous linewidth γ_{hom} of a single molecule line is much smaller than the inhomogeneous bandwidth Γ_{inh} of their distribution.[23] This hybrid method of spectral-spatial imaging is useful at low temperatures, when the homogeneous width is small enough. The inhomogeneous broadening of the line is then exploited so as to efficiently reduce the concentration of resonant molecules in the sample.

3. Main Host–Guest Systems

Not all combinations of host and guest molecules are suited to high-resolution spectroscopy. The intensity of ZPLs (i.e., their Debye–Waller factor, see Section 2) decays exponentially with electron–phonon coupling. A weak coupling to phonons requires rigid guest molecules, for which no large reorganization follows excitation, and also a good fit

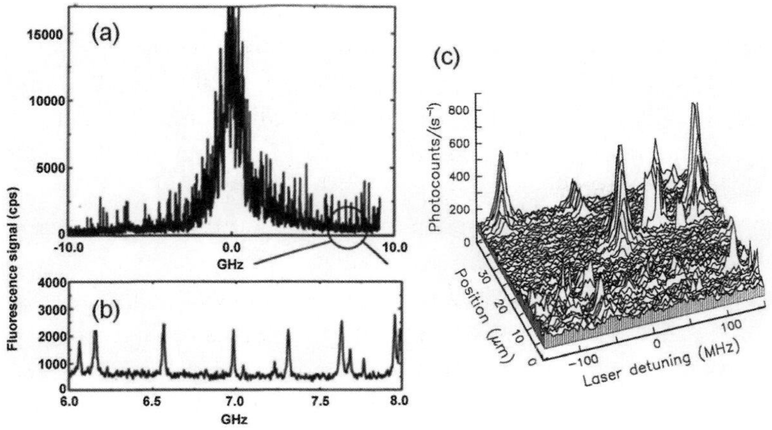

Fig. 3. Spectroscopy of single molecules of pentacene in *p*-terphenyl in frequency and space. Reprinted with permission from Ref. 44. Copyright 1991, American Institute of Physics.

between the guest and the cavity it occupies in the host lattice. The rigidity of the cage ensures that the frequency of local libration modes of the guest is not too low. Molecular rigidity is also one of the required features for high fluorescence yield. The above conditions are met for many rigid aromatic guests in well-ordered molecular crystals, and for several inorganic systems as well. Hereafter, we review the main systems used in the literature.

3.1. *Organic crystals*

Mixed molecular crystals offer many advantages for SMS. They can be fine-tuned through the host chemical structure to fit the guest molecule, they often present a small number of well-defined insertion sites, and the defects responsible for inhomogeneous broadening of the optical lines are often static. Single-molecule lines are therefore usually very stable in molecular crystals, at least at cryogenic temperatures. Mixed crystals can be grown by well-established methods.[45] In the Bridgman method, crystals are grown from a melt.[46] Sublimation is often used for larger molecules which would tend to react or decompose in the melt.[3] Recently, spin-coating of a solution in a suitable solvent has been

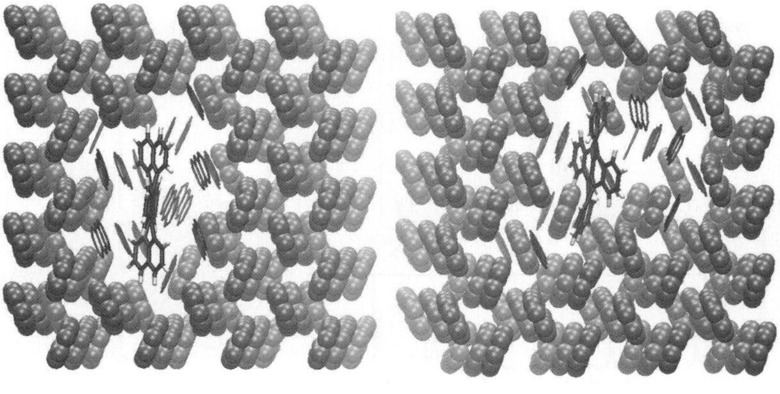

Fig. 4. Molecular mechanics calculations of the two main insertion sites of dibenzoterrylene in the crystal lattice of anthracene. On the left the "main" site, on the right the "red" site.[41]

successfully applied to grow thin crystalline films on solid substrates.[47] The samples must then be annealed to eliminate most defects and reduce spectral diffusion.

Good examples of stable single-molecule lines in a mixed crystal are terrylene in naphthalene,[48,49] or dibenzoterrylene in anthracene, whose insertion sites are illustrated in Fig. 4.[40,41] In crystals undergoing phase transitions upon cooling, numerous defects can be created and present residual dynamics even at liquid helium temperatures. An example is *p*-terphenyl, which presents a ferro-elastic phase transition. The symmetry breaking in the low-temperature phase leads to mosaic structures with different domains separated by walls. Pentacene guest molecules in the crystal probe the conformational changes of host molecules at domain walls, leading to spectral diffusion.[23,24,50] Terrylene as a guest behaves in a very different way in this matrix. Of the four possible insertion sites, only two are photostable.[51] The other two give rise to large photoinduced jumps, which can be reversible.[52] More recently, single molecules were used to probe the dynamics of more complex crystals, e.g., perylene[39] or terrylene[53] in the incommensurate phase of biphenyl, or terrylene for the dipolar disorder in a 2,3-dimethylnaphthalene crystal.[54]

Shpol'skii matrixes are quickly quenched crystals of linear n-alkanes, and can serve as hosts for many kinds of aromatic molecules.[55,56] In the crystal, the alkane molecules assemble into layers with their long axis perpendicular to the layer plane. Aromatic guests can be accommodated by substituting one or more alkane molecules. Being saturated and chemically inert, alkanes are perfectly transparent host molecules in the visible and near-UV, and do not perturb the singlet or triplet levels of guest molecules. n-Hexadecane ($C_{16}H_{34}$), e.g., is a much used matrix because it is fluid at ambient temperature, yet presents a low vapor pressure, making its manipulation very convenient. The orientation of single guest molecules in Shpol'skii matrixes has been found to be spread around the average orientation by some 20°,[57,58] i.e., significantly more than what might have been expected from the quality of the crystals. Although this observation is not completely understood yet, we propose that the fluctuations of the transition dipole vectors are actually much larger than those of the molecular axes. Small fluctuations in the relative positions of the molecules can lead to large variations of van der Waals interactions between host and guest, which are mixing guest electronic levels and altering the directions of associated transition dipoles. Similar small transient variations of intermolecular distances were estimated to induce significant fluctuations in the magnitude of the transition dipole of single fluorophores in polymers hosts close to the glass transition, leading to fluorescence lifetime fluctuations[59] and should obviously be accompanied by fluctuations of the dipole's direction. Correlation between the inhomogeneous width and the static dipoles responsible for the Stark effect of single molecules was discussed in Refs. 41 and 60. The sensitivity of electronic wavefunctions to small perturbations confirms optical spectroscopy as a sensitive but indirect indicator of the structure of the insertion site at the molecular scale, a precise description of which is still elusive.

3.2. *Polymers and other organic guests*

From the early days of SMS, polymers were used as matrixes for their ease of processing. The first experiments on perylene in the semi-crystalline polymer polyethylene (PE)[61] showed relatively narrow ZPLs subject to spontaneous and light-induced spectral jumps, phenomena known earlier from ensemble experiments as spectral diffusion and photophysical spectral hole-burning, respectively. The first Stark effect experiments on terrylene in PE showed the large distribution of orientations and of magnitudes of the dipole moments induced by polymer disorder.[62] Single terrylene molecules were later studied in several other polymers: poly-styrene, -methyl methacrylate, -vinyl butyral,[63] and -isobutylene.[64] The structure of these polymers is more disordered, leading to broader lines and larger and more frequent spectral jumps.[63,65–67] Polymers are convenient to study the low-temperature dynamics of glasses and amorphous materials,[65,66] which can be qualitatively explained with the model of the two-level systems (TLSs).[68] Single molecules, however, allow for very detailed comparisons to the model showing clear deviations from the general assumptions.[69] This line of work has been continued in the past decade and will be reviewed in Section 4.2.

Single bacterial antenna complexes were detected and studied for the first time at cryogenic temperatures in 1999.[70] This study showed the spectacular difference between the narrow lines in the ring of weakly interacting bacteriochlorophyll molecules and the broad excitonic bands of the strongly interacting ring. Much information has been gained from high-resolution spectra of these and related systems in the last years, and the progress has been recently reviewed.[71,72] Single fluorescent proteins have been studied at low temperature and present narrower transitions than at room temperature.[73]

Conjugated polymers are complex absorbers and emitters whose photophysical properties depend strongly on their conformations. They are therefore subject to a large inhomogeneity, which becomes immediately clear at the single-molecule level. The first pioneering experiments of Barbara's group[74] have now been complemented by low-temperature spectroscopy from several groups.[75–79] Red chains of polydiacetylene in the crystalline matrix of monomers are good model systems of a single polymer molecule in a perfectly ordered environment. Careful spectroscopy of these single chains shows exciton delocalization, electron–phonon coupling, and exciton transport along the chain.[80]

3.3. *Inorganic Systems*

Fluorescence spectroscopy of single emitters has so far been applied mostly to organic molecules. However, it has also been performed on a few inorganic systems, which we enumerate hereafter.

The most studied molecule-like inorganic emitters are color centers in wide-gap insulators, particularly in diamond. The rigidity of the diamond lattice and the lightness of carbon give rise to very high phonon frequencies (Debye temperature higher than 1000 K) and to ZPLs still observable at room temperature, for some types of defects. The nitrogen-vacancy (NV) center consists of a nitrogen substituting a carbon atom, with a carbon vacancy in the first neighbor shell.[81] It exists in two forms, neutral and negatively charged. The presence of spin in the ground and optically excited states makes it attractive for optically detected spin resonance and spin manipulations. Although much of the work on single NV centers was done at room temperature, the concepts are the same as at low temperature, and we mention some low-temperature experiments in Section 4.3. More optically active centers have been observed individually in diamond, in particular the nickel–nitrogen (NE8),[82] and a promising NiSi center which appears to involve elements nickel and silicon.[83] Many semiconductors or insulators would certainly present favorable hosts for color centers for single-center studies, e.g., SiC or LiF,[84] but no successful detection of single centers was reported yet.

Semiconductor nanostructures are fertile ground for single-emitter spectroscopy, with colloidal semiconductor nanoparticles[85,86] and self-assembled epitaxially grown quantum dots.[87] Emission lines appear to become lifetime-limited at low temperature in the case of self-assembled epitaxial QDs, but usually not for colloidal nanocrystals, where charge rearrangements and spectral diffusion appear to broaden the excitonic lines. Only special kinds of quantum dots give rise to narrow lines at low temperatures[88] although these lines remain much broader than the lifetime limit. Optical studies of single semiconductor quantum dots reviewed in Ref. 86 have expanded considerably in the last years, and will not be reviewed in depth in this chapter.

Many other emitters could possibly be detected individually through their fluorescence, e.g., rare earth ions. Few papers have reported successful experiments with these systems, but none of the early results has ever been independently confirmed.[89] To our knowledge, only one report of single ion detection with the current single-molecule methods has appeared.[90] The main experimental obstacle appears to be the extremely long lifetime of rare earth ions, which makes detection slow, and therefore difficult to discriminate against detector dark counts. A possible solution to this problem would be to enhance the strength of the optical transition by the near-field of a plasmonic nanoantenna.

4. Examples of Experiments

Many kinds of experiments have been demonstrated with single emitters at low temperatures, and the possibilities are certainly not exhausted. The narrow, often lifetime-limited linewidth of single systems presents several attractive features and properties.

- Because of the sharpness and intensity of the ZPL, and because the only adjustable parameter is its position, it becomes possible to compare experiment to theory carefully, and to reach deeper understanding of dephasing or spectral diffusion processes than with ensemble measurements.
- The sharp ZPL of a single molecule is very sensitive to local dynamics and to external perturbations. This provides excellent opportunities for nanoprobing, i.e., the direct measurement of a local parameter by a (nearly) point probe. A similar idea is exploited in scanning probe microscopies, where a tunneling current or a force exerted on a cantilever replace the total fluorescence intensity as the macroscopically accessible parameter.
- Lifetime-limited lines are very interesting in quantum optics, because the irreversibility due to decoherence processes is limited by spontaneous emission to a minimum. When the lifetime has a purely radiative origin, and is due to spontaneous emission towards one level only, coherent effects such as the absorption of a single molecule are optimized.[91,92]

- The ratio of inhomogeneous to homogeneous widths is often very large, particularly in systems presenting frozen disorder such as orientational disorder of noncentrosymmetric host molecules in their lattice.[54] This means that thousands to hundreds of thousands of single molecules can often be excited separately, and therefore distinguished, in each single focal spot. This feature makes it straightforward to achieve superresolution, by tuning the excitation laser into resonance with each molecule in turn.[15] With the resulting frequency-position map, many processes can in principle be mapped with sub-wavelength resolution.[93] This is a very simple and direct variant of the more recent superresolution schemes at room temperatures, which all rely on a controlled switching mechanism.[16,94]

The limited space in this chapter does not allow us to review all kinds of experiments done to this day (for a review of the early work, see, e.g., Ref. 17, and references therein). Instead of reviewing the whole of the low-temperature work done since 2000, we choose a few examples which are particularly illustrative of the advantages of SMS at low temperatures.

4.1. *Molecular photophysics and spectroscopy*

Much information can be deduced from the fluorescence spectra of single molecules, which reveal their isotopic composition[95] and their vibration modes in the ground electronic state.[40,54,96–99] It is also possible with broadband lasers to scan the blue wing of the absorption spectrum and to record excitation spectra of molecular vibronic lines, showing vibrations in the excited state.[49,100–102] If other properties can be also measured on the same molecules, e.g., optically detected paramagnetic resonance (ODMR), a very powerful method is open for the study of individual isotopomers[103] or of molecules in particular configurations. It is important to realize that these correlations all but suppress the need for a special synthesis of each isotopomer, or for a full control of the molecular environment. One just lets nature randomly fabricate all isomers or all micro-configurations, and one selects the molecules of interest by a suitable choice of frequency and spatial position.

4.1.1. *Molecular photophysics*

Of particular interest are the intramolecular relaxation pathways, which determine the photophysical properties of each individual molecule. The intersystem crossing rate, which controls the branching ratio from the excited singlet to the triplet manifold, is especially important for the brightness and detectability of single molecules.[38,95,104]

4.1.2. *Blinking*

One of the most striking of the early observations on single molecules has been blinking, the sudden switching on and off of their fluorescence brightness. These sudden intensity changes reflect transitions between quantum states, and relate to early discussions on quantum jumps. Transitions between quantum states are described statistically by the evolution of a density matrix for large ensemble of quantum systems, but appear as discontinuous events, or quantum jumps, in observations of single quantum systems, in agreement with the collapse postulate of quantum mechanics. A quantum measurement projects a system into an eigenstate of the measured observable, and the continuous observation or measurement of the fluorescence intensity, therefore, leads to projection and to the associated discontinuous jumps. The first indications of intensity changes, due to intersystem crossing between singlet and triplet manifolds, have been the bunching measurements by autocorrelation of intensity fluctuations,[3] but the quantum jumps due to intersystem crossing into triplet states were subsequently observed directly.[52] A well-studied mechanism of blinking at low temperature arises from the low-energy excitations of amorphous materials, known as TLSs, and related to tunneling motions of atoms or groups of atoms (see Section 4.2). While relatively common in amorphous polymeric hosts,[105] this process can even occur in crystals with defects,[106] where phonon-driven transitions between configurations of the nearby host crystal cause the

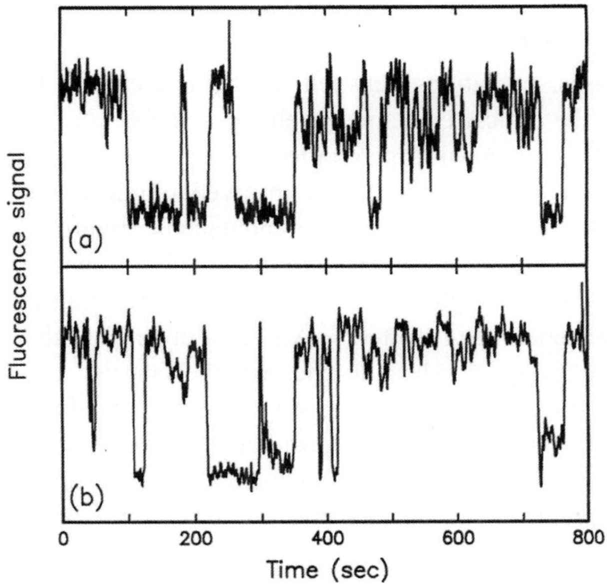

Fig. 5. Early example of blinking for a single molecule of pentacene in *p*-terphenyl at 2 K, where the emission pumped by a single-frequency laser turns on and off as the molecule jumps into and out of resonance. For low (a) and high (b) powers, the rate is unchanged, signifying the spontaneous nature of the effect.[106]

optical absorption frequency of the single molecule to spontaneously jump in and out of resonance with the pumping laser (Fig. 5).

A third type of blinking is induced by charge separation consecutive to optical excitation, with one or more photons. The dark states in this type of intermittency can be extremely long-lived, and often give rise to power-law distributions of off- and on-times. Although this type of blinking has been discovered in semiconductor nanocrystals and mostly studied at room temperature,[107] it appears also active for molecular systems.[108–110] Low temperatures do not seem to qualitatively change the nature of the observations. This observation strongly suggests that tunneling of charges is responsible for this effect. The mechanism of power-law blinking is still hotly debated,[111–113] and low-temperature experiments would certainly help much to clarify the current situation.

4.1.3. *Light-induced blinking or switching*

In some cases, transitions between nearby low-energy TLS states are induced by the optical excitation itself, presumably due to nonradiative processes. This effect was first reported for perylene single molecules in a poly(ethylene) host, where light was observed to induce a shift to another resonance frequency, followed by a spontaneous return to the original frequency.[61] By measuring the waiting time before a spectral jump many times for the same single molecule, the first observation of exponential (Poisson) statistics was reported.[114] A similar effect was also reported for terrylene molecules in n-hexadecane, and the fact that the second state was only on the order of 100 MHz away in frequency allowed a molecule to be optically pushed back and forth between two electronic resonance frequencies.[115] These light-driven switching effects were thought to be useful for optical storage, but they also would be useful for superresolution imaging of molecular positions.[94]

4.1.4. *Bichromophoric systems*

A major requirement of SMS is that the distance between two molecules resonantly excited by the laser is larger than the diffraction-limited resolution (typically 300 nm, more often 1000 nm in low-temperature experiments). Therefore, the probability of finding two resonant molecules in the same laser spot is very small. It is nonetheless possible to study two interacting molecules, either by identifying rare cases of close-by molecules in a random distribution, or by attaching two chromophores, e.g., covalently, to the same molecule, as we discuss hereafter.

Basché's group have studied a variety of multichromophoric molecules in polymeric environments and determined the transfer rates via the lifetime-limited homogeneous width of the donors' ZPLs. For heteropairs of perylene-diimide (PDI) donor and terrylene-diimide (TDI) acceptor, they correlated the transfer rate with the (independently measured) spectral overlap and found evidence for a non-Förster contribution,[116,117] which is thought to be mediated by the aromatic bridge between the chromophores. They also studied PDI–PDI homopairs and consistently found transfer towards the lower excited state at low temperature.[118]

In disordered environments, inhomogeneous broadening makes it nearly impossible to find a resonant pair of molecules, even for the homopairs discussed above. However, in a crystalline solution, even with a low concentration of guest, there is a small but finite probability to find two resonant guest molecules at a short enough distance that their coupling becomes important. Sandoghdar's group have found such a pair of molecules and studied their spectrum as a function of laser intensity or applied electric field.[119] They were able to determine the intermolecular distance, and to show the presence of excitonic coupling between them. Such pairs of molecules open interesting schemes for the quantum treatment of information (see Section 4.3). However, as these experiments are seriously limited by the short dephasing (or decoherence) time, mainly due to spontaneous emission, longer lived coherences are desirable.

4.2. *Probing at nanoscales*

4.2.1. *Low-temperature lattice dynamics*

We have mentioned the dynamics of TLSs as one of the sources of blinking in ill-ordered matrixes (glasses, crystals with defects, and polymers) at low temperatures. Through their study, much can be learned about those excitations, their coupling to acoustic phonons,[65,66] or with other TLSs,[69] and about their distributions.

The problem of a single molecule coupled to a single TLS is by no means trivial when no simplifying assumption is made. Following an experimental demonstration by Bauer and Kador,[120] Hofmann *et al.*[121] have proposed a theoretical treatment of a single molecule coupled to a single TLS, for arbitrary driving field intensity and arbitrary differences of the TLS potential between the ground and excited electronic states. Their simulations, based on the density matrix formalism, account for experimental observations so far.[119,122–124] This theory predicts complex spectral profiles and dependences on laser intensity, which can be sought

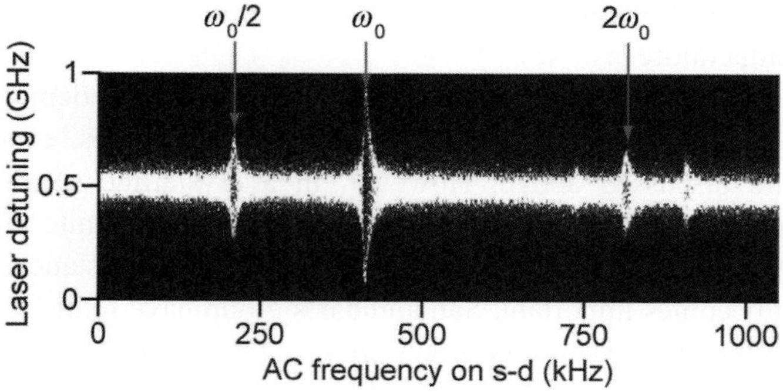

Fig. 6. Example of low-frequency acoustic resonances observed in the spectrum of a single dibenzoterrylene molecule (vertical axis) in an anthracene crystal. The frequency of the applied ac-voltage frequency is scanned (horizontal axis). The resonances at the double and half frequency are absent for low-driving voltages.[126] See also Colour Insert.

in the spectra of molecules coupled to TLSs of different kinds, including electronic ones based on charge hopping.[125]

Several studies showed clear evidence that a significant portion of the molecules cannot be adequately described in the standard TLS framework.[69,127–129] In particular, a wide distribution of low-energy oscillators is found in the study of individual molecules, with various frequencies in the 1–100 cm^{-1} range, various couplings to the optical electron,[130] and clear correlation between them.[131,132] These soft modes contribute to the excess of vibrational degrees of freedom known as "boson peak" in glassy and disordered systems.[133]

In addition to those modes, ultralow frequency (kHz–MHz) oscillators were discovered in crystal matrixes contacted to metal electrodes.[126] An example of resonances of the Stark shift amplitude for certain driving frequencies is presented in Fig. 6. These modes appear related to crystal defects, which concentrate stress and anharmonicity in small and heavily deformed areas. Other types of dynamics were identified in disordered crystals. In *p*-terphenyl, the domain walls appearing at the ferroelastic order–disorder transition cause a specific dynamics affecting the energies of single pentacene molecules (Fig. 5),[44] which has been explored by theoretical studies.[24,50] Upon cooling, a biphenyl crystal undergoes two phase transitions, leading to an incommensurate phase which persists down to a few Kelvin. The large residual width of single dopant molecules in biphenyl may be caused by coupling to phasons, low-energy excitations still significantly populated at low temperatures.[39,53] A last example is the dipolar disorder of 2,3-dimethylnaphthalene. Although the structure of this crystal resembles that of the parent crystal of naphthalene, the orientation of the methyl groups at one end of the molecule is largely random. The associated dipole points in one or the other of two possible directions. This configuration is static at low temperatures, giving rise to an enhanced inhomogeneous broadening, characteristic of an amorphous system, but without the associated dynamics. At higher temperature, however, the molecules start flipping between their two possible orientations, giving rise to strong thermally and light-induced spectral diffusion.[54]

Electronic states presenting spin are sensitive to a particular type of spectral diffusion, which arises from the flips of nuclear spins (and sometimes also electronic spins), which are still active at liquid helium temperatures. We discuss those in Section 4.3.2.

4.2.2. *Methyl group tunneling*

Methyl groups are well-known tunneling systems. Provided their environment allows for nearly free rotation, tunneling can be detected optically by spectral hole-burning.[134,135] The rotational potential is often flat enough for methyl substituted aromatic hydrocarbons in suitable matrices. Terrylene molecules with methyl substituents on the corner benzo rings (1- and 2-methyl terrylene) were synthesized in the hope of observing individual jumps of single methyl groups.[136] Precise measurements of the distributions of tunneling barriers in the ground and excited electronic states could confirm and quantify their sensitivity to the slight fluctuations of intermolecular distances responsible for the inhomogeneous broadening in the optical domain. Spectral hole-burning and single-molecule traces showed that the two isomers behaved differently in the Shpol'skii matrixes hexadecane and deuterated hexadecane. Beta(2)-methyl-terrylene in hydrogenated hexadecane appeared to be the best system for the observation of single methyl group tunneling.[136] Further work is needed to determine the distributions of tunneling potentials and of temperature dependences of the tunneling rates.

4.2.3. *External field effects and charge transport*

The sharp ZPLs of single molecules are easily shifted by external perturbations, and can be used to probe local fields. For example, electric fields can be applied with external electrodes, leading to a Stark shift of electronic levels, which depends on static dipole moments and polarizabilities of the ground and excited electronic states.[137] The Stark effect is routinely measured to probe possible symmetry breaking of a guest upon insertion into the host lattice.[41,52,60]

However, if charges can move within the environment of the guest molecule, or if charges are injected from the electrodes into the sample, very different electric field effects can appear. First, the local field on the molecule under study can be much stronger than the nominal applied field, because of the proximity of field sources. Second, because the local field is often created by charges trapped in the insulating or poorly conducting material, and because this distribution often relaxes extremely slowly, the dynamics of the electric field effects can extend over many orders of magnitude, from picoseconds to days and longer. The electric field effects become then much more complicated, but their study with single molecules opens an unique window on injection and transport of charges at nanometer scales. Early examples of single molecules in Shpol'skii matrixes deposited on semiconductor films can be found in Refs. 138 and 139. Later experiments on similar samples on a silicon carbide (SiC) film demonstrated the very slow charging of capacitors through extremely high resistors,[140,141] giving rise to time constants as long as minutes or hours. The combination of charge injection with single-molecule optical detection opens the interesting possibility of single-molecule electroluminescence. With an electrically driven single-photon source, e.g., background from the excitation laser would be completely eliminated. However, it is much more difficult to control charge transport in organic and disordered materials, than it is in semiconductors. Nevertheless, Vacha's group have demonstrated the feasibility of single-molecule electroluminescence.[142]

The effect of a magnetic field on singlet states is very weak, because these states are diamagnetic (they undergo a diamagnetic Zeeman effect), but it is still measurable thanks to the sharp single-molecule lines, as demonstrated by Bauer and Kador.[143] The Zeeman shift of terrylene is quadratic and weak, of the order of a few MHz/T^2, because contrary to porphyrins and other disk-like molecules, terrylene does not have two closely spaced, orthogonally polarized singlet levels.

4.3. *Quantum optics*

Single molecules behave in many ways as simple electronic TLSs that can be used to test laser–matter interactions, i.e., for nonlinear and quantum optics. The requirements on a host–guest system for this purpose are as follows. The single-molecule lines must be very stable and the branching ratio to the triplet manifold should be as small as possible, because the triplet severely limits the magnitude of the observed effects.[144] In addition, the ZPLs should be narrow, ideally lifetime-limited as they are in many crystal matrixes at liquid helium temperatures. To reach a high frequency stability, one may scan small frequency intervals with a single frequency-stabilized laser and with acousto-optical modulators.[145] Many convenient systems have been found for such measurements, in particular pentacene and terrylene in pterphenyl, and dibenzanthanthrene in naphthalene,[30] or dibenzoterrylene in anthracene.[40]

4.3.1. *Nonlinear spectroscopy*

The simplest nonlinear effect on an optical TLS is its saturation, due to the finite lifetime of the excited state. The optical line broadens and the fluorescence signal saturates when the laser power increases.[23] The antibunching of fluorescence photons is a quantum-mechanical effect that appears because a single molecule emits only one photon at a time. Immediately after one fluorescence photon has been observed, the probability of observing a second photon is zero. The measurement of the first photon has projected the molecule into its ground state, which cannot emit a second photon. The experiment runs like a fluorescence autocorrelation measurement (Section 4.1), but on a shorter timescale, with a setup of the Hanbury–Brown–Twiss type,[21,146] in order to avoid distortions from dead time and afterpulses of the detector. In addition to the antibunching dip at short times, the correlation function shows Rabi oscillations at high enough excitation intensities.

Several nonlinear effects can be observed with single molecules, e.g., the response to a weak probe beam in the presence of a strong (pump) beam at a different frequency. Under the strong pump illumination, the molecular levels are shifted and new resonances appear. The shift of the levels (called light shift) leads to a shift of the molecular transition, which can be measured as a function of the pump intensity and frequency.[144] The molecule in the pump field can be described in the dressed atom formalism and displays three resonances: (i) the light-shifted molecular transition, (ii) the Rayleigh or two-wave mixing structure at the pump frequency, and (iii) the hyper-Raman resonance corresponding to the absorption of two pump photons and to the stimulated emission of one probe photon.[145] When both pump and probe beams are strong, more complex, multiphoton structures appear due to a transient motion of the Bloch vector of the molecule in the beating of pump and probe beams.[147] All these effects and structures are in very good agreement with optical Bloch equations and show that single molecules are very well described by a simple two-level model.[148]

4.3.2. *Single photons on command*

The good control of the quantum state of a TLS in a laser field makes it possible to produce single photons on command.[149] In this experiment, the molecule was brought into its excited state with nearly unity probability by a rapid adiabatic passage, and spontaneous emission produced one photon per excitation cycle. Instead of a rapid passage, a laser pulse can be tailored with acousto-optical modulators to prepare the molecule in the desired coherent state.[150] A different scheme was used at room temperature by exciting a vibronic transition of the excited state.[151] The vibronic state relaxed quickly to the emitting excited state, producing the desired photon, and detected photon count rates as large as 310 kHz were observed. However, at room temperature, the emission is considerably broadened by dephasing. An extensive review of the possible schemes and uses of single emitters of different kinds to generate single photons on command can be found in Ref. 152.

Careful optimization of the microscope point-spread function allows for maximal interaction of a single molecule with light, leading to an extinction larger than 10% for a single molecule.[153] Zumofen *et al.* have recently proven that the theoretical limit for a single molecule with only one transition and for a correctly patterned incident light mode is actually total extinction.[154]

The two coupled molecules discussed in Section 4.1.3 offer possibilities to emit two photons in a correlated manner, as well as to manipulate quantum states.[119] For example, one could shift the frequency of one of the molecules by an inhomogeneous electric field produced by a coated pointed metal tip,[155] or a local strain field. By an adiabatic sweep, one could then create a delocalized exciton state from a localized state on one of the molecules.[156] One could thus create entangled excited states on two or more molecules. However, such coherent states will never live longer than some nanoseconds, the short excited state lifetime. Therefore, much effort is devoted to transferring optical coherence to longer lived states, e.g., electron or nuclear spin states, as discussed hereafter.

The optical control of a single emitter's states open the way to all-optical information processing, e.g., a single-photon transistor.[157] In their scheme, the plasmonic near-field of a silver nanowire concentrates the incoming photons to enhance the interaction with the emitter. Hwang *et al.*[158] have recently demonstrated a similar idea with careful focusing of the light, and achieved all-optical modulation of a light beam with the ZPL of a single molecule.

4.3.3. *Systems with spins*

Most of the optical experiments on single molecules (and more generally single emitters) require strong transitions, i.e., ones in which the spin does not change. All the organic molecules discussed in this chapter are cycled between singlet ground and excited states. This means that the optical spectra are insensitive to spin and to magnetic field (except for the very weak diamagnetic Zeeman effect discussed in Section 4.2.3). For many applications, in particular magnetic resonance on a nanometer scale and quantum nanophysics, it would be very interesting to have optical transitions sensitive to magnetic fields and to nearby spins. This can be achieved in two different ways. In the first, direct scheme, the optical transitions take place between two states, of which at least one carries a spin. In the second (indirect) scheme, the emitter is cycled between spinless states, but visits a spin-carrying state during the optical cycle, e.g., via relaxation. Examples of the second scheme have been demonstrated early on with the ODMR spectroscopy of the triplet states of single pentacene molecules.[159,160] Exploiting the sensitivity of the optical signal to the electron spin, and the coupling of the electron spin to nuclear spins, Wrachtrup *et al.* demonstrated nuclear magnetic resonance on a single proton spin.[161]

The latter demonstration, however, requires favorable circumstances and only applies to a nucleus within the molecule under study. Moreover, the experiment is limited by the triplet yield and the triplet lifetime. The yield should be weak and the lifetime short to allow for strong optical signals. Therefore, it is desirable to find single emitters with spin-carrying ground and excited states. The NV color center in diamond introduced in Section 3.3 and discussed in the next section has triplet ground and excited states, and is the only example so far of a bright, single molecule-like emitter with spin.

4.3.4. *Spectral diffusion and dephasing from spin dynamics*

Optical transitions between spin-carrying states are subject to a particular type of spectral dynamics, induced by spin flips in the emitter's environment. The random magnetic fields induced by these dynamics give rise to spectral diffusion and/or to dephasing.

Spin-induced dynamics has been recently studied on the optical transition of the negatively charged NV center in diamond. In the early experiments on single NV centers,[162] optical pumping between the spin states would limit the brightness of single centers at low temperature. Therefore, much of the initial spectroscopy was done at room temperature, including spin-echo studies showing the coupling of a

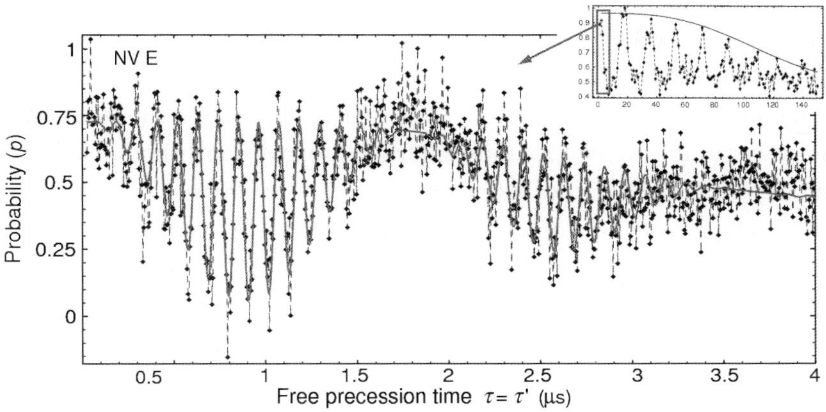

Fig. 7. Two-pulse echo signal of the electron spin of a single NV center in diamond, as the delay between the pulses is varied. The vertical axis indicates the fluorescence intensity, itself a function of the spin state. The modulations are due to isotopic impurities in the diamond lattice, and vary from center to center. Adapted with permission from Ref. 163. See also Colour Insert.

single NV spin to ^{13}C nuclear spins, as illustrated by Fig. 7.[163] These studies show the discrete levels due to nearby spins and the dephasing contributions of the outer shell of spins. A second problem with the low-temperature studies of NV centers was the large concentration of spin-carrying nitrogen impurities in regular diamond. These would give rise to large spectral diffusion through charge transport and through spin flips. Progress in material control has allowed Tamarat *et al.*[164] to practically eliminate spectral diffusion in ultrapure diamond (type IIa). They reached the lifetime-limited width of the optical lines, and could achieve high count rates by repumping the other spin states towards the optical system, either by optical recycling or by the application of microwaves. The application of electric[165] or magnetic[166] fields makes a fine spectroscopy of the excited state possible. High magnetic fields can freeze the dynamics of the electron spin bath, thereby reducing the decoherence of the NV spin.[167,168]

The manipulation of single spins open many interesting applications in the quantum treatment and storage of information, as well as in nanotechnology. For example, magnetometry in the microtesla range has already been demonstrated.[169,170]

5. Conclusions and Outlook

In this chapter, we have reviewed the main advances in SMS at cryogenic temperatures made in the last 10 years. After the early demonstrations of the power of the method for molecular spectroscopy and photophysics, the emphasis has shifted towards probing the nanoenvironment of individual molecules, by electric or magnetic field effects. A number of types of low-temperature dynamics have been studied in molecular crystals and in disordered organic materials. The advantages of single-molecule methods are well known, as they can probe any processes in condensed matter at low temperature, free from any averaging: tunneling, phonons, vibration modes, and conduction. The main development, however, has been in the direction of quantum optics, particularly for systems which possess spin in the states involved in the optical transition.

The combination of optical reading with coherent spin dynamics provides a unique way towards the manipulation, treatment, and storage of quantum information. The reduced size of molecules or nanoparticles makes it in principle possible to integrate them into larger structures. Several recent papers have demonstrated all the power of the optical single-molecule approach with experiments on NV centers in diamond. Although this system is convenient because of its stability and ruggedness, it has marked disadvantages. Diamond is difficult to prepare and modify. For example, the usual method to place individual centers at well-defined relative positions, or close to other structures is by implantation, which entails a low spatial resolution. Spin-carrying molecular states would enable similar experiments to color centers, but they could be engineered and tuned thanks to chemical synthesis capabilities.

Another fascinating perspective is the combination of single-molecule methods with nano-optics. Plasmon excitations in metal nanostructures can enhance the optical fields by orders of magnitude. Plasmon resonances can be utilized in nanoantennas, which can adapt single molecules to better interact with propagating photons.[171] In this way, light–matter interactions can be tailored at will. Similarly, photonic crystal cavities could be coupled to single molecules to store or manipulate exciting or emitting photons. Nano-optical devices interfaced with single emitters would spell tremendous advances for the all-optical treatment of information. More generally, single molecules and single localized electronic states can be seen as local ports for reading and writing information at nanometer scale, while the communication between them could be ensured by photons in different spectral ranges.

Acknowledgments

This work was supported by many funding agencies and organizations over the years, and the authors are thankful for this support. For W. E. M., these include the U. S. Office of Naval Research, the U. S. National Science Foundation, the IBM Research Division, the University of California, San Diego, and Stanford University. For M. O., supporting agencies include NWO and the FOM Foundation, the European Community (IST program), and the European Research Council.

References

1. W. E. Moerner and L. Kador, *Phys. Rev. Lett.* **62**, 2535 (1989).
2. L. Kador, D. E. Horne and W. E. Moerner, *J. Phys. Chem.* **94**, 1237 (1990).
3. M. Orrit and J. Bernard, *Phys. Rev. Lett.* **65**, 2716 (1990).
4. W. E. Moerner and Th. Basché, *Angew. Chem. Int. Ed.* **105**, 537 (1993); *Angew. Chem. Int. Ed.* **32**, 457 (1993).
5. M. Orrit, J. Bernard and R. Personov, *J. Phys. Chem.* **97**, 10256 (1993).
6. W. M. Itano, J. C. Bergquist and D. J. Wineland, *Science* **237**, 612 (1987).
7. F. Diedrich, J. Krause, G. Rempe, M. O. Scully and H. Walther, *IEEE J. Quant. Elect.* **24**, 1314 (1988), and references therein.
8. H. Dehmelt, W. Paul and N. F. Ramsey, *Rev. Mod. Phys.* **62**, 525, (1990).
9. G. Binnig, H. Rohrer and C. Gerber, *Phys. Rev. Lett.* **49**, 57 (1982).
10. G. Binnig, C. F. Quate and C. Gerber, *Phys. Rev. Lett.* **56**, 930 (1986).
11. G. Binnig and H. Rohrer, *Rev. Mod. Phys.* **59**, 615 (1987).
12. E. Betzig and R. J. Chichester, *Science* **262**, 1422 (1993).
13. W. P. Ambrose, P. M. Goodwin, J. C. Martin and R. A. Keller, *Phys. Rev. Lett.* **72**, 160 (1994).
14. X. S. Xie and R. C. Dunn, *Science* **265**, 361 (1994).
15. A. M. van Oijen, J. Koehler, J. Schmidt, M. Mueller and G. J. Brakenhoff, *Chem. Phys. Lett.* **292**, 183 (1998).
16. S. W. Hell, *Science* **316**, 1153 (2007).

17. W. E. Moerner and M. Orrit, *Science* **283**, 1670 (1999).
18. S. Weiss, *Science* **283**, 1676 (1999).
19. P. Tinnefeld and M. Sauer, *Angew. Chem. Int. Ed.* **44**, 2642 (2005).
20. P. V. Cornish and T. Ha, *ACS Chem. Biol.* **2**, 53 (2007).
21. T. Basché, S. Kummer and C. Bräuchle, *Nature* **373**, 132 (1995).
22. W. E. Moerner, *Science* **277**, 1059 (1997).
23. W. P. Ambrose and W. E. Moerner, *Nature* **349**, 225 (1991).
24. P. D. Reilly and J. L. Skinner, *Phys. Rev. Lett.* **71**, 4257 (1993).
25. E. Barkai and R. M. Silbey, *Annu. Rev. Phys. Chem.* **55**, 457 (2004).
26. W. E. Moerner, *Acc. Chem. Res.* **29**, 563 (1996).
27. M. Orrit, J. Bernard, R. Brown and B. Lounis, In: E. Wolf (editor), *Progress in Optics*, vol. 35, Elsevier (1996), p. 61.
28. T. Plakhotnik, E. A. Donley and U. P. Wild, *Annu. Rev. Phys. Chem.* **48**, 181 (1996).
29. T. Basché, W. E. Moerner, M. Orrit and U. P. Wild (editors), *Single Molecule Optical Detection, Imaging, and Spectroscopy,* Verlag-Chemie, Weinheim (1997).
30. P. Tamarat, F. Jelezko, B. Lounis and M. Orrit, In: C. Gooijer, F. Ariese, J. W. Hoofstraat (editors), *Shpol'skii Spectroscopy and Other Site-Selection Methods*, *Chemical Analysis*, vol. 156. Wiley (2000).
31. D. A. Wiersma, In: J. Jortner, R. D. Levine and S. A. Rice (editors), *Photoselective Chemistry, Part 2* (*Adv. Chem. Phys. vol.* 47).Wiley, New York (1981), p. 421.
32. W. M. Yen and P. M. Selzer (editors), *Laser Spectroscopy of Solids (Topics Appl. Phys., vol.* **49**), Springer Berlin (1981).
33. W. E. Moerner (editor), *Persistent Spectral Hole-Burning: Science and Applications (Topics Curr. Phys. vol.* **44**), Springer, Berlin, Heidelberg (1988).
34. A. M. Stoneham, *Rev. Mod. Phys.* **41**, 82 (1969).
35. K. K. Rebane, *Impurity Spectra of Solids*, Plenum, New York (1970), p. 99.
36. K. K. Rebane, *Chem. Phys.* **189**, 139 (1994).
37. W. E. Moerner, R. M. Dickson and D. J. Norris, *Adv. Atom. Molec. Opt. Phys.* **38**, 193 (1997).
38. A. A. L. Nicolet, M. A. Kol'chenko, B. Kozankiewicz and M. Orrit, *J. Chem. Phys.* **124**, 164711 (2006).
39. P. J. Walla, F. Jelezko, P. Tamarat, B. Lounis and M. Orrit, *Chem. Phys.* **233**, 117 (1998).
40. A. A. L. Nicolet, C. Hofmann, M. A. Kol'chenko, B. Kozankiewicz and M. Orrit, *Chem. Phys. Chem.* **8**, 1215 (2007).
41. A. A. L. Nicolet, P. Bordat, C. Hofmann, M. A. Kol'chenko, B. Kozankiewicz, R. Brown and M. Orrit, *Chem. Phys. Chem.* **8**, 1929 (2007).
42. T. Funatsu, Y. Harada, M. Tokunaga, K. Saito and T. Yanagida, *Nature* **374**, 555 (1995).
43. R. M. Dickson, D. J. Norris, Y.-L. Tzeng and W. E. Moerner, *Science* **274**, 966 (1996).

44. W. P. Ambrose, Th. Basché and W. E. Moerner, *J. Chem. Phys.* **95**, 7150 (1991)
45. N. Karl, In: H. C. Freyhardt (editor), *Crystals: Growth, Properties and Applications* (*Organic Crystals,* vol. 4). Springer (1980).
46. A. M. Boiron, F. Jelezko, Y. Durand, B. Lounis and M. Orrit, *Mol. Cryst. Liq. Cryst.* **291**, 41 (1996).
47. R. J. Pfab, J. Zimmermann, C. Hettich, I. Gerhardt, A. Renn and V. Sandoghdar, *Chem. Phys. Lett.* **387**, 490 (2004).
48. T. Plakhotnik, T. Nonn and V. Palm, *Chem. Phys. Lett.* **357**, 397 (2002).
49. D. Wiacek and B. Kozankiewicz, *Chem. Phys. Lett.* **462**, 280 (2008).
50. P. D. Reilly and J. L. Skinner, *J. Chem. Phys.* **102**, 1540 (1995).
51. S. Kummer, T. Basché and C. Bräuchle, *Chem. Phys. Lett.* **229**, 309 (1994); *Chem. Phys. Lett.* **232**, 414 (1995).
52. F. Kulzer, S. Kummer, R. Matzke, C. Brauchle and T. Basché, *Nature* **387**, 688 (1997).
53. V. Palm, M. Pars, J. Kikas, M. Nilsson and S. Kroll, *J. Lumin.* ***127***, 218 (2007).
54. M. Banasiewicz, D. Wiacek and B. Kozankiewicz, *Chem. Phys. Lett.* **425**, 94 (2006).
55. L. Nakhimovsky, M. Lamotte and J. Joussot-Dubien, *Handbook of Low Temperature Electronic Spectra of Polycyclic Aromatic Hydrocarbons (Phys. Scie. Data* **40**), Elsevier (1989).
56. C. Gooijer, F. Ariese and J. W. Hoofstraat (editors), S*hpol'skii Spectroscopy and Other Site-Selection Methods (Chemical Analysis, vol.* **156**), Wiley (2000).
57. A. Bloess, Y. Durand, M. Matsushita, J. Schmidt and E. J. J. Groenen, *Chem. Phys. Lett.* **344**, 55 (2001).
58. M. Matsushita, A. Bloess, Y. Durand, J. Y. P. Butter, J. Schmidt, E. J. J. Groenen, *J. Chem. Phys.* **117**, 3383 (2002).
59. R. A. L. Vallée, M. Van Der Auweraer, F. C. De Schryver, D. Beljonne and M. Orrit, *Chem. Phys. Chem.* **6**, 81 (2005).
60. P. Bordat, M. Orrit, R. Brown and A. Würger, *Chem. Phys.* **258**, 63 (2000).
61. T. Basché and W. E. Moerner, *Nature* **355**, 335 (1992).
62. M. Orrit, J. Bernard, A. Zumbusch and R. I. Personov, *Chem. Phys. Lett.* **196**, 595 (1992).
63. B. Kozankiewicz, J. Bernard and M. Orrit, *J. Chem. Phys.* **101**, 9377 (1994).
64. J. Tittel, R. Kettner, T. Basché, C. Bräuchle, H. Quante and K. Müllen, *J. Lumin.* **64**, 1 (1995).
65. L. Fleury, A. Zumbusch, M. Orrit, R. Brown and J. Bernard, *J. Lumin.* **56**, 15 (1993).
66. A. Zumbusch, L. Fleury, R. Brown, J. Bernard and M. Orrit, *Phys. Rev. Lett.* **70**, 3584 (1993).
67. R. Brown and M. Orrit, In: T. Basché, W. E. Moerner, M. Orrit and U. P. Wild (editors), *Single Molecule Optical Detection, Imaging and Spectroscopy.* VCH, Weinheim (1997), Chapter 1.4.

68. W. A. Phillips (editor), *Amorphous Solids Low-Temperature Properties, Topics in Curr. Phys. vol.* **24**, Springer, Berlin (1981).
69. A. M. Boiron, P. Tamarat, B. Lounis, R. Brown and M. Orrit, *Chem. Phys.* **247**, 119 (1999).
70. A. M. van Oijen, M. Ketelaars, J. Köhler, T. J. Aartsma and J. Schmidt, *Science* **285**, 400 (1999).
71. Y. Berlin, A. Burin, J. Friedrich and J. Köhler, *Phys. Life Rev.* **4**, 64 (2007).
72. M. Brecht, *Mol. Phys.* **107**, 1955 (2009).
73. S. Fujiyoshi, M. Fujiwara and M. Matsushita, *Phys. Rev. Lett.* **100**, 168101 (2008).
74. P. F. Barbara, W. S. Chang, S. Link, G. D. Scholes and A. Yethiraj, *Annu. Rev. Phys. Chem.* **58**, 565 (2007).
75. J. G. Mueller *et al.*, *Appl. Phys. Lett.* **84,** 1183 (2004).
76. R. Hildner, U. Lemmer, U. Scherf, M. VanHeel and J. Köhler, *Adv. Mat.* **19**, 1978 (2007).
77. F.A. Feist, G. Tommaseo and T. Basché, *Phys. Rev. Lett.* **98**, 208301 (2007).
78. F. A. Feist, G. Tommaseo and T. Basché, *J. Phys. Chem.* **C113**, 11484 (2009).
79. Y. Ebihara and M. Vacha, *J. Phys. Chem.* **B112**, 12575 (2008).
80. F. Dubin, J. Berrehar, R. Grousson, T. Guillet, C. Lapersonne-Meyer, M. Schott and V. Voliotis, *Phys. Rev.* **B66**, 113202 (2002).
81. N. B. Manson and R. L. McMurtrie, *J. Lumin.* **127**, 98 (2007).
82. J. R. Rabeau, Y. L. Chin, S. Prawer, F. Jelezko, T. Gaebel and J. Wrachtrup, *Appl. Phys. Lett.* **86**, 131926 (2005).
83. I. Aharonovich, C. Y. Zhou, A. Stacey, J. Orwa, S. Castelletto, D. Simpson, A. D. Greentree, F. Treussart, J. F. Roch and S. Prawer, *Phys. Rev.* **B79**, 235316 (2009).
84. A. van Duijn-Arnold, R. Zondervan, J. Schmidt, P. G. Baranov and E. N. Mokhov, *Phys. Rev.* **B64**, 085206 (2001).
85. A. P. Alivisatos, *Science* **271**, 933 (1996).
86. S. A. Empedocles, R. Neuhauser, K. Shimizu and M. G. Bawendi, *Advanced Materials* **11**, 1243 (1999).
87. D. Gammon and D. G. Steel, *Phys. Today* **55**, 36 (2002).
88. L. Biadala, Y. Louyer, P. Tamarat and B. Lounis, *Phys. Rev. Lett.* **103**, 037404 (2009).
89. R. Lange, W. Grill and W. Martienssen, *Europhys. Lett.* **6**, 499 (1988).
90. R. Rodrigues-Herzog, F. Trotta, H. Bill, J. M. Segura, B. Hecht and H. Guntherodt, *Phys. Rev.* **B62** 11163 (2000).
91. G. Wrigge, I. Gerhardt, J. Hwang, G. Zumofen and V. Sandoghdar, *Nature Phys.* **4**, 60 (2008).
92. G. Wrigge, J. Hwang, I. Gerhardt, G. Zumofen and V. Sandoghdar, *Opt. Expr.* **16,** 17358 (2008).
93. T. Plakhotnik, *Chem. Phys. Chem.* **7**, 1699 (2006).
94. W. E. Moerner, *Proc. Natl. Acad. Sci. USA* **104**, 12596 (2007).

95. A. C. J. Brouwer, J. Köhler, E. J. J. Groenen and J. Schmidt, *J. Chem. Phys.* **105**, 2212 (1996).
96. P. Tchénio, A. B. Myers and W. E. Moerner, *J. Phys. Chem.* **97**, 2491 (1993).
97. L. Fleury, Ph. Tamarat, B. Lounis, J. Bernard and M. Orrit, *Chem. Phys. Lett.* **236**, 87 (1995).
98. S. Kummer, F. Kulzer, R. Kettner, T. Basche, C. Tietz, C. Glowatz and C. Kryschi, *J. Chem. Phys.* **107**, 7673 (1997).
99. M. Banasiewicz, D. Wiacek and B. Kozankiewicz, *Chem. Phys. Lett.* **425**, 289 (2006).
100. T. Nonn and T. Plakhotnik, *Chem. Phys. Lett.* **336**, 97 (2001).
101. A. Kiraz, M. Ehrl, Ch. Bräuchle and A. Zumbusch, *J. Chem. Phys.* **118**, 10821 (2003).
102. A. Kiraz, M. Ehrl, C. Hellriegel, C. Bräuchle and A. Zumbusch, *Chem. Phys. Chem.* **6**, 919 (2005).
103. A. C. J. Brouwer, J. Köhler, A. M. van Oijen, E. J. J. Groenen and J. Schmidt, *J. Chem. Phys.* **110**, 9151 (1999).
104. J. Bernard, L. Fleury, H. Talon and M. Orrit, *J. Chem. Phys.* **98**, 850 (1993).
105. P. Tchénio, A. B. Myers and W. E. Moerner, *J. Lumin.* **56**, 1 (1993).
106. W. E. Moerner and W. P. Ambrose, *Phys. Rev. Lett.* **66**, 1376 (1991).
107. M. Kuno, D. P. Fromm, H. F. Hamann, A. Gallagher and D. J. Nesbitt, *J. Chem. Phys.* **112**, 3117 (2000).
108. M. Haase, C. G. Hübner, E. Reuther, A. Herrmann, K. Müllen and T. Basché, *J. Phys. Chem.* **B108**, 10445 (2004).
109. J. P. Hoogenboom, E. M. H. P. van Dijk, J. Hernando, N. F. van Hulst and M. F. Garcia-Parajo, *Phys. Rev. Lett.* **95**, 097401 (2005).
110. J. Schuster, F. Cichos and C. von Borczyskowski, *Appl. Phys. Lett.* **87**, 051915 (2005).
111. F. Cichos, C. von Borczyskowski and M. Orrit, *Curr. Opin. Colloid Interf. Sci.* **12**, 272 (2007).
112. P. Frantsuzov, M. Kuno, B. Janko and R. A. Marcus, *Nature. Phys.* **4**, 519 (2008).
113. F. D. Stefani, J. P. Hoogenboom and E. Barkai, *Phys. Today* **62**, 34 (2009).
114. T. Basché, W. P. Ambrose and W. E. Moerner, *J. Opt. Soc. Am.* **B9**, 829 (1992).
115. W. E. Moerner, T. Plakhotnik, T. Irngartinger, M. Croci, V. Palm and U. P. Wild, *J. Phys. Chem.* **98**, 7382 (1994).
116. R. Métivier, F. Nolde, K. Müllen and T. Basché, *Phys. Rev. Lett.* **98**, 047802 (2007).
117. G. Hinze, R. Métivier, F. Nolde, K. Müllen and T. Basché, *J. Chem. Phys.* **128**, 124516 (2008).
118. R. Métivier, T. Christ, F. Kulzer, T. Weil, K. Müllen and T. Basché, *J. Lumin.* **110**, 217 (2004).
119. C. Hettich, C. Schmitt, J. Zitzmann, S. Kuhn, I. Gerhardt and V. Sandoghdar, *Science* **298**, 385 (2002).
120. M. Bauer and L. Kador, *J. Chem. Phys.* **118**, 9069 (2003).

121. C. Hofmann and M. Orrit, *J. Phys. Chem.* **B110**, 18925 (2006).
122. T. Nonn and T. Plakhotnik, *Phys. Rev. Lett.* **85** 1556 (2000).
123. T. Plakhotnik, *J. Chem. Phys.* **114**, 5631 (2001).
124. M. Bauer and L. Kador, *J. Lumin.* **107**, 32 (2004).
125. C. Hofmann, A. Nicolet, M. A. Kol'chenko and M. Orrit, *Chem. Phys.* **318**, 1 (2005).
126. M. A. Kol'chenko, A. A. L. Nicolet, M. D. Galouzis, C. Hofmann, B. Kozankiewicz and M. Orrit, *New J. Phys.* **11**, 023037 (2009).
127. Y. G. Vainer, A. V. Naumov, M. Bauer and L. Kador, *Opt. Spectrosc.* **94**, 864; *Opt. Spectrosc.* **94**, 873 (2003).
128. A. V. Naumov, Y. G. Vainer, M. Bauer and L. Kador, *J. Chem. Phys.* **119**, 6296 (2003).
129. A. V. Naumov, Y. G. Vainer and L. Kador, *Phys. Rev. Lett.* **98**, 145501 (2007).
130. Y. G. Vainer, A. V. Naumov, M. Bauer and L. Kador, *J. Chem. Phys.* **122**, 244705 (2005).
131. Y. G. Vainer, *J. Lumin.* ***125***, 279 (2007).
132. A. V. Naumov, Y. G. Vainer and L. Kador, *Phys. Rev.* **B79**, 132201 (2009).
133. Y. G. Vainer, A. V. Naumov, M. Bauer and L. Kador L, *Phys. Rev. Lett.* **97**, 185501 (2006).
134. C. von Borczyskowski, A. Oppenländer, H. P. Trommsdorff and J. C. Vial, *Phys. Rev. Lett.* **65**, 3277 (1990).
135. M. Johnson, K. Orth, J. Friedrich and H. P. Trommsdorff, *J. Chem. Phys.* **105**, 9762 (1996).
136. A. Sigl, C. Scharnagl, J. Friedrich, J. A. Gourdon and M. Orrit, *J. Chem. Phys.* **128**, 044508 (2008).
137. C. Brunel, P. Tamarat, B. Lounis, J. C. Woehl and M. Orrit, *J. Phys. Chem.* **A103**, 2429 (1999).
138. J. M. Caruge and M. Orrit, *Phys. Rev.* **B64**, 205202 (2001).
139. J. M. Caruge and M. Orrit, *J. Lumin.* **98**, 1 (2002).
140. M. Bauer and L. Kador, *J. Lumin.* **98**, 75 (2002).
141. M. Bauer and L. Kador, *J. Phys. Chem.* **B107,** 14301 (2003).
142. Y. Sekiguchi, S. Habuchi and M. Vacha, *Chem. Phys. Chem.* **10**, 1195 (2009).
143. M. Bauer and L. Kador, *Chem. Phys. Lett.* **407**, 450 (2005).
144. P. Tamarat, B. Lounis, J. Bernard, M. Orrit, S. Kummer, R. Kettner, S. Mais and T. Basché, *Phys. Rev. Lett.* **75**, 1514 (1995).
145. B. Lounis, F. Jelezko and M. Orrit, *Phys. Rev. Lett.* **78**, 3673 (1997).
146. T. Basché, W. E. Moerner, M. Orrit and H. Talon, *Phys. Rev. Lett.* **69**, 1516 (1992).
147. C. Brunel, B. Lounis, P. Tamarat and M. Orrit, *Phys. Rev. Lett.* **81**, 2679 (1998).
148. F. Jelezko, B. Lounis and M. Orrit, *J. Chem. Phys.* **107**, 1692 (1997).
149. C. Brunel, B. Lounis, P. Tamarat and M. Orrit, *Phys. Rev. Lett.* **83**, 2722 (1999).
150. I. Gerhardt, G. Wrigge, G. Zumofen, J. Hwang, A. Renn and V. Sandoghdar, *Phys. Rev.* **A79**, 011402 (2009).
151. B. Lounis and W. E. Moerner, *Nature* **407**, 491 (2000).

152. B. Lounis and M. Orrit, *Rep. Progr. Phys.* **68**, 1129 (2005).
153. I. Gerhardt, G. Wrigge, P. Bushev, G. Zumofen, M. Agio, R. Pfab and V. Sandoghdar, *Phys. Rev. Lett.* **98**, 033601 (2007).
154. G. Zumofen, N. M. Mojarad, V. Sandoghdar and M. Agio, *Phys. Rev. Lett.* **101**, 180404 (2008).
155. W. E. Moerner, T. Plakhotnik, T. Irngartinger and U. P. Wild, *Phys. Rev. Lett.* **73**, 2764 (1994).
156. M. Orrit, *Science* **298**, 369 (2002).
157. D. E. Chang, A. S. Sorensen, E. A. Demler and M. D. Lukin, *Nature Phys.* **3**, 807 (2007).
158. J. Hwang, M. Pototschnig, R. Lettow, G. Zumofen, A. Renn, S. Götzinger and V. Sandoghdar, *Nature* **460**, 76 (2009).
159. J. Köhler, J. A. J. M. Disselhorst, M. C. J. M. Donckers, E. J. J. Groenen, J. Schmidt and W. E. Moerner, *Nature* **363**, 242 (1993).
160. J. Wrachtrup, C. von Borczyskowski, J. Bernard, M. Orrit and R. Brown, *Nature* **363**, 244 (1993).
161. J. Wrachtrup, A. Gruber, L. Fleury and C. von Borczyskowski, *Chem. Phys. Lett.* **267**, 179 (1997).
162. A. Gruber, A. Dräbenstedt, C. Tietz, L. Fleury, J. Wrachtrup and C. von Borczyskowski, *Science* **276**, 2012 (1997).
163. L. Childress, M. V. Gurudev Dutt, J. M. Taylor, A. S. Zibrov, F. Jelezko, J. Wrachtrup, P. R. Hemmer and M. D. Lukin, *Science* **314**, 281 (2006).
164. P. Tamarat, T. Gaebel, J. R. Rabeau, M. Khan, A. D. Greentree, H. Wilson, L. C. L. Hollenberg, S. Prawer, P. Hemmer, F. Jelezko and J. Wrachtrup, *Phys. Rev. Lett.* **97**, 083002 (2006).
165. P. Tamarat, N. B. Manson, J. P. Harrison, R. L. McMurtrie, A. Nizovtsev, C. Santori, R. G. Beausoleil, P. Neumann, T. Gaebel, F. Jelezko, P. Hemmer and J. Wrachtrup, *New J. Phys.* **10**, 045004 (2008).
166. G. D. Fuchs, V. V. Dobrovitski, R. Hanson, A. Batra, C. D. Weis, T. Schenkel and D. D. Awshalom, *Phys. Rev. Lett.* **101** 117601 (2008).
167. R. Hanson, V. V. Dobrovitski, A. E. Feiguin, O. Gywat and D. D. Awschalom, *Science* **320**, 352 (2008).
168. S. Takahashi, R. Hanson, J. van Tol, M. S. Sherwin and D. D. Awschalom, *Phys. Rev. Lett.* **101**, 047601 (2008).
169. J. R. Maze, P. L. Stanwix, J. S. Hodges, S. Hong, J. M. Taylor, P. Cappellaro, L. Jiang and M. V. Gurudev Dutt, *Nature* **455**, 644 (2008).
170. G. Balasubramanian, P. Neumann, D. Twitchen, M. Markham, R. Kolesov, N. Mizuochi, J. Isoya, J. Achard, J. Beck, J. Tissler, V. Jacques, P. R. Hemmer, F. Jelezko and J. Wrachtrup, Nat. *Materials* **8**, 383 (2009).
171. A. Kinkhabwala, Z. Yu, S. Fan, Y. Avlasevich, K. Müllen, and W. E. Moerner, Nature Photon. **3**, 654 (2009).

CHAPTER 13

NOBLE-GAS CHEMISTRY

Wojciech Grochala

ICM, University of Warsaw, Pawińskiego 5a, 02106 Warsaw Poland and Faculty of Chemistry, University of Warsaw, Pasteur 1, 02093 Warsaw Poland
E-mail: wg22@cornell.edu

Leonid Khriachtchev and Markku Räsänen

Department of Chemistry, P.O. Box 55, FIN-00014 University of Helsinki, Finland
E-mails: leonid.khriachtchev@helsinki.fi; markku.rasanen@helsinki.fi

Noble-gas chemistry was started in 1962 with the discovery of xenon hexafluoroplatinate followed with a number of compounds binding xenon or krypton. We highlight the classical and more exotic noble-gas compounds and discuss the nature of their bonding starting with strongly bound systems and progressing to weak interactions. Noble-gas hydrides with the common formula HNgY were found in 1995, which led later to the identification of the first argon compound HArF. The formation mechanism of noble-gas hydrides at low temperatures is described in detail followed with a model of bonding. The interactions of the noble-gas hydrides with their surroundings and with complexing molecules are discussed. The chapter ends with known and potential applications of noble gases and with challenges encountered.

1. Introduction

The chemistry of noble gases began with a room temperature synthesis of the first xenon compound in the solid state, xenon hexafluoroplatinate ($XePtF_6$) by Neil Bartlett.[1] The true and indeed very complex nature of the "$XePtF_6$" compound was confirmed by experiments conducted nearly

40 years after the initial preparation.[2] The history of the breaking of the inertness of the noble gases is certainly complex and fascinating, full of many misleading reports, surprises, and ingenuity.[3]

The chemical bonds between noble gas and other atoms are usually quite fragile due to various redox (electron-transfer) reactions. In consequence of that, the exploration of noble-gas chemistry constitutes a very good probe of the capabilities offered by low-temperature experimental techniques and also serves as an important incentive for their development. Low-temperature solid hosts have shown to be excellent to synthesize and identify novel species. For noble-gas containing species, soon after the breakthrough of Neil Bartlett, matrix isolation methods were applied to produce noble-gas halides like KrF_2, $XeCl_2$, and XeClF.[4,5]

Matrix photolysis and thermal annealing of different small hydrides resulted in a puzzle by producing extremely strong unidentified absorptions in the mid-IR region. One characteristic example was the photolysis of HCl in Kr and Xe matrices.[6] The first suggestion on the structure of these species came from the diatomics-in-ionic-systems (DIIS) simulation of the HCl–Xe_n system made by Last and George.[7] They have found that a charge-transfer structure of HXeCl appears to be stable. Such a structure was then verified by quantum chemical calculations, and a number of noble-gas hydrides of the common structure HNgY, where Ng = Ar, Kr, or Xe and Y is an electronegative atom or fragment, were predicted and experimentally characterized. The latest member of this family of molecules is doubly "xenonated" water, HXeOXeH.[8] By the year 2009, the number of these molecules amounts 23 (see Table 1), including the first neutral chemical compound of argon, HArF.[9]

In this chapter on noble-gas chemistry, we present our understanding of the properties of noble-gas compounds made at ambient or cryogenic temperatures and describe some interesting predictions and challenges. As a shining example, we discuss in detail the noble-gas hydrides, their synthesis, properties, and complexation with other molecules in low-temperature matrices.

Table 1. Experimentally identified noble-gas hydride molecules by the year 2009.

		HXeH
		HXeI
		HXeBr
	HKrCl	HXeCl
HArF	HKrF	
	HKrCN	HXeCN
		HXeNC
		HXeOH
		HXeO
		HXeOXeH
		HXeSH
		HXeNCO
	HKrCCH	HXeCCH
	HKrCCCN	HXeCCCN
		HXeCCXeH
		HXeCC
	HKrCCCCH	HXeCCCCH

2. Noble-Gas Chemistry from Cryogenic to Ambient Temperatures

2.1. *The ambient temperature regime: "Classical" noble-gas compounds and the nature of chemical bonding*

The seminal 1962 discoveries that noble gases are indeed reactive enough to form chemical bonds have resulted in an avalanche of novel compounds during the next 50 years. It is estimated that the number of noble-gas compounds reaches today about five hundred. The barely noticeable seed immediately sprouted and the myth of inertness collapsed. The chemical bonding in the multitude of novel species is hardly different from those observed in the first synthesized compounds. Indeed, it is mostly the fluoro- and oxo-connections of the heavier noble gases (Kr and Xe), which predominate this beautiful chemistry.[10,11]

Xenon shows reactivity similar to iodine and it forms compounds at oxidation states from (II), via (IV) and (VI), to (VIII); these are isoelectronic and most often isostructural with connections of I(I), I(III),

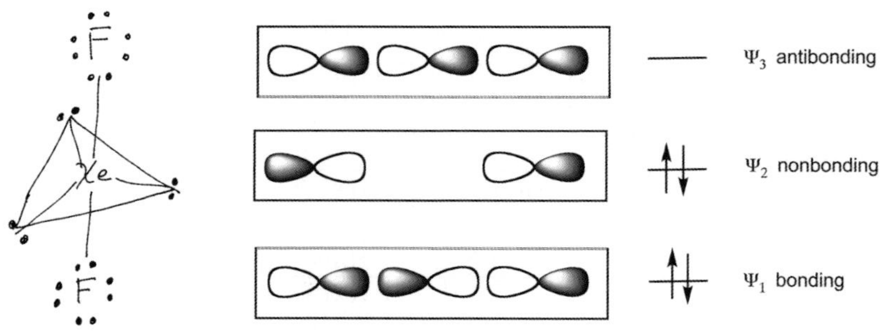

Fig. 1. Classical picture of chemical bonding in XeF_2 molecule. The Lewis (dot) diagram (left) and the molecular orbital picture of three-center four-electron bonding neglecting the *s/p* mixing (right). Similar diagrams may be drawn for tetragonal-planar XeF_4 and for octahedral XeF_6.

I(V), and I(VII), respectively. The binary fluorides of xenon (XeF_2, XeF_4, and XeF_6) are probably the best characterized of all the compounds of this element by both experiment and theory.[12] Meticulous studies of Xe(VI)O_3 and Xe(VIII)O_4 have been limited by the explosive nature of these compounds, but the oxo-salts of xenon (perxenathes) such as Ba_2Xe(VIII)O_6 are thermally quite stable.

XeF_2 is one of the most stable chemical compounds of the noble gases; it forms soft molecular crystals (space group *I*4/*mmm*) and sublimes easily (even at room temperature) without decomposition. The Xe–F bond length is 1.974 Å in the gas phase and increases only slightly to 2.00 Å in the solid state. The most frequently drawn picture of the chemical bonding in XeF_2 is that of 3-center 4-electron bonding (molecular orbital diagram in Fig. 1 accounts only for 5*p* orbitals of Xe and neglects 5*s* contribution). Both Xe–F bonds are equivalent in the linear symmetric XeF_2 molecule. However, the occupied Xe–F bonding HOMO-1 orbital (σ_u) and the unoccupied LUMO (also σ_u) are prone to mixing with the occupied Xe–F nonbonding HOMO (σ_g) when a molecule is found in an asymmetric environment (otherwise mixing is symmetry-forbidden). Thus, the Xe–F bonds have substantially different lengths in the case of ionic [A–F^-]···[XeF^+] salts, where A represents a strong Lewis acid (SbF_5, SO_3, etc.). For example, the closest Xe–F separations are 1.84 and 2.35 Å for the [XeF][Sb_2F_{11}] salt.

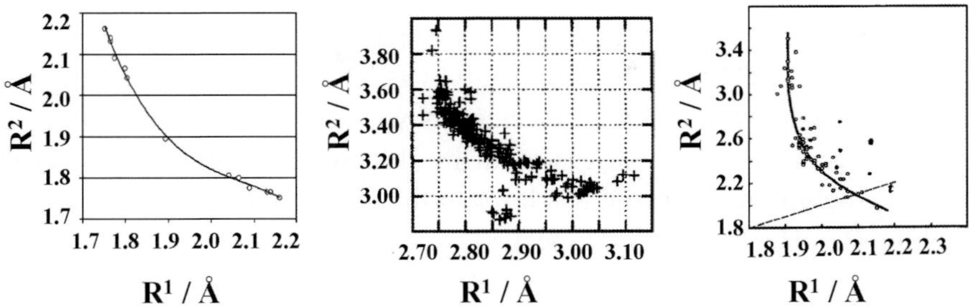

Fig. 2. Illustration of the plasticity of the first coordination sphere of (left) Kr(II) for a few fluoro-connections, (middle) hypervalent Sb(III) and Sb(V) linked to iodine, and (right) Cu(II) in an oxide environment. The first- and second-order Jahn–Teller effects are responsible to various degrees for deformation of the first coordination sphere of the central cations. Reproduced with permission from Refs. 13–15.

One interesting observation is that if one Xe–ligand bond strengthens, the other weakens and *vice versa*. All known chemical connections of noble gases exhibit the same characteristic feature as illustrated in Fig. 2 for selected fluoro-connections of krypton. The electron-rich Ng(II) cations exhibit substantial flexibility of the first coordination sphere in analogy with other "hypervalent" species as Sb(III), Sb(V), etc.[14] Such behavior, sometimes referred to as distortion isomerism, is due to the second-order Jahn–Teller effect; it renders noble-gas species analogous to the odd–electron transition-metal cations such as Cu(II) (susceptible to the first-order Jahn–Teller effect, see Fig. 2).[15]

It is remarkable that the entire route from ideally symmetric to highly asymmetric hypervalent bonding may be encompassed by deliberate variation of one parameter only: the Lewis acidity.[16] Thus, at least five distinct well-characterized phases exist in the phase diagram of the XeF_2/XeF_5AsF_6 mixture, and each of them exhibits a more or less pronounced dissociation of XeF_2 into the ionic $[A{-}F^-]\cdots[XeF^+]$ form, consistent with the acid/base ratio, for example:

$$2XeF_2 + XeF_5\,AsF_6 \rightarrow 2[XeF_2]\cdot XeF_5\,AsF_6, \tag{1}$$

$$2[XeF_2]\cdot[XeF_5\,AsF_6] + XeF_5\,AsF_6 \rightarrow 2\{XeF_2\cdot[XeF_5\,AsF_6]\}, \tag{2}$$

$$\{XeF_2\cdot[XeF_5\,AsF_6]\} + XeF_5\,AsF_6 \rightarrow [XeF_2]\cdot 2[XeF_5\,AsF_6]. \tag{3}$$

The more covalent bond between Xe and F in the XeF^+ cation is of course appreciably stronger than the average hypervalent Xe–F bond for symmetric XeF_2.[16] This observation gives a hint as to one possible strategy toward new stable noble-gas species. As we see later, attempts to avoid hypervalence of the noble-gas atom by designing a highly asymmetric ionic structure in fact bring us closer to the first chemical connections of helium ($F^-\cdots HeO$ and related species).

Finally, we note that an entire family of compounds is known that contain XeF_2 or XeF_4 as molecular ligands for "naked" metal cations.[17]

2.2. *It gets colder: Nonclassical noble-gas compounds with the Xe–N, Xe–C, Xe–Cl, Kr–O, Kr–N, and other similar bonds*

The first genuine Xe–N bond saw daylight in 1974, with the preparation of $[FXe^+][^-N(SO_2F)_2]$ by LeBlond and DesMarteau,[18] the crystal structure of which has been obtained a few years later by Sawyer *et al.*[11] This white compound is isolable in gram quantities and is stable to decomposition up to +70 °C, but its high-yield preparation is usually carried out at *ca.* −200 °C. All other compounds with the Xe–N bond are much less thermally stable than the first synthesized species in this family.

It was also quite difficult to produce compounds with real Xe–C bonds presented in Fig. 3. The precedents, derivatives of the perfluorophenylxenon(II) cation, $[(F_5C_6)Xe^+]$, were independently synthesized by Naumann and Tyrra and by Frohn and Jacobs as late as 1988. Modest cooling of the reaction mixtures to −30 °C was applied during synthesis and solutions of these colorless solids in acetonitrile proved to be stable at *ca.* 0 °C.[19,20] Further developments in this field paved the way to preparation of $[(F_5C_6)Xe^+][AsF_6^-]$, which can actually be melted at +102 °C with negligible decomposition. Due to its remarkable stability, the fluoroarsenate salt has become an important reagent in the emerging field of organoxenon chemistry.

The formation of the Xe–Cl bond at temperatures close to ambient was first seen one decade ago.[22] This new chemical bond was brought to life in two novel crystalline species, $(F_5C_6)XeCl$ and $\{[(F_5C_6)Xe]_2Cl^+\}(AsF_6)$, which showed reasonable kinetic stability at

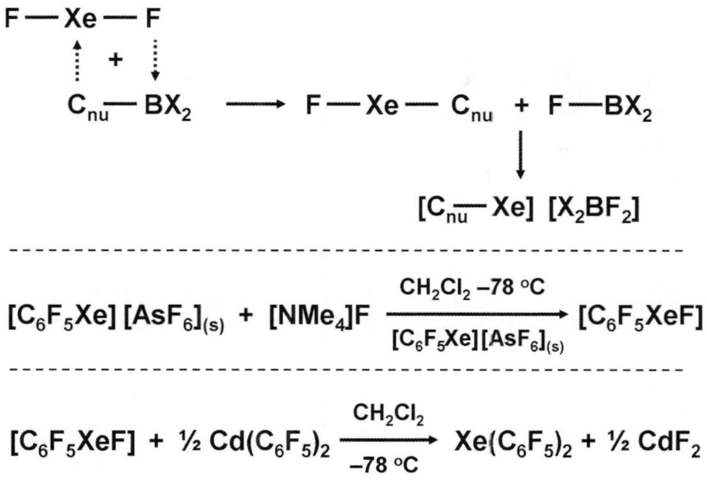

Fig. 3. Preparation of the Xe–C bonds via the xenodeborylation reaction (top) and the subsequent utilization of the $[(F_5C_6)Xe^+][AsF_6]$ precursor for synthesis of $[(F_5C_6)Xe^+]F$ (middle) and $Xe(F_5C_6)_2$ (bottom). Reproduced with permission from Ref. 21.

ambient temperature but their preparation required cooling of the reaction vessels to −78 °C. Since then only one more compound, $(XeCl^+)(Sb_2F_{11})$, has been added to the list of compounds isolable "in the flask."[23] Cooling turns out to be a necessary precondition in the synthesis of the less-kinetically stable species. Indeed, the formation of Xe–Cl bonds in low-temperature matrices has been achieved.[5]

Krypton, which is much less prone to chemical bond formation than Xe, has been successfully linked to N and O by Schrobilgen and co-workers, but this occurred only below −60 °C (for O: −90 °C) with BrF_5 (for O: SO_2ClF) as solvent.[24,25]

The formation of chemical bonds between noble gases and nonmetals, which are less electronegative than C, requires the use of very low temperatures. This has been shown by the formation of a multitude of novel chemical bonds (e.g., Xe–Br, Xe–S, Kr–C, Kr–H, etc.) in hydride molecules synthesized in matrices[9] and by the successful preparation of the first compound of argon, HArF.[26]

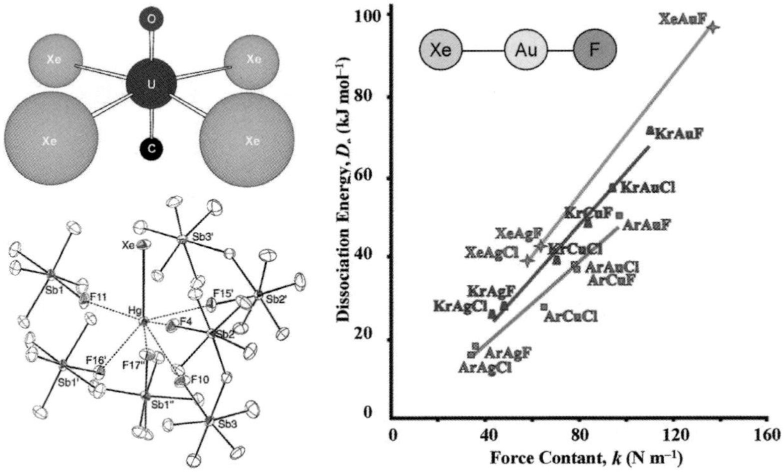

Fig. 4. Selected examples of chemical connections of xenon to metal cations in molecules and solids: $U(C)(O)(Xe)_4$ and $XeHgSb_3F_{16}$ (left). The XeAuF molecule and the calculated bond energy versus the experimental force constant for the Ng–M–X family of molecules where Ng = Ar, Kr, Xe; M = Cu, Ag, Au; and X = F, Cl, Br (right). Data reproduced with permission from Refs. 32, 35, and 38. See also Colour Insert.

2.3. *Very cold and exotic compounds with noble gas–metal bonds, and an unexpected warming*

It is now known that noble-gas atoms do not bind exclusively to nonmetals, but they may also form bonds to metals (see Fig. 4). This was realized, however, only in 1983 despite the many preceding inorganic and organometallic syntheses performed in a liquid or supercritical xenon environment, and many previous experimental efforts in the field of low-temperature matrix isolation. The inherent weakness of the majority of the Xe–metal cation bonds should be held responsible for these late yet fascinating discoveries, which are described below.

The first hints that xenon can bind to positively charged metal centers were obtained by Wells and Weitz and by Simpson *et al.* in 1983, when $XeM(CO)_5$ species (M = Cr, Mo, W) were observed for the first time at temperatures from −100 °C to −122 °C.[27,28] Further studies of these systems were conducted by Weiller in 1992[29] and more systematically by

Sun *et al.* in 1996.[30] In 1994, Thompson and Andrews synthesized novel adducts of noble gases to a metal center, NgBe(II)O, where Ng = Ar, Kr, and Xe.[31] The case of XeBeO is rather straightforward: a coordinatively unsaturated Be(II) cation exposes its empty (*sp*) hybrid, and is ready to bind whatever Lewis base you provide. Hence, it will bind a Xe atom with a surprisingly high-binding energy of over 0.3 eV. Electron-rich Cr(0) should of course be much less prone to auxiliary bonding than Be(II) (even if coordinatively unsaturated) but five CO π-acceptors are capable of withdrawing a large share of the electron density from Cr(0). The bonding of xenon to M(0) in $XeM(CO)_5$ is weaker than the respective bonding of H_2 or N_2 molecules yet it may be considered moderately strong, with M–Xe bond dissociation energies of (0.37 ± 0.04) eV.[27] Another interesting example of bonding between noble-gas atoms and neutral transition metal-containing molecules exhibiting a low coordination number of a metal center, has been provided by *trans*-$U(C)(O)Ng_4$ complexes (Ng = Ar, Kr, Xe).[32,33]

Another impressive piece of research came from the field of supersonic jets. In a series of landmark papers appearing since 2000, Gerry and co-workers have reported that noble-gas atoms bind to isolated MX molecules (where M = Cu, Ag, and Au; X = F, Cl, and Br), and the binding energy was estimated to be as large as a quarter an eV for the Ar···AgF derivative.[34] The work started from the most inert element (Ar) of the Ar–Kr–Xe set, and Kr and Xe were conquered in the following steps. To date, most of the molecules in the NgMX series (where M = Cu, Ag, and Au; X = F, Cl, and Br; Ng = Ar, Kr, and Xe) have been studied, and their fundamental properties measured and calculated. The XeAuF molecule synthesized only recently (Ref. 35) has proved to be the most strongly bound of all the complexes, and the Xe–Au(I) bond energy was estimated to exceed 1 eV.[36] In this important family of triatomics, the metal–Xe bonding smoothly traverses the frontier between weak interactions and a regular chemical bond.

All examples of metal–noble-gas bonding discussed so far are provided by molecules which exist only at very low temperatures and in isolation. However, these seemingly necessary preconditions were omitted in the 2000 study by Seidel and Seppelt. They were able to isolate the first complex containing the Xe–Au bond, which is kinetically stable at 20 °C at small overpressure of xenon (*ca.* 4 atm).[37] Studies for similar complexes and for bound Hg(II)–Xe species have followed.[38] Here the promising stage of the Ng–metal bonding stable at ambient (*p*, *T*) conditions has been opened and is now waiting for its new explorers.

2.4. *Species exhibiting very weak interactions to noble-gas atoms*

It has been long known that the heavier noble gases (Ar, Kr, Xe) can form [Ng · $6H_2O$] hydrates as well as clathrates with hydroquinone and toluene.[39] These cage compounds are formed due to beneficial packing of Ng atoms with host molecules and they are held together by very weak polarization forces; these compounds are by no means unique to noble gases.

Noble gases can also form endohedral Buckminster fullerene compounds, where the Ng atom is trapped inside a fullerene molecule. In 1993, Saunders *et al.* discovered that when C_{60} is exposed to a pressure of around 3 bar of He or Ne, a tiny amount of the complexes He@C_{60} and Ne@C_{60} are formed.[40] With higher pressures (3 kbar), it is possible to achieve a yield of up to 0.1%. Endohedral complexes with argon, krypton, and xenon have also been prepared. It is not the chemical character of these clathrates which is fascinating but rather the mechanism of their formation since it necessitates insertion of quite large Ne atoms through the walls of a fullerene molecule. A transition state for this genuine chemical reaction must exhibit chemical bonds between C atoms and the lightest noble gases.

3. Theoretical Predictions of Novel Compounds of Noble Gases: Toward Bonding to the Lightest Noble Gases

Theoretical calculations have played a vital role in the development of noble-gas chemistry, often providing a stimulus for experiments. It is impossible to discuss here all the important theoretical contributions exemplified by the inspiring works of Pyykkö and co-workers.[41,42] We briefly mention here some recent theoretical predictions of hypothetical compounds of the lightest noble gases, notably helium (see Fig. 5).

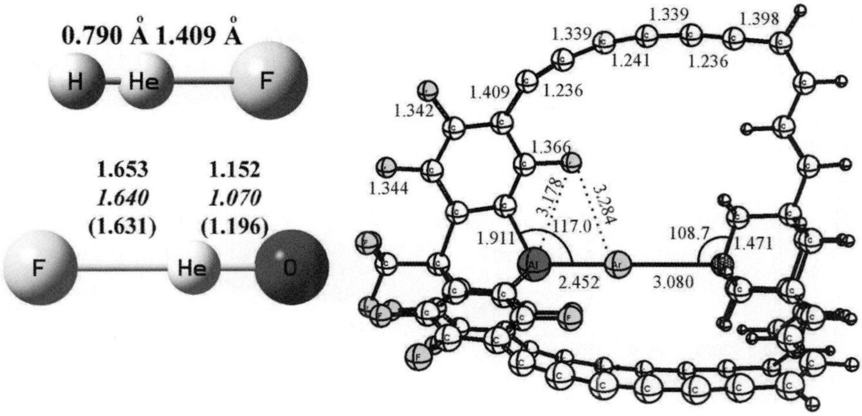

Fig. 5. Examples of exotic species containing noble-gas atoms according to recent theoretical predictions: HHeF,[43,44] $OHeF^-$, and a cage push–pull compound of Ar. Reproduced with permission from Refs. 47 and 50. Bond distances are in Å. See also Colour Insert.

In a direct analogy to the known HArF,[26] HHeF has been theorized.[43,44] It turns out, however, that the estimated lifetime of this interesting molecule is as short as 120 ps at 0 K. In agreement with expectations, deuteration brings elongation in lifetime up to 14 ns. The anions $OHeF^-$,[45–47] $SHeF^-$,[48] and $NBHeF^-$,[49] and two neutral salts $MOHeF^-$ [M = Cs, $N(CH_3)_4$][47] are a few other interesting examples of as-yet unknown exotic species which contain chemically bound helium atom. All these species suffer from relatively small barriers for chemical bond rupture along the stretching and/or bending coordinate. The consequences of tunneling to the lifetime can also be serious.

One possible tactic to stabilize the bonds between light noble gases and other elements might be to enclose them in a carefully prepared molecular cavity[50] or to apply external pressure.[13] The second approach, however, may be unsuccessful due to the opening of novel intermolecular decomposition channels of the fragile species. Unfortunately, the aspects of experimental and theoretical high-pressure chemistry and physics of noble gases and their compounds remain largely unexplored.[13]

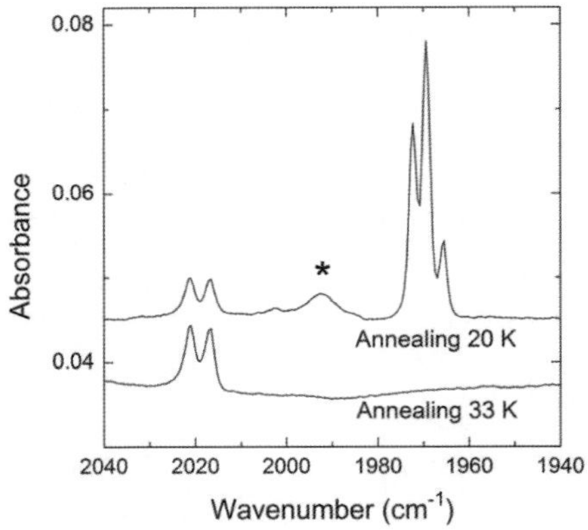

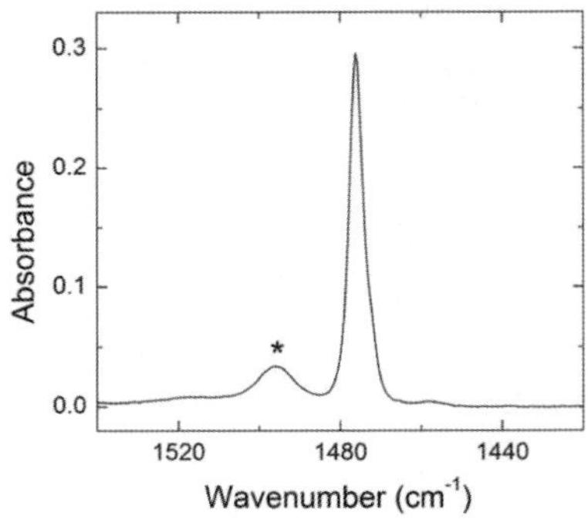

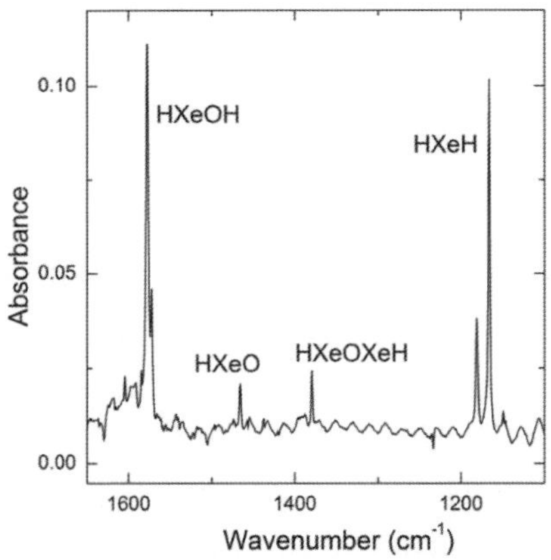

Fig. 6. IR spectra of selected noble-gas hydrides: HArF in an Ar matrix (upper left panel), HKrCl in a Kr matrix (upper right panel), and hydride molecules prepared from H_2O in a Xe matrix (lower panel). The observed absorptions are from the H–Ng stretching modes. The HY/Ng matrices were first photolyzed and then annealed. The bands from HArF and HKrCl librational motion are marked with asterisk.

4. Noble-Gas Hydrides (HNgY)

One interesting part of noble-gas chemistry is constituted by noble-gas hydrides with the general formula HNgY.[9] In the experiments, the HNgY molecules have been prepared in most cases by using UV photolysis and annealing of low-temperature HY/Ng (~1/1000) solid matrices (see Fig. 6).[51] Photodissociation of hydrogen-containing HY precursors with UV light can lead to isolation of H and Y fragments in the noble-gas lattice. The Y fragment usually remains in the parent cage whereas the H atom escapes it.[52] The main formation of HNgY molecules is achieved upon thermal annealing of UV photolyzed HY/Ng matrices. The HNgY formation presumably originates from the H + Ng + Y reaction of the neutral atoms and fragments. The high yields of the HNgY formation obtained upon thermal annealing, which are often tens of percent of dissociated HY molecules, suggest relatively low losses of H atoms during UV photolysis and thermal annealing.[53] In other words, the amount of annealing-induced HNgY molecules is comparable in favorite cases to the amount of photodissociated HY molecules.

4.1. *Direct formation and locality of photolysis*

In some cases (e.g., HArF, HKrCl, HXeNCO), the formation of noble-gas hydrides has been detected in relatively small amounts already during UV photolysis of HY/Ng matrices.[26,54,55] These experimental results suggest that permanent photodissociation of HY fragments in solid matrices is essentially a local process. The locality of photolysis means that dissociated H atoms are principally separated from the parent cage by a relatively short distance that is comparable with the lattice parameter (a few Å for noble-gas crystals). In this situation, the atoms can react with the parent Ng-Y center after the dissociation event. This principal concept was studied in detail for formation of HKrCl in an HCl/Kr matrix.[55] Laser-induced fluorescence of Cl atoms in solid krypton was used to estimate the proportion of different channels of solid-state photolysis. It appears that the formation of HKrCl intermediates is the major channel (about 60%) for permanent dissociation of HCl in solid krypton with excess energy of 1.8 eV. The direct stabilization of H and Cl atoms occurs only with a probability of 40%. These values do not account in-cage recombination of HCl if no cage exit occurs. The low photostability of HKrCl explains the relatively small steady-state HKrCl concentration observed during photolysis.

For more photolabile molecules, the intermediate channel is very difficult to detect; however, it can be dynamically very important. For example, an indirect evidence of such direct formation was also reported for UV photolysis of HI in solid xenon.[56] The HNgY intermediates are decomposed by further photolysis producing H and Y fragments trapped in a matrix.

The low-temperature formation of HArF in solid argon found experimentally further highlights the concept of local solid-state photolysis.[57] The HArF concentration slowly (on scale of hours) increases in a photolyzed HF/Ar matrix at 8 K. The formation of the deuterated molecule DArF at 8 K is much slower, by a factor of ~ 50, than the HArF formation. A probable mechanism for this low-temperature formation process is quantum tunneling of a H atom to the Ar-F neutral center through a matrix-induced barrier, and the large H/D isotope effect is a good fingerprint of quantum tunneling. It is clear at once that a short separation distance is required for efficient H-atom tunneling, which supports a short light-induced distance in the primary HF photolysis. Bihary *et al.* estimated the barrier of ~0.3 eV and the barrier thickness of ~1.3 Å for the H + Ar + F reaction from the closest separation of the fragments in an Ar matrix.[58] This calculated geometry allows a relatively high penetration probability through the barrier making the quantum tunneling process quite realistic. The H + Ar + F reaction barrier calculated by Bihary *et al.* is similar to the barrier for rotational isomerization of higher energy conformers of some carboxylic acids to the ground-state conformers controlled by H-atom quantum tunneling,[59] which supports the possibility of the tunneling mechanism for the HArF formation at the lowest experimental temperature.

These conclusions on local solid phase photolysis contradict some literature discussions on a long-range light-induced flight of H atoms in noble-gas matrices.[52] The analysis of IR absorption and laser-induced luminescence data has shown that some artifact reported in the literature can be caused by a large optical thickness of the matrices, which leads to major effects connected with self-limited photolysis and rising light absorbers.[60,61] Neglecting such important optical processes can also cause strong overestimates of cage exit probability for H atoms. In practical photolysis conditions, the cage exit of H atom seems to be relatively small, and in-cage recombination and/or in-cage reactions dominate. Systematic studies of cage-exit probabilities are not available; however, the values are most probably well below 10% in most cases of relevant matrix isolation studies.[62]

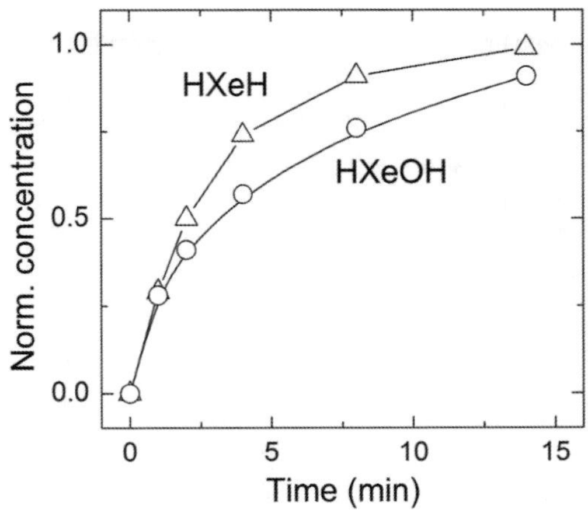

Fig. 7. Kinetics of the HXeOH and HXeH concentrations in a Xe matrix at 40 K.[63] The H_2O/Xe matrix was first photolyzed at 193 nm. The integrated band intensities are normalized by the values obtained after additional annealing at 45 K, which saturates the processes connected with H atom mobility. The data were obtained by integrating the H–Xe stretching absorptions.

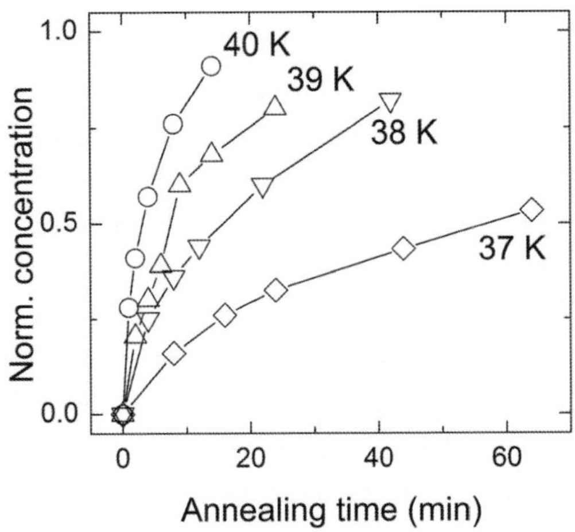

Fig. 8. Kinetics of the HXeOH concentration at different temperatures in a Xe matrix.[63] The H_2O/Xe matrix was first photolyzed at 193 nm. The integrated band intensities are normalized by the result obtained after additional annealing at 45 K.

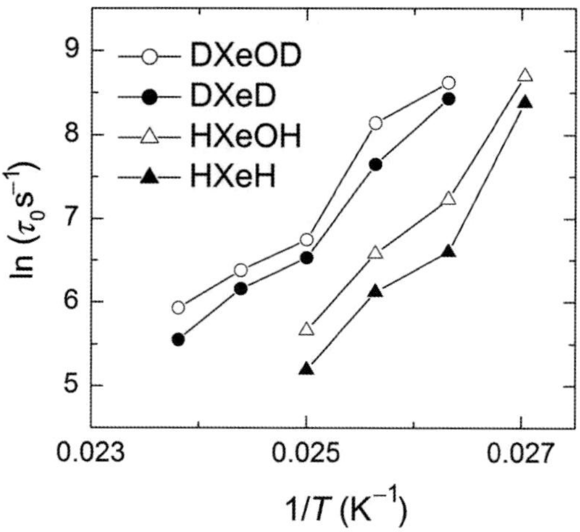

Fig. 9. H/D isotope effect on the characteristic formation time (in seconds) of noble-gas hydrides HXeH and HXeOH.[63] The formation time was measured at the 0.63 level of the product amount obtained after additional annealing at 45 K.

4.2. *Annealing-induced formation*

The main formation of noble-gas hydrides is observed upon thermal annealing of photolyzed HY/Ng matrices, which is connected with H-atom mobility. The rich chemistry found upon photolysis of HY/Ng matrices with subsequent annealing is demonstrated in Fig. 6.

When the matrix temperature is sufficiently high (*ca.* 30 and 40 K for Kr and Xe matrices), the HNgY concentration increases as presented for HXeOH and HXeH in Fig. 7. In the experimental formation kinetics, the HXeH concentration typically saturates faster compared to other HXeY molecules. The formation rate typically increases by a factor *ca.* 2 for a temperature increase by 1 K as seen in Fig. 8.[63] The deuterated molecules DNgY (Ng = Kr and Xe) form more slowly or at higher temperatures compared to HNgY, which means an isotope effect (see Fig. 9). The experimental activation energy in a Xe matrix is ~110 meV, and it is ~4 meV higher for the deuterated species. The experiments in Kr matrices yielded activation energies of ~64 and 68 meV for formation of HKrCl and DKrCl.[64] These activation energies correspond to global mobility of H and D atoms in noble-gas matrices as discussed later. The activation energies were extracted assuming the same preexponential factors in the Arrhenius plots for H and D atoms meaning dominating matrix effects in the mobility. If the preexponential factors for H and D atoms differ by $(2)^{1/2}$, similarly to the vibrational frequency in the cavity, the H/D isotope-induced difference becomes ~3 meV. Most probably, the activation energy differs for H and D atom mobility by a value from 3 to 4 meV for both Kr and Xe matrices.

In general, the thermally activated formation of HNgY molecules is a combination of global and local mobility processes.[56] Primary photolysis of HY precursors is presumably an essentially local event as described earlier. This fact means that the H atom may react upon annealing with the parent Ng-Y center with a relatively high probability. Local mobility is a short-range effect occurring in a perturbed matrix morphology, which generally changes the activation energy of the process in favor of geminate recombination.[65] In contrast, global mobility means random motion of atoms across the matrix lattice over large distances compared to the lattice parameter when the position of reaction is not essentially determined by the initial position of the H atom after photolysis. In other words, globally mobile H atoms loose the memory of the position of the parent cage.

Global mobility leads to reaction of H atoms mainly apart from the parent cage. In this case, the HNgY formation time should become longer for smaller H and Y concentrations because in average more jumps are required for H atoms to meet a Ng-Y center. In contrast, the formation time should be concentration-independent in the case of local (geminate) formation process when a H atom reacts with the parent Ng-Y center. The features of global mobility were shown experimentally for HKrCl and HXeCCH because the formation became slower for smaller precursor concentrations.[64,66] However, Pettersson *et al.* have shown that local thermal mobility of H atoms contributes to some extent to the formation of HXeI molecules in a Xe matrix, and the contribution of the local process decreases for more intense photolysis.[56] They suggested that the randomization of H atom distribution is produced by light-induced neutralization of solvated protons $XeHXe^+$.

A dependence of the product formation on the matrix deposition temperature was found and explained by different losses of mobile H atoms.

These losses are contributed by lattice defects (traps) that can permanently capture globally mobile H atoms. The short-range mobility of H atoms should not be influenced much by matrix defects. The amount of defects can be probed by the matrix optical properties due to the fact that matrices with lower amounts of defects are less optically scattering.[67,68] The matrix morphology depends on the deposition temperature, and the best optical properties suggest the lowest defect concentration for deposition of HBr/Xe matrices at 30 K.[66] The HXeBr amount was maximum for deposition at 30 K and it decreased for higher and lower deposition temperatures.[66] Thus, the HXeBr formation is more efficient in matrices with lower amounts of defects. Moreover, the higher HXeBr formation efficiency correlates with the narrowing of the HBr precursor band, which is also a fingerprint of a regular matrix structure. The long-range hydrogen mobility is substantially suppressed in Xe matrices with large amounts of defects (deposited at 10 K) and the product concentration is strongly decreased in this case. These experimental results suggest a dominating role of the global mobility for formation of noble-gas hydrides in matrices with a low amount of traps after extensive photolysis.

In the kinetic model of global H-atom mobility in noble-gas matrices, the HNgY formation time is a nearly linear function of the Ng/H ratio when losses are not taken into consideration.[66] However, the corresponding experimental dependences clearly exhibit saturation for small HY/Ng concentrations, and this fact can be also caused by losses of globally mobile H atoms. If the losses of H atoms are included into the kinetic model, the formation time becomes saturating for low precursor concentrations, in agreement with the experimental data.

4.3. *Modeling of the HNgY formation*

Many experimental observations can be interpreted based on simple kinetic considerations. A kinetic model can be constructed for the annealing-induced formation of HNgY molecules in noble-gas matrices. It has been experimentally shown that these molecules are formed via the neutral H + Ng + Y mechanism mainly upon thermally activated global mobility of H atoms. It is assumed for simplicity that photolysis fully decomposes the HY precursor producing isolated H and Y fragments. Annealing mobilizes H atoms, and the following thermal reactions are considered for the annealing-induced process in solid xenon:[66,69,70]

$$H + Xe + Y \rightarrow HXeY, \quad (4)$$

$$H + Xe + H \rightarrow HXeH, \quad (5)$$

$$H + HXeY \rightarrow H_2 + Xe + Y \text{ or } H + Xe + HY, \quad (6)$$

$$H + HXeH \rightarrow H_2 + Xe + H, \quad (7)$$

$$H + T \rightarrow HT. \quad (8)$$

Reactions (4) and (5) describe the formation of HXeH and HXeY whereas reactions (6) and (7) describe the decomposition reactions of these molecules that has been also demonstrated experimentally.[64,71] The (H + T) loss channel for H atoms is included [reaction (8)], it describes the capture of mobile H atoms by various traps such as matrix defects, impurities, and other reactive species formed during photolysis, and HT denotes the concentration of the trapped (immobile) H atoms. The xenon atoms participating in reactions (4) and (5) are always available because the process occurs in a Xe matrix. The same reaction cross sections and the absence of reaction barriers are assumed in this model and reactions between mobile and immobile H atoms are neglected. In particular, the negligible reaction barriers are suggested by the formation of HXeBr and HXeCl observed in Ne matrices at 10 K.[72] By solving the corresponding differential equations, one gets the product concentration as a function of time. The described approach has been successfully used to simulate formation of such noble-gas molecules as HXeH, HXeOH, HKrCl, HXeCC, and HXeCCXeH.[63,64,66,71,73] The situation with HArF in an argon matrix seems to be different, and local processes dominate in the HArF formation mechanism.[57,74]

The described model allows studying the kinetics of the product amounts for different reaction constants and precursor concentrations in Xe and Kr matrices. In particular, it shows that the HXeH concentration in a Xe matrix increases and saturates quicker as compared to other HXeY, which is due to a faster decrease of the H concentration compared to the Y concentration. This kinetic feature was found experimentally as mentioned earlier and seen in Fig. 7. The developed model suggests that losses of H atoms due to formation of H_2 and HY molecules [reactions (6) and (7)] are probably of a lesser importance. On the other hand, losses by various traps [reaction (8)] can be significant depending on the trap concentration. Losses of mobile H atoms limit both formation efficiency and formation time of HNgY species.

A microscopic model of global hydrogen mobility in a perfect Xe lattice has been developed by Shestakov and co-workers.[63] In this model, Xe atoms constitute a face-centered-cubic lattice, and the harmonic elastic forces are considered between the nearest Xe neighbors only. The H-Xe interaction potential is taken in Tang–Toennies form with the parameters obtained from experiment,[75] which provides more reliable potential at small H-Xe distances compared to the computational data. From two possible interstitial positions of a H atom (octahedral and tetrahedral), the octahedral position is preferable energetically. The minimal model describing a H atom jump from an octahedral to tetrahedral cavity through the common Xe_3 triangle contains seven flexible Xe atoms around both cavities. A somewhat larger model with additional optimization of 34 Xe atoms which are the closest neighbors of the first 7 Xe atoms, 41 Xe atoms overall, was used. The interaction of the hydrogen atom with the rest of Xe lattice atoms was included for their fixed (not relaxed) positions.

The calculations performed by Shestakov and co-workers for a H atom in a Xe cluster yield at 40 K total energies of 61.9, 233.9, and 238.8 meV in the octahedral site, the tetrahedral site, and the transition state, respectively. It is shown that H atoms cannot be stabilized in tetrahedral sites, and they relax to octahedral sites after passing the transition state. The zero-point energy was included to estimate the activation energy of an elementary jump. As a result, the energies of H in the octahedral and transition states at 40 K are 130.6 and 298.3 meV, and the corresponding values for D system are 109.5 and 279.9 meV. This gives the activation energies of 167.7 and 170.3 meV for the elementary jumps of H and D atoms at 40 K. The theoretical activation energy is somewhat larger than the experimental value (~110 meV). The obtained difference between the H and D activation energies (2.6 meV) agrees reasonably well with the experimental estimate (3–4 meV). The jump rate was obtained using the transition-state theory. As a result, the k_H/k_D ratio was found to be 2.15 at 40 K, which is somewhat smaller than the experimental value of ~4 (see

Fig. 9). The calculated temperature leading to "experimental" H mobility ($k_H = 1\ s^{-1}$) is 60 K, which is somewhat higher than the experimental value. These discrepancies might originate from larger anharmonicity of the vibrations in the transition state, which would decrease the zero-point energy in the transition state considerably, and some additional processes probably contribute. For instance, the effects of tunneling and coupling between vibrations were not taken into account in this model, and the number of matrix atoms is probably insufficient. Evidently there is a lot to do to improve this model.

4.4. *Bonding of HNgY compounds*

The bonding of the HNgY compounds can be described as $(HNg)^+Y^-$ where $(HNg)^+$ is mainly covalently bonded and the interaction of the $(HNg)^+$ and Y^- is strongly ionic. It is interesting to note that the $(HNg)^+$ fragments are isoelectronic with the corresponding hydrogen halides, and should resemble them at the charge limit of $+1e$. Typically, the positive charge of the $(HNg)^+$ fragment is between $+0.5e$ and $+0.8e$, and external influences to this value have marked spectroscopic consequences. On the basis of this main bonding structure, one can easily design new noble-gas hydrides by reacting H and Ng atoms with strongly electronegative fragments Y. Interesting examples can be found in acetylenic systems, where the electron withdrawing effect of the $(CC)_nH$ fragment increases with n.[76]

4.5. *Complexes of HNgY molecules*

The ionic nature of HNgY molecules and their relatively weak bonding makes them amenable to remarkably strong external effects, and a number of 1:1 complexes have been experimentally studied in matrices ($HArF\cdots N_2$, $HKrCl\cdots N_2$, $HXeCl\cdots HCl$, $HXeCCH\cdots CO_2$, etc.).[77] The complexation-induced effects can easily be detected experimentally due to the strong H–Ng stretching intensity. As examples of complex formation with a nonpolar and polar molecule, we can take HKrCl complexed with N_2 or HCl.

For HKrCl complexes with N_2, the linear structure with N_2 interacting with HKrCl from the H end is the most stable.[77] Computationally the interaction shortens the H–Ng bond from its monomeric value of 1.500 to 1.465 Å and increases the Cl partial charge from $-0.67e$ to $-0.72e$. The computed H–Kr stretching absorption shifts from its monomeric value of 1828 cm^{-1} by +146 cm^{-1} to higher frequencies, and its intensity decreases from 2400 to 900 km mol^{-1}. Experimentally, in solid krypton the H–Kr stretching band for the linear complex was found blue-shifted by +113 cm^{-1} from its monomeric value. The bent structure also shows a substantial blue shift of +32.4 cm^{-1} of the H–Kr stretching band.

Even a more striking effect was found for HKrCl complexed with HCl where computationally three structures were found, and a bent structure with an interaction energy of about −40 kJ mol^{-1} was the most stable.[78] In this interaction, the H atom of the HCl molecule points toward the Cl atom of HKrCl at a distance of about 2.05 Å and at an angle of 78°. This interaction increases the positive charge of the $(HKr)^+$ unit by about $+0.1e$. Experimentally this 1:1 complexation was found to shift the H–Kr stretching band by up to +306 cm^{-1}, which seems to be the largest vibrational blue shift found for molecular complexes to date.

4.6. *Excited states of HNgY*

As mentioned earlier, the direct formation of HNgY molecules upon photolysis of the HY precursor is observed only in special cases. This limitation is due to the fact that the HNgY molecules possess very strong dissociative electronic absorptions where the photosensitivity originates from strong charge-transfer absorptions onto a repulsive excited electronic state.[79–81] Correlation of the IR spectra with the broad UV absorptions allows identification of the individual UV spectra of different HNgY molecules. The transition dipoles involved exceed 7 D as predicted by multireference configuration interaction calculations.[79] The experimental estimates agree with the theoretical predictions.[55,81]

4.7. *Thermal stability and enhanced site effects*

Many of the prepared noble-gas hydrides are stable at elevated temperatures up to the matrix degradation. As an exception, HXeOH is found to decay at *ca.* 55 K.[8,81]

An interesting case was found for HArF where annealing a photolyzed HF/Ar matrix at 16–20 K produces the H−Ar stretching triplet with components at 1965.7, 1969.4, and 1972.3 cm^{-1} (see Fig. 6).[26] These bands disappear at about 30 K, and the H−Ar stretching doublet absorption rises at 2016.3 and 2020.8 cm^{-1}.[74,82] The blue-shifted doublet belongs to the stable HArF which disappears only with the degradation of the Ar matrix. These configurations computationally are described in terms of local matrix morphology.[83] An additional broad band between the doublet and triplet absorptions (marked with asterisk in Fig. 6) is described in terms of librational motion of HArF in solid argon.[84] The extensive theoretical simulations have resulted in a good agreement between theory and experimental observations as described in Chapter 14.

5. The Known and Possible Applications and Challenges of Noble Gases and Their Compounds

Currently the majority of practical applications of noble gases utilize the elements and not the chemically bound species, the glow-discharge bulbs, ion and excimer lasers, providing a protective atmosphere (Ar) and cryogenic refrigeration (He) being the most important examples.

The heaviest noble gas, Xe, offers an impressive range of isotopes (124, 126, 128, 129, 130, 131, 132, 134, and 136) and practically all of them are available from commercial vendors. Since seven out of nine known nuclides of xenon have a zero nuclear spin, it must be considered fortunate that the natural abundance of the only spin-½ nucleus (^{129}Xe) is pretty high (+26%). This fact has been utilized in ^{129}Xe NMR spectroscopy which is now routinely performed without expensive isotopic enrichment.[85] Using NMR, xenon may be utilized as a probe of binding strength to sites of varying acidity on the surfaces of solids.

Let us now turn to the scarce and hypothetical applications of noble-gas compounds:

(i) Radioactive isotopes of krypton and xenon (which find uses in medicine and in archeological dating) are difficult to store and dispose, and compounds of these elements may be more easily handled than the gaseous forms. Thus, clathrates have been used for the separation of lighter He and Ne from heavier Ar, Kr, and Xe, and also for the transportation of the latter. In addition, ^{85}Kr hydrate provides a safe source of beta particles, while ^{133}Xe hydrate is a useful source of gamma rays.

(ii) Xenon fluorides are good fluorinating agents. A measure of the success of the XeF_2 is that it is the only chemical noble-gas compound which is commercially available (e.g., *Fluorochem* sells it as cheap as £5.60 per gram). Of the noble-gas compounds, XeF_2 is probably the most frequently used in chemical synthesis; as far as fundamental research is concerned, noble-gas chemistry is ultimately linked to the chemistry of fluorine and that of high-valent metal species. Another valuable use of XeF_2 is that of etching surfaces of silicon and other semiconductors (*cf.* US patent No. 4, 761, 302).

(iii) Perxenic acid (H_4XeO_6) is a valuable oxidizing agent because it has no potential for introducing impurities: the xenon is simply liberated as a gas. This reagent is rivaled only by ozone in this respect.

(iv) Xenon is used in the atomic energy field in bubble chambers, probes, and other applications where its high molecular weight is of value. For the same reason and due to relativistic effects, xenon accelerates the rate of the singlet–triplet intersystem crossing which is sometimes used to quench fluorescence in optical spectroscopy.

(v) It has been suggested that xenon might be utilized as a catalyst for the formation of as-yet unsynthesized compounds, such as AuF,[86,87] and for the preparation of novel cage polymorphs of various elements and compounds, utilizing "soft" adducts different from the known clathrates and hydrates.[13]

(vi) The noble-gas hydrides are examples of very high-energy compounds. For example, HXeOXeH is about 8 eV above the thermodynamical minimum of H_2O + 2Xe.[8] The perspective of this remarkable property is fully uncertain at the moment.

(vii) The "missing Xe-problem" is still without explanation, despite several attempts.[88] Connected with this the noble-gas hydrides are of particular interest and especially their stability dependence on solvation and pressure.

(viii) Xenon acts as an anesthetic substance. In fact, quite little is known about the molecular level mechanisms of the anesthesia. The Xe compounds where xenon is inserted to various functional groups may play some role and their studies may help in understanding the physiological effects.

Acknowledgments

The work at the University of Warsaw was supported by the Polish Ministry of Science and Higher Education via the BST 120000-501/641-32655 grant. W.G. thanks ICM and the Faculty of Chemistry for continuing sustenance, P. Leszczyński for screening patent databases, and A. J. Churchard for never-failing technical assistance. The work at the University of Helsinki was supported by the Academy of Finland and the Finnish Center of Excellence in Computational Molecular Science. L. K. and M. R. thank all the colleagues who have contributed to the described results on noble-gas hydrides.

References

1. N. Bartlett, Proc. Chem. Soc. (London) 218 (1962).
2. L. Graham, O. Graudejus, N. K. Jha and N. Bartlett, *Coord. Chem. Rev.* **197**, 321 (2000).
3. P. Laszlo and G. J. Schrobilgen, *Angew. Chem. Int. Ed. Engl.* **27**, 479 (1988), and references therein.
4. J. J. Turner and G. C. Pimentel, *Science* **180**, 974 (1963).
5. W. F. Howard and L. Andrews, *J. Amer. Chem. Soc.* **96**, 7864 (1974).
6. M. Pettersson, J. Lundell and M. Räsänen, *J. Chem. Phys.* **102**, 6423 (1995).
7. I. Last and T. F. George, *J. Chem. Phys.* **89,** 3071, (1988).
8. L. Khriachtchev, K. Isokoski, A. Cohen, M. Räsänen and R. B. Gerber, *J. Am. Chem. Soc.* **130,** 6114 (2008).
9. L. Khriachtchev, M. Räsänen and R. B. Gerber, *Acc. Chem. Res.* **42**, 183 (2009).
10. J. F. Lehmann, H. P. A. Mercier and G. J. Schrobilgen, *Coord. Chem. Rev.* **233**, 1 (2002).

11. J. F. Sawyer, G. J. Schrobilgen and S. J. Sutherland, *J. Chem. Soc., Chem. Commun.* 210 (1982).
12. Theory is now able to reproduce even the NMR shielding and spin–spin coupling constants of xenon fluorides despite the large relativistic effects for Xe: A. Antušek, M. Pecul and J. Sadlej, *Chem. Phys. Lett.* **427**, 281 (2006).
13. W. Grochala, *Chem. Soc. Rev.* **36**, 1533 (2007).
14. G. A. Landrum and R. Hoffmann, *Angew. Chem. Int. Ed. Engl.* **37**, 1887 (1998).
15. J. Gažo, I. B. Bersuker, J. Garaj, M. Kabešová, J. Kohout, H. Langfelderová, M. Melnik, M. Serator and F. Valach, *Coord. Chem. Rev.* **19**, 253 (1976).
16. B. Žemva, A. Jesih, D. H. Templeton, A. Zajkin, A. K. Cheetham and N. Bartlett, *J. Amer. Chem. Soc.* **109**, 7420 (1987).
17. M. Tramšek and B. Žemva, *J. Fluor. Chem.* **127**, 1275 (2006), and references therein.
18. R. D. LeBlond and D. D. DesMarteau, *J. Chem. Soc., Chem. Commun.* 555 (1974).
19. D. Naumann and W. Tyrra, *J. Chem. Soc., Chem. Commun.* 47 (1989).
20. H. J. Frohn and S. Jakobs, *J. Chem. Soc., Chem. Commun.* 625 (1989).
21. H. Frohn and V. V. Bardin, *Organometallics* **20**, 4750 (2001).
22. H. J. Frohn, T. Schroer and G. Henkel, *Angew. Chem. Int. Ed. Engl.* **38**, 2554 (1999).
23. S. Seidel and K. Seppelt, *Angew. Chem. Int. Ed. Engl.* **40**, 4225 (2001).
24. G. J. Schrobilgen, *J. Chem. Soc., Chem. Commun.* 863 (1988).
25. J. C. P. Sanders and G. J. Schrobilgen, *J. Chem. Soc., Chem. Commun.* 1576 (1989).
26. L. Khriachtchev, M. Pettersson, J. Lundell, N. Runeberg and M. Räsänen, *Nature* (*London*) **406**, 874 (2000).
27. J. R. Wells and E. Weitz, *J. Amer. Chem. Soc.* **114**, 2783 (1992).
28. M. B. Simpson, M. Poliakoff, J. J. Turner, W. B. Maier and J. G. J. McLaughlin, *J. Chem. Soc., Chem. Commun.* 1355 (1983).
29. B. H. Weiller, *J. Amer. Chem. Soc.* **114**, 10910 (1992).
30. X. Z. Sun, M. W. George, S. G. Kazarian, S. M. Nikiforov and M. Poliakoff, *J. Amer. Chem. Soc.* **118**, 10525 (1996). Many reactions involving photolysis with UV light can be performed in liquid xenon without involving a noble gas in the reaction.
31. C. A. Thompson and L. Andrews, *J. Am. Chem. Soc.* **116**, 423 (1994).
32. L. Andrews, B. Liang, J. Li and B. E. Bursten, *Angew. Chem. Int. Ed. Engl.* **39**, 4565 (2000).
33. J. Li, B. E. Bursten, B. Liang and L. Andrews, *Science* **295**, 5563 (2002).
34. C. J. Evans and M. C. L. Gerry, *J. Chem. Phys.* **112**, 1321 (2000).
35. S. A. Cooke and M. C. L. Gerry, *J. Am. Chem. Soc.* **126**, 17000 (2004).
36. C. C. Lovallo and M. Kłobukowski, *Chem. Phys. Lett.* **368**, 589 (2003).
37. S. Seidel and K. Seppelt, *Science* **290**, 117 (2000).
38. I. C. Hwang, S. Seidel and K. Seppelt, *Angew. Chem. Int. Ed. Engl.* **42**, 4392 (2003).
39. J. A. Glasel, *Proc. Nat. Acad. Sci. Am.* **55**, 479 (1966).

40. M. Saunders, H. A. Jiménez-Vázquez, R. J. Cross and R. J. Poreda, *Science* **259**, 1428 (1993).
41. P. Pyykkö, *J. Am. Chem. Soc.* **117**, 2067 (1995).
42. N. Runeberg, M. Seth and P. Pyykkö, *Chem. Phys. Lett.* **246**, 239 (1995).
43. N. W. Wong, *J. Am. Chem. Soc.* **122**, 6289 (2000).
44. G. M. Chaban, J. Lundell and R. B. Gerber, *J. Chem. Phys.* **115**, 7341 (2001).
45. T.-H. Li, C.-H. Mou, H.-R. Chen and W.-P. Hu, *J. Am. Chem. Soc.* **127**, 9241 (2005).
46 Y.-L. Liu, Y.-H. Chang, T.-H. Li, H.-R. Chen and W.-P. Hu, *Chem. Phys. Lett.* **439**, 14 (2007).
47. W. Grochala, *Pol. J. Chem.* **83**, 87 (2009).
48. S. Borocci, N. Bronzolino and F. Grandinetti, *Chem. Phys. Lett.* **458**, 48 (2008).
49. P. Antoniotti, S. Borocci, N. Bronzolino, P. Cecchi and F. Grandinetti, *J. Phys. Chem.* **A111**, 10144 (2007).
50. L. A. Mück, A. Y. Timoshkin, M. von Hopffgarten and G. Frenking, *J. Am. Chem. Soc.* **131**, 3942 (2009).
51. Feldman and co-workers have prepared some noble-gas hydrides by irradiating HY/Ng matrices with fast electrons. Buck and co-workers have performed photolysis experiments in Xe clusters in the gas phase.
52. V. A. Apkarian and N. Schwentner, *Chem. Rev.* **99**, 1481 (1999).
53. J. Lundell, L. Khriachtchev, M. Pettersson and M. Räsänen, *Low. Temp. Phys.* **26**, 680 (2000).
54. M. Pettersson, L. Khriachtchev, J. Lundell, S. Jolkkonen and M. Räsänen, *J. Phys. Chem.* **A104**, 3579 (2000).
55. L. Khriachtchev, M. Pettersson, J. Lundell and M. Räsänen, *J. Chem. Phys.* **114**, 7727 (2001).
56. M. Pettersson, L. Khriachtchev, R.-J. Roozeman and M. Räsänen, *Chem. Phys. Lett.* **323**, 506 (2000).
57. L. Khriachtchev, A. Lignell and M. Räsänen, *J. Chem. Phys.* **123**, 064507 (2005).
58. Z. Bihary, G. M. Chaban, R. B. Gerber, *J. Chem. Phys.* **119**, 11278 (2003).
59. L. Khriachtchev, *J. Mol. Struct.* **880**, 14 (2008).
60. L. Khriachtchev, M. Pettersson, and M. Räsänen, *Chem. Phys. Lett.* **288**, 727 (1998).
61. L. Khriachtchev, M. Pettersson, S. Jolkkonen and M. Räsänen, *Chem. Phys. Lett.* **316**, 115 (2000).
62. M. Dickgiesser and N. Schwentner, *Nucl. Instrum. Meth.* **B168**, 252 (2000).
63. L. Khriachtchev, H. Tanskanen, M. Pettersson, M. Räsänen, V. Feldman, F. Sukhov, A. Orlov and A. F. Shestakov, *J. Chem. Phys.* **116**, 5708 (2002).
64. L. Khriachtchev, M. Saarelainen, M. Pettersson and M. Räsänen, *J. Chem. Phys.* **118**, 6403 (2003).
65. L. Khriachtchev, M. Pettersson, S. Pehkonen, E. Isoniemi and M. Räsänen, *J. Chem. Phys.* **111**, 1650 (1999).

66. H. Tanskanen, L. Khriachtchev, A. Lignell, M. Räsänen, S. Johanson, I. Khyzhniy and E. Savchenko, *Phys. Chem. Chem. Phys.* **10**, 692 (2008).
67. D. H. Fairbrother, D. LaBrake and E. Weitz, *J. Phys. Chem.* **100**, 18848 (1996).
68. I. Ya Fugol, *Adv. Phys.* **27**, 1 (1978).
69. These equations describe processes in solid xenon. In Kr matrices, reactions (5) and (7) should be replaced by direct formation of hydrogen molecules in the H + H reaction. It seems that the HArF formation in solid argon is dominated by local recombination without global mobility.
70. HXeCCXeH and HXeOXeH are formed upon H atom mobility in reactions involving noble-gas radicals HXeCC and HXeO: HXeCC + Xe + H and HXeO + Xe + H. HXeO radicals are formed upon O-atom mobility in the H + Xe + O reaction. The O-atom mobility occurs in solid xenon at about 30 K, i.e., at lower temperatures than the H-atom mobility. This chapter mostly discusses the mobility of H atoms; however, some other fragments can also move in the lattice.
71. L. Khriachtchev, M. Pettersson, H. Tanskanen and M. Räsänen, *Chem. Phys. Lett.* **359**, 135 (2002).
72. M. Lorenz, M. Räsänen and V. E. Bondybey, *J. Phys. Chem.* **A104**, 3770 (2000).
73. H. Tanskanen, L. Khriachtchev, J. Lundell and M. Räsänen, *J. Chem. Phys.* **121**, 8291 (2004).
74. L. Khriachtchev, A. Lignell and M. Räsänen, *J. Chem. Phys.* **120**, 3353 (2004).
75. K. T. Tang and J. P. Toennies, *Chem. Phys.* **156**, 413 (1991).
76. H. Tanskanen, L. Khriachtchev, J. Lundell, H. Kilijunen and M. Räsänen, *J. Am. Chem. Soc.* **125,** 16361 (2003).
77. A. Lignell and L. Khriachtchev, *J. Mol. Struct.* **889,** 1 (2008).
78. A. Corani, A. Domanskaya, L. Khriachtchev, M. Räsänen and A. Lignell, *J. Phys. Chem.* **A113**, 10687 (2009).
79. J. Ahokas, K. Vaskonen, J. Eloranta and H. Kunttu, *J. Phys. Chem.* **A104,** 9506 (2000).
80. J. Ahokas, H. Kunttu, L. Khriachtchev, M. Pettersson and M. Räsänen, *J. Phys. Chem* **A106,** 7743 (2002).
81. L. Khriachtchev, H. Tanskanen, M. Pettersson, M. Räsänen, J. Ahokas, H. Kunttu and V. Feldman, *J. Chem. Phys.* **116**, 5649 (2002).
82. L. Khriachtchev, M. Pettersson, A. Lignell and M. Räsänen, *J. Am. Chem. Soc.* **123**, 8610 (2001).
83. A. V. Bochenkova, D. A. Firsov and A. V. Nemukhin, *Chem. Phys. Lett.* **405**, 165 (2005).
84. A. V. Bochenkova, L. Khriachtchev, A. Lignell, M. Räsänen, A. A. Granovsky and A. V. Nemukhin, *Phys. Rev.* **B77** 094301 (2008).
85. D. Raftery, *Ann. Rep. NMR Spectr.* **57**, 205 (2006).
86. P. Schwerdtfeger, J. S. McFeaters, M. J. Liddell, J. Hrušák and H. Schwarz, *J. Chem. Phys.* **103**, 245 (1995).
87. D. Kurzydłowski and W. Grochala, *Z. Anorg. Allg. Chem.* **634**, 1082 (2008).
88. C. Sanloup, B. C. Schmidt, E. M. Perez, A. Jambon, E. Gregoryanz and M. Mezouar, *Science* **310**, 1174 (2005).

CHAPTER 14

MODELING STRUCTURES AND SPECTRA OF TRAPPED SPECIES IN LOW-TEMPERATURE MATRICES

Alexander Nemukhin and Bella Grigorenko

Department of Chemistry, M.V. Lomonosov Moscow State University
1/3 Leninskie Gory, Moscow 119991, Russia
E-mails: anem@lcc.chem.msu.ru; bell_grig@yahoo.com

We discuss development and applications of the hybrid quantum mechanical-molecular mechanical (QM/MM) methods for modeling properties of species trapped in low-temperature rare-gas matrices. The main emphasis is on calculations on potential energy surfaces of guest–host systems assuming the guest molecule to be the QM part. We show advantages in using the diatomics-in-molecules (DIM)-based approaches for coupling the QM and MM regions in these applications. The results of simulations of metal clusters, diatomic molecules, and rare-gas compounds are considered.

1. Introduction

The low-temperature matrix helps to capture and to temporary stabilize reactive chemical particles thus providing conditions to examine their properties what is often unattainable in the gas phase or in solutions. Although experimental studies in chemistry are of greater priority than computational based modeling, the role of molecular simulations is also very important. The most valuable theoretical results may refer to those accomplishments which are difficult or too expensive to achieve experimentally. Traditionally, molecular modeling is applied to elucidate details of the structure of molecules, molecular associates (with the recent emphasis on nanoparticles) as well as to describe mechanisms of chemical reactions possibly accompanied by excitations. Modeling structures and spectra of trapped species in low-temperature matrices lies within the traditional boundaries, but has peculiar features some of which will be discussed in this chapter.

Several excellent reviews, including those in the present book, address questions related to simulations of low-temperature matrix phenomena.[1–4] It should be noted that it is hardly possible to review numerous recent papers in which quantum chemical calculations are performed for the bare molecules or intermolecular complexes in order to help assignment of experimental spectra of these species isolated in low-temperature matrices. A special emphasis in this chapter is primarily on development and applications of a particular tool for molecular simulations in condensed phases which is based on the hybrid quantum–classical approaches to construct potential energy surfaces of model guest–host systems. This technique is perfectly suited for modeling properties of trapped species which presumably weakly interact with the host species — the guest molecule is assigned to the subsystem described by the rigorous quantum equations while surrounding matrix particles are described by molecular mechanics. In the case of matrix phenomena, interaction of so selected quantum and classical subsystems accounts, in particular, for the so-called matrix shifts of spectral bands which always are of interest in matrix studies.

2. Hybrid QM/MM Methods in Molecular Simulations

The idea to combine approaches of different theoretical levels to obtain a practical calculation technique is widely explored in modeling properties and reactivity of molecular systems. A motivation for use of such combinations is clear: one can seldom select a model system small enough to describe its properties at a uniform high-accuracy level based on the consistent quantum theory. Modeling structures and reactions in the gas phase is probably one of such areas. However, for treating condensed phase phenomena it is difficult to separate out a subsystem of a primary interest as a fraction of a large piece of substance and to ignore completely the influence of the environmental particles on the properties of the central part. As such, the thought to use a multilayer description for the large model system comes to the surface. In the electronic structure theory, the ONION (our own *N*-layer integrated molecular orbital molecular mechanics) model of Morokuma and co-authors presents an example of the multilayer description.[5] Another widely known theory is the quantum mechanical — molecular mechanical (QM/MM) approach for modeling biomolecular phenomena,[6,7] in particular, simulations of enzymatic mechanisms[8] or electronic excitations in photoreceptor proteins.[9]

The use of the QM/MM-related methodology for simulations molecular events in low-temperature rare-gas matrices may be explained by considering the following expression for the energy of the system which includes a guest molecule (GM) trapped in the low-temperature matrix composed of the rare-gas atoms (Rg):

$$E = E\{\text{GM}\} + \sum_{i=1}^{n} \sum_{j>i}^{n} E\{\text{Rg}_i\text{Rg}_j\} + \sum_{i=1}^{n} E\{\text{Rg}_i \cdots \text{GM}\}. \quad (1)$$

This approach assumes that the number, n, of atoms in the rare-gas crystal lattice is finite and the rare-gas atoms interact via the pair-wise potentials. The latter assumption may be generalized by including many-body interaction potentials, which are sometimes considered as important contributions for rare-gas solids. The key issue of this approach [Eq. (1)] is that one applies quantum mechanical description for the guest molecule and MM-type interaction potentials for pairs of rare-gas atoms. Probably the most difficult obstacle presents a treatment of the Rg···GM interaction, and different approaches have been suggested to approximate the corresponding potentials.[10,11] We advocate the use of the diatomics-in-molecules (DIM) theory which turn out to be very successful approach in this field.

The DIM method was firstly formulated by Ellison[12] and generalized by Tully.[13] According to this methodology the potential energy of a polyatomic molecular system may be expressed as a proper combination of potential curves of all diatomic constituents following the rules of the quantum theory.[13] It was quite popular some time ago to estimate potentials describing collisions of atoms with small molecules[14] or structures and spectra of small molecules and clusters of small molecules.[15–21] Nowadays, modern methods of computational quantum chemistry can provide inexpensive procedures for estimates energies and energy gradients for a separated molecular system composed of few atoms (say, less than 10) even for dynamics calculations on the fly, thus diminishing the role of DIM in this field. The DIM-based approaches keep their significance in modeling condensed phase phenomena since they present a certain generalization of standard molecular mechanical interaction potentials, in particular, accounting for many-body contributions.

Important steps in developing the DIM-type approximations for describing molecular systems with rare-gas atoms were due to Last and George who introduced the so-called "diatomics-in-ionic-system" (DIIS) method to study electronic structure of the rare-gas compounds.[22] It was found that the charge-transfer effects treated by such extension of DIM are very important for the description of van der Waals minima on potential energy surfaces. On the contrary, Gersonde and Gabriel implemented the DIM model, ignoring the ionic configurations, for simulating the HCl photodissociation in xenon crystals.[23] Krylov and Gerber improved the latter model by taking into account the spin–orbit coupling.[24] We also should mention the development of Buchachenko and Stepanov[25] who formulated the DIM-based perturbation theory which turned out to be very efficient for calculations of vibrational predissociation dynamics of rare-gas complexes.[26] Grigorenko, Nemukhin, and Apkarian[27] suggested the analytical DIIS-type approximation for the Ar−HF interaction potential which incorporated the most important neutral and ionic contributions mixed in the spirit of the first-order DIM perturbation theory.[25] These authors also underlined importance of the ionic contributions in the DIIS model for the systems $He\cdots Cl_2$ and $Ar\cdots Cl_2$,[28] following the earlier DIM considerations of $He\cdots Cl_2$.[29] All these calculation schemes were intensively exploited when simulating phenomena in low-temperature rare-gas matrices.

3. First Applications of QM/MM to Low-Temperature Matrices: Metal Clusters in Rare-Gas Matrices

We return to Eq. (1) for the energy of the guest–host molecular system. More generally, we can consider this equation as relating the corresponding Hamiltonian operators of the guest molecule, of the host subsystem, and the interaction terms. Application of the QM/MM methodology implies that we compute contributions from the environmental particles (here, the rare-gas atoms) directly to the Hamiltonian operator of the guest molecule and take into account interaction between environmental particles with the empirical potentials.

Such approach was used in Ref. 30, in which charge fluctuations in small lithium clusters embedded to argon environment were modeled. The guest system (lithium cluster) was considered at the Hartree–Fock level with the Fock operator F expressed in the MO LCAO approximation. Surrounding rare-gas atoms donated external potential V_{ext}, and electronic density of the embedded metal cluster was computed from this equation:

$$\sum_{\nu} [(F + V_{ext})_{\mu\nu} - \lambda_i (S')_{\mu\nu}] c_{i\nu} = 0. \quad (2)$$

For V_{ext} we used the following approximation

$$V_{ext} = \sum_{a} \left[V_a^{Ar} - \sum_{j} \varepsilon_j^a \, |Y_j^a\rangle \langle Y_j^a| \right], \quad (3)$$

where the orbitals were computed in the density functional theory approach with the Vosko–Wilk–Nisair functional

$$\left[-\frac{1}{2}\Delta + \Phi^a(r) + V_{ext} \right] Y_j^a(r) = \varepsilon_j^a Y_j^a(r) \quad (4)$$

and initial one-electron potential from each argon atom was approximated as follows

$$V_a^{Ar} = \left(\frac{Z_{Ar}}{r} \right) e^{-\alpha r^2} + \sum_{k} d_k e^{-\beta_k r^2}. \quad (5)$$

The computed electronic density matrix

$$P_{\mu\nu} = \sum_{i} o_i c_{i\mu} c_{i\nu} \quad (6)$$

of the embedded metal cluster corrected for the presence of the matrix atoms was analyzed in terms of natural atomic orbitals.[31] Positions of the rare-gas atoms were varied along the classical molecular dynamics (MD) trajectories at temperatures between 2 and 35 K. As usual, the temperature in MD simulations is considered as mean kinetic energy of the particles. The computer program used in simulations was created as a combination of slightly modified GAMESS(US)[32] code and original MD code. The main conclusion drawn from these simulations was that charge alteration in small lithium clusters may be as high as 0.2 atomic units, reflecting local changes in the matrix environment. These charge fluctuations may be of importance when considering low-temperature chemical reactions of metal clusters even in a presumably inert environment. Another observation of these simulations was that the electronic density of the metal cluster on the surface of argon matrix should noticeably be different from that inside the matrix.[30]

A close QM/MM-type approach was applied when modeling matrix shifts in electronic spectra of the Na_2 molecule in krypton matrix.[33] According to the results of experimental studies on absorption spectroscopy of sodium atoms and molecules deposited from a supersonic nozzle beam in krypton matrices at 4 K a blue shift of 12 nm for the B–X ($^1\Pi_u - {}^1\Sigma_g^+$) transition and a red shift of 17 nm for the A–X ($^1\Sigma_u^+ - {}^1\Sigma_g^+$) transition in Na_2 due to matrix effects were observed.[34] In simulations, the matrix was modeled by a heterocluster Na_2Kr_{62}: the guest molecule Na_2 was surrounded by first five solvation shells of the face-centered-cubic (*fcc*) krypton lattice. Positions of rare-gas atoms in the heterocluster were adjusted in MD calculations at temperatures lower than 6 K with the pair-wise Lennard–Jones potentials for Kr–Kr and Na–Kr interactions. Electronic structure of the guest molecule Na_2 was described by the Hartree–Fock equations in the MOLCAO approximation. One-electron contributions to the Fock matrix from the surrounding matrix atoms were computed following preliminary calculations of the krypton atom electron densities in the density functional theory approximation [Eqs. (2)–(5)]. Positions of bands in the electronic spectrum of the trapped Na_2 molecule were estimated via differences of the corresponding orbital energies: $\pi_u - \sigma_g$ for the B–X transition, and $\sigma_u - \sigma_g$ for the A–X transition.

Table 1 shows some of the results illustrating dependence of the calculated shifts on temperature.[33] Apparently, this QM/MM-based model was capable to reproduce the major qualitative effect — the matrix-induced spectral shifts in opposite directions for two

Table 1. Calculated[33] and experimental[34] matrix shifts (cm^{-1}) for the Na_2 molecule in a krypton matrix.

	A–X	B–X
Experimental	−406	+523
Na_2Kr_{62}, $T = 0$ K	−461	+636
Na_2Kr_{62}, $T = 4$ K	−658	+549
Na_2Kr_{62}, $T = 5$ K	−395	+944

experimentally studied transitions in Na_2. These matrix shifts are reproduced almost quantitatively.

4. Diatomic Molecules in Low-Temperature Rare-Gas Matrices

We discuss in this section the results of simulations for diatomic molecules Cl_2 and SH in low-temperature rare-gas matrices performed with the hybrid modeling technique QM/DIM. This means that the potential curves of the guest molecules, Cl_2 or SH, were taken from quantum chemical calculations, while the guest–host interaction was modeled by using the DIM method considered as a generalization of conventional molecular mechanics. In both cases the goal of simulations was to provide interpretation of the experimental spectral studies.

4.1. *Spectrum of Cl_2 in neon matrices*

Experimental spectroscopy of the chlorine molecule in the gas phase and in low-temperature matrices has a rich history.[35–40] The results of the study[38] illustrate a solvation effect of the rare-gas environment on the spectral bands corresponding to the $^3\Pi_g$(ion-pair) – $^3\Pi_u$(valence) transition: the red shift of about 0.1 eV in neon matrices compared to the gas phase. The fine structure shown in Ref. 38 refers to the fluorescence from the $v' = 0$ level of the $^3\Pi_g$ ion-pair state to the successive v'' vibrational levels of the valence $^3\Pi_u$ state. In the gas phase the most intense peak corresponds to $v'' = 6$ of the A′ $^3\Pi_u$ state, while in the matrix the most intense line refers to $v'' = 8$. Thereby, according to experiments a matrix noticeably modifies the view of the spectrum.

In a theoretical work,[41] four excited state potential curves of chlorine molecule, correlating to the valence-like states $1^3\Pi_g$, $1^3\Pi_u$ of Cl_2, which in turn correspond to the dissociation limit $Cl(^2P) + Cl(^2P)$, and the ion-pair states $2^3\Pi_g$, $2^3\Pi_u$ related to the ionic limit $Cl^+(^3P) + Cl^-(^1S)$, were considered. These potential curves were constructed on the base of experimental data and *ab initio* quantum chemical calculations. The guest molecule was surrounded by a large number ($n = 498$) of neon atoms. In simulations, the corresponding potential energy surfaces of the $Cl_2@Ne_n$ heterocluster computed with the DIM theory were used for MD simulations "on the fly" to estimate transitions between vibronic levels of Cl_2 buried inside neon solvation shells.

The polyatomic Hamiltonian for the cluster $Cl_2@Ne_n$ was written in the usual DIM (or DIIS) manner:

$$\mathrm{H} = \mathrm{H}_{\mathrm{Cl}^{(1)}\mathrm{Cl}^{(2)}} + \sum_{i=1}^{2}\sum_{j=1}^{n}\mathrm{H}_{\mathrm{Cl}^{(i)}\mathrm{Ne}^{(j)}} - n\left(\sum_{i=1}^{2}\mathrm{H}_{\mathrm{Cl}^{(i)}} + \sum_{j=1}^{n}\mathrm{H}_{\mathrm{Ne}^{(j)}}\right). \tag{7}$$

With a chosen set of atomic basis functions, which constituted the $Cl(^2P)$, $Ne(^1S)$, $Cl^+(^3P)$, and $Cl^-(^1S)$ states, the fragment Hamiltonian matrices were constructed as described elsewhere.[23,28,29] Inclusion of ionic states was essential for correct description of the shapes of potential energy surfaces. All required potential curves of diatomic fragments were approximated by results of *ab initio* quantum chemistry calculations or following experimental data.

MD simulations were performed for the model system for the fraction of the *fcc* rare-gas lattice composed of 500 neon atoms. Two neon atoms in the center were replaced by the guest Cl_2 molecule. Classical MD trajectories with the periodical boundary conditions and the integration step of 0.3 fs were computed for temperatures 3–5 K. The energies E have been computed directly from the DIM Hamiltonian matrix specified by Eq. (7)

$$\mathrm{HC} = \mathrm{EC}. \tag{8}$$

The forces have been calculated half-analytically by using the following strategy. The eigenvectors C are recomputed at every integration step, and the energy gradients with respect to each geometry coordinate q are estimated by the formula

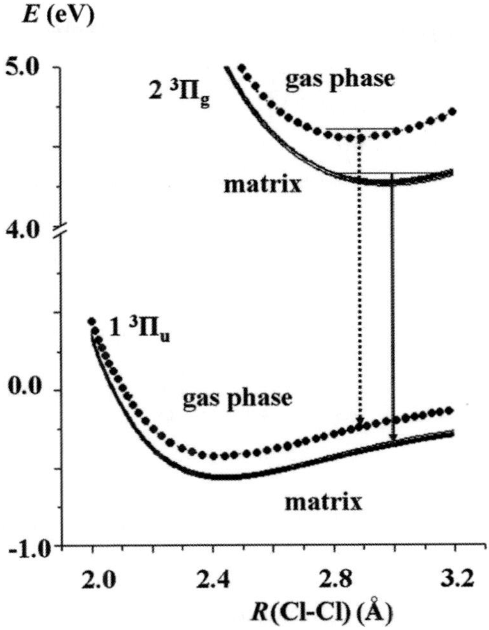

Fig. 1. Potential energy curves of the Cl_2 molecule in the gas phase and in a neon matrix.

$$\frac{\partial E}{\partial q} \approx C^{+} \frac{\partial H}{\partial q} C. \tag{9}$$

The derivatives of the Hamiltonian [see Eq. (7)] matrix elements over the set of coordinates q were expressed analytically. At every integration step the cuts through the multidimensional potential energy surfaces corresponding to the Cl_2 diatomic fragment were evaluated, thus obtaining effective potential curves of the Cl_2 molecule inside the rare-gas shells (see Fig. 1).

By numerical solution of the one-dimensional Schrödinger equation the vibrational levels supported by these effective potential curves were calculated. The positions of vibronic bands in the spectrum and the corresponding Franck–Condon factors could be estimated.

In simulations shift of the bands in the spectrum $2^3\Pi_g(v' = 0) - 1^3\Pi_u(v'')$ to the red was reproduced almost precisely despite no adjustable parameters were introduced to the computational protocol. The usual uncertain parameter of the DIIS construct, the mixing parameter of

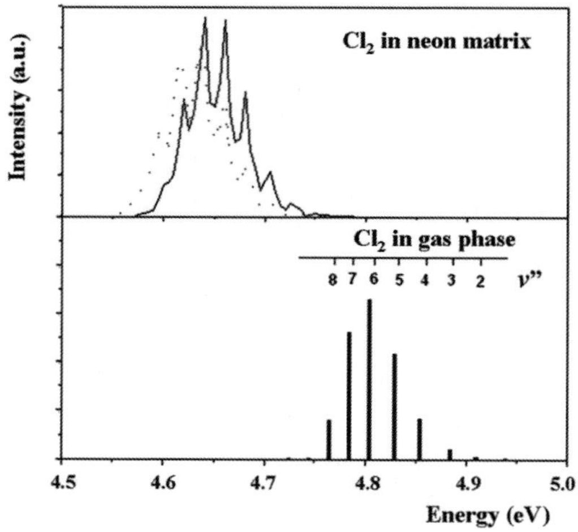

Fig. 2. Computed vibronic bands of the Cl_2 molecule in the gas phase (bottom) and in the neon matrix (top).

neutral and ionic states, was estimated on the base of quantum calculations.[41] The modeling also reproduced slight differences in distribution of intensities of the bands: in the gas phase the $v'' = 6$ peak clearly dominated in the spectrum,[38] while in the matrix the peaks $v'' = 8$ and $v'' = 7$ competed for the role of the most intense line (see Fig. 2).

4.2. *Spectroscopy of SH radicals in krypton matrices*

The spectroscopy of SH in solid rare gases was found to resemble the general tendencies known for the OH radical. However, an outstanding feature of SH emission was noticed in solid krypton: two bands at 375 and 413 nm possessing very different lifetimes, 350 and 750 ns, respectively, were observed while exciting SH radicals.[42] Simulations of the $A^2\Sigma^+$ ($v' = 0$) – $X^2\Pi$ (v'') emission of SH in solid krypton performed by using mixed quantum–classical dynamics allowed the authors to assign these two observed bands to different trapping sites of electronically excited SH in the matrix.[43,44]

These simulations included construction of the DIM potential energy surfaces for the cluster SH@Kr_n in the ground [correlating to the

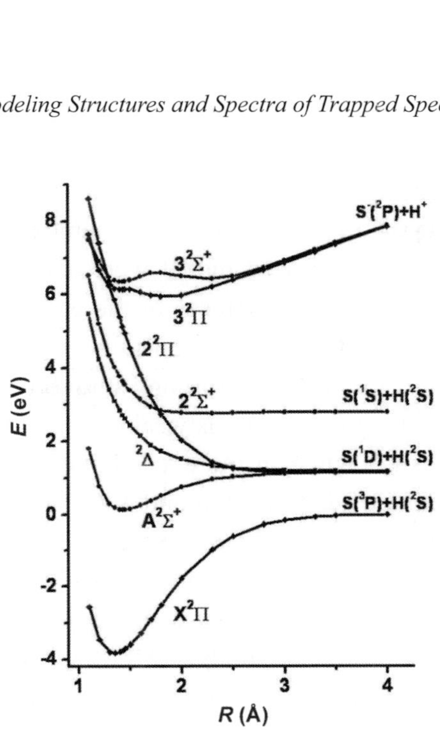

Fig. 3. Computed potential energy curves of the SH molecule.

SH(X) + *n*Kr combination] and excited [SH(A) + *n*Kr] states followed by mixed quantum–classical MD calculations to model the emission spectrum $A^2\Sigma^+$ ($v'' = 0$) – $X^2\Pi$ (v''). The polyatomic Hamiltonian for the SH@Kr_n cluster was represented as a sum of various diatomic contributions:

$$H = H_{SH} + \sum_{i=1}^{n} H_{SKr^{(i)}} + \sum_{i=1}^{n} H_{HKr^{(i)}} + \sum_{i=1}^{n-1}\sum_{j=i+1}^{n} H_{Kr^{(i)}Kr^{(j)}} - n\,(H_H + H_S + \sum_{i=1}^{n} H_{Kr^{(i)}}). \tag{10}$$

The $^2S(H)$, $^1S(H^+)$, $^3P(S)$, $^1D(S)$, $^2P(S^-)$, $^1S(Kr)$, and $^2P(Kr^+)$ states were used to construct polyatomic basis functions as the products of the corresponding atomic functions. Inclusion of ionic constituents was dictated by previous experience, e.g., when considering the chlorine molecule in neon. The matrix elements of the Hamiltonian operator [Eq. (10)] contained energies of the state-specific diatomic fragments, spin coupling parameters, parameters of angular dependencies of atomic basis functions, and mixing parameters of the diatomic states of the same spin and spatial symmetry. Potential energy curves required for the calculation scheme were computed by using high-level *ab initio* methods of quantum chemistry.[45] Figure 3 shows the scheme of the SH potentials.

Parameters of this DIM-based model were additionally adjusted by comparing the potential energy surfaces of the gas-phase van der Waals complex SH···Kr in the ground and electronically excited states which were available from high-resolution spectroscopy studies of Miller *et al.*[46] The SH···Kr complex in the ground state is weakly bound with a dissociation energy about 100 cm^{-1}, while in the excited state the potential surface possesses two minima of depths 1700 and 1400 cm^{-1}, corresponding to the linear geometry configurations SH···Kr and Kr···SH, separated by a saddle point.

With the potential energy surfaces of the SH@Kr_n clusters at hand, it was possible to compute and analyze MD trajectories.[43,44] The guest molecule SH was inserted into the fragment of the *fcc* rare-gas lattice containing 256 krypton atoms. The double-substituted sites were initiated for the majority of trajectories. The classical trajectories were propagated on the ground state potential energy surface at effective temperatures close to 7–12 K. The periodical boundary conditions were imposed. The time integration step of 0.2–0.5 fs was used, and the total length of each trajectory amounted up to 1 ps. At the second stage, the trajectory was switched to the upper electronic surface correlating with the channel SH($A^2\Sigma^+$) + *n*Kr and started from positions of all atoms achieved on the lower potential surface, corresponding to the limit SH($X^2\Pi$) + *n*Kr. The length of each trajectory on the upper surface was about 10 ps. The current data were selected for calculations of the spectrum, lifetimes, and spatial correlation functions.

As in the case of the model system described in Section 4.1, the effective one-dimensional potentials along the distance in diatomic guest molecule (here, S–H) inside the host subsystem were constructed. By numerical integration of the one-dimensional Schrödinger equation corresponding to the ground and excited states the vibrational levels for both states were estimated. The energy differences and overlap integrals of the vibrational wavefunctions (the Frank–Condon factors) were computed along with the radiative lifetimes in the upper electronic state.

Table 2. Experimental[42] and theoretical[43,44] characterization of bands I and II in the emission spectrum of SH in krypton matrices.

	Emission wavelength (nm)		**Lifetime (ns)**		**Features in MD spatial correlation functions**	
	I	**II**	**I**	**II**	**I**	**II**
Experiment	375	413	350	750	–	–
Theory	340– 350	390– 400	550– 600	800– 840	Rotation of SH in Kr	Tight SH···Kr complex

Table 2 summarizes the essential features of both theoretical and experimental findings on the emission spectrum of SH in krypton. The simulations show that both bands in the emission spectra refer to the same $A^2\Sigma^+$ ($v' = 0$) – $X^2\Pi$ (v'') transition and originate from two different trapping sites in the excited state. The band at 375 nm corresponds to the site with a relatively free movement of SH(A) inside the lattice while the adjacent more intense band at 413 nm is associated with the site where one of the matrix atoms is attached to the guest molecule forming a transient complex SH(A)•Kr inside the solvation shell.

5. Rare-Gas Compounds in Low-Temperature Rare-Gas Matrices

In this section, we describe applications of the QM/MM approach to model properties of the intriguing rare-gas compounds that can be formed in low-temperature matrices. A series of stable hydrides HRgY, where Y is an electronegative fragment have been characterized by spectroscopic tools.[47] The most important experimental and theoretical findings are summarized in the recent review of Khriachtchev, Räsänen, and Gerber.[48] We discuss here only the works relevant to the central topic of the present chapter — simulations with the QM/MM technique.

Theoretical studies of the $HXeOH\cdots(H_2O)_n$ structures described in Ref. 49 stay slightly apart from other QM/MM applications for modeling low-temperature phenomena. In simulations, the rare-gas compound, HXeOH, was considered as a guest species (treated at the QM level) inside the cages of water molecules (treated at the MM level). However, the most important conclusions with respect of stability of the $HXeOH\cdots(H_2O)_n$ complexes were drawn only for the systems with the number of attached water molecules not exceeding $n = 3$.

Properties of another rare-gas compound, HArF, in low-temperature argon matrices were modeled in a series of recent papers of Bochenkova and co-authors.[50–52] Among other findings for HArF in solid argon, the experimental studies showed an existence of two trapping sites of the guest species with different argon environment.[53] These sites were characterized by different thermal stability and different spectral bands referring to the H–Ar stretching vibrations. Theoretical modeling aimed to reproduce the experimentally observed features and to describe transitions between unstable to thermally stable configurations of HArF in solid argon.[10,11,50–52]

The DIM-based QM/MM methodology used in Refs. 50–52 relied on previous simulations of structures and spectra of the gas-phase clusters $Ar_k(HF)_m$.[27,54–56] The energy and forces acting on each atom in the HArF@Ar_n model system, mimicking a fragment of the solid argon with the trapped HArF species, were calculated as a sum of three types of contributions from the Ar···HArF, Ar···Ar, and HArF interaction potentials:

$$E = E\{\mathrm{HArF}\} + \sum_{i=1}^{n} \sum_{j>i}^{n} E\{\mathrm{Ar}_i \cdots \mathrm{Ar}_j\} + \sum_{i=1}^{n} E\{\mathrm{Ar}_i \cdots \mathrm{HArF}\}. \qquad (11)$$

The electronic structure of the HArF molecule was calculated in the *ab initio* MP2/6–311++G(2d, 2p) approximation of quantum chemistry. To describe the Ar···Ar interaction, the conventional Aziz-Chen multiproperty adjusted pair-wise interatomic potential[57] was used. The key issue was the application of the DIM (DIIS) approximation for the interaction of the guest molecule with each Ar atom of the host subsystem:

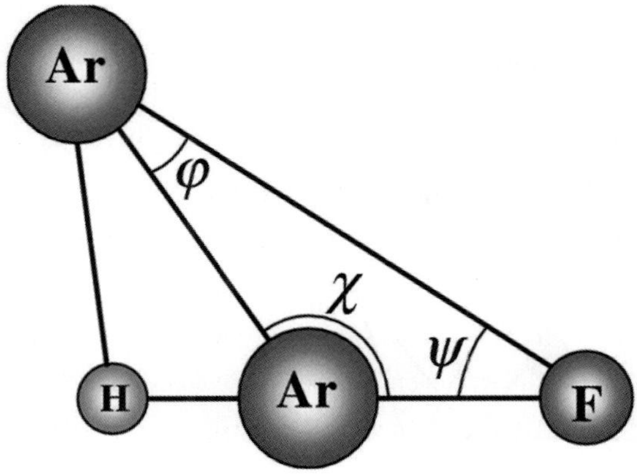

Fig. 4. Geometry parameters of the Ar···HArF complex.

$$V(\mathrm{Ar-HArF}) = \begin{bmatrix} V(\mathrm{ArH}) + V(\mathrm{ArAr}) + \\ V(\mathrm{ArF}, \Sigma)\cos^2(\mathrm{Y}) + \\ V(\mathrm{ArF}, \Pi)\sin^2(\mathrm{Y}) \end{bmatrix} \cos^2(\beta) + \begin{bmatrix} V(\mathrm{ArAr}^+, \Sigma)\cos^2(\chi) + \\ V(\mathrm{ArAr}^+, \Pi)\sin^2(\chi) + \\ V(\mathrm{ArF}^-) + \dfrac{\alpha(\mathrm{Ar})\cos\varphi}{R^2_{\mathrm{ArAr}} R^2_{\mathrm{ArF}}} \end{bmatrix} \sin^2(\beta). \quad (12)$$

In Eq. (12), $V(\mathrm{AB})$ denotes a state-specific diatomic interaction potential as a function of the corresponding internuclear distance. $V(\mathrm{ArH})$ refers to the $^2\Sigma$ potential of ArH, $V(\mathrm{ArAr})$ and $V(\mathrm{ArF}^-)$ correspond to the $^1\Sigma$ potentials of Ar_2 and ArF^-. For the ArF and Ar_2^+ fragments, two electronic states, $^2\Sigma$ and $^2\Pi$ contribute to the interaction energy depending on specific geometry configuration of the Ar–HArF complex through the angles Y and χ (see Fig. 4). The last term in Eq. (12), containing the argon atom polarizability $\alpha(\mathrm{Ar})$, accounts for the nonadditive charge-induced dipole contribution due to the ionic structure Ar···(HAr^+F^-), also depends on specific geometry configuration of the complex — the angle φ between the Ar–F and Ar–Ar directions.

Parameter β accounts for the mixing of neutral and ionic electronic configurations in the system. Its value, $\beta = 40°$, was adjusted by mimicking valence bond description of HArF at the equilibrium geometry. All contributions (QM for HArF, MM for $Ar_i \cdots Ar_j$, and DIM

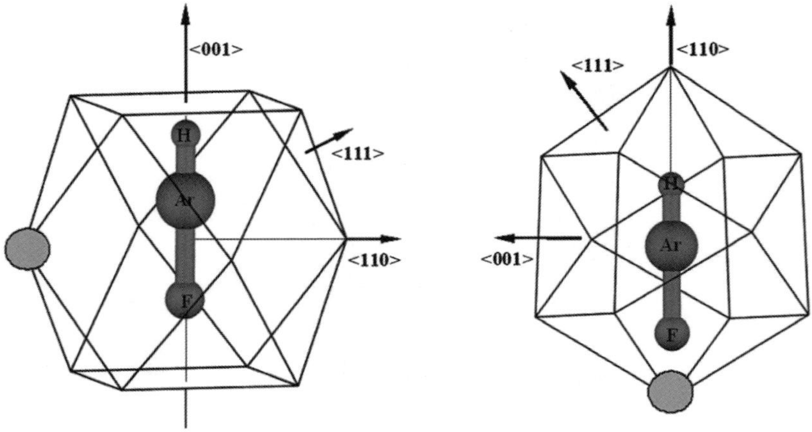

Fig. 5. Thermally unstable (left) and stable (right) solid-state configurations of HArF in an argon matrix.

for $Ar_i \cdots HArF$) to the energy of $HArF@Ar_n$ were summed up for use within the frame of PC GAMESS computer package.[58]

In simulations,[51,52] the cage of host atoms constituted 363 or 364 argon atoms initially arranged according to the *fcc* lattice. The guest species, HArF, was inserted in the center of the atomic cluster. After the QM/MM energy optimization of geometry parameters, three types of structures for trapping sites were found. They were assigned to the double substitutional (DS) trapping sites and classified according to local symmetries: DS-I (C_{4v}), DS-II (C_{2v}), and DS-III (C_{3v}). Vibrational analysis performed for these minimum energy points on the potential energy surface allowed the authors to attribute the DS-III structure to the experimentally observed thermally stable configuration of HArF in solid argon, and the DS-I structure to the unstable configuration (see Fig. 5).

Comparison of the computed[50] and measured[53] shifts in vibrational frequencies upon transition from the unstable to the thermally stable structures (Table 3) provides support to the assignment.[50]

Development of this methodology led to better interpretation of the experimental IR absorption spectra of HArF in low-temperature argon matrices.[51] The DIM-based QM/MM potential energy surface was used to analyze rotation of the trapped molecule inside the matrix cavities. To this goal more than 260 000 grid points on the two-dimensional

Table 3. Differences in frequencies (cm^{-1}) between the unstable (DS-I) and thermally stable (DS-III) configurations of HArF in solid argon.

Mode	Theory[50]	Experiment[53]
ν(H–Ar)	+48.3	+49.5
δ(H–Ar–F)	+10.2	+8.4
ν(Ar–F)	+3.7	–

rotational potential surface were computed for spherical angle coordinates Θ, Φ, defining orientation of the rigid rotor HArF in the matrix.

The rotational eigenvalues and the corresponding eigenvectors of the hindered rotor were computed by diagonalization of the Hamiltonian matrix in a basis set of real valued spherical harmonics Y_{lm}:

$$(\hat{H})_{(jm)(j'm')} = Bj(j+1)\delta_{jj'}\delta_{mm'} + V_{(jm)(j'm')}. \tag{13}$$

Here B is a rotational constant of the HArF molecule ($B = 0.3053\ cm^{-1}$). The potential energy matrix elements $V_{(jm)(j'm')}$ were calculated by numerical integration over the surface of a unit sphere:

$$V_{(jm)(j'm')} = \int_0^{\pi} \sin\Theta d\Theta \int_0^{2\pi} Y_{lm}(\Theta,\Phi)V(\Theta,\Phi)Y_{l'm'}(\Theta,\Phi)d\Phi. \tag{14}$$

The total dimension of the basis set was 5041, and the maximum value of angular momentum j was 70. The lowest energy minimum on the computed rotational surface at Θ = 90° and Φ = 45° corresponded to the stable HArF configuration, while another minimum at Θ = 0° and Φ = (0°, 360°) was attributed to the unstable configuration of HArF in solid argon.

The computed adiabatic rotational potential surface possesses fairly high barriers compared to the rotational constant of the HArF molecule, and therefore the motion of the trapped species resembles a librational limit of the rotational problem, which means hindered or restricted rotation. The lowest energy librational spacing is fairly similar for both sites, namely, $\Delta E_{10\leftarrow 00} = 42\ cm^{-1}$ and $\Delta E_{10\leftarrow 00} = 33\ cm^{-1}$ for the unstable and stable configurations, respectively. By comparing these data with the observed satellite lines which might originate from thermally excited hindered rotation of HArF coupled to the H−Ar stretching vibration, the authors of paper[51] suggested an assignment of the detected earlier the 1992 cm^{-1} absorption band in the IR spectrum to hindered rotation of HArF in the thermally unstable configuration in solid argon.

In the most recent work of this series, Bochenkova, Bochenkov, and Khriachtchev considered the thermal conversion from unstable to stable configurations.[52] In their simulations, the authors used the methodology based on a combination of the accelerated MD and the QM/DIM/MM potential for the HArF@Ar_n system. Temperature accelerated dynamics (TAD)[59] is one of the accelerated MD techniques. This approach is applicable to the systems that are characterized by "infrequent events," in which the system makes an occasional transition from one potential basin to another. The transitions are forced to occur more rapidly by increasing the temperature of the system. In each basin, the system is evolved at high temperature during a certain time. Each time when transition out of the basin is detected, the saddle point is determined, the low-temperature waiting time for this transition is estimated, and the trajectory is reflected back into the basin. Such a basin-constrained dynamics provides a list of different pathways with corresponding activation barriers and waiting times at low temperature. When this list can be considered as complete, the transition with the lowest waiting time is accepted and the system clock is advanced by this time. The saddle points are located using the climbing image Nudged Elastic Band method[60] and the microscopic rate constants are determined using the harmonic transition state theory. Finally, the time evolution of the system, i.e., the sequence of the visited basin minima connected by the saddle point configurations, is obtained. The conversion of the unstable HArF configuration in low-temperature argon matrices should occur at the typical timescale of seconds or longer. This approach used the energies and forces computed "on the fly" for the QM/DIM/MM potential.

The computed values of the activation energy for the unstable–stable transition, 64 meV, as well as the corresponding pre-exponential factor, exp(28) s^{-1} are well consistent with the experimental estimates of 70 meV and exp(30 ± 3) s^{-1}, respectively.

6. Conclusions

This chapter has focused almost exclusively on the use of the quantum mechanical — molecular mechanical (QM/MM) type theoretical approach to simulate structure, spectra, and molecular transformations in low-temperature rare-gas matrices. This technique is perfectly suited for modeling properties of trapped (guest) species which presumably weakly interact with the host particles. The guest molecule is assigned to the subsystem described by the rigorous quantum equations while surrounding matrix particles are described by molecular mechanics. We show certain advantages in application of the DIM-based methods for coupling the QM and MM regions of a large guest–host system. Fairly long ago we noticed that interpretation of the matrix-induced shifts in vibrational spectra of trapped molecules could not be successfully accomplished if only pair-wise interaction potentials were used in molecular modeling.[61] We believe that the DIM-based approaches will keep their significance in modeling condensed phase phenomena since they present a certain generalization of standard MM pair-wise interaction potentials, accounting, in particular, for many-body contributions to the energy.

Acknowledgments

We would like to warmly thank our colleagues and collaborators, firstly, G. B. Sergeev, V. A. Apkarian, M. Räsänen, L. Khriachtchev, and A. V. Bochenkova.

References

1. L. M. Raff, In: D. L. Thompson (editor), *Modern Methods for Multidimensional Dynamics Computations in Chemistry,* p. 266 World Scientific, Singapore (1998).
2. Y. Haas and U. Samuni, *Prog. React. Kinet.* **23**, 211 (1998).
3. V. A. Apkarian and N. Schwentner, *Chem. Rev.* **99**, 1481 (1999).
4. R. B. Gerber, *Annu. Rev. Phys. Chem.* **55**, 55 (2004).
5. S. Dapprich, I. Komaroni, K. S. Byun, K. Morokuma and M. J. Frish, *J. Mol. Struct. (THEOCHEM)* **461–462**, 1 (1999).
6. A. Warshel, *Computer Modeling of Chemical Reactions in Enzymes and Solutions*, John Wiley, New York (1991).
7. A. Warshel, *Annu. Rev. Biophys. Biomol. Struct.* **32**, 425 (2003).

8. H. M. Senn and W. Thiel, *Topics Curr. Chem.* **268**, 173 (2007).
9. A. Altun, S. Yokoyama and K. Morokuma, *Photochem. Photobiol.* **84**, 845 (2008).
10. Z. Bihary, G. M. Chaban and R. B. Gerber, *J. Chem. Phys.* **116**, 5521 (2002).
11. S. Jolkkonen, M. Pettersson and J. Lundell, *J. Chem. Phys.* **119**, 7356 (2003).
12. F. O. Ellison, *J. Amer. Chem. Soc.* **85**, 3540 (1963).
13. J. C. Tully, In: G. A. Segal (editor), *Modern Theoretical Chemistry, Semiempirical Methods of Electronic Structure Calculations, V. 7A*, Plenum, New York (1977).
14. P. J. Kuntz, In: R. B. Bernstein (Editor), *Atom-Molecule Collision Theory,* Plenum, New York (1979).
15. A. V. Nemukhin and N. F. Stepanov, *J. Mol. Struct.* **67**, 81 (1980).
16. A. V. Nemukhin, A. V. Zaitsevsky and N. F. Stepanov, *Mol. Phys.* **41**, 377 (1980).
17. A. V. Nemukhin, S. V. Lyudkovsky and N. F. Stepanov, *J. Mol. Struct. (THEOCHEM)* **104**, 403 (1984).
18. B. L. Grigorenko, A. V. Nemukhin and V. A. Apkarian, *Chem. Phys.* **232**, 321 (1998).
19. B. L. Grigorenko, A. V. Nemukhin and V. A. Apkarian, *J. Chem. Phys.* **108**, 4413 (1998).
20. A. V. Nemukhin, B. L. Grigorenko, E. Ya. Skasyrskaya, I. A. Topol and S. K. Burt, *J. Chem. Phys.* **112**, 513 (2000).
21. B. L. Grigorenko, A. V. Nemukhin, I. A. Topol and S. K. Burt, *J. Chem. Phys.* **112**, 2638 (2000).
22. I. Last and T. F. George, *J. Chem. Phys.* **87**, 1183 (1987).
23. I. H. Gersonde and H. Gabriel, *J. Chem. Phys.* **98**, 2094 (1993).
24. A. I. Krylov and R. B. Gerber, *J. Chem. Phys.* **106**, 6574 (1997).
25. A. A. Buchachenko and N. F. Stepanov, *J. Chem. Phys.* **104**, 9913 (1996).
26. A. A. Buchachenko, N. F. Stepanov, B. L. Grigorenko and A. V. Nemukhin, *J. Chem. Phys.* **111**, 2470 (1999).
27. B. L. Grigorenko, A. V. Nemukhin and V. A. Apkarian, *J. Chem. Phys.* **104**, 5510 (1996).
28. B. L. Grigorenko, A. V. Nemukhin and V. A. Apkarian, *Chem. Phys.* **219**, 161 (1997).
29. B. L. Grigorenko, A. V. Nemukhin, A. A. Buchachenko, N. F. Stepanov and S. Ya. Umansky, *J. Chem. Phys.* **106**, 4575 (1997).
30. B. L. Grigorenko, A. V. Nemukhin, G. B. Sergeev, V. S. Stepanyuk and A. Szasz, *Phys. Rev.* **A50**, 18666 (1994).
31. A. E. Reed, R. B. Weinstock and F. Weinhold, *J. Chem. Phys.* **83**, 735 (1985).

32. M. W. Schmidt, K. K. Baldridge, J. A. Boatz, S. T. Elbert, M. S. Gordon, J. H. Jensen, S. Koseki, N. Matsunaga, K. A. Nguyen, S. J. Su, T. L. Windus, M. Dupuis and J. A. Montgomery, *J. Comput. Chem.* **14**, 1347 (1993).
33. A. V. Nemukhin, B. L. Grigorenko and G. B. Sergeev, *Canad. J. Phys.* **72**, 909 (1994).
34. M. Hofmann, S. Leutwyler and W. Schultze, *Chem. Phys.* **40**, 154 (1979).
35. B. S. Ault, W. F. Howard and L. Andrews, *J. Mol. Spectrosc.* **55**, 217 (1975).
36. V. E. Bondybey and C. Fletcher, *J. Chem. Phys.* **64**, 3615 (1976).
37. J. A. Coxon, *J. Mol. Spectrosc.* **82**, 264 (1980).
38. P. Guertler, H. Kunz and J. Le Calve, *J. Chem. Phys.* **91**, 6020 (1989).
39. T. Ishiwata, Y. Kasai and K. Obi, *Chem. Phys. Lett.* **261**, 175 (1996).
40. M. H. Hill and V. A. Apkarian, *J. Chem. Phys.* **105**, 4023 (1996).
41. B. L. Grigorenko, A. V. Nemukhin and N. V. Ozhegova, *Chem. Phys. Lett.* **296**, 84 (1998).
42. L. Khriachtchev, M. Pettersson, E. Isoniemi and M. Räsänen, *J. Chem. Phys.* **108**, 5747 (1998).
43. B. Grigorenko, L. Khriachtchev, A. Nemukhin, M. Pettersson, E. Isoniemi and M. Räsänen, *J. Chem. Phys.* **110**, 5836 (1999).
44. D. A. Firsov, B. L. Grigorenko, A. V. Nemukhin, L. Yu. Khriachtchev and M. Räsänen, *Chem. Phys. Lett.* **338**, 317 (2001).
45. A. V. Nemukhin, B. L. Grigorenko and A. A. Granovsky, *Chem. Phys. Lett.* **301**, 287 (1999).
46. P. P. Korambath, X. T. Wu, E. F. Hayes, C. C. Carter and T. A. Miller, *J. Chem. Phys.* **107**, 3460 (1997).
47. L. Khriachtchev, M. Pettersson, R. Runeberg, J. Lundell and M. Räsänen, *Nature (London)* **406**, 874 (2000).
48. L. Khriachtchev, M. Räsänen and R. B. Gerber, *Acc. Chem. Res.* **42**, 183 (2009).
49. A. V. Nemukhin, B. L. Grigorenko, L. Khriachtchev, H. Tanskanen, M. Pettersson and M. Räsänen, *J. Am. Chem. Soc.* **124**, 10706 (2002).
50. A. V. Bochenkova, D. A. Firsov and A. V. Nemukhin, *Chem. Phys. Lett.* **405**, 165 (2005).
51. A. V. Bochenkova, L. Khriachtchev, A. Lignell, M. Räsänen, H. Lignell, A. A. Granovsky and A. V. Nemukhin, *Phys. Rev.* **B77**, 0943011 (2008).
52. A. V. Bochenkova, V. E. Bochenkov and L. Khriachtchev, *J. Phys. Chem.* **A113**, 7654 (2009).
53. L. Khriachtchev, M. Pettersson, A. Lignell and M. Räsänen, *J. Am. Chem. Soc.* **123**, 8610 (2001).
54. A. V. Nemukhin and B. L. Grigorenko, *Intern. J. Quant. Chem.* **62**, 55 (1997).
55. B. L. Grigorenko and A. V. Nemukhin, *Chem. Phys. Lett.* **270**, 103 (1997).
56. A. V. Bochenkova, M. A. Suhm, A. A. Granovsky and A. V. Nemukhin, *J. Chem. Phys.* **120**, 3732 (2004).
57. R. A. Aziz and H. H. Chen, *J. Chem. Phys.* **67**, 5719 (1977).

58. A. A. Granovsky, PC GAMESS, *http://www.classic.chem.msu.su/gran/gamess.*
59. A. F. Voter, F. Montalenti and T. C. Germann, *Annu. Rev. Mater. Res.* **32**, 321 (2002).
60. G. Henkelman, B. P. Uberuaga and H. Jónsson, *J. Chem. Phys.* **113**, 9901 (2000).
61. A. I. Krylov, A. V. Nemukhin and N. F. Stepanov, *J. Molec. Struct.* (*THEOCHEM*) **262**, 55 (1992).

CHAPTER 15

SPECTROSCOPY OF BIOLOGICAL MOLECULES AT VERY LOW TEMPERATURES: THEORETICAL STUDIES

R. Benny Gerber

Institute of Chemistry and The Fritz Haber Research Center, The Hebrew University, Jerusalem 91904, Israel and Department of Chemistry, University of California Irvine, CA 92697, USA
E-mails: benny@fh.huji.ac.il; bgerber@uci.edu

Jiri Sebek

Institute of Chemistry and The Fritz Haber Research Center, The Hebrew University, Jerusalem 91904, Israel
E-mails: jirka@fh.huji.ac.il; jirka.sebek@seznam.cz

There has been remarkable progress in recent years in experimental studies of high-resolution vibrational spectroscopy of biological molecules. Such studies generally require low-temperature conditions. The experiments carry in principle a wealth of information on the potential energy surfaces, and the structures and other properties of the molecules. The quantitative interpretation of the spectroscopic experiments is a challenge to theory, since an anharmonic level of treatment is often required, and since the number of vibrational degrees of freedom is large for the systems of interest. This chapter presents an overview of recent progress in theoretical calculations of vibrational spectroscopy of small biological molecules, and examines the lessons of comparison with experiment. The emphasis is on anharmonic calculations with the vibrational self-consistent field (VSCF) method, using directly potential energy surfaces from *ab initio* methods. Based on comparison with experiments several issues are examined including the role of anharmonic effects; the merits of different types of potential energy surfaces; the sensitivity of the spectra to specific conformers; and the interaction between biological molecules and a selected number of water molecules. Several outstanding challenges and important directions for future progress are discussed.

1. Introduction

Biological molecules function at ambient temperatures in life processes. Nevertheless, there are important reasons for studying biological molecules also at very low temperatures, in the cryogenic range. Above all, low temperatures make possible the measurement of certain physical properties of the molecules that may not be accessible in other conditions. This is indeed the case of vibrational spectroscopy studies at high resolution of biomolecules. The sensitivity of the vibrational spectrum to the potential energy surface of the system is well recognized, and vibrational spectroscopy is a powerful tool for exploring also other properties of molecular systems. Biological molecules, including the smallest ones, have multiple conformations separated by low barriers. At room temperatures, several of these conformers are significantly populated. In addition, all but the very smallest biological molecules have soft vibrational modes, excited vibrational states of which are expected to be populated at room temperatures. These two factors alone exclude in many cases the possibility of measuring, say, an extensive range of the fundamental infrared (IR) spectrum of such a molecule at high resolution in the regime of ambient temperatures. Both of the above difficulties can be eliminated by measurements at low temperatures. Specifically, in these conditions it is often possible to obtain a spectrum from the lowest-energy conformer only or from a small number of conformers as well. This enormously simplifies the observed spectroscopic features, and greatly helps in the interpretation. There has been remarkable progress in the development of experimental techniques for measuring vibrational spectra of biological molecules at low temperatures. The oldest approach, and still very fruitful, involves measuring the IR spectrum of biological molecules in noble-gas matrices (and in several other cryogenic host solids).[1–7] In a number of cases, the spectra of different conformers of a biomolecule could be identified in matrix isolation spectroscopy. Not all biological molecules can be introduced in a matrix and usefully studied by vibrational spectroscopy. A major limitation is that many molecules occupy multiple sites in the matrix. The larger and the more flexible the molecule, the more severe this problem seems to be. Quite obviously, the matrix environment does affect the spectrum of the guest molecule. The matrix shifts and other effects are believed to be relatively small. However, despite partial evidence from experiments, knowledge on this issue is still very incomplete. An approach, or a group of methods that have greatly expanded the field in very recent years involve spectroscopic studies of biological molecules in molecular beams, in which case essentially perfect isolation of the species of interest is attainable, and there are no environmental effects.[8–27] In many of the cases, the molecular beam spectroscopic methods require the insertion of a suitable chromophore into the molecule of interest. However, on the whole, the gas-phase approach has remarkably versatile possibilities in applicability to different families of biological molecules, including amino acids, peptides, nucleic acid bases, sugars, and hydrogen-bonded complexes of all the latter with water and with other species of interest. Finally, we mention the approach involving spectroscopy of the molecule in superfluid helium droplets.[28,29] This method is still in its infancy, and so far only few organic or biological molecules were explored using this approach. However, the method is novel, and seems to be very promising for the future. One advantage is that superfluid helium is expected to have very small "environment" effects on the spectrum. It should prove less of a perturber than the "usual" noble-gas matrices.

This chapter describes recent developments in computational studies of vibrational spectroscopy of small biological molecules. Section 2 briefly mentions the most widely used and the very familiar approach, namely the harmonic approximation, improved in most practical use by empirical scaling "rules", and describes anharmonic methods in use for biological molecules, focusing on the vibrational self-consistent field (VSCF) approach. The properties, advantages and limitations of this approach are discussed. Section 3 describes applications to different types of small biological molecules, discusses comparisons of calculations with experiments, and draws conclusions on the state of the art, and what has been learned on the properties of biological molecules from the spectroscopic studies. Section 4 presents concluding remarks.

2. Methodology

A wide range of computational methods for vibrational spectroscopy of polyatomic molecules are available. However, only very few methods have been actually applied systematically to the spectroscopy of biological molecules. The obvious obstacles for such calculations are the large number of degrees of freedom of even relatively small biomolecules, and the difficulty of obtaining sufficiently accurate potential energy surfaces for the purpose of spectroscopic calculations. We will focus here on the principles of the methods in use, on their advantages and limitations, but will not discuss equations, or the derivation of the approximations involved. An extensive recent review of both experimental and computational methods has been written by de Vries and Hobza.[27] Two previous brief overviews deal with computational methods that treat anharmonic effects.[30,31] These methods, and their applications, will also be the subject matter of this chapter.

2.1. *Harmonic approximation and anharmonic scaling*

Since its introduction in the earliest stages of molecular quantum mechanics, the harmonic approximation has been the main tool for analyzing vibrational spectroscopy of polyatomic systems.[32] In harmonic calculations, the first step is to determine the equilibrium geometry of the molecule. The potential function is expanded in the series of displacements of the atoms from equilibrium, while the terms of the first order are zero and the terms of order higher than the second are neglected. Then, normal modes of displacements are computed, for which the Hamiltonian of the molecular vibrations can be expressed simply as a sum of independent oscillators — each of them corresponding to one of the normal modes.[32] All the vibrational wavefunctions and energy levels for this model are analytically obtained. This very simple framework is very useful in the qualitative analysis and assignment of the spectra.[33–35] Codes for harmonic calculations using *ab initio* methods for the potential surfaces are implemented in all the important suites of programs for electronic structure calculations, and their use has become a matter of routine.[27]

Unfortunately, the harmonic approximation is generally insufficient in accuracy for quantitative comparisons with experiment. This is certainly the case for part of the vibrational transitions in nearly all, if not all, biological molecules for which experimental data are available. To bring the computed frequencies into closer agreement with experiment, an *empirical* correction is often employed. This is referred to as "anharmonic scaling." Based on tests of harmonic calculations against experimental data for a set of molecules, correction factors are determined for frequencies of different types of transitions.[36] These "anharmonic scaling factors" can then be applied to computed harmonic frequencies of molecules of interest. Experience shows that the carefully determined "scaling factors" do indeed greatly improve agreement with experiment. Anharmonic scaling calculations involve very little computational effort, and are therefore very convenient to use. Such calculations have, however, several limitations that should be kept in mind:

(i) The objective of many spectroscopic studies comparing theory with experiment is to learn about the potential energy surface of the molecule, including the anharmonic interactions. The empirical anharmonic scaling approach provides no information on the very interesting anharmonic terms of the true potential function.

(ii) The scaling approach was not tested as to whether it improves *intensities of vibrational transitions*. At least for biological molecules, this issue remains to be established. To our knowledge, no argument has been given as to why such scaling should substantially improve intensities.

(iii) Scaling factors were not determined for overtone and combination mode frequencies. The suitability for intensity calculations of such transitions seems unknown.

We conclude that anharmonic scaling is a very efficient, practical tool for improving agreement with experiment for fundamental transition frequencies. For other transitions, the potential usefulness of such scaling is unknown. Since the approach is not a "first principles" one, it cannot be used to characterize anharmonic properties of potential energy surfaces.

2.2. *Vibrational Self-Consistent Field (VSCF) methods*

These methods have been extensively used in recent years for anharmonic calculations of the vibrational spectra of biological molecules. Possibly, VSCF is currently the most widely used tool for this purpose. The first form of the VSCF approach to vibrational spectroscopy of polyatomic molecules was in approximations proposed by Bowman[37,38] and by Gerber and Ratner,[39,40] and was initially applied to model systems. Subsequent developments brought the basic method, and subsequently several higher-level variants, into forms that are computationally effective even for relatively large (and realistic) polyatomic molecules.[41,42] The basic VSCF approach begins with the determination of the normal modes, describing vibrational displacements from the equilibrium structure, as in standard harmonic calculations. Then, the VSCF approximation is introduced, which assumes that the full vibrational wavefunction is factorizable into single-mode wavefunctions corresponding to the different normal modes. Equations for the wavefunctions and energy levels of the system are then derived on this basis.[41,42] The full vibrational wavefunctions of the system in basic VSCF are separable, as in the harmonic approximation, but are not harmonic. Unlike the harmonic approximation, the basic VSCF equations are not analytically solvable, but they can be treated quite efficiently numerically. Early applications of basic VSCF for biological molecules include calculations of small peptides[43] and of monosaccharides,[44] using empirical force fields for the potentials. The results showed a clear improvement over the harmonic approximation. In fact, a demonstration calculation with basic VSCF was carried out for a small protein, BPTI.[45,46] There are no experiments to compare with, and it is questionable if the accuracy of the force field used is sufficient for comparison with spectroscopic measurements. Nevertheless, the protein application is important for providing qualitative insights (e.g., anharmonic properties of vibrational wavefunctions of proteins), and for indicating that realistic anharmonic calculations of the spectra of biological macromolecules are likely to become feasible.

2.2.1. *Treatment of potentials in VSCF*

To carry out practical VSCF calculations, it is necessary to simplify, or approximate, the full potential energy surface of a real molecule. Two approaches have been used so far: (i) Carrying out a Taylor expansion of the potential function in the normal modes, which has to be truncated after some order.[41] Quartic force fields were used in most applications of VSCF for biological molecules in this approach.[43–46] (ii) Inclusion of interactions between pairs of normal modes only in the calculation, thus neglecting effects of direct interaction of three or more modes.[42] Both approaches, (i) and (ii) above, greatly simplify the evaluation of integrals in the VSCF equations. Which of the approaches is preferable is still an open question, and the answer may depend on the type of molecule. However, so far the experience for biological molecules points to superiority of the pairwise interaction approximation of the potential, over the perhaps more expected truncated Taylor expansion. Quartic expansion seems in numerous cases to be insufficient, and higher-order expansions seem computationally costly (especially for *ab initio* potentials). The pairwise approximation must be examined for validity in principle in each case, but has yielded very good results so far. Extensions to include a selected number of triple-mode interactions should be feasible at least for relatively small biomolecules.

2.2.2. *VSCF-PT2*

Basic VSCF provides improvement over the harmonic approximation, but a quest for higher accuracy is obviously desirable. This means to go beyond the approximation of separability of the wavefunction. Using the second-order perturbation theory for introducing the contributions of the effects of inseparability has resulted in a method that is currently the main "work-horse" of the VSCF approach.[42] Unfortunately, several different and confusing titles were used at different times for this method: correlation-corrected VSCF (CC-VSCF),[42] and vibrational Møller–Plesset method,[47] since it is analogous to MP2 from electronic structure theory. We choose to accept here the acronym VSCF-PT2 that seems to describe best the nature of the method. The method seems to perform very well in tests against experiments for a wide range of systems, including biological molecules where many satisfactory applications are at hand. Some of these will be discussed later.

Using properties of the single-mode VSCF wavefunctions, Pele, Brauer and Gerber were able to algebraically recast the VSCF-PT2 equations into a form that is computationally much more efficient for large systems than the original version.[48] Specifically, the PBG algorithm for VSCF-PT2 scales only as N^3 with the number of modes N, much superior to the N^6 scaling of the computational effort for the old version. This opens the way for calculations at the VSCF-PT2 level for systems as large as tetrapeptides and beyond.[49] This acceleration of the VSCFPT2 algorithm is for the pairwise mode–mode coupling for the potential surface, and does not hold if triplets or higher-order interactions between the normal modes are included. Another very useful "acceleration" of VSCF-PT2 was introduced by Benoit,[50] for cases where a single transition only in a large polyatomic system is of interest. The clue to efficient calculations for large systems depends not only on the algorithm for the vibrational wavefunctions but also on the number of (important) mode–mode interaction terms that have to be included. Physical considerations can be used to predict which mode–mode couplings are unlikely to contribute significantly and in this way substantially reduce the required computational effort.[49–52] As experiments increasingly provide high-resolution spectroscopic data for very large biomolecules, the challenge of providing improved algorithms, that scale better with the number of vibrational modes N, becomes more important.

2.2.3. *VSCF methods for vibrationally degenerate states*

Results seem to indicate that the consequences of vibrational degeneracy on the accuracy of VSCF calculations are quite often not very severe. Certainly, the effects are of much weaker consequence than is the case for electronic degeneracy in *ab initio* calculations. However, there are cases, even for large molecules, where treatment of vibrational degeneracies is necessary for the interpretation of spectroscopic data. Two variants of VSCF were introduced for computing vibrational transitions when 1:1 resonances occur.[53] The simpler of the two algorithms involves diagonalizing the vibrational Hamiltonian in a VSCF basis, over the degenerate subspace. This method by Matsunaga *et al.* was originally named VCI.[53] In retrospect, a more adequate terminology would probably be degenerate VSCF (D-VSCF). A second, more accurate algorithm, also due to Matsunanga, is VSCF-DPT2. It involves, beyond diagonalization over the space of the 1:1 resonances, also interactions with nondegenerate states, treated by second-order perturbation theory. This method is, however, inapplicable, for example, to Fermi resonances. An extension of VSCF for such more demanding types of degeneracy was provided by Yagi, Hirata, and Hirao.[54,55] On the whole, the treatment of vibrational degeneracy in the spectroscopy of biomolecules is still at an early state of development, perhaps because so far there seem relatively few experimental cases that require computational treatment of such effects. More extensive high-resolution studies in the future are quite likely to require much increased attention to degeneracy.

2.2.4. *VSCF calculations for ab initio potentials*

An algorithm for first-principles calculations of anharmonic spectra of polyatomic molecules by direct application of VSCF to potential surface points from *ab initio* codes was introduced by Chaban, Jung, and Gerber.[56,57] Soon afterwards, the algorithm was introduced in the GAMESS suit of programs,[58] and later also in MOLPRO.[59] *Ab initio* potentials in such calculations are computed for a grid of points. It is of great computational convenience to use VSCF in this direct way, in which the potential is generated over a grid of normal-mode points and used directly in the vibrational algorithm. An alternative approach is to fit the potential surface, then use it in the vibrational calculations. For very small molecules such an approach is indeed very effective. In the present state of the art, for the highly dimensional potentials of even small biological molecules, it seems that fitting is not yet competitive with the direct calculations. The great majority of anharmonic calculations for biomolecules were done to date by the direct approach. However, one must keep an open mind in such a rapidly evolving area.

Increasingly sophisticated, powerful analytical methods have been developed recently. It is possible that the balance, so far heavily tilted toward the direct algorithm, may become more favorable to analytical fitting in the future. We note that direct *ab initio* calculations of anharmonic spectra can be carried out, e.g., with GAMESS, for basic VSCF, and also VSCF-PT2, D-VSCF (called also VCI), and VSCFDPT2.

2.3. *Second-order perturbation theory for anharmonic vibrations*

This is another widely used computational approach to anharmonic spectroscopy. The method has a long history. A modern code implemented for *ab initio* potentials and already with a record of applications for several small biological molecules is due to V. Barone.[60,61] In this method, the harmonic potential is used in the zeroth-order vibrational Hamiltonian. The anharmonic potential, assumed very weak compared with the harmonic one, is treated by quantum-mechanical perturbation theory. Second-order perturbation theory (PT2) seems to provide sufficient accuracy in most cases, and is generally used. For molecules relevant to the present review (organic molecules, biological building blocks), good accord with experiment is generally reported. Systematic comparison with VSCF results for biological molecules is not available. In the case of glycine, where both methods were applied for similar potentials to three low-lying conformers, agreement between the two methods seems very good.[62,63] In this case, there is considerable confidence in both methods, since the completed frequencies agree very well with matrix isolation spectra. For several nonbiological systems, e.g., the water dimer $(H_2O)_2$, comparisons between the perturbation-theoretic algorithm and VSCF variants were made,[64] with the conclusion that each of the two approaches has some advantages for different transitions. From the theoretical point of view, VSCF-PT2 treatment described here should start from a better initial point for corrections, since in this method the zeroth-order approximation already includes anharmonic effects. There is, however, no systematic evidence on the actual relative performance of the two methods.

It can, however, be said that the VSCF methods have the important advantage of computing explicitly also anharmonic wavefunctions and wavefunction properties, the most relevant of which are intensities. Thus, the basic VSCF approximation allows anharmonic calculations of intensities of fundamental transitions, and more recently expressions have been given also for combination transitions, where the anharmonic effects are critical.[65,66] The anharmonic vibrational wavefunctions provided by VSCF can also be useful for calculations of electronic excitation properties. The PT2 method does not provide explicit anharmonic wavefunctions.

2.4. *Higher-level methods and quantum-dynamical approaches*

Very powerful coupled-cluster methods were recently introduced by Christiansen and co-workers for anharmonic vibrations of a polyatomic molecule.[67,68] The coupled-cluster algorithms can attain accuracies well beyond those of VSCF-PT2, for example. For relatively small polyatomic molecules, this may well be the emerging method of choice. However, vibrational coupled-cluster calculations are not yet applicable for the systems addressed here. Improving the efficiency of the approach for larger systems seems a very promising goal.

Vibrational energy redistribution dynamics in polyatomic systems involves effects and mechanisms that are not included at all in VSCFPT2, or in the simple perturbation-theoretic approach PT2. A high degree of correlation between the different vibrational degrees of freedom, multiple resonances as well as nonresonance effects must generally be incorporated for a realistic treatment. This topic, still in very early stages of development, lies outside the scope of the present review. We mention very interesting studies by Yagi *et al.*[69] and by Fujisaki *et al.*[70] that address IR spectrum and vibrational dynamics in systems such as guanine–cytosine base pair[69] and *N*-methylacetamide.[70]

A rapidly developing spectroscopic field, which bears a strong connection to some of the issues discussed in this review, is 2D IR spectroscopy.[71–73] 2D IR is already an extensive discipline in its own right, with aims and approach that differ greatly from frequency-domain IR and Raman spectroscopies that are the context of this review. 2D IR experiments and the corresponding theoretical developments are already having a significant impact on the study of biological molecules in condensed phases. Despite differences between the experimental and theoretical approaches used for the two spectroscopic fields, there are important issues that are common to the two areas, such as the nature and role of anharmonic interaction in biomolecules. Hopefully, such issues and the connection between the two spectroscopic fields will receive increased attention in the near future.

3. Applications and Comparisons with Experiment

3.1. *Spectroscopic testing of potentials for biological molecules*

Vibrational spectroscopy is a sensitive probe of molecular properties. Perhaps most important in this respect is the sensitivity of vibrational spectroscopy to the potential energy landscape of the molecule. Thus, spectroscopy has already served to assess the validity and accuracy of different force fields for biological molecules. It seems to us that there is a great scope for more extensive use of spectroscopy as a tool for testing force fields for biological molecules, and probably also for guiding the development of new potentials.

One example is the comparison of the spectroscopic accuracy of force fields such as AMBER, OPLS-AA, and CHARMM with that of *ab initio* potentials. Such comparisons were carried out for amino acids[63] for *N*-methylacetamide[74] and for several other biological building blocks. The comparisons show that good electronic structure methods such as MP2 or DFT/B3LYP are considerably superior to the empirical force fields in predicting the vibrational spectroscopy. It must be kept in mind, of course, that force fields such as AMBER, OPLS-AA, and CHARMM were not calibrated for spectroscopy. Moreover, these force fields are generic for each class of molecules, and it is certain that fitting a force field for a specific molecule is bound to give improved agreement for many or all properties. While the result does not seem surprising, it does strongly support the use of *ab initio* potentials over empirical force fields for systems where this is feasible. Such applications of *ab initio* potentials need not be necessarily for vibrational properties. It suffices that the spectroscopy validates the potential. It seems desirable to pursue, e.g., molecular dynamics simulations of small biological molecules (number of atoms $\leq$100) using directly good *ab initio* methods, as this is quite feasible at the present state of the art. For large biological molecules, the use of *ab initio* potentials is not feasible at present, and unlikely to be realistic in the near future. Thus, empirical force fields, which are now almost the only tools of biomolecular modeling, are likely to remain in extensive use. A potentially useful strategy could be to pursue improvements of empirical force fields through spectroscopic testing and fitting. Several examples suggest that fitting on the basis of the present functional forms of the standard force yields some improvement, but only a limited one.[63,64] It is very helpful that the spectroscopic studies of, e.g., monosaccharides,[44] amino acids,[63] small peptides,[43] and others provide insights into some of the flaws of the standard force fields, thus suggesting targets for improvements. Almost obvious examples are stretching vibrations, since these are modeled in standard force fields by harmonic interactions.[75] An important inadequacy of the empirical force fields revealed in flaws of their spectroscopic predictions is with regard to the description of hydrogen bonding. This is the case of intermolecular hydrogen bonds, while the inadequacy can be quite dramatic for complexes of biomolecules with water, e.g., glycine–H_2O.[76] Anharmonic calculations can be used also to test different electronic structure methods and the potentials they provide against spectroscopic data. Chaban and Gerber compared the spectroscopic predictions of MP2, DFT/BLYP, and DFT/B3LYP with experiment for several small organic and biological molecules, using VSCF-PT2 in computing the spectra.[77] MP2 was found to give better agreement with experiment than B3LYP, and much better agreement than BLYP. The results can be useful in guiding choices of the *ab initio* methods for modeling small biomolecules. The computationally more convenient B3LYP is extensively used, and the findings suggest that this consideration should be balanced against the superior accuracy of MP2 potential surfaces for molecules of this type. In at least one case has validation by spectroscopy served as the basis for developing a new variant of potential energy surfaces for biological molecules.

Semiempirical electronic structure methods, such as PM3, are computationally very efficient and feasible for applications to quite large molecules. However, the accuracy of such potentials is not satisfactory, being far inferior to MP2 or B3LYP. Brauer *et al.* have introduced an adaptation of PM3 that seems to greatly improve the spectroscopic accuracy of the method.[78] The standard PM3 potentials are modified by scaling factors applied to the coordinates, so chosen that the modified PM3 reproduces the harmonic frequencies of MP2. In essence, this "fixes" PM3 for the harmonic level, for which also MP2 calculations are easy, and then the modified PM3 is used for anharmonic calculations. VSCF-PT2 calculations for glycine, alanine, and proline with the improved PM3 gave excellent results.[78] Since its introduction, the method was used for spectroscopic calculations for a range of different biological molecules, with very good results, e.g., for the complex G···C of two nuclear acid bases.[79] The results seem, in several applications, to be in accuracy close to that obtainable from the far more costly MP2 or B3LYP. However, the validity of the method must be examined for each type of molecule quite carefully, since there is no systematic theoretical basis for this approach. For systems where the method is valid, its computational efficiency is great, making it applicable even for large peptides.

3.2. *Significance of anharmonic effects*

The contributions of the anharmonic effects to the vibrational spectrum are of much interest, since these have implications also to dynamical properties of the molecules. Within the harmonic approximation, there is no energy transfer between the normal modes, and thus intramolecular vibrational energy flow cannot be accounted for. The study of energy flow in proteins and in other biological macromolecules is of major interest, and knowledge of anharmonic coupling even in small biomolecules should be very useful for progress on this.[80,81]

We discuss first the quantitative importance of anharmonic effects on spectroscopy. As examples, we take protonated imidazole (ImH^+) and its complexes with water, $(ImH^+)(H_2O)_n$, for $n = 1, 2$. The results shown are from Adesokan *et al.*,[82] in which VSCF calculations were used to

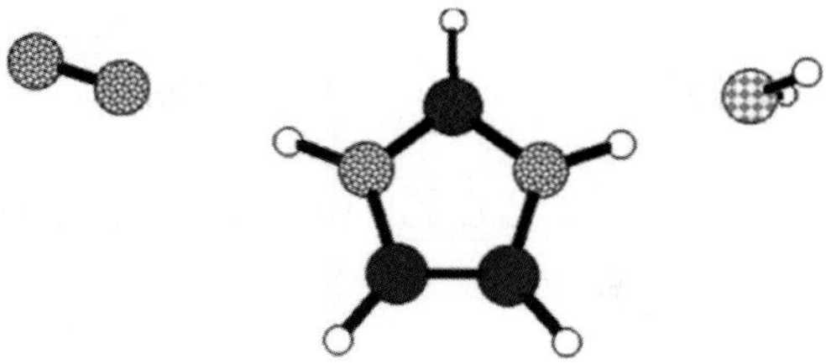

Fig. 1. Equilibrium structure of $(ImH^+)(H_2O)N_2$. Adapted with permission from Ref. 82. Copyright 2007 American Chemical Society.

interpret gas-phase infrared photodissociation (IRPD) spectroscopy. In the experiments, the ionic species actually studied were $(ImH^+)(H_2O)_nN_2$, in which the N_2 serves as a convenient "tag" for the mass spectrometric detection of the photoabsorption. The N_2 molecule is included in the theoretical calculations, but its bonding to the $(ImH^+)(H_2O)_n$ species is relatively weak, and the calculations indicate that it only has a very small effect on the computed transitions. Figure 1 shows the equilibrium structure of $(ImH^+)(H_2O)N_2$.[82]

Figure 2 shows the percentage of deviation of calculations from experiment for $(ImH^+)(H_2O)_nN_2$, $n = 1, 2$, and for ImH^+ (measured without an N_2 "tag"). In each part of the figure, an MP2 harmonic calculation is compared with a VSCF calculation. In Fig. 2a, the VSCF calculations for $(ImH^+)(H_2O)N_2$ are for an MP2 potential; in Fig. 2b for $(ImH^+)(H_2O)_2$, the improved PM3 potentials are used in the VSCF calculations; in Fig. 2c improved PM3 is used for $(ImH^+)(H_2O)$; and in Fig. 2d the comparison shown is between harmonic MP2 and VSCF for improved PM3, for (ImH^+). In all these cases, the deviation from experiment is drastically smaller for the anharmonic calculations. In the case of $(ImH^+)(H_2O)N_2$ in Fig. 2a, where full MP2 was used for the anharmonic potential surface, the accuracy of the VSCF results is astonishing. The anharmonic calculations with improved PM3 potentials are somewhat less accurate, but also provide a very large reduction of the deviation from experiment compared to the harmonic calculations. In all cases, harmonic approximation errors that can be as large as ~12% for some transitions, are reduced to no more than about 2% in the anharmonic calculations. This example is admittedly a case of very large anharmonicity. Many hydrogen stretching transitions in other systems also show large anharmonic effects. For other types of

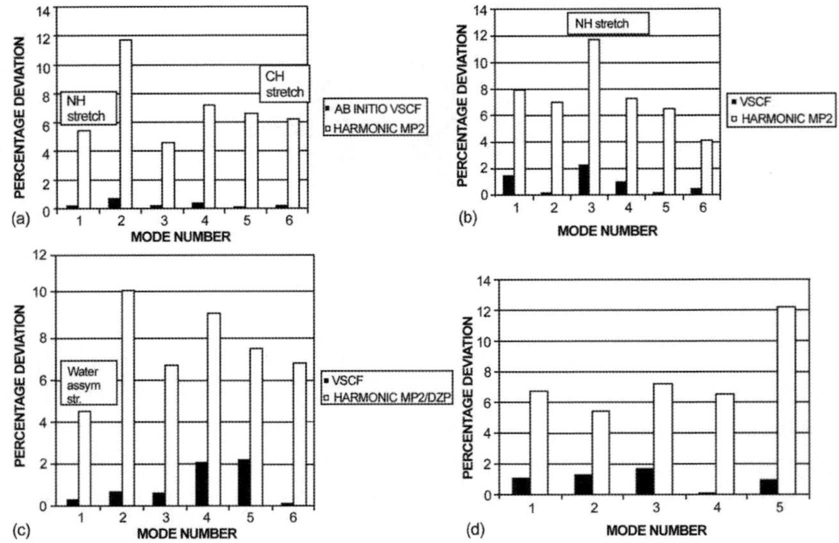

Fig. 2. (a) Comparison of percentage deviations of MP2 VSCF frequencies of $(ImH^+)(H_2O)N_2$ and MP2 harmonic frequency calculations from experiment. (b) Comparison of percentage deviations of improved PM3 VSCF frequencies of $(ImH^+)(H_2O)_2$ and MP2 harmonic frequency calculations from experiment. (c) Comparison of percentage deviations of improved PM3 VSCF frequencies of $(ImH^+)H_2O$, and MP2 harmonic frequency calculations from experiment. (d) Comparison of percentage deviations of improved PM3 VSCF frequencies of ImH^+ and MP2 harmonic frequency calculations from experiment. Adapted with permission from Ref. 82. Copyright 2007 American Chemical Society.

transitions the anharmonic effects are typically smaller. Many transitions in the mid-IR of biological molecules have anharmonic effects of not more than 2%–3% in the frequencies.[83] The conclusion is, however, that at least part of the full vibrational spectrum of any biological molecule is quite generally subject to large anharmonic effects.

3.3. *Single-mode anharmonicity and mode-mode anharmonic coupling*

The anharmonic interactions in the potential energy can be classified into two types. The first and simplest is the intrinsic anharmonicity of each mode: consider displacing the molecule away from equilibrium along a

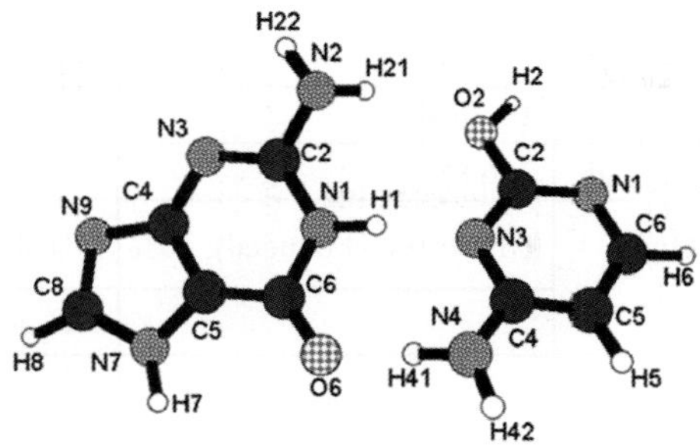

Fig. 3. Equilibrium structure of the enol tautomer of G···C. Adapted with permission from Ref. 79. Copyright 2005 American Chemical Society.

Table 1. Strong mode–mode couplings in G···C.[79]

Mode Type	Mode 1	Mode 2
All	NH_2 sym str (G)	NH_2 group tors (G)
	OH str	COH tors
	NH str (no H bond, G)	NH oopb (no H bond, G)
All, coord. suitable	NH_2 sym str (G)	NH_2 asym str (G)
	NH str (H bonded, G)	NH oopb, (H bonded,G)
	NH_2 sym str (C)	shear (G, C)
All inter-ring	NH_2 group tors (G)	buckle
	COH tors	stagger
	NH_2 group tors (G)	inter-ring str
Inter-ring, coord. suitable	NH str (H bonded, G)	Opening
	Shear (G, C)	Stagger
	NH_2 asym str (C)	Inter-ring str

single normal mode. The deviation from harmonic behavior in such a displacement reveals the intrinsic anharmonicity of that mode. All other anharmonic interactions are due to coupling between different normal modes. We adhere in this part to the pairwise coupling approximation for the potential in terms of the normal modes, discussed in Section 1.[42] In this framework, the second type of anharmonicity is due to mode–mode coupling. Though both types of anharmonicity contribute to the vibrational spectrum, the difference between the two is important. Only

Table 2. Intrinsic and coupling anharmonicities.[79]

Description	Value	Intrinsic	Coupling
Hydrogenic stretches	Typical	2–6.8%	1–6.4%
	High	NH str (G, H bonded), 10%	NH_2 asym str (G), 8.9%
			NH_2 asym str (C), 7.5%
	Low	NH_2 asym str (G), 0.5%	NH str (G, no H bond), 0.9%
C=O stretch	–	0.20%	1.50%
Bends 1000 to 1700 cm^{-1}	Typical	0–1.2%	1.3–3.1%
	High	CH b (G), 3.1%	NH b (G, no H bond), 3.9%
		NH b (G, no H bond), 1.8%	CH b (G), 3.7%
	Low	–	COH b, NCH b C 0.8%
Oop bends, shear, torsion	Typical	0–4.2%	0.30–4.2%
	High	NH oopb (G, H bond), 10.5%	NH oopb (G, H bond), 7.4%
		CH oopb (G), 7.5%	CH oopb (G), 5.6%
	Low	–	CCC oopb (G), 0%
Intermol. modes	Range	0–11.2%	0.4–24.7%

the anharmonic coupling between modes can cause, in dynamics, intramolecular vibrational energy flow.

Analysis of the anharmonic coupling between modes can conveniently be carried out in the framework of VSCF calculations. As an example, we present here results from the combined theoretical and experimental study by Brauer *et al.*[79] on the vibrational spectroscopy of G···C, a complex of the two nucleobases G and C. The computed equilibrium structure of this complex, which has the enolic form, is shown in Fig. 3. Following the definition of Ref. 79 that provides a measure of the coupling strength, it is found that only a limited set of pairs of normal models have strong mutual interactions. Table 1 presents the strongly coupled pairs of normal modes in this case.

These are the pairs that are expected to be involved most efficiently in intramolecular vibrational energy flow. In considering the relation between these couplings and the vibrational spectrum, one must keep in

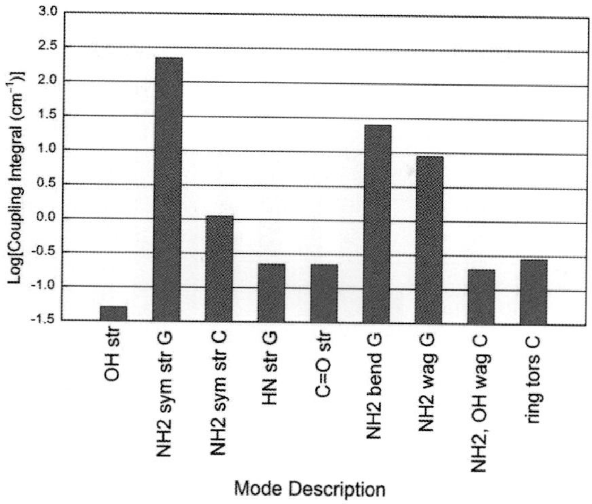

Fig. 4. Guanine asymmetric stretching mode: coupling to other modes in G···C. Data from Ref. 79.

mind that also the intrinsic anharmonicity of modes contributes to the spectrum (though not to the dynamics of vibrational energy flow). Intrinsic and coupling anharmonicities can act in the same direction, shifting both the harmonic frequencies to the red or to the blue, or they may act in opposite directions, in some cases even cancelling each other. The fact that a certain frequency is found to be in exact accord with harmonic prediction does not prove, therefore, that the corresponding mode will not be involved in vibrational energy flow.

The intrinsic and coupling anharmonicities for different types of modes in G···C are given in Table 2. The percentage figure in the table corresponds, roughly, to the fraction of the frequency effect of the interaction, from the value of the anharmonic frequency contribution included. (A more accurate definition can be found in Ref. 79). As Table 2 shows, for hydrogenic stretches in this complex, the intrinsic and the coupling anharmonicity effects on the corresponding frequency is, roughly, of the same order. The same is true for the out-of-plane bends, showing frequencies and torsions. On the other hand, for the intermolecular modes the coupling anharmonicity is significantly more important, and for the C=O stretch it is far more important than the intrinsic anharmonicity. The strength of the couplings between different

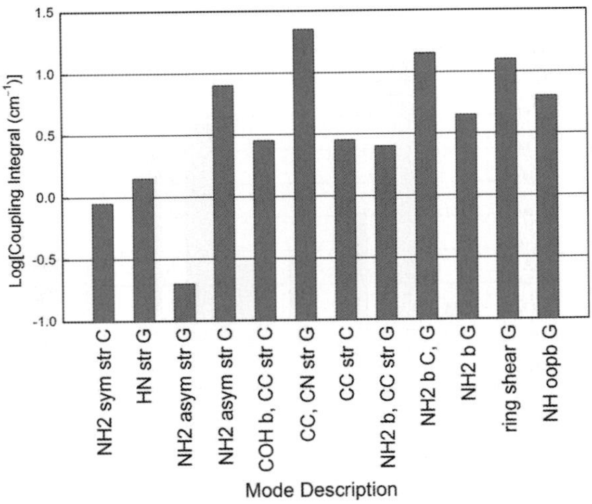

Fig. 5. C=O stretch mode coupling to other modes in G···C. Data from Ref. 79.

modes is in no way "democratic" in general. Figure 4 shows the strength of the couplings between the guanine NH_2 asymmetric stretching mode and the other modes. In this case the couplings vary greatly, depending on the nature of the other mode. On the other hand, consider the couplings of the C=O stretching mode with other modes, depicted in Fig. 5. In this case, the different couplings are (mostly) of the same approximate strength.

Such analysis should be useful for predicting pathways of vibrational energy flow in the system.

3.4. *Vibrational calculations for biological molecules in an environment*

In the previous sections, we discussed spectra of biological molecules treated as isolated species. This corresponds to the conditions for molecular beam experiments, but for molecules in a host such isolated-molecule calculations are at best an approximation. It can be argued that for molecules in noble-gas matrices, the isolated-molecule approximation is expected to be a very reasonable one. Indeed, this is one of the great merits of matrix isolation spectroscopy, compared with experiments in

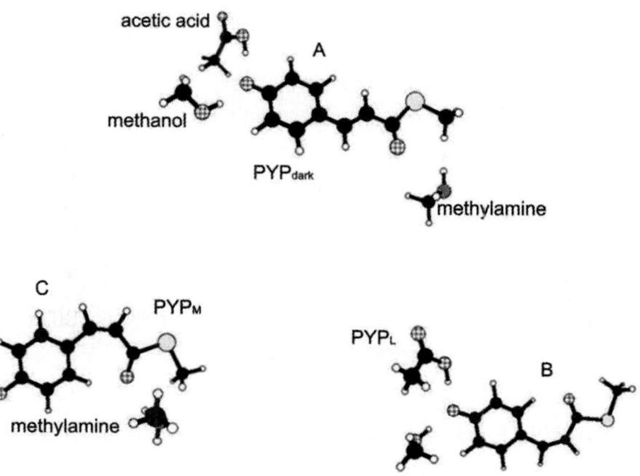

Fig. 6. Structures of the mimic systems used to compute the Raman spectroscopy of three intermediates of the chromophore of PYP. Adapted with permission from Ref. 84. Copyright 2007 American Chemical Society.

other types of media. In the present section, we would like to illustrate how realistic vibrational spectroscopy calculations can be carried out for a molecule in a host medium that interacts quite strongly with the guest molecule. As for isolated molecules, the desired goal is performing of anharmonic calculations using potentials from an adequate level of electronic structure theory. We are aware of only a few such calculations for molecules in relatively strong interactions with the host medium.[84,85]

As an example illustrating a possible approach for such systems, we consider anharmonic calculations for the Raman spectra of intermediates in the photocycle of photoactive yellow protein (PYP).[84] The Raman-active atmosphere in this system is embedded in the protein. The experiments were done at ambient temperature. The Raman spectra of three intermediates were explored: the initial "dark" state prior to photoexcitation, and two short-lived intermediates, the *M*-intermediate (blue-shifted in electronic adsorption with respect to the "dark" state), and the *L*-intermediate (red-shifted). Vibrational spectroscopy of these three states, or structures, can offer understanding of three regions of the potential energy surface of the chromophore, and of the evolution of the potential along the photocycle process. The calculations of Adesokan

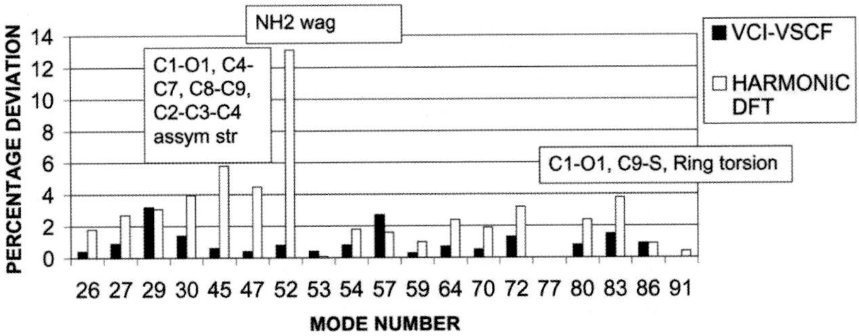

Fig. 7. Deviations of VSCF and of harmonic frequencies from Raman experiments for the "dark" state of PYP. Adapted with permission from Ref. 84. Copyright 2007 American Chemical Society.

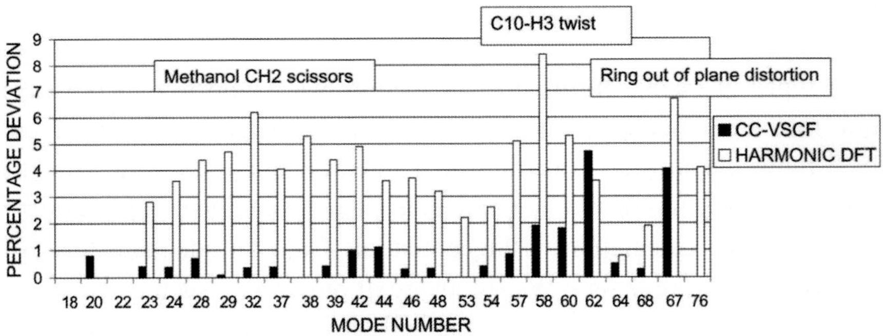

Fig. 8. Deviations of VSCF and of harmonic calculations from the measured frequencies for the *L*-state of PYP. Adapted with permission from Ref. 84. Copyright 2007 American Chemical Society.

et al.[84] treated the chromophore molecule explicitly, but replaced for each intermediate the protein environment around the chromophore *small-molecule mimics* that provide hydrogen-bonding interactions similar to those due to the real protein. Due to the structural changes in the photocycle process, the proper mimics for the different intermediates are different. The mimic systems for the three intermediates are shown in Fig. 6. At this level the system becomes computationally feasible. The harmonic spectrum was computed by DFT/B3LYP. Then, improved

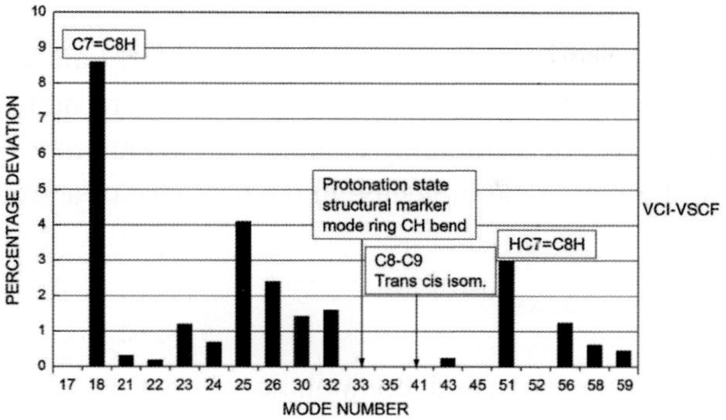

Fig. 9. Deviations of VSCF calculations from the measured frequencies for the *M*-intermediate of PYP. Adapted with permission from Ref. 84. Copyright 2007 American Chemical Society.

PM3 potentials were constructed on the basis of the harmonic DFT frequencies, as discussed in Section 2.[78]

Figure 7 shows the deviation percentages of the calculations from the Raman experiments for the dark state. Both VSCF and harmonic deviation are shown versus mode number (where the modes were numbered in decreasing value of the harmonic frequency). The VCI (or degenerate) variant of VSCF was used in this case, as there are a number of important resonances. Two important conclusions emerge from Fig. 7. First, the anharmonic calculation results in much lower deviations from experiment, and thus anharmonic effects are important also in condensed-phase spectroscopy. Second, the VCI-VSCF calculations deviate very little from the experimental value, supporting the hope that quantitative calculations can be done also for condensed-phase spectroscopy.

In Fig. 8, results of the deviations from the measured Raman frequencies of both VSCF and harmonic calculations are shown for the *L*-intermediate. Again, the anharmonic calculations give much better accord with experiment, emphasizing again the essential role of anharmonic effects in vibrational spectroscopy also in an interacting environment. Finally, Fig. 9 shows the deviations of VSCF calculations from experiment for the *M*-intermediate. In this case, there is at least one large deviation from experiment at the VSCF level. The origins of this are not yet clear. We believe this could be due to a flaw of the improved PM3 potential for this state. Nevertheless, for nearly all transitions, the deviations from experiment, also for the *M*-state, are quite modest. All this adds up to a very encouraging example of quantitative spectroscopic calculations of a biological molecule in the presence of a protein host.

4. Concluding Remarks

In this chapter, we discussed computational methods for quantitative calculations of the vibrational spectroscopy of biological molecules at low temperatures. The examples presented allow for evaluation of the methods as tools for quantitative interpretation of experiments in the field. While other methods are available, we focused here on the VSCF approach, in several variants. Several conclusions emerge from the examples presented. First, the treatment of anharmonic effects is generally essential. The role of anharmonicity can differ greatly for different transitions, but for essentially all the molecules studied, at least some of the fundamental transitions were subject to important anharmonic contributions. Second, methods of electronic structure theory seem to be the only available ones of accuracy sufficient for vibrational spectroscopy calculations. Obviously, these must be of the appropriate level for providing the required accuracy. Methods such as MP2 meet the state-of-the-art requirements for spectroscopy of biomolecules, but are computationally limited to small systems. We argued in this case for using adapted or improved variants of standard semiempirical electronic structure methods such as PM3. These were found to yield good results in several applications, and certainly can be used to address much larger molecules. However, the improved semiempirical electronic structure methods for the purpose of large-molecule spectroscopy are at an early stage of development. This is definitely more of a promising direction for future research than a well-established working tool. Further progress on this is essential. Finally, it is very encouraging that the methodology of anharmonic calculations with direct use of *ab initio* potentials seems to work to remarkable degree at least in some cases of molecules in strongly interacting (i.e., hydrogen bonding) host environments. This was achieved in the presented example through the use of small mimics, to model the effect of a host such as a protein. Clearly, much more general representations of a condensed-phase environment are desirable. This defines another major challenge for the future.

Acknowledgments

Research by the authors presented in this chapter was carried out under the auspices of the Saerree K. and Louis P. Fiedler Chair in Chemistry (RBG). The research by RBG presented here was also partly supported by the US-Israel Binational Science Foundation (BSF Grant 2004009).

References

1. A. Y. Ivanov, G. Sheina and Y. P. Blagoi, *Spectrochim. Acta.* **A55**, 219 (1999).
2. S. G. Stepanian, I. D. Reva, E. D. Radchenko and L. Adamowicz, *J. Phys. Chem.* **A102**, 4623 (1998).
3. A. Y. Ivanov, A. M. Plokhotnichenko, V. Izvekov, G. G. Sheina and Y. P. Blagoi, *J. Mol. Struct.* **408**, 459 (1997).
4. I. D. Reva, S. G. Stepanian, A. M. Plokhotnichenko, E. D. Radchenko, G. G. Sheina and Y. P. Blagoi, *J. Mol. Struct.* **318**, 1 (1994).
5. S. Breda, I. Reva and R. Fausto, *Vib. Spectrosc.* **50**, 57 (2009).
6. S. Breda, I. Reva and R. Fausto, *J. Mol. Struct.* **887**, 75 (2008).
7. A. Kovacs and A. Y. Ivanov, *J. Phys. Chem.* **B113**, 2151 (2009).
8. G. Meijer, G. Berden, W. L. Meerts, H. E. Hunziker, M. S. Devries and H. R. Wendt, *Chem. Phys.* **163**, 209 (1992).
9. M. G. H. Boogaarts, G. von Helden and G.Meijer, *J. Chem. Phys.* **105**, 8556 (1996).
10. H. Piest, G. von Helden and G. Meijer, *J. Chem. Phys.* **110**, 2010 (1999).
11. E. Nir, L. Grace, B. Brauer and M. S. de Vries, *J. Am. Chem. Soc.* **121**, 4896 (1999).
12. K. Remmers, R. G. Satink, G. von Helden, H. Piest, G. Meijer and W. L. Meerts, *Chem. Phys. Lett.* **317**, 197 (2000).
13. L. C. Snoek, R. T. Kroemer and J. P. Simons, *Phys. Chem. Chem. Phys.* **4**, 2130 (2002).
14. L. Grace, R. Cohen, T. M. Dunn, D. M. Lubman and M. S. de Vries, *J. Mol. Spectrosc.* **215**, 204 (2002).
15. R. A. Jockusch, F. O. Talbot and J. P. Simons, *Phys. Chem. Chem. Phys.* **5**, 1502 (2003).
16. J. M. Bakker, R. G. Satink, G. von Helden and G. Meijer, *Phys. Chem. Chem. Phys.* **4**, 24 (2002).

17. E. Nir, I. Hunig, K. Kleinermanns and M. S. de Vries, *Phys. Chem. Chem. Phys.* **5**, 4780 (2003).
18. C. Plutzer, I. Hunig, K. Kleinermanns, E. Nir and M. S. de Vries, *Chem. Phys. Chem* **4**, 838 (2003).
19. J. M. Bakker, I. Compagnon, G. Meijer, G. von Helden, M. Kabelac, P. Hobza and M. S. de Vries, *Phys. Chem. Chem. Phys.* **6**, 2810 (2004).
20. E. Nir, I. Hunig, K. Kleinermanns and M. S. de Vries, *Chem. Phys. Chem* **5**, 131 (2004).
21. B. Lucas, G. Gregoire, F. Lecomte, B. Reimann, J. P. Schermann and C. Desfrancois, *Mol. Phys.* **103**, 1497 (2005).
22. J. M. Bakker, C. Plutzer, I. Hunig, T. Haber, I. Compagnon, G. von Helden, G. Meijer and K. Kleinermanns, *Chem. Phys. Chem* **6**, 120 (2005).
23. A. Abo-Riziq, J. E. Bushnell, B. Crews, M. Callahan, L. Grace and M. S. de Vries, *Chem. Phys. Lett.* **431**, 227 (2006).
24. M. P. Callahan, B. Crews, A. Abo-Riziq, L. Grace, M. S. de Vries, Z. Gengeliczki, T. M. Holmes and G. A. Hill, *Phys. Chem. Chem. Phys.* **9**, 4587 (2007).
25. E. J. Cocinero, E. C. Stanca-Kaposta, E. M. Scanlan, D. P. Gamblin, B. G. Davis and J. P. Simons, *Chem. Eur. J.* **14**, 8947 (2008).
26. E. E. Baquero, W. H. James, S. H. Choi, S. H. Gellman and T. S. Zwier, *J. Am. Chem. Soc.* **130**, 4784 (2008).
27. M. S. de Vries and P. Hobza, *Ann. Rev. Phys. Chem.* **58**, 585 (2007).
28. M. Y. Choi, F. Dong, S. W. Han and R. E. Miller, *J. Phys. Chem.* **A112**, 7185 (2008).
29. D. S. Skvortsov and A. F. Vilesov, *J. Chem. Phys.* **130**, 151101 (2009).
30. R. B. Gerber, B. Brauer, S. K. Gregurick and G. M. Chaban, *Phys. Chem. Comm.* **5**, 142 (2002).
31. R. B. Gerber, G. M. Chaban, S. K. Gregurick and B. Brauer, *Biopolymers* **68**, 370 (2003).
32. E. B. Wilson, J. C. Decius and P. C. Cross, *Molecular Vibrations,* McGraw-Hill, New York (1955).
33. N. G. Mirkin and S. Krimm, *J. Am. Chem. Soc.* **113**, 9742 (1991).
34. A. G. Csaszar, *J. Mol. Struct.* **346**, 141 (1995).
35. A. G. Csaszar, *J. Phys. Chem.* **100**, 3541 (1996).
36. A. P. Scott and L. Radom, *J. Phys. Chem.* **100**, 16502 (1996).
37. J. M. Bowman, *J. Chem. Phys.* **68**, 608 (1978).
38. J. M. Bowman, *Acc. Chem. Res.* **19**, 202 (1986).
39. R. B. Gerber and M. A. Ratner, *Chem. Phys. Lett.* **68**, 195 (1979).
40. R. B. Gerber and M. A. Ratner, *Adv. Chem. Phys.* **70**, 97 (1988).
41. J. O. Jung and R. B. Gerber, *J. Chem. Phys.* **105**, 10682 (1996).
42. J. O. Jung and R. B. Gerber, *J. Chem. Phys.* **105**, 10332 (1996).
43. S. K. Gregurick, E. Fredj, R. Elber and R. B. Gerber, *J. Phys. Chem.* **B101**, 8595 (1997).

44. S. K. Gregurick, J. H.-Y. Liu, D. A. Brant and R. B. Gerber, *J. Phys. Chem.* **B103**, 3476 (1999).
45. A. E. Roitberg, R. B. Gerber, R. Elber and M. A. Ratner, *Science* **268**, 1319 (1995).
46. A. E. Roitberg, R. B. Gerber and M. A. Ratner, *J. Phys. Chem.* **B101**, 1700 (1997).
47. L. S. Norris, M. A. Ratner, A. E. Roitberg and R. B. Gerber, *J. Chem. Phys.* **105**, 11261 (1996).
48. L. Pele, B. Brauer and R. B. Gerber, *Theor. Chem. Acc.* **117**, 69 (2007).
49. L. Pele and R. B. Gerber, *J. Chem. Phys.* **128**, 165105 (2008).
50. D. M. Benoit, *J. Chem. Phys.* **125**, 244110 (2006).
51. Y. Scribano and D. M. Benoit, *J. Chem. Phys.* **127**, 164118 (2007).
52. D. M. Benoit, *J. Chem. Phys.* **129**, 234304 (2008).
53. N. Matsunaga, G. M. Chaban and R. B. Gerber, *J. Chem. Phys.* **117**, 3541 (2002).
54. K. Yagi, S. Hirata and K. Hirao, *Phys. Chem. Chem. Phys.* **10**, 1781 (2008).
55. S. Hirata and K. Yagi, *Chem. Phys. Lett.* **464**, 123 (2008).
56. G. M. Chaban, J. O. Jung and R. B. Gerber, *J. Chem. Phys.* **111**, 1823 (1999).
57. G. M. Chaban, J. O. Jung and R. B. Gerber, *J. Phys. Chem.* **A104**, 2772 (2000).
58. M. W. Schmidt, K. K. Baldridge, J. A. Boatz, S. T. Elbert, M. S. Gordon, J. H. Jensen, S. Koseki, N. Matsunaga, K. A. Nguyen, S. J. Su, T. L. Windus, M. Dupuis and J. A. Montgomery, *J. Comp. Chem.* **14**, 1347 (1993).
59. H.-J. Werner, P. J. Knowles, R. Lindh, F. R. Manby, M. Schütz, P. Celani, T. Korona, A. Mitrushenkov, G. Rauhut, T. B. Adler, R. D. Amos, A. Bernhardsson, A. Berning, D. L. Cooper, M. J. O. Deegan, A. J. Dobbyn, F. Eckert, E. Goll, C. Hampel, G. Hetzer, T. Hrenar, G. Knizia, C. Köppl, Y. Liu, A. W. Lloyd, R. A. Mata, A. J. May, S. J. McNicholas, W. Meyer, M. E. Mura, A. Nicklass, P. Palmieri, K. Pflüger, R. Pitzer, M. Reiher, U. Schumann, H. Stoll, A. J. Stone, R. Tarroni, T. Thorsteinsson, M. Wang and A. Wolf, in *MOLPRO, Version 2008.1, a Package of ab initio Programs* (2006).
60. V. Barone, *J. Chem. Phys.* **122**, 014108 (2005).
61. P. Carbonniere, T. Lucca, C. Pouchan, N. Rega and V. Barone, *J. Comp. Chem.* **26**, 384 (2005).
62. O. Bludský, J. Chocholoušová, J. Vacek, F. Huisken and P. Hobza, *J. Chem. Phys.* **113**, 4629 (2000).
63. G. M. Chaban, J. O. Jung and R. B. Gerber, *J. Phys. Chem.* **A104**, 10035 (2000).
64. H. G. Kjaergaard, A. L. Garden, G. M. Chaban, R. B. Gerber, D. A. Matthews and J. F. Stanton, *J. Phys. Chem.* **A112**, 4324 (2008).
65. Y. Miller, G. M. Chaban and R. B. Gerber, *Chem. Phys.* **313**, 213 (2005).
66. B. Brauer, F. Dubnikova, Y. Zeiri, R. Kosloff and R. B. Gerber, *Spectrochim. Acta* **A71**, 1438 (2008).
67. P. Seidler, M. B. Hansen and O. Christiansen, *J. Chem. Phys.* **128**, 154113 (2008).
68. P. Seidler, E. Matito and O. Christiansen, *J. Chem. Phys.* **131**, 034115 (2009).
69. K. Yagi, H. Karasawa, S. Hirata and K. Hirao, *Chem. Phys. Chem* **10**, 1442 (2009).

70. H. Fujisaki, K. Yagi, J. E. Straub and G. Stock, *Int. J. Quantum Chem.* **109**, 2047 (2009).
71. Y. S. Kim and R. M. Hochstrasser, *J. Phys. Chem.* **B113**, 8231 (2009).
72. R. M. Hochstrasser, *Adv. Chem. Phys.* **132**, 1 (2006).
73. W. Zhuang, T. Hayashi and S. Mukamel, *Angew. Chem. Int. Ed.* **48**, 3750 (2009).
74. S. K. Gregurick, G. M. Chaban and R. B. Gerber, *J. Phys. Chem.* **A106**, 8696 (2002).
75. A. R. Leach, *Molecular Modelling. Principles and Applications,* 2nd Ed., Essex (2001).
76. G. M. Chaban and R. B. Gerber, *J. Chem. Phys.* **115**, 1340 (2001).
77. G. M. Chaban and R. B. Gerber, *Theor. Chem. Acc.* **120**, 273 (2008).
78. B. Brauer, G. M. Chaban and R. B. Gerber, *Phys. Chem. Chem. Phys.* **6**, 2543 (2004).
79. B. Brauer, R. B. Gerber, M. Kabeláč, P. Hobza, J. M. Bakker, A. G. A. Riziq and M. S. de Vries, *J. Phys. Chem.* **A109**, 6974 (2005).
80. D. M. Leitner, *Phys. Rev. Lett.* **87**, 188102 (2001).
81. D. M. Leitner, *Ann. Rev. Phys. Chem.* **59**, 233 (2008).
82. A. A. Adesokan, G. M. Chaban, O. Dopfer and R. B. Gerber, *J. Phys. Chem.* **A111**, 7374 (2007).
83. G. von Helden, I. Compagnon, M. N. Blom, M. Frankowski, U. Erlekam, J. Oomens, B. Brauer, R. B. Gerber and G. Meijer, *Phys. Chem. Chem. Phys.* **10**, 1248 (2008).
84. A. A. Adesokan, D. Pan, E. Fredj, R. A. Mathies and R. B. Gerber, *J. Am. Chem. Soc.* **129**, 4584 (2007).
85. D. Shemesh, J. Mullin, M. S. Gordon and R. B. Gerber, *Chem. Phys.* **347**, 218 (2008).

Index

*Many chapters of this book describe this subject, and other relevant discussions of this topic appear in the text.

*Many chapters of this book describe this subject, and other relevant discussions of this topic appear in the text.

*Many chapters of this book describe this subject, and other relevant discussions of this topic appear in the text.

*Many chapters of this book describe this subject, and other relevant discussions of this topic appear in the text.

Color Insert

Chapter 4

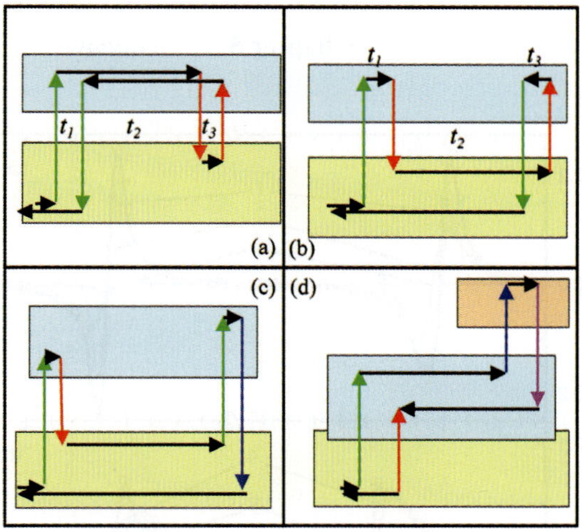

Fig. 1. Time-circuit diagrams define the time correlations responsible for various observables: (a) absorption followed by emission, which is part of the RR response if there is no dephasing during t_2; (b) the nonclassical RR response contribution in which the fields of the excitation and scattering photons are intertwined; (c) CARS diagram designed to interrogate the ground state vibrational coherence, (d) four-color four-wave mixing designed to interrogate quantum coherences in the excited electronic state in real time. Note, in (a) and (b) a photon is represented by its two fields, E and E^*. In (c) and (d) the complex conjugate diagrams are not shown, now each up/down arrow corresponds to interaction with a different photon. Time progresses along the horizontal, forward on ket states and backward on bra. The background colors designate different electronic states to highlight the resonant processes. For the case of the I_2 experiments discussed, the X, B, and E states are represented by the colored bands.

Chapter 5

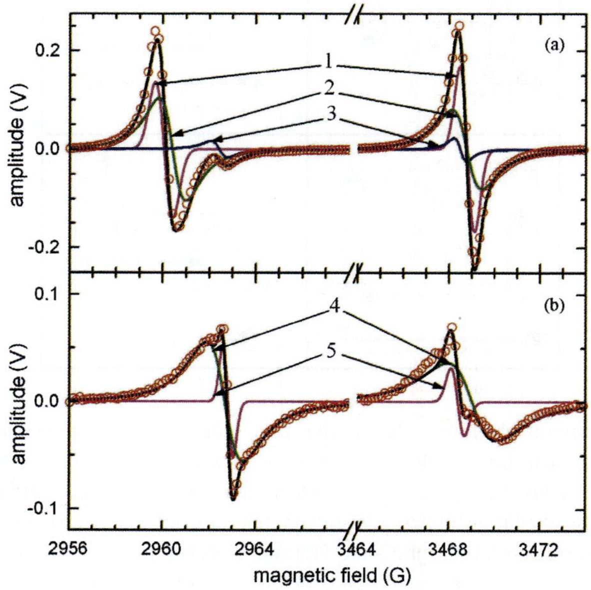

Fig. 14. (a) Experimental ESR spectra of H atoms for as-prepared H–Kr–He sample (open red circles) shown in Fig. 13a and three pairs of fitting lines (1, magenta are Gaussian lines; 2, green and 3, blue are Lorentzian lines). (b) Experimental ESR spectra of H atoms for annealed sample (open red circles) shown in Fig. 13c and two pairs of fitting lines (4, green are Lorentzian lines and 5, magenta are Gaussian lines). The sums of fitting lines are shown as black lines for both (a) and (b).[66]

Chapter 8

Fig. 4. Cryostat used for the study of solutions in LNe, LAr, LKr, and LXe. The cell is equipped with wedged Si windows and has a pathlength of 40 mm.

Fig. 5. Cryostat used for the study of solutions in LAr, LKr, and LXe. The cell has a pathlength of 10 mm and is equipped with ZnSe windows.

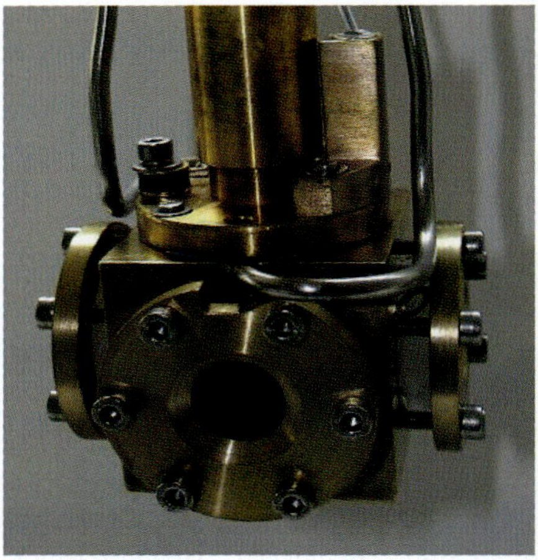

Fig. 6. Liquid cell used to study the Raman spectra of cryosolutions.

Chapter 10

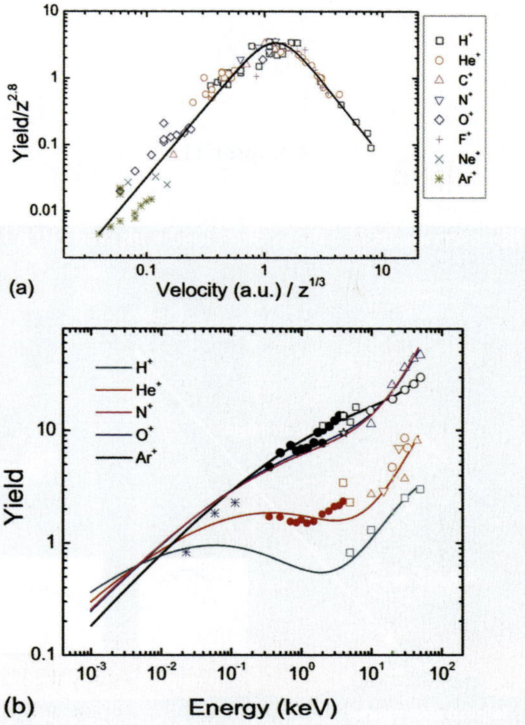

Fig. 1. (a) Yield, Y, in equivalent water molecules ejected from ice per incident fast ion versus the ion velocity in atomic mass units (2.18×10^8 cm s^{-1}) $T < 100$ K: ejection is primarily due to electronic excitations and ionizations. Yield and velocity are scaled by nuclear charge, Z, of ion from Johnson *et al.*[14]; line gives fit from http://www.people.virginia.edu/~rej. (b) Y in lower energy regime includes electronic and knock-on sputtering processes: summary of data, see Fama *et al.*[30]; lines are analytic expressions in that paper: electronic contribution indicted by upturn above ~1 keV amu^{-1}. Lower energies dominated by knock-on sputtering.

Chapter 11

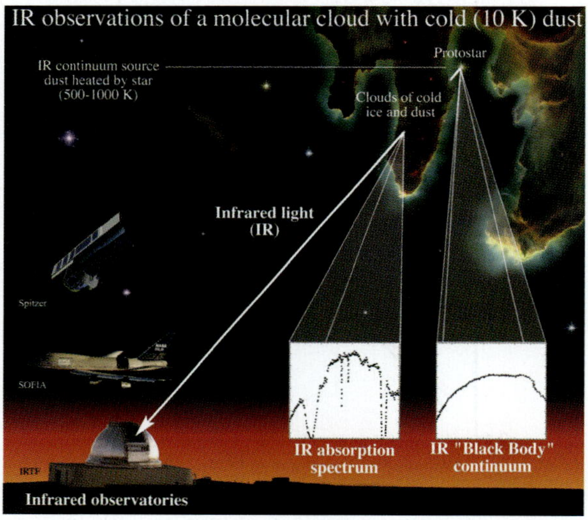

Fig. 1. Measuring the IR absorption spectra of interstellar clouds. A background star (protostar) serves as the IR source, the cloud is the sample, and the telescope gathers the light and sends it to a monochromator or spectrometer. The advent of airborne IR telescopes in the 1970s and orbiting telescopes in the 1980s made it possible to avoid IR absorptions by atmospheric H_2O, CO_2, and so on, opening a new window into the Cosmos. Figure courtesy of Dr. Jason Dworkin.

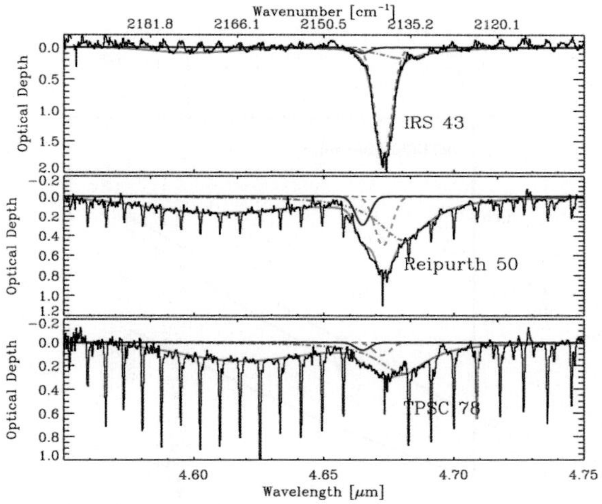

Fig. 3. The 4.5–4.75 μm spectra of three low-mass protostars (IRS 43, Reipurth 50, TPSC 78) show the relatively broad solid CO absorption at about 4.67 μm and a forest of gas-phase CO rotational–vibrational transitions. The profile of the interstellar solid CO band consists of several independent components: A narrow band at 2139.9 cm^{-1} (4.67 μm) dominates the top spectrum. The profile in the bottom panel is dominated by a much broader band peaking at 2136.5 cm^{-1} (4.68 μm). The middle panel is a combination of these two extreme profiles. In addition, the solid CO band is accompanied by a weak band at somewhat shorter wavelengths (2143 cm^{-1}; 4.66 μm). Finally, the much broader and weaker band at ~ 4.61 μm actually consists of two components at 4.62 μm (2165 cm^{-1}) and 4.60 μm (2175 cm^{-1}). The figure is adapted from Ref. 18.

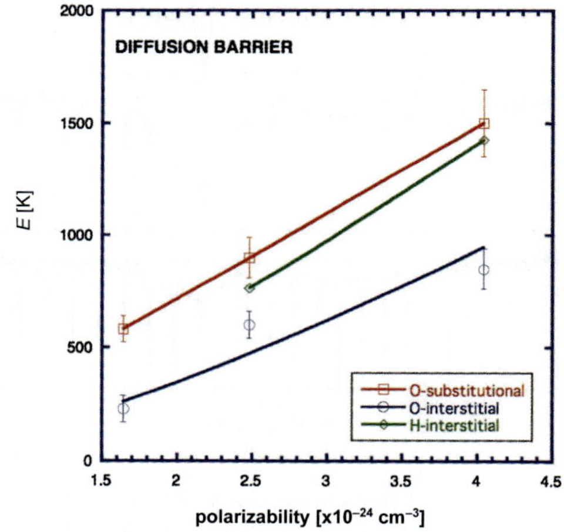

Fig. 6. Barriers against migration for oxygen and hydrogen atoms trapped in noble-gas matrices as a function of the polarizability of the matrix. Migration barriers have been estimated from molecular thermoluminescence, laser-induced fluorescence, or electron paramagnetic resonance. The data are taken from Refs. 50–53.

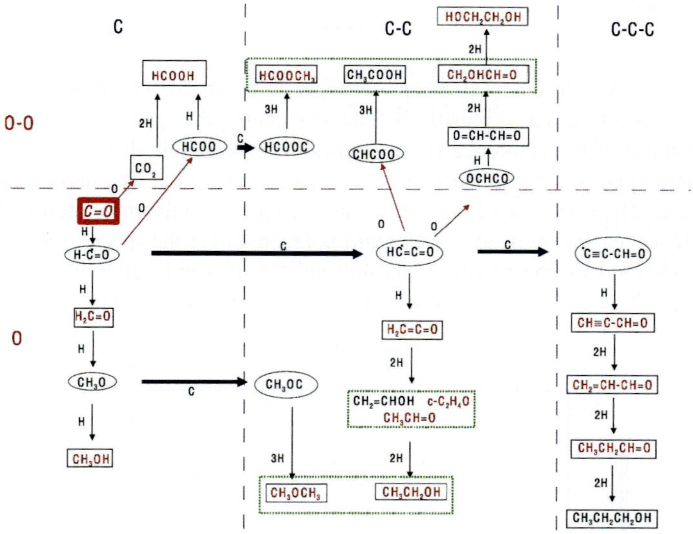

Fig. 8. Hydrogenation of CO enables a rich organic chemistry. Hydrogenation, oxidation, and carbon addition reactions are indicated with black (vertically), red (diagonally), and blue (horizontally) arrows, respectively. See text for details. Taken from Ref. 72 after Ref. 6.

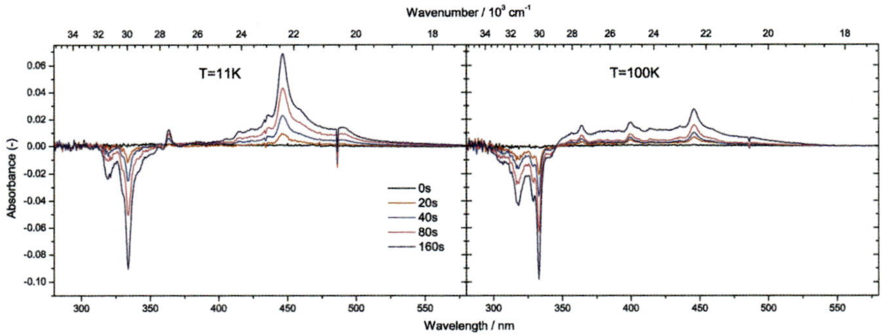

Fig. 15. The VUV-induced spectroscopic changes in pyrene: H_2O ice for two different temperatures as a function of photolysis time. Comparing the spectra from the 11-K ice (left) with those of the 100-K ice (right) shows the critical role that temperature plays in determining photochemical pathways in a PAH-containing ice. The loss of the neutral pyrene (Py, $C_{14}H_{12}$) band complex near 330 nm is similar at both temperatures. However, in the 11-K ice, cation formation (~ 450 nm) is favored over the species responsible for the broad, 350–500 nm absorption and the 400 and 405 nm band carriers formed at higher temperatures. The opposite holds for the 100-K ice. Taken from Ref. 116.

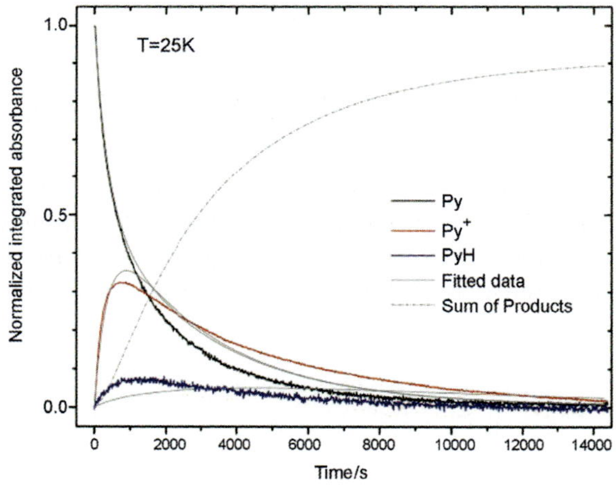

Fig. 16. Behavior of the Py 334 nm, Py^+ 445 nm, and PyH 400 nm bands as a function of VUV irradiation in $Py:H_2O$ ices at 25 K. Taken from Ref. 116. The curves labeled fits are produced by the chemical network described in Ref. 116.

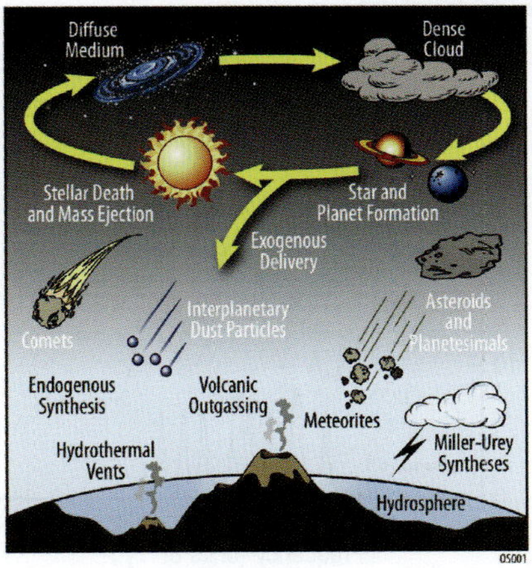

Fig. 18. The cycle of interstellar material leading to the production and evolution of complex organic compounds that can be delivered intact to, and mixed with, those produced on the newly formed planets and moons. There they participate in the primordial chemistry on these young objects. Figure courtesy of Jason Dworkin, NASA Goddard Space Flight Center, taken from Ref. 123.

Chapter 12

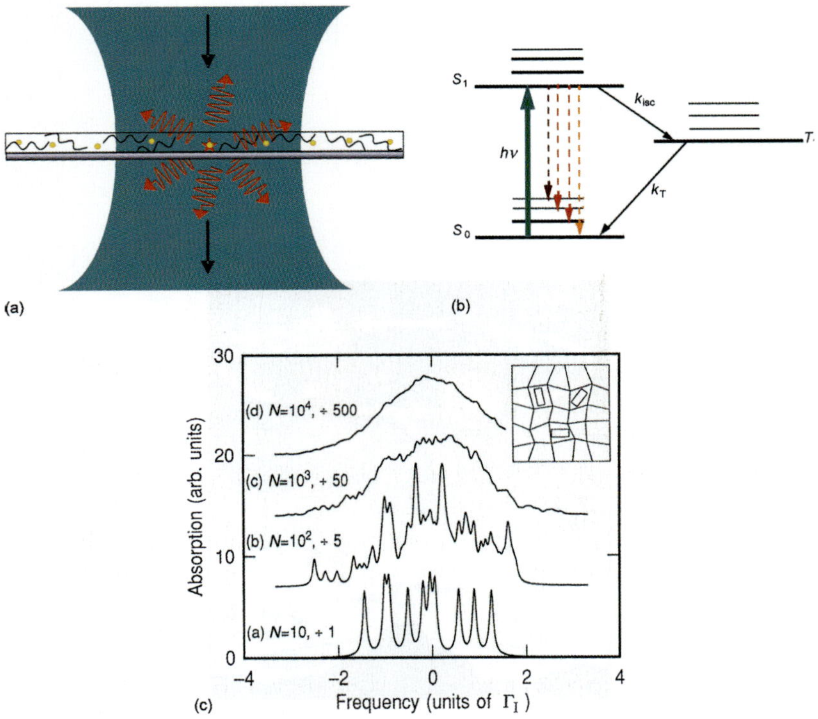

Fig. 1. (a) Schematic of the selection of a single molecule out of many in the same focal volume, (b) typical energy level scheme, and (c) illustration of inhomogeneous broadening.

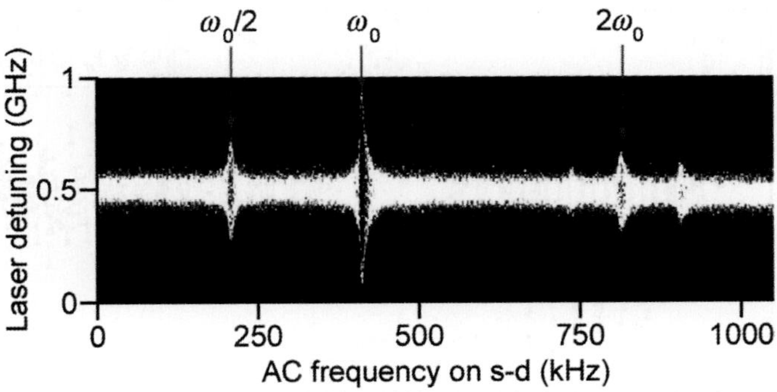

Fig. 6. Example of low-frequency acoustic resonances observed in the spectrum of a single dibenzoterrylene molecule (vertical axis) in an anthracene crystal. The frequency of the applied ac-voltage frequency is scanned (horizontal axis). The resonances at the double and half frequency are absent for low-driving voltages.[126]

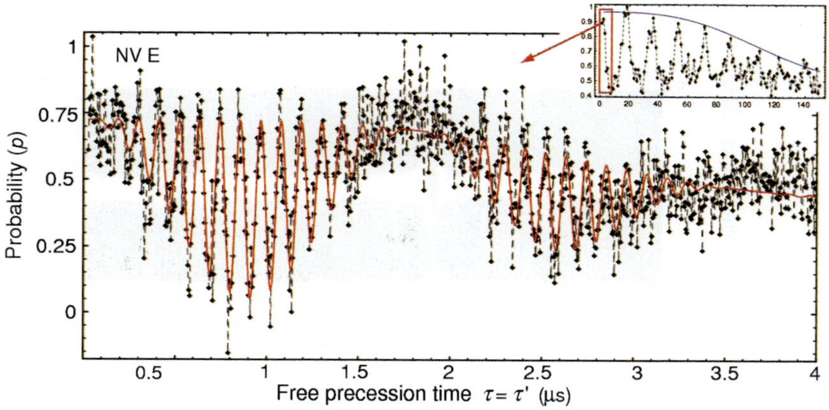

Fig. 7. Two-pulse echo signal of the electron spin of a single NV center in diamond, as the delay between the pulses is varied. The vertical axis indicates the fluorescence intensity, itself a function of the spin state. The modulations are due to isotopic impurities in the diamond lattice, and vary from center to center. Adapted with permission from Ref. 163.

Chapter 13

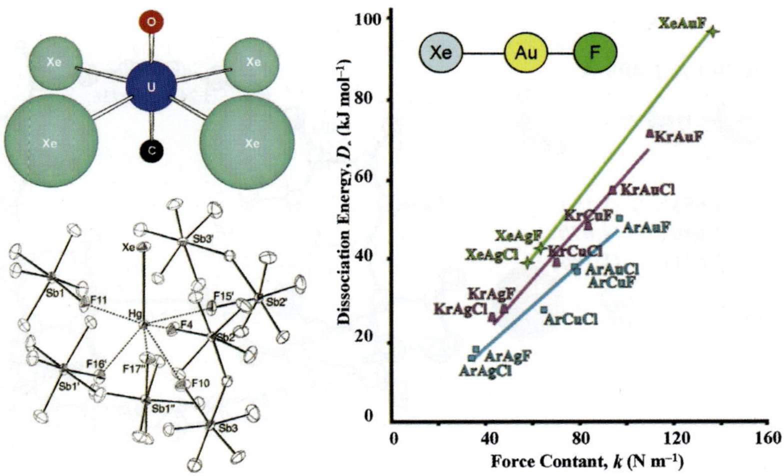

Fig. 4. Selected examples of chemical connections of xenon to metal cations in molecules and solids: U(C)(O)$(Xe)_4$ and $XeHgSb_3F_{16}$ (left). The XeAuF molecule and the calculated bond energy versus the experimental force constant for the Ng–M–X family of molecules where Ng = Ar, Kr, Xe; M = Cu, Ag, Au; and X = F, Cl, Br (right). Data reproduced with permission from Refs. 32, 35, and 38.

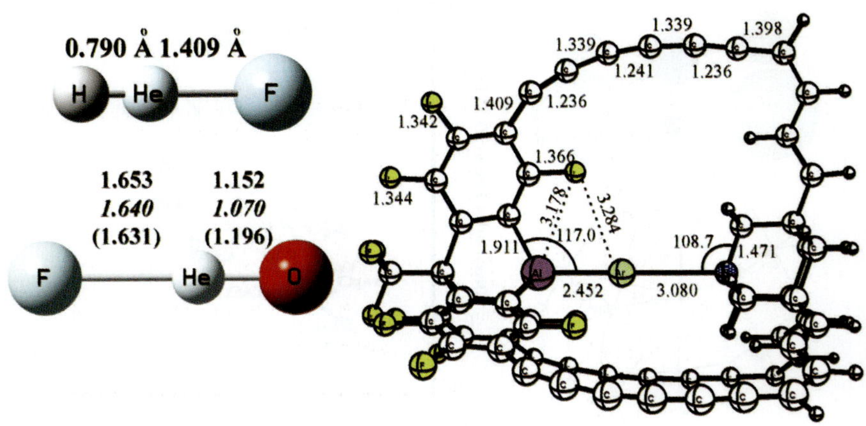

Fig. 5. Examples of exotic species containing noble-gas atoms according to recent theoretical predictions: HHeF,[43,44] $OHeF^-$, and a cage push–pull compound of Ar. Reproduced with permission from Refs. 47 and 50. Bond distances are in Å.